COLLOID
INTERFACE
CHEMISTRY FOR
NANOTECHNOLOGY

PROGRESS IN
COLLOID AND INTERFACE SCIENCE

Series Editors
Reinhardt Miller and Libero Liggieri

COLLOID AND INTERFACE CHEMISTRY FOR NANOTECHNOLOGY

Edited by

Peter Kralchevsky

Reinhard Miller

Francesca Ravera

CRC Press
Taylor & Francis Group
Boca Raton London New York

CRC Press is an imprint of the
Taylor & Francis Group, an **informa** business

CRC Press
Taylor & Francis Group
6000 Broken Sound Parkway NW, Suite 300
Boca Raton, FL 33487-2742

First issued in paperback 2019

ISBN-13: 978-1-4665-6905-8 (hbk)
ISBN-13: 978-0-367-37975-9 (pbk)

Library of Congress Cataloging-in-Publication Data

Colloid and interface chemistry for nanotechnology / editors, Peter A. Kralchevsky,
   Reinhard Miller, Francesca Ravera.
      pages cm. -- (Progress in colloid and interface science)
   Includes bibliographical references and index.
   ISBN 978-1-4665-6905-8 (hardback)
   1. Colloids. 2. Surface chemistry. 3. Nanoparticles. I. Kralchevsky, Peter A.

QD549.C558 2013
541'.345--dc23   2013018867

Visit the Taylor & Francis Web site at
http://www.taylorandfrancis.com

and the CRC Press Web site at
http://www.crcpress.com

Contents

SECTION I Nanoparticle Synthesis and Characterization

SECTION II   New Experimental Tools and Interpretations

SECTION III   Interfaces and Nanocolloidal Dispersions

Preface

The present book is a collection of manuscripts reflecting the activities of research teams that have been involved in the networking project *Colloid and Interface Chemistry for Nanotechnology* (2006–2011), Action D43, the European Science Foundation. The project was a part of the intergovernmental framework for Cooperation in Science and Technology (COST), allowing the coordination of nationally funded research on a European level. COST contributes to reducing the fragmentation in European research investments and opening the European research area to worldwide cooperation. In particular, 91 research teams and partner institutions from 28 European countries, as well as from Australia, Egypt, and Israel, took part in COST Action D43. Among the Action's main deliverables is the increased networking in colloid chemistry through the organization of 19 conferences and workshops, 4 training schools, and 52 short-term scientific missions for early-stage researchers, representing important steps toward their future careers.

Colloid and interface science had been dealing with nanoscale objects for almost a century before the term *nanotechnology* was coined. Colloid science is interdisciplinary in nature, because it bridges between our macroscopic world and the small world of atoms and molecules. The colloids appeared on the frontier of science after the fundamental studies by Einstein on the Brownian motion (1905) and Perrin on the determination of the Avogadro number and Boltzmann constant (1908), which led to the acceptance of the molecular theory by the scientific community. A cornerstone of colloid chemistry is the theory of interactions and stability of colloids developed by Derjaguin, Landau, Verwey, and Overbeek in 1941–1948. A strong boost of colloid and interface science was observed after the 1980s with the growing interest in nanoscience. Modern nanotechnology emerged in the 1990s out of the boom in microfabrication techniques representing the *top–down* approaches of the modern semiconductor industry, for example, beam lithography, sputtering, and etching. In addition, objects on the scale of nanometres (e.g., nanowires and nanoparticles) were fabricated, the most promising approach for their utilization being the *bottom–up* assembly in a controlled fashion.

This book contains contributions from six basic research directions in colloid and interface science, which have been the thematic areas of corresponding working groups within COST Action D43:

- The synthesis of nanostructured materials of well-defined size and functionalities, for example, uniform inorganic and organic nanoparticles of controlled size, shape, and uniformity, is the material basis of the bottom–up approaches.
- The control and characterization of the synthesized nanostructured materials demand the development, standardization, and automation of appropriate analytical methods, experimental tools, and procedures.

- The self-assembly is a basic route to the design of responsive soft materials. Examples for such materials are the microemulsions, polyelectrolyte capsules, and copolymer micelles, with their potential pharmaceutical and life science applications.
- The bio-inspired nanostructured materials represent another important research area. They are mainly based on nanoparticles and polymers, on the engineering of their structure and properties, and on the investigation of their potential for applications in materials and life sciences.
- Another important area is related to the design of active and soft functional interfaces with nanoscale structures and specific coatings that provide unique properties as sensors, catalysts, or as particular biomedical assays.
- Finally, nanoscale elements are combined into soft nanoscale devices and machines, in particular lab-on-a-chip devices, nanomotors, or nanopipettes for applications in analytical and biomedical sciences.

The great contribution of Michal Borkovec should be acknowledged for formulating the research strategy of the project and chairing the COST Action D43 during its first two years. The important contributions of the working group leaders, Ger Koper, Ceco Dushkin, Regine von Klitzing, Thomas Zemb, and Zbigniew Adamczyk, who devoted much of their time to coordinate the international teams, must be also thankfully acknowledged. The editors of the present book served to the Action as chair, 2008–2011 (P.K.), working group leader (R.M.), and coordinator for the exchange of researchers (F.R.).

The chapters of the book are summarizing recent results obtained by members of the COST Action D43 in the aforementioned research directions. The papers are grouped into three thematic sections, viz., "Nanoparticle Synthesis and Characterization," "New Experimental Tools and Interpretation," and "Nanocolloidal Dispersions and Interfaces." We hope that the book will represent an interesting and useful reading for researchers working in the diverse fields of colloid and interface science and nanotechnology.

Peter Kralchevsky
Reinhard Miller
Francesca Ravera

Contributors

Zbigniew Adamczyk
J. Haber Institute of Catalysis and
  Surface Chemistry
Polish Academy of Science
Cracow, Poland

Silvia Ahualli
Department of Applied Physics, Faculty
  of Science
University of Granada
Granada, Spain

Nikola A. Alexandrov
Department of Chemical Engineering
Sofia University
Sofia, Bulgaria

Mickaël Antoni
Centre Scientifique de St. Jérôme
Aix-Marseille Université
Marseille, France

Lise Arleth
Niels Bohr Institute
University of Copenhagen
Frederiksberg, Denmark

Francisco J. Arroyo
Department of Physics, Faculty of
  Experimental Sciences
University of Jaén
Jaén, Spain

Naz Zeynep Atay
Department of Chemistry
Bogazici University
Istanbul, Turkey

Carlo Baldisserri
National Research Council of Italy
  (CNR)
Institute of Science and Technology for
  Ceramics (ISTEC)
Faenza, Italy

Elka S. Basheva
Department of Chemical Engineering
Sofia University
Sofia, Bulgaria

Marija Bešter-Rogač
Faculty of Chemistry and Chemical
  Technology
University of Ljubljana
Ljubljana, Slovenia

Magali Boutonnet
School of Chemistry, Division of
  Chemical Technology
Kungliga Tekniska Högskolan (KTH)
Stockholm, Sweden

Seyda Bucak
Department of Chemical Engineering
Yeditepe University
Istanbul, Turkey

Rafael Contreras-Caceres
Complex Fluids Physics Group and
  Nano Laboratory
University of Almería
Almería, Spain

Romina Cuculayef
Department of Chemical Engineering
Yeditepe University
Istanbul, Turkey

Mustafa Culha
Department of Genetics and
    Bioengineering
Yeditepe University
Istanbul, Turkey

Maria Dąbkowska
J. Haber Institute of Catalysis and
    Surface Chemistry
Polish Academy of Science
Cracow, Poland

Abhijit Dan
Max Planck Institute of Colloids and
    Interfaces
Potsdam-Golm, Germany

A. V. Delgado
Department of Applied Physics, Faculty
    of Science
University of Granada
Granada, Spain

Jean-François Dufrêche
Institut de Chimie Séparative de
    Marcoule (ICSM)
CEA-CNRS-Université Montpellier
    2-ENSCM
Bagnols-sur-Ceze, France

Victoria Dutschk
Faculty of Engineering Technology
University of Twente
Enschede, The Netherlands

Magali Duvail
Institut de Chimie Séparative de
    Marcoule (ICSM)
CEA-CNRS-Université Montpellier
    2-ENSCM
Bagnols-sur-Ceze, France

Ozlem Erol
Science Faculty, Smart Materials
    Research Laboratory,
    Chemistry Department
Gazi University
Ankara, Turkey

Mustafa Ersoz
Science Faculty, Chemistry Department
Selcuk University
Konya, Turkey

Jordi Esquena
Instituto de Quimica Avanzada
    de Cataluña,
    Consejo Superior de Investigaciones
    Cientificas (IQAC-CSIC)
Barcelona, Spain

and

CIBER en Biotecnologia, Biomateriales
    y Nanomedicina (CIBER-BBN)
Barcelona, Spain

Antonio Fernandez-Barbero
Complex Fluids Physics Group and
    Nano Laboratory
University of Almería
Almería, Spain

and

NanoPhotonics Centre, Cavendish
    Laboratory
Department of Physics
University of Cambridge
Cambridge, United Kingdom

Carmen Galassi
National Research Council of Italy
    (CNR)
Institute of Science and Technology for
    Ceramics (ISTEC)
Faenza, Italy

Davide Gardini
National Research Council of Italy (CNR)
Institute of Science and Technology for Ceramics (ISTEC)
Faenza, Italy

Wolfgang Gaschler
BASF SE
Ludwigshafen am Rhein, Germany

Mihail T. Georgiev
Department of Chemical Engineering
Sofia University
Sofia, Bulgaria

Gospodinka Gicheva
Department of Chemistry
University of Mining and Geology "St. Ivan Rilski"
Sofia, Bulgaria

Georgi Gochev
Max Planck Institute of Colloids and Interfaces
Potsdam-Golm, Germany

Dmitry Grigoriev
MPI für Kolloid- und Grenzflächenforschung
Golm, Germany

Theodor D. Gurkov
Department of Chemical Engineering
Sofia University
Sofia, Bulgaria

Eduardo Guzmán
CNR—Istituto per l'Energetica e le Interfasi
UOS Genova
Genoa, Italy

Aliyar Javadi
Max Planck Institute of Colloids and Interfaces
Potsdam-Golm, Germany

Maria L. Jiménez
Department of Applied Physics, Faculty of Science
University of Granada
Granada, Spain

Mohsen Karbaschi
Max Planck Institute of Colloids and Interfaces
Potsdam-Golm, Germany

and

Sharif University of Technology
Tehran, Iran

Elena K. Kostova
Department of Chemical Engineering
Sofia University
Sofia, Bulgaria

Nina M. Kovalchuk
Institute of Biocolloid Chemistry
Kiev, Ukraine

Volodymyr I. Kovalchuk
Institute of Biocolloid Chemistry
Kiev, Ukraine

Jürgen Krägel
MPI für Kolloid- und Grenzflächenforschung
Golm, Germany

Peter A. Kralchevsky
Department of Chemical Engineering
Sofia University
Sofia, Bulgaria

Ana Kroflič
Faculty of Chemistry and Chemical
   Technology
University of Ljubljana
Ljubljana, Slovenia

Marta Kujda
J. Haber Institute of Catalysis and
   Surface Chemistry
Polish Academy of Science
Cracow, Poland

Kyriacos C. Kyriacou
The Cyprus Institute of Neurology and
   Genetics
Nicosia, Cyprus

Marco Laurenti
Departamento de Química Física II,
   Facultad de Farmacia
Universidad Complutense de Madrid
Madrid, Spain

Adil Lekhlifi
Centre Scientifique de St. Jérôme
Aix-Marseille Université
Marseille, France

Epameinondas Leontidis
Department of Chemistry
University of Cyprus
Nicosia, Cyprus

Libero Liggieri
CNR—Istituto per l'Energetica e le
   Interfasi
UOS Genova
Genoa, Italy

Guiseppe Loglio
Dipartimento di Chimica Organica
Università degli Studi di Firenze
Firenze, Italy

Enrique López-Cabarcos
Departamento de Química Física II,
   Facultad de Farmacia
Universidad Complutense de Madrid
Madrid, Spain

Marzieh Lotfi
Max Planck Institute of Colloids and
   Interfaces
Potsdam-Golm, Germany

and

Sharif University of Technology
Tehran, Iran

Armando Maestro
Departamento de Química Física I
Universidad Complutense de Madrid
Madrid, Spain

Alexander V. Makievski
SINTERFACE Technologies
Berlin, Germany

Krastanka G. Marinova
Department of Chemical Engineering
Sofia University
Sofia, Bulgaria

Abraham Marmur
Chemical Engineering Department
Technion—Israel Institute of
   Technology
Haifa, Israel

Reinhard Miller
MPI für Kolloid- und
   Grenzflächenforschung
Golm, Germany

Boryana Nenova
Department of Chemical Engineering
Sofia University
Sofia, Bulgaria

Tugce Ozdemir
Department of Chemical Engineering
Yeditepe University
Istanbul, Turkey

Kelly Pemartin
Instituto de Química Avanzada
    de Cataluña, Consejo Superior
    de Investigaciones Científicas
    (IQAC-CSIC)
Barcelona, Spain

and

CIBER en Biotecnología, Biomateriales
    y Nanomedicina (CIBER-BBN)
Barcelona, Spain

Jorge Perez-Juste
Departamento de Química-Física y
    Unidad Asociada CSIC
Universidad de Vigo
Vigo, Spain

Francesca Ravera
CNR—Istituto per l'Energetica e le
    Interfasi
UOS Genova
Genoa, Italy

Raúl A. Rica
Department of Applied Physics, Faculty
    of Science
University of Granada
Granada, Spain

Ramón G. Rubio
Departamento de Química Física I
Universidad Complutense de Madrid
Madrid, Spain

Jorge Rubio-Retama
Departamento de QuímicaFísica II,
    Facultad de Farmacia
Universidad Complutense de Madrid
Madrid, Spain

Margarita Sanchez-Dominguez
Centro de Investigacion en Materiales
    Avanzados, S. C. (CIMAV)
Unidad Monterrey, GENES—Group of
    Embedded Nanomaterials for Energy
    Scavenging
Nuevo Leon, Mexico

Deniz Sandal
Department of Genetics and
    Bioengineering
Yeditepe University
Istanbul, Turkey

Eva Santini
CNR—Istituto per l'Energetica e le
    Interfasi
UOS Genova
Genoa, Italy

Amitav Sanyal
Department of Chemistry
Bogazici University
Istanbul, Turkey

Bojan Šarac
Faculty of Chemistry and Chemical
    Technology
University of Ljubljana
Ljubljana, Slovenia

Kamila Sofińska
J. Haber Institute of Catalysis and
    Surface Chemistry
Polish Academy of Science
Cracow, Poland

Conxita Solans
Instituto de Química Avanzada
    de Cataluña, Consejo Superior
    de Investigaciones Científicas
    (IQAC-CSIC)
Barcelona, Spain

and

CIBER en Biotecnología, biomateriales
    y Nanomedicina (CIBER-BBN)
Barcelona, Spain

Rumyana D. Stanimirova
Department of Chemical Engineering
Sofia University
Sofia, Bulgaria

Vamseekrishna Ulaganathan
Max Planck Institute of Colloids and
    Interface
Potsdam/Golm, Germany

Halil Ibrahim Unal
Science Faculty, Smart Materials
    Research Laboratory,
    Chemistry Department
Gazi University
Ankara, Turkey

Kamil Wojciechowski
Faculty of Chemistry
Warsaw University of Technology
Warsaw, Poland

Jooyoung Won
Max Planck Institute of Colloids and
    Interfaces
Potsdam-Golm, Germany

Georgi Yordanov
Faculty of Chemistry and Pharmacy
Sofia University "St. Kliment Ohridski"
Sofia, Bulgaria

Pierre-Léonard Zaffalon
Department of Chemistry
University of Geneva
Geneva, Switzerland

Thomas Zemb
Institut de Chimie Séparative de
    Marcoule (ICSM)
CEA-CNRS-Université Montpellier
    2-ENSCM
Bagnols-sur-Ceze, France

Andreas Zumbuehl
Department of Chemistry
University of Fribourg
Fribourg, Switzerland

Section I

Nanoparticle Synthesis and Characterization

1 Advanced Strategies for Drug Delivery in Nanomedicine

Georgi Yordanov

CONTENTS

1.1 INTRODUCTION

The idea for targeted chemotherapy of diseases has been proposed in the 19th century by the German scientist Dr. Paul Ehrlich (1854–1915), a Nobel laureate (in 1908) for achievements in immunology (Krause 1999; Tan and Grimes 2010; Turk 1994). He is noted also for his contributions in histology, treatment of infectious diseases, and discovering Salvarsan 606—the first drug targeted against a specific pathogen (trypanosome, which causes syphilis) (Gensini et al. 2007; Kaufmann 2008). Dr. Ehrlich popularized the concept of a "magic bullet"—a compound that could selectively target a disease-causing organism conjugated with a toxin for that organism; then, the toxin could be delivered along with the agent of selectivity, and this magic bullet would kill only the targeted organism (Winau et al. 2004). Nowadays, about a century later, this concept became the main aim of nanomedicine that emerged from the

3

converging of medicine with modern nanotechnology and colloid science (Moghimi et al. 2005). The early concepts of nanomedicine sprang from the idea that tiny nano-robots (nanobots) and nanomachines could be designed, manufactured, and introduced into the human body to perform cellular repairs at the cellular and molecular level (Datta and Jaitawat 2006; Freitas 2005). From the viewpoint of colloid and pharmaceutical sciences, the current concept of a magic bullet is interpreted as a submicron colloidal system (usually called a nanocarrier) loaded with therapeutic and diagnostic agents intended for targeted delivery of these agents to a desired location in the body (Caruthers et al. 2007; Rannard and Owen 2009). There are many different colloidal systems for drug delivery (some of which are described shortly in Section 1.4), such as polymer nanoparticles, micelles, liposomes, lipid nanoparticles, albumin nanoparticles, microemulsions, polymeric vesicles, niosomes, dendrimers, and magnetic nanoparticles (Farokhzad and Langer 2006).

Nanomedicine provides a variety of tools for control of drug pharmacokinetics (bioavailability, biodistribution profile, area under the curve, clearance, etc.) without changing the drug's molecular structure and mechanism of action (Vizirianakis 2011). In the ideal case, nanocarriers can provide increased bioavailability of drugs, decreased effective dose, protection of unstable drug molecules from rapid degradation, achievement of high drug concentration in infected or abnormal cells and low drug concentration in normal cells, low drug toxicity and decrease in undesirable side effects, and improved therapeutic index. A comprehensive understanding of the body characteristics seems to be absolutely necessary in order to predict correctly the fate of colloidal systems in the body and to prevent possible inadequate biodistribution and undesirable side effects (Bertrand and Leroux 2012).

In this chapter, we consider the pathophysiological elements that should be taken into account when designing new carriers for delivery or imaging purposes. Nanocarrier systems hold promise for applications in targeted treatment of cancer (Brannon-Peppas and Blanchette 2004; Brigger et al. 2002; Kawasaki and Player 2005; Shapira et al. 2011), infections (Briones et al. 2008; Lisziewicz and Töke 2013; Pinto-Alphandary et al. 2000), vascular diseases (Gupta 2011), neurological diseases (Kabanov and Gendelman 2007), and other severe diseases. The increasing use of nanomaterials raises concerns about the possible toxic effects of nanocarriers. Nanotoxicology has emerged only recently, when various nanomaterials had already been introduced into a number of industrial processes and products (Elsaesser and Howard 2012; Kagan et al. 2005; Nyström and Fadeel 2012). Different materials for the preparation of nanocarriers (synthetic polymers, proteins, lipids, etc.) have different properties, which result in different combinations of advantages and disadvantages. In any case, colloidal drug formulations (also referred to as nanomedicines) should not cause acute and chronic toxicity when applied in therapeutic concentrations, inflammation, blood clot, and instability of blood cells, as well as production of toxic and difficult-to-eliminate products of biodegradation. The following paragraphs contain a short description of the interactions between nanocarrier systems and living organisms. The chapter then continues with consideration of the basic principles for targeting of colloidal drug carriers to pathological locations (mainly focusing on drug delivery to solid tumors), as well as some illustrative examples for drug delivery by colloidal systems. Finally, the chapter concludes with a brief summary of the advantages of nanomedicines, as well as the basic problems facing the development of colloidal systems for targeted drug delivery.

1.2 INTERACTIONS OF COLLOIDAL DRUG CARRIERS WITH PROTEINS AND CELLS

1.2.1 INTERACTIONS BETWEEN PROTEINS AND COLLOIDAL DRUG CARRIERS

Colloidal drug carriers are frequently designed for parenteral administration. Nanocarriers become coated with plasma proteins after their introduction into the bloodstream, forming the so-called protein corona—a complex and variable protein coating around the nanocarrier (Lynch and Dawson 2008). Approximately 50 different proteins from human plasma (which contains as many as 3700 proteins) have been identified to interact with various nanoparticles (Aggarwal et al. 2009). Some of these proteins are opsonins—compounds (such as complement C3b and immunoglobulins) that enhance the uptake of the coated nanocarrier by phagocytic cells (Patel 1992). Therefore, the nanoparticle interactions with plasma proteins relate to particle biodistribution, biocompatibility, and therapeutic efficacy (Aggarwal et al. 2009). The physicochemical properties of nanocarriers (size, shape, zeta potential, surface modifications) as well as the route of administration are known to determine how nanocarriers are distributed in the body (Chithrani et al. 2006; Dobrovolskaia et al. 2008; Moghimi and Davis 1994; Moghimi et al. 2001; Patel 1992; Poznansky and Juliano 1984). However, there are still not enough quantitative data to clearly reveal how the physicochemical properties of nanocarriers affect the binding of plasma proteins. One should keep in mind that the physicochemical properties and the fate of actual colloidal formulation of a drug (a nanomedicine) in the body may differ from that of the drug-free colloidal carrier. There are different methods used to separate (centrifugation, gel filtration, polyacrylamide gel electrophoresis) and identify (mass spectroscopy, protein sequencing, immunoblotting) proteins bound to nanocarriers, which are reviewed in detail elsewhere (Aggarwal et al. 2009).

Most probably, the interactions between nanocarriers and proteins are nonspecific hydrophobic/hydrophilic and electrostatic interactions (including van der Waals) (Bousquet et al. 1999; Liu and Wang 1995; Patil et al. 2007; Sant et al. 2008). Differences among the many types of nanocarriers can result in different protein binding profiles, although the proteins that have been identified on virtually all nanocarriers include mainly albumin, immunoglobulins (IgG, IgM), fibrinogen, and apolipoproteins. It has been suggested that these proteins are found bound to nanocarriers because of their large abundance in blood plasma (Cedervall et al. 2007); however, their affinity to nanocarriers and the kinetics of binding to the nanocarrier surface are not studied enough to clearly understand the dynamic composition of the protein corona in time. The largely abundant proteins may dominate the surface initially but over time may dissociate and be replaced by proteins of lower concentrations, slower adsorption kinetics, or higher affinities (Lynch et al. 2007). For example, kinetic studies on solid lipid nanoparticles have demonstrated that albumin initially predominates and is replaced over time by fibrinogen, which is then replaced by apolipoproteins and other proteins (Göppert and Müller 2005c). Also, it should be taken into account that when the nanocarriers distribute from the blood to various locations (interstitial space, intracellular compartments, lymphatic vessels, etc.), the differences in protein composition and concentrations may affect the evolution of the protein corona and therefore may influence the interactions between nanocarriers and cells.

It has been shown that neutrally charged nanocarriers have a distinctively slower opsonization rate than charged nanocarriers (Owens and Peppas 2006). An increase in plasma protein adsorption has been observed with an increase in the surface charge density but has shown no significant difference in the profile of detected proteins (Gessner et al. 2002). Positively charged colloids have been found to adsorb mainly proteins with isoelectric points less than 5.5 (such as albumin), while negatively charged colloids tend to adsorb proteins with isoelectric points larger than 5.5 (such as IgG) (Gessner et al. 2003). There are many studies that have demonstrated that hydrophobicity may influence the amount of adsorbed proteins, as well as the protein identities (Cedervall et al. 2007; Göppert and Müller 2005a,b; Labarre et al. 2005). Generally, hydrophobic nanocarriers are opsonized more quickly than hydrophilic ones (Carstensen et al. 1992; Owens and Peppas 2006). However, it has been demonstrated that properties such as size, shape, and morphology may influence only the amount of adsorbed protein but not identities of bound proteins (Cedervall et al. 2007).

Protein adsorption can change nanocarrier size and zeta potential (Dutta et al. 2007; Nagayama et al. 2007). These changes may affect the internalization of the nanocarriers into phagocytic cells and the overall biodistribution profile. Some adsorbed proteins (such as IgG, complement factors, and fibrinogen) may enhance the phagocytosis of nanocarriers by macrophages of the reticuloendothelial system (RES), leading to concentration of these nanocarriers in the phagocytes of the liver and spleen (Camner et al. 2002; Göppert and Müller 2005c; Leroux et al. 1995; Nagayama et al. 2007; Owens and Peppas 2006). However, adsorption of proteins like albumin is known to prolong blood circulation times (Göppert and Müller 2005c; Ogawara et al. 2004). Sometimes, it is useful to have some proteins bind as they can target the nanocarriers to a particular location in the body. For example, adsorption of certain apolipoproteins on the nanocarrier surface can be useful for transportation of the nanocarriers across the blood–brain barrier (BBB) (Göppert and Müller 2005b; Kreuter 2001; Olivier 2005). Therefore, drugs like doxorubicin, tubocurarine, and dalargin, which normally have poor transportation through the BBB, have been delivered to the brain using polysorbate 80-coated nanoparticles, which preferentially adsorbed ApoE (Gulyaev et al. 1999; Olivier 2005). It has been shown that covalent attachment of apolipoproteins (mainly ApoE, ApoA-I, and ApoB-100) to nanoparticles may enable drug transport through the BBB (Kreuter et al. 2007). However, it should be taken into account that adsorption of these proteins has also been shown to correlate with rapid uptake into the RES organs, which is a negative effect if increase of circulation time is desired (Owens and Peppas 2006). On the other hand, nanocarriers that are rapidly opsonized (nanocarriers with relatively hydrophobic surfaces or with high absolute value of the zeta potential) and phagocytosed by macrophages can be utilized for delivery of antibiotics to infected phagocytes (Briones et al. 2008; Lisziewicz and Töke 2013; Pinto-Alphandary et al.

2000), for delivery of vaccines to antigen-presenting cells (Peek et al. 2008), or for delivery of drugs and diagnostic agents to the lymph nodes (Hawley et al. 1995).

Modification of the nanocarrier surface with hydrophilic polymers has widely been used to prevent protein binding (or at least to decrease their amount) and change the profile of adsorbed proteins (Moghimi and Hunter 2001). Poly(ethylene glycol) (PEG) is usually used as a hydrophilic polymer for modification of nanocarrier surfaces and therefore such nanocarriers are known as PEGylated. These nanocarriers are less recognizable for phagocytic cells and have prolonged circulation in the bloodstream, for which reason they are usually known as "stealth" carriers or long-circulating carriers (Gref et al. 2000; Owens and Peppas 2006). PEGylation can be achieved by covalently linking, entrapping, or adsorbing PEG chains onto the surface of nanocarriers (Owens and Peppas 2006). The PEG chains can prevent interactions between the nanocarrier surface and the plasma proteins by providing hydrophilicity and steric repulsion of the protein molecules (Jeon et al. 1991; Owens and Peppas 2006). Protein adsorption can be decreased by increasing the molecular mass of PEG (from 2000 to 20,000 Da) as well as by increasing the surface density of PEG (Gref et al. 2000). PEGylation cannot lead to a completely avoided protein adsorption (albumin, fibrinogen, IgG, and apolipoproteins were still detected) but can remarkably decrease the amount of adsorbed proteins. PEGylation also can decrease the uptake of nanocarriers by polymorphonuclear cells, proportionally to the surface density of PEG (Gref et al. 2000). Increasing the molecular mass of PEG chains can also decrease the cellular uptake and increase the blood circulation half-life of nanocarriers (Owens and Peppas 2006; Zahr et al. 2006). It can be concluded that engineering a nanocarrier to specifically adsorb certain proteins for targeting purposes, or engineering nanocarriers with these proteins covalently bound to the carrier surface, can provide many possibilities for the development of nanocarriers for targeted drug delivery. On the other hand, nanocarriers that are designed to bind less proteins (such as PEGylated long-circulating nanocarriers) can be used to deliver drugs to solid tumors via the so-called enhanced permeability and retention (EPR) effect, which involves accumulation of macromolecular and colloidal structures in the tumor interstitium as a result of passage of the colloids through the abnormal and fenestrated blood vessels in the tumor and their retention there because of non-functional lymphatic drainage (see Section 1.3.2).

1.2.2 Endocytosis of Colloidal Drug Carriers

Nanocarrier systems often require site-specific intracellular localization to deliver their drug payload to subcellular locations beneath cell membranes. Nanocarriers can employ multiple pathways for cellular entry, which are currently insufficiently understood. Endocytosis of nanocarriers represents the process of internalization of nanocarriers into living cells and is an important research topic in nanomedicine (Iversen et al. 2011; Sahay et al. 2010c). The different mechanisms of endocytosis are as follows (classified on the basis of cellular proteins involved in the process): phagocytosis, clathrin-dependent endocytosis, and clathrin-independent endocytosis (macropinocytosis, caveolae-mediated endocytosis, caveolae- and clathrin-independent endocytosis) (Sahay et al. 2010c). The intracellular traffic and localization of nanoparticles (Figure 1.1) have been found to depend on the mechanism of nanoparticle endocytosis.

Phagocytosis occurs mainly in phagocytes (macrophages, monocytes, neutrophils, and dendritic cells), although some other types of cells (fibroblasts and epithelial and endothelial cells) may also possess phagocytic activity (Hillaireau and Couvreur 2009). Phagocytosis consists of three steps: (1) opsonization of the nanocarriers in the bloodstream (adsorption of plasma proteins like laminin, fibrinogen, complement factors, and immunoglobulins), (2) recognition (via Fc receptors, complement receptors, etc.) of the opsonized nanocarriers by the phagocytic cell, and (3) internalization of the nanocarrier by the phagocytic cell and formation of a membrane-enclosed vesicle, called the phagosome (that may have different sizes depending on the size of the internalized object, which can range from as little as a few hundred nanometers to dozens of microns). The fast clearance of easily opsonized nanocarrier systems from the bloodstream is due to phagocytosis of the nanocarriers by the macrophages of the RES. Nanocarriers with hydrophilic surfaces are less opsonized and therefore are recognized by phagocytes to a lower extent, which results in longer circulation times of these systems in the bloodstream (see Section 1.2.1).

Clathrin-mediated endocytosis involves engulfment of receptors associated with the respective ligands and formation of a clathrin-coated pit (its formation is due to polymerization of the cytosolic protein clathrin-1 and other assembly proteins). A clathrin-coated vesicle of average size ~120 nm is formed and then the clathrin coat is shed off (Pucadyil and Schmid 2009). The vesicle is then usually fused with the endosomes or could be transported to the trans-Golgi network or could be transported back to plasma membrane for exocytosis (Rappoport 2008). Nanoparticles of poly(ethylene glycol)-polylactide (PEG-*co*-PLA) (Harush-Frenkel et al. 2007, 2008; Lahtinen et al. 2003), poly(lactic-co-glycolic acid) (PLGA) (Panyam and Labhasetwar 2003; Qaddoumi et al. 2004), and chitosan (Huang et al. 2002; Ma and Lim 2003) have been found to enter into cells via clathrin-mediated endocytosis. Nanocarriers can be specially designed with surface modification with specific

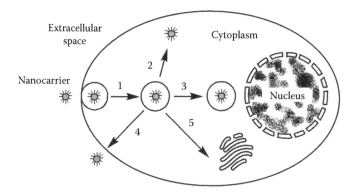

FIGURE 1.1 Illustration of the major intracellular pathways of nanocarriers: (1) endocytosis of the nanocarrier and formation of endosome, (2) endosomal escape and release of the nanocarrier in the cytoplasm, (3) endosome accumulation, (4) exocytosis, and (5) localization to other intracellular compartments (endoplasmic reticulum, Golgi complex).

ligands that can trigger clathrin-mediated endocytosis by interaction with cell surface receptors (Bareford and Swaan 2007). However, one should take into account that the intracellular trafficking of such modified nanocarriers may differ from the trafficking of the respective free ligands (Tekle et al. 2008).

The caveolae are a subset of lipid rafts, cholesterol-rich plasma membrane regions with the presence of the membrane protein caveolin-1 (Simons and Toomre 2000). After formation, the caveolae vesicles transport and fuse with caveosomes or multi-vesicular bodies that have neutral pH (Parton and Simons 2007), although some nanocarriers (like polymeric micelles and PEGylated liposomes) (Sahay et al. 2010c) that utilize the caveolae-mediated endocytosis have been found to localize in lysosomes. It appears that at least in some cases, the caveolae-mediated pathway can bypass lysosomes and this pathway is believed to be beneficial for intracellular delivery of proteins and DNA (Rejman et al. 2006). Paclitaxel-loaded albumin nanoparticles can bind to gp60 (receptor for albumin present in caveolae of endothelial cells) and transport across the vascular walls to the tumor interstitium (Schnitzer 1992). The process of endocytosis of a nanocarrier followed by exocytosis is known as transcytosis, which is most important for transport of nanocarriers through the endothelium (like in the case of albumin nanoparticles described above). Transcytosis through the endothelial cells of brain capillaries could be quite an important strategy to overcome the BBB for successful delivery of bioactive compounds to the brain (Gulyaev et al. 1999; Kreuter et al. 2007; Olivier 2005). This could be achieved via modification of the nanocarrier surface with apolipoproteins that are recognized by low-density lipoprotein receptors on the surface of brain endothelial cells (Göppert and Müller 2005b; Kreuter 2001; Olivier 2005). Different clathrin- and caveolae-independent pathways of endocytosis of nanocarriers also exist (Doherty and McMahon 2009). Nanoparticles and polymers modified with folate can bind to the folate receptor that is usually overexpressed on the membrane of many cancer cells, which could be utilized in drug targeting (Lu and Low 2002). The entry of folate-modified nanocarriers in cells is complex, and along with clathrin- and caveolae-independent endocytosis, it can also involve clathrin-mediated pathway in some cell types (Doherty and McMahon 2009). Multiple mechanisms of entry into cells can be utilized also by other types of nanocarriers, such as micelles of amphophilic triblock copolymer of poly(ethylene oxide) (PEO) and poly(propylene oxide) (PPO) (Batrakova and Kabanov 2008), polystyrene nanoparticles (Lai et al. 2008), dendrimers (Kitchens et al. 2007), and nanocarriers modified with cell-penetrating peptides (Kaplan et al. 2005). Endocytosed nanocarriers can be designed to escape from endosomal compartments and release the loaded drugs in the cytoplasm (from where the drugs may diffuse to the nucleus and other cell organelles). For example, to increase the endosomal escape, PLGA nanoparticles have been modified with a cationic polymer, poly(L-lysine) (Vasir and Labhasetwar 2008). Such nanoparticles that absorb protons in response to the acidification of endosomes (after fusion with lysosomes) can be used to disrupt the endosomes via the so-called proton sponge effect (swelling and increased osmotic pressure) and are thus able to deliver bioactive compounds to the cytosol (Hu et al. 2007).

Factors that determine the mechanism of endocytosis and intracellular fate of nanocarriers are nanocarrier size, charge, shape, and the type of cells used for tests

(Sahay et al. 2010c). For example, positively charged nanomaterials predominantly internalize through clathrin-mediated endocytosis and macropinocytosis and in some cases multiple pathways are utilized (Harush-Frenkel et al. 2007, 2008). Negatively charged nanocarriers (such as PEGylated liposomes and some polymeric micelles) may utilize the caveolae-mediated pathway (Sahay et al. 2010c). Nanocarriers with sharp edges are known to be phagocytosed faster than those with smooth surfaces (Champion and Mitragotri 2009). There are some nanocarriers (such as carboxy-lated polystyrene and PLGA nanoparticles) that can enter cells through caveolae-independent mechanisms (Panyam and Labhasetwar 2003; Qaddoumi et al. 2004). Importantly, it appears that there are different endocytic pathways in normal and cancer cells that may be explored for selective targeting of nanomaterials into tumors. This is the case of using cross-linked polymeric micelles that selectively enter cancer cells but do not enter normal epithelial cells owing to their ability to target differences in endocytic mechanisms (Sahay et al. 2010a). These micelles enter cancer cells via caveolae-mediated endocytosis, which is absent at the apical site of confluent normal epithelial cells.

1.3 BASIC PRINCIPLES OF COLLOIDAL DRUG DELIVERY TO PATHOLOGICAL LOCATIONS

1.3.1 DELIVERY OF BIOACTIVE COMPOUNDS TO PHAGOCYTES

Phagocytic cells (such as polymorphonuclear leucocytes, monocytes, and tissue macrophages) are an essential component of the immune system and their main function is to recognize microorganisms, ingest them by phagocytosis, and destroy them inside phagolysosomes (Ernst and Stendahl 2006). Nanocarriers with relatively hydrophobic surfaces and sizes larger than 100 nm can be easily opsonized, recognized, and engulfed by phagocytic cells (see Section 1.2.2) and therefore can be utilized for delivery of different bioactive compounds (antibiotics, antiviral agents, nucleic acids, and immunoacive substances) to these cells.

Intracellular pathogens (such as *Salmonella* spp., *Listeria monocytogenes, Mycobacterium tuberculosis, Brucella abortus*, and *Legionella pneumophila*) present a major problem because they are especially difficult to eradicate via conventional antibiotic therapies. These microorganisms are able to survive and reproduce after they have been ingested by phagocytic cells, utilizing different survival strategies, such as inhibition of the phagosome–lysosome fusion and resistance to attack by lysosomal enzymes, oxygenated compounds, and other defenses of the host macrophages, and are able to escape from the phagosome into the cytoplasm (Pinto-Alphandary et al. 2000). The intracellular location of these microorganisms protects them also from the host defense mechanisms (antibodies, complement factors, etc.) and antibiotics, which may encounter difficulties in reaching the phagolysosomes inside phagocytes. The intracellular activity of antibiotics is dependent on many factors (Silverstein and Kabbash 1994), although the poor penetration into the infected cells and decreased intracellular activity are the major limitations for most antibiotics (penicillins, cephalosporins, aminoglycosides) in intracellular infections. Another problem with antibiotic therapy of intracellular infections is that many intracellular microorganisms are dormant and can persist for extended periods (Kaprelyants et al. 1993), which results in a dramatic decrease of their susceptibility to antibiotics (Gilbert et al. 1990). By loading the antibiotics into colloidal carrier systems, one can expect improved delivery to infected cells (Briones et al. 2008; Pinto-Alphandary et al. 2000). Antibiotic carrier systems vary in nature (liposomes, micro- and nanoparticles, nanosuspensions and conjugates with water-soluble polymers and with lipoproteins, etc.). The utilization of various antimicrobial agents formulated with different colloidal delivery systems is reviewed in detail elsewhere (Briones et al. 2008). Drug-loaded colloidal systems can be phagocytosed by phagocytic cells, thus leading to targeted delivery of antibiotics to phagocytes and increased intraphagocytic drug accumulation. Moreover, the ingestion of such colloidal systems may involve macrophage activation, increasing the immune response of the host (Prior et al. 2002). Also, it should be taken into account that phagocytes are also able to transport drugs to the site of infection by chemotactic mechanisms. Therefore, antibiotics that are concentrated in phagocytes could be released at the site of infection, which may act against extracellular microorganisms located there (Gladue et al. 1989). The colloidal systems used to target drugs to phagocytes must be biocompatible and biodegradable, stable enough to deliver the drug load to phagocytes and should also have some specific characteristics (Briones et al. 2008): they should be able to solve technical problems with drug solubility, stability, and loading efficiency; once in the bloodstream, they must be rapidly opsonized, recognized, and ingested by phagocytes, leading to elevation of the drug concentration in the target cells (phagocytes), where the pathogen is located; they should prevent drug release until phagocytes are reached and allow sustained drug release inside phagolysosomes; they should allow increased drug retention in the infected location; drug permeation in bacteria should be high enough to avoid development of drug resistance; if possible, they should minimize undesirable side effects. The colloidal nanocarriers can be actively targeted to phagocytes by binding of ligands, which are recognized by specific scavenger receptors of phagocytic cells, such as glycoproteins or polysaccharides ending in mannose or fucose residues, and polyanionic macromolecules (such as acetylated low-density lipoproteins) (Hu et al. 2000; Vasir et al. 2005). Once the phagolysosomes are reached, the drug can be released from the carrier by various mechanisms, which depend on the nature of the delivery system. It can involve passive diffusion from the nanocarrier or release after degradation and bioerosion of the carrier material, chemical cleavage from the nanocarrier (in the cases of drug–carrier conjugates), osmotic pressure effects, stimuli-responsive drug release (such as induced by the change in pH), and others (see Section 1.3.4). Examples of antibiotic-loaded colloidal nanocarriers used to treat intracellular pathogens include liposomes, nanoparticles, nanosuspensions, lipoproteins, dendrimers, polymer conjugates, and lipoproteins (reviewed by Briones et al. 2008).

Colloidal drug delivery systems can be utilized also for the treatment of fungal infections. The intracellular location of the fungal pathogens requires high doses of drugs in order to reach therapeutic levels inside the cell, leading to toxic side effects. For example, drugs like amphotericin B can be accumulated in phagocytes, increasing the antifungal activity of macrophages and polymorphonuclear cells (Jahn et al. 1998; Vyas et al. 2000). The colloidal delivery system for delivery of amphotericin

B, based on liposomes, allows administration of higher doses and decreased nephrotoxicity of the drug (Gibbs et al. 2005).

Intracellular delivery of antiviral drugs to phagocytes seems to be important for improvement of treatment of the HIV infection (Bender et al. 1996; Briesen et al. 2000; Löbenberg et al. 1998). The monocyte–macrophage system is the first to be infected by HIV, which supports the intracellular replication of the virus and protects the virus against antiretroviral drugs. The use of colloidal carriers allows sustained drug release and lower drug doses (Briesen et al. 2000). The ability of nanocarriers to target drugs to infected cells is increased because macrophages infected by HIV show greater phagocytic activity than healthy phagocytes (Lanao et al. 2007; Schafer et al. 1992). Nanoparticles can be used to increase the antiviral effect of acyclovir against herpes simplex virus type 1 (Cavalli et al. 2009), as well as for delivery of antiviral drugs to the liver (Zhang and He 1999), which may be beneficial for the treatment of liver-located viral infections.

Colloidal carrier systems pursued as vaccine delivery systems and immunopotentiators (especially for stimulation of cell-mediated immunity) are the focus of intensive research (Park 2010; Peek et al. 2008). The idea behind using nanocarriers is to increase the delivery of vaccine components to antigen-presenting cells (such as dendritic cells and macrophages). The antigens (peptides) or plasmid DNA (encoding for an antigenic peptide) can be loaded in nanocarriers that are designed to be phagocytosed by antigen-presenting cells. Liposomes have already been approved as vehicles for vaccine delivery (Altin and Parish 2006; Gregoriadis et al. 2006); however, other colloidal systems are also studied as potential carriers for peptides/proteins (Calvo et al. 1997; Waeckerle-Men and Groettrup 2005) and nucleic acids (Gregoriadis et al. 2006). Dendritic cells can capture the nanocarriers with loaded antigen (or DNA) by phagocytosis and transport the antigens from epithelia and tissues to lymph nodes, where antigens are presented to naïve T-cells. T-cells are thus activated to proliferate and differentiate into effector and memory cells that serve various functions in cell-mediated immunity.

1.3.2 EPR EFFECT

In 1986, a new concept for macromolecular therapeutics in cancer chemotherapy called the EPR effect (Figure 1.2) has been demonstrated (Matsumura and Maeda 1986), which has been later recognized as the greatest breakthrough leading to more general targeted antitumor therapy (Torchilin 2011). These authors showed that macromolecular anticancer drugs could be selectively accumulated in solid tumors owing to the unique pathological microenvironment of most solid tumors, characterized mainly by highly permeable vasculature and impaired lymphatic drainage in comparison with normal healthy tissues (Danhier et al. 2010). The blood vessels in normal tissues are continuous with no gaps between the endothelial cells, which are bound together by tight junctions and covered by smooth muscle cells. These tight junctions stop large particles in the blood from leaking out of the vessel, while smooth muscle cells may change the size of the lumen and participate in the regulation of blood pressure.

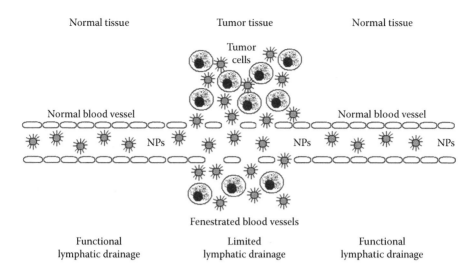

FIGURE 1.2 Illustration of the EPR-based accumulation of nanomedicines in solid tumors.

Most solid tumors have blood vessels with defective architecture and usually produce extensive amounts of various vascular permeability factors, which make the tumor vasculature leaky, exhibiting enhanced permeability to ensure a sufficient supply of nutrients and oxygen to rapidly growing tumor tissues (Maeda 2012). In order to be effectively targeted to solid tumors, macromolecules must have molecular mass larger than 40 kDa, which is the renal clearance threshold. For example, albumin (67 kDa) is easily concentrated in solid tumors via the EPR effect, which makes this protein interesting for development of anticancer chemotherapeutics (Kratz 2008). The first drug described to accumulate in solid tumors via the EPR effect was the antitumor agent SMANCS (neocarzinostatin conjugated with synthetic copolymer) (Matsumura and Maeda 1986), which has been shown to be highly effective in liver and renal cancers (Maeda 2012). Currently, the EPR effect is the gold standard in anticancer drug design and anticancer strategies in nanomedicine (Torchilin 2011). However, there are still many problems with that strategy. Most importantly, the nanocarriers should not be recognized by phagocytic cells in order to ensure long circulating times of the nanocarriers in the bloodstream. At least a few hours of circulation are required to achieve effective accumulation of the nanocarriers in solid tumor via the EPR effect. This can be achieved by modification of the nanocarrier surface with hydrophilic compounds (polymers like PEG) or adsorption of anti-opsonins (like albumin) (see Section 1.2.1). One interesting example is the alternating copolymer poly(styrene-*co*-maleic acid), which may associate with albumin to form a complex suitable for EPR-based targeting to solid tumors. This strategy has been utilized for the development of SMANCS (Matsumura and Maeda 1986), as well as for the development of a macromolecular drug by conjugation of anthracyclines with poly(styrene-*co*-maleic acid) (Greish et al. 2004, 2005). A remarkable result with pirarubicin conjugated with poly(styrene-*co*-maleic acid) has been achieved during in vivo experiments with a model of mouse sarcoma, where complete tumor eradication in 100% of tested animals has been reported (with 20 mg/kg pirarubicin equivalent) and mice survived for more than 1 year after treatment with the highest micellar drug doses tested (100 mg/kg pirarubicin equivalent) (Greish et al. 2005). It must be noted that similar results with an analogous formulation of doxorubicin have been achieved at a fivefold higher concentration of the drug (Greish et al. 2004), clearly indicating the higher activity of the pirarubicin-based formulation. Pathophysiological barriers that represent limitations against the EPR-based strategy for drug delivery also exist (Fang et al. 2011; Maeda 2010). Solid tumors are usually hypoxic and may contain necrotic areas and embolized blood vessels, which makes their structure highly heterogeneous and represents a serious limitation for the successful delivery of nanomedicines to all areas of the tumor. Increased interstitial fluid pressure in the tumor is also a limiting factor. Investigations of Maeda and coworkers have demonstrated that these limitations can be at least partially overcome and the efficiency of the EPR-based drug targeting of nanomedicines can be improved by controlled increase of the systemic blood pressure (by administration of angiotensin II or ACE inhibitors), as well as by administration of nitroglycerin or other NO donors (Fang et al. 2011; Maeda 2001, 2010, 2012). In both cases, the permeability of the tumor blood vessels is increased, thus favoring the accumulation of nanomedicines in the tumor interstitium. To achieve increased tumor accumulation and delivery to local lymph metastasis, nanomedicines could be applied by injection into the tumor-feeding artery via a catheter (Nagamitsu et al. 2009). It should be taken into account that the EPR-based accumulation of nanomedicines in solid tumors is just the first stage toward the successful treatment of the tumor. The delivery of enough amounts of drug inside the cancer cells for a prolonged period depends on the properties and stability of the drug and the carrier, the characteristics of the tumor microenvironment, the strength of drug association with the carrier system (the kinetics of drug release), and the efficiency of drug penetration into cancer cells. The efficiency of EPR-based delivery of nanomedicines probably could be further increased by combining the EPR effect with active targeting of the nanomedicines, as described in Section 1.3.3.

A serious complication during development of solid tumors is the occurrence of metastasis, usually in the regional lymph nodes and distant metastasis at later stages. Therefore, targeting of anticancer drugs to lymph node metastasis becomes highly important for prevention and eradication of metastasis. The route of administration of colloidal drug carriers largely determines their biodistribution profile. For achievement of higher lymphatic accumulation of nanomedicines, they should be administered by subcutaneous or intraperitoneal injection. The factors that influence the lymphatic localization of colloidal drug carriers and illustrative examples are considered in detail elsewhere (Cai et al. 2011; Hawley et al. 1995; Nishioka and Yoshino 2001). Smaller nanoparticles (<100 nm) are favored for effective lymph node targeting because they may diffuse faster through the intercellular space. Most of the colloids are actually captured by the fixed macrophages in the lymph nodes (therefore, nanocarriers with relatively hydrophobic surfaces that are easily opsonized are usually effectively accumulated in lymph nodes), but these macrophages can become like drug reservoirs from where the drug may diffuse to the local cancer cells. It should be taken into account that the lymph flow can be decreased or blocked in nodes that are invaded by tumor metastasis, which may decrease accumulation of colloids that enter the lymph nodes via the afferent lymphatic vessels.

1.3.3 NANOCARRIERS FOR ACTIVE TARGETING

The concept of active targeting in nanomedicine is based on the idea for attaching targeting ligands on the surface of nanocarriers, thus enabling them to specifically recognize the respective target (Byrne et al. 2008). This concept is actually based on the pioneering ideas of Dr. Paul Ehrlich for using magic bullets during the dawn of immunology in the beginning of the 20th century (Winau et al. 2004). The surface of nanocarrier systems could be modified with a targeting ligand (such as a substrate or antibody that can recognize specifically a receptor on the cell surface), thus increasing the selectivity of nanocarriers toward specific target cells that express receptors for binding with the targeting ligand. The active targeting concept gained great popularity in nanomedicine during recent decades. Modification of the nanocarrier surface with apolipoproteins to target cells that express lipoprotein receptors (Kreuter et al. 2007) and modifying the nanocarrier surface with opsonins to target them to phagocytic cells (Hu et al. 2000; Vasir et al. 2005) are examples for utilization of the active targeting strategy. The active targeting concept became quite popular in the development of nanomedicines for cancer treatment, but it seems that many problems still exist with its application in real cancer chemotherapy. Both cancer cells and tumor endothelium can be used as potential targets (Danhier et al. 2010). One strategy is to increase the intracellular accumulation (e.g., by endocytosis) of the nanomedicine by designing delivery systems targeted to endocytosis-prone surface receptors that are overexpressed in cancer cells (Kirpotin et al. 2006), such as folate receptor (Low and Kularatne 2009), transferrin receptor (Daniels et al. 2006), epidermal growth factor receptor (EGFR) (Acharya et al. 2009; Lurje and Lenz 2009; Scaltriti and Baselga 2006), some glycoproteins recognizable by lectins (Minko 2004), and so forth. The concept of targeting nanomedicines to these receptors in order to achieve improved cancer chemotherapy on a surface appears to be simple, but actually it may be quite complicated if not impossible to obtain improved therapeutic outcome with in vivo induced cancers. There are many reports that demonstrate the superior efficiency of actively targeted nanomedicines when tested in vitro on cancer cell lines, and in most cases, such delivery systems increase the intracellular accumulation of the drug, thus leading to a higher cytotoxicity. However, few, but serious, problems with the in vivo application of these nanomedicines may arise. First, cancer cells in induced tumors (e.g., like those induced by carcinogens and pathogens) are quite heterogeneous in antigen presentation and most probably not all of them overexpress the targeted receptor (which is different from tumors obtained by implantation of cancer cells from a single cancer cell line into animals; in this case, the implanted cells and their progeny express the same receptors and could be easily targeted). Second, healthy cells that are normally actively dividing (such as bone marrow) also express more of the same receptors for growth factors, folate, and transferrin; therefore, the nanomedicines could also target these healthy cells. Third, nanomedicines decorated with antibodies or other targeting ligands may interact nonspecifically with plasma opsonins and can be recognized by phagocytes, which will reduce their circulation lifetime in blood and therefore may decrease their effective accumulation at the tumor site. An interesting approach that may overcome some of these problems could be modification of the nanocarrier surface with a targeting ligand and long PEG chains to hide the ligand and to reduce opsonization, while the PEG chains are conjugated with the nanocarrier by covalent bonds unstable in acidic medium. For example, this strategy has been utilized for the development of double-targeted PEGylated liposomes (Sawant et al. 2006). Inside the acidic microenvironment of the tumor, the long PEG chains of these liposomes could be detached from the nanocarrier, resulting in exposure of the targeting ligand hidden inside the PEG layer. The ligand can then interact with the cell membrane and facilitate endocytosis of the drug-loaded nanocarrier into the cancer cells. However, it should always be taken into account that there are clear indications that PEG (although considered as the gold standard in the design of long-circulating nanomedicines) attached to nanomaterials can activate the complement, and otherwise engage immune response (Arima et al. 2008).

Another strategy for active targeting of nanomedicines to solid tumors is by utilizing ligands that target the tumor endothelium (Lammers et al. 2008). This strategy is based on that the tumor growth could be inhibited by preventing blood and nutrient supply to the tumor by inhibiting the formation of new blood vessels (angiogenesis). Simple embolization of tumor-feeding arteries seems to be ineffective because soon new blood vessels are developed laterally around the tumor. Therefore, a more effective strategy seems to be the development of ligand-targeted nanocarriers that bind to and kill angiogenic blood vessels and, indirectly, the tumor cells that these vessels supply with nutrients and oxygen. Targeting the tumor endothelium has more advantages than targeting individual cancer cells: (i) nanocarriers do not need to pass through vessel walls in order to reach their targets, (ii) decreased risk of development of drug resistance (endothelial cells are more genetically stable than cancer cells), and (iii) the target endothelial receptors are expressed independently of the tumor type, which makes this strategy more universal. Examples for endothelial targets include the following: vascular endothelial growth factor receptors (VEGFR-1 and VEGFR-2) (Carmeliet 2005), $\alpha_v\beta_3$ integrin (endothelial cell receptor for extracellular matrix proteins; it can bind to arginine–glycine–aspartic acid [RGD] tripeptide sequence, and RGD modification, therefore, has been used to direct nanocarriers to capillary endothelial cells of the angiogenic blood vessels) (Desgrosellier and Cheresh 2010), vascular cell adhesion molecule-1 (VCAM-1) (Dienst et al. 2005), membrane type 1 matrix metalloproteinase (MT1-MMP) (Genis et al. 2006), and others.

1.3.4 STIMULI-RESPONSIVE NANOCARRIERS

The use of stimuli-responsive nanocarriers offers an interesting opportunity for drug and gene delivery in optimization of therapy (Ganta et al. 2008; Torchilin 2009). Various stimuli, such as changes in the pH, temperature, and redox potential, could be used to control the physicochemical properties of nanocarriers. The pH in pathological areas (inflammation, infection, and cancer) is significantly different from that of the normal tissue. The pH is on average lower in the tumor tissue (pH ~6.5

or less) than in normal tissue (pH ~7.4), which is a clear indication of the different metabolic activity of the normal and cancer cells (Wike-Hooley et al. 1984). This is a result from the insufficient oxygen in tumors (hypoxia) and therefore the intensive anaerobic glycolysis in the tumor cells leading to the formation of lactic acid. The intracellular pH in cancer cells is close to the pH inside normal cells and therefore the transmembrane gradient in pH can be exploited for the intracellular accumulation of weakly acidic anticancer drugs (Gerweck and Seetharaman 1996). Such drugs would be in their non-ionized form in the acidic extracellular space and may diffuse freely across the cell membrane. Upon reaching a relatively basic intracellular compartment, such drugs may become anionic and therefore trapped within the cell. Nanocarriers can be specially designed to change their physicochemical properties and to release the entrapped drugs upon reaching the acidic tumor microenvironment, thus increasing the drug accumulation in the cancer tissue (Schmaljohann 2006).

Temperature differences between pathological and normal areas may also be exploited for stimuli-responsive drug delivery. Nanocarrier-mediated delivery of anticancer drugs can be positively influenced by localized increase of temperature (hyperthermia) (Ponce et al. 2006). There are at least two factors that may contribute for enhanced delivery of anticancer drugs to tumor cells as a result of local temperature increase: (i) drug nanocarriers can be designed to be temperature responsive and to change their physicochemical properties or drug release rate upon local change in temperature, and (ii) tumor cells seem to be more sensitive to heat-induced damage than normal cells. Recently, the majority of clinical studies of hyperthermia have utilized super-paramagnetic iron oxide–containing nanocarriers (Gupta et al. 2007). Changes in the redox potential could also be used as a stimulus for inducing changes in the properties of nanocarrier systems, taking into account the fact that a high redox potential difference may exist between the reducing intracellular space and oxidizing extracellular space (Schafer and Buettner 2001). For example, gene-delivery systems containing disulfide linkages can be taken up by endocytosis and may undergo disulfide cleavage in the lysosomal compartments (Collins et al. 1991).

1.4 ILLUSTRATIVE EXAMPLES FOR TARGETED DRUG DELIVERY BY NANOCARRIER SYSTEMS

Here, some basic types of colloidal systems for drug delivery are described, focusing mainly on the different strategies that can be utilized to make these systems suitable for targeted delivery to pathological locations. There are already drug formulations approved for clinical use on the basis of nanocarrier systems (like liposomes and albumin nanoparticles), although most of the known nanocarrier systems are still under development and in preclinical or clinical trials.

1.4.1 LIPOSOMES

Liposomes are vesicular systems with an aqueous interior surrounded by one or more phospholipid bilayers with a diameter ranging from ~30 nm to several microns

(Barenholz 2001). The phospholipid bilayer is usually composed of phospholipids (such as phosphatidylcholine and PEGylated phospholipids) and usually also contains cholesterol, which makes liposomes more stable in the bloodstream. Hydrophobic compounds can be dissolved into the phospholipid membrane, and hydrophilic ones can be dissolved in the aqueous interior. The major types of liposomes are the multilamellar vesicles (up to several microns in diameter), the small unilamellar vesicles (30–70 nm in diameter), and the large unilamellar vesicles (70–120 nm in diameter). The loaded drugs can be released as the liposomes are broken down by enzymes or by the fusion of the liposome bilayer with other bilayers such as the cell membrane or endosomal membrane. There are four major types of liposomes that differ in their surface chemistry (Figure 1.3): (i) classical non-modified liposomes, (ii) PEGylated liposomes with PEG chains attached to the surface of liposomes that makes them more stable and long circulating, (iii) immunoliposomes with antibodies attached to the liposome surface intended to specifically recognize target receptors and antigens, and (iv) liposomes modified with stimuli-responsive materials, such as temperature- and pH-sensitive polymers. Current research is focused on the development of multifunctional nanocarriers that combine the advantages of the basic types leading to improved efficacy (Torchilin 2009).

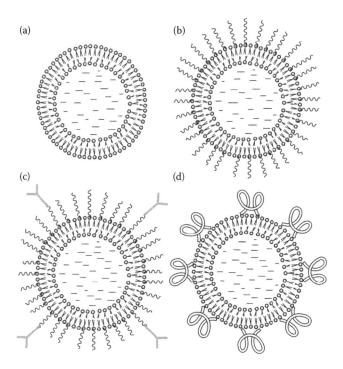

FIGURE 1.3 Schematic diagram of various liposomes: (a) non-modified liposome, (b) PEGylated liposome, (c) immunoliposome, and (d) liposome modified with stimuli-responsive polymers.

Liposomes, which do not have hydrophilic polymers attached to their surface, are rapidly recognized by phagocytic cells and are concentrated in endosomal compartments. Liposomes can be made in a particular size range that makes them targets for phagocytosis. These liposomes may be digested in the phagosome, thus releasing its drug content. Liposomes can be decorated with opsonins and other ligands to activate endocytosis in other cell types. This makes them suitable candidates for delivery of antibiotics (Pinto-Alphandary et al. 2000) and vaccines (Gregoriadis et al. 2006; Peek et al. 2008). Liposomes can be targeted to lymph nodes via subcutaneous injection, which may improve drug delivery to the lymphatic system (Oussoren and Storm 1998). This strategy may be useful in the treatment of diseases with lymphatic involvement such as tumor metastases, viral and bacterial infections, and immunization. Long-circulating liposomes are prepared by using PEGylated phospholipids and can be used for passive targeting of drugs to solid tumors via the EPR effect (such as the PEGylated liposomal formulation of doxorubicin, which is approved for clinical use, Doxil) (Allen et al. 1995; Barenholz 2001, 2012; Maruyama 2011; Slingerland et al. 2012). Different anticancer drugs such as doxorubicin, camptothecin, and daunorubicin are currently being marketed as liposome formulations. Liposomal cisplatin is used to treat pancreatic cancer (Yang et al. 2011). Indeed, PEGylated liposomal formulations resulted in increased half-life in plasma, decreased drug clearance, increased tumor accumulation, and decreased toxicity (mainly decreased cardiotoxicity of anthracyclines). There are some indications that PEGylated liposomes may enter cells via caveolae-mediated endocytosis and are localized in endosomes (Sahay et al. 2010c). Efficiency of doxorubicin formulated in PEGylated liposomes has been found to be equivalent to that of the free drug in cases of metastatic breast cancer and multiple myeloma, but drug toxicity has been decreased. Despite the overall tolerability of Doxil over free doxorubicin, two side effects not typical of what is observed for the free drug were observed for Doxil. The first and more dominant one, which may appear as a dose-limiting factor, is grade 2 or 3 of desquamating dermatitis and is referred to as palmar–plantar erythrodysesthesia (PPE) or "foot and hand syndrome," which appears as redness, tenderness, and peeling of the skin (Solomon and Gabizon 2008). The PPE is dose dependent and so far there is no complete solution to this effect, except increasing the interval between treatments (from 3 to 4 weeks). Another adverse effect of the PEGylated liposomal doxorubicin is an infusion-related reaction that shows up as flushing and shortness of breath. This is an adverse immune phenomenon that many other nanosystems can provoke and is actually a complement activation–related pseudo-allergy, which can be reduced by slowing the infusion rate and by premedication (Szebeni et al. 2011).

Liposomes can also be designed to deliver bioactive compounds in other ways (Andresen et al. 2005). Liposomes intended for the active targeting of cancer cells have been developed by conjugation of phospholipids with various targeting ligands, such as folate, antibodies (usually conjugated with liposomes via PEG linker), and cell-penetrating peptides. Interesting double-targeted liposomes have been developed by modification of PEGylated liposomes with targeting antibodies exposed at the surface (above the PEG layer) and with cell-penetrating peptides hidden within the PEG layer (Sawant et al. 2006). The bond between the PEG and the liposome surface is designed to be labile in acidic medium. The PEG chains are designed to be detached from liposomes in the acidic microenvironment of a solid tumor, thus revealing the hidden cell-penetrating peptides, which can then interact with the cell membrane of target cells.

Liposomes can be modified with pH-sensitive components to achieve pH sensitivity. Such pH-sensitive liposomes can be endocytosed and the acidic pH in endosomes may trigger liposomes to fuse with the endosomal membrane and release its bioactive load into the cytoplasm (Simoes et al. 2004). Such pH sensitivity of liposomes can be achieved by changing the lipid composition or by modifying liposomes with various pH-sensitive polymers. For example, pH-sensitive phosphatidylethanolamine can be incorporated into the liposomal bilayer resulting in the formation of liposomes that are stable in the blood but undergo phase transition at acidic endosomal pH, which facilitates cytoplasmic delivery of bioactive substances (Litzinger and Huang 1992). It is supposed that the head group of phosphatidylethanolamine becomes protonated in the acidic endosomes, thus destabilizing the liposomal bilayer and causing the release of liposomal contents (Ellens et al. 1984). Polymers, such as *N*-isopropylacrylamide copolymers, have been demonstrated to be suitable for the preparation of pH-sensitive liposomes (Leroux et al. 2001). Temperature-sensitive liposomes (Yatvin et al. 1978) that release their contents at temperatures above physiological levels are widely investigated (Grüll and Langereis 2012). By using such liposomes, targeted delivery of bioactive substances can be achieved by local hyperthermia. Such liposomes can be prepared by utilizing a proper ratio between phospholipid (dipalmitoylphosphatidylcholine and distearoylphosphatidylcholine) components or by using thermosensitive polymers (Kono 2001). These polymers undergo a coil-to-globule transition as the temperature changes, which results in alteration of the conformation of the polymer chains and their hydrophobicity, thus changing the surface properties of the modified liposomes. In such a way, liposomes can be destabilized and their contents released or the interactions of these liposomes with cells can be changed. Disulfide (–S–S–) bonds can be utilized as linkers to prepare liposome conjugates where the disulfide bond is critical to liposomal stability. Such liposomes can be destabilized upon reaching a reducing environment that reduces the disulfide bonds, disrupting the liposomal membrane and releasing the liposomal contents (Kevin and Otto 2005). For example, redox-sensitive liposomes have been prepared by conjugation of liposomes with PEG coating via disulfide links (Kirpotin et al. 1996). Such liposomes are expected to be long circulating, to accumulate in solid tumors via the EPR effect, and then to release their contents into target cells upon a redox stimulus.

1.4.2 Albumin Nanoparticles

Albumin is recognized as a protein carrier for drug targeting and for improving the pharmacokinetic profile of various drugs, prodrugs, drug conjugates, and nanoparticles (Kratz 2008). Albumin is the most abundant protein in human plasma, synthesized in the liver, with a molecular mass of 66.5 kDa. Its half-life in blood plasma is 19 days and performs many different functions in the organism (Peters 1985). It binds to fatty acids, bilirubin, metal ions in blood, and a great number of therapeutic drugs; it regulates the colloid osmotic pressure of the blood and may provide nutrition to peripheral tissue by breaking down into amino acids. Its stability, preferential uptake in solid tumors and inflamed tissue, availability, biodegradability, and lack of toxicity and immunogenicity make it an ideal biomaterial for drug delivery (Kratz 2008). Albumin-binding drug derivatives have been developed to target solid tumors. Prodrug derivatives intended to conjugate with endogenous albumin by targeting the cysteine-34 position in albumin seem to be quite perspective (Kratz 2008). The proof of concept has been obtained with acid-sensitive doxorubicin prodrugs—like the (6-maleimidocaproyl)hydrazone derivative of doxorubicin (DOXO-EMCH)— that have been found to rapidly and selectively bind to circulating albumin within a few minutes and dramatically improve the efficacy of doxorubicin in preclinical tumor models (Kratz et al. 2002, 2007). The DOXO-EMCH formulation resulted in increased tumor accumulation in time, but the major difference in comparison with the free doxorubicin seems to be the decreased cardiotoxicity and decreased mitochondrial damage that allows a two- to fivefold increase in the maximum tolerated dose. Another interesting example of a complex of anticancer prodrug with albumin that results in dramatically increased therapeutic efficacy is a formulation containing pirarubicin conjugated with poly(styrene-*co*-maleic acid) (Greish et al. 2005). Remarkably, during in vivo experiments with a model of mouse sarcoma, complete tumor eradication in 100% of tested animals has been observed with this formulation.

Attempts to develop albumin nanoparticles (100–300 nm in diameter, usually prepared by desolvation and cross-linking of albumin) loaded with anticancer drugs also resulted in successful formulations. A good example is the Food and Drug Administration–approved drug Abraxane, which is a nanoparticle albumin-bound form of paclitaxel (Desai et al. 2006). It has been shown that this formulation may target solid tumors, enhance tumor penetration, and minimize toxicity compared to classical formulations of paclitaxel. Drug-loaded albumin nanoparticles may reach solid tumors via the EPR effect, as well as via transcytosis and uptake into endothelial cells, mediated by binding of albumin to a cell surface 60-kDa glycoprotein (gp60) receptor. Treatment with albumin-bound paclitaxel resulted in increased response rates compared with the Cremophor-based formulation of paclitaxel and prolonged time to tumor progression. In addition, survival has been increased and incidence of neuropathy has been decreased in patients receiving nanoparticle albumin-bound paclitaxel (Gradishar et al. 2005). These results motivated further investigation aimed to increase the efficacy of albumin nanoparticles as carriers of anticancer drugs, mainly by modification with targeting ligands, such as folate, transferrin, and antibodies (Bae et al. 2012; Ulbrich et al. 2011; Wagner et al. 2010). Indeed, most of these experiments demonstrated increased cellular uptake of drug and increased cytotoxicity *in vitro*, but the therapeutic efficacy of these formulations should be further examined by in vivo tumor models. It has been shown that attachment of apolipoproteins on the surface of albumin nanoparticles may be used to target these carriers to the brain (Kreuter et al. 2007). This strategy can be utilized to increase the transport of drugs through the BBB that may be beneficial for the treatment of neurological and other brain diseases, such as brain cancers.

1.4.3 POLYMERIC MICELLES

Polymeric micelles are considered as among the most promising modalities of drug carriers (Nishiyama and Kataoka 2006b). It is well known that block amphophilic copolymers at concentrations above the critical micellar concentration (CMC) spontaneously assemble into polymeric micelles with a diameter of several tens of nanometers in aqueous media. Polymeric micelles can be easily prepared, reaching efficient drug loading and controlled drug release, as well as easily sterilized by filtration taking advantage of their small size. These micelles are more kinetically stable than micelles from low-molecular surfactants, have a core–shell structure (with an inner hydrophobic core serving as a nano-container of hydrophobic drugs, surrounded by an outer hydrophilic shell, usually composed of PEG), and have demonstrated long circulation lifetime in the bloodstream, thus being suitable for effective tumor accumulation by the EPR-based mechanism (Bae et al. 2005; Nishiyama et al. 2005; Yokoyama et al. 1999). Also, core–shell nanostructures termed polyion complex (PIC) micelles can be prepared by electrostatic interaction between charged block copolymers and oppositely charged macromolecules/drugs that are potentially useful for delivery of nucleic acids (Harada and Kataoka 1999; Kataoka et al. 2001). Different types of polymeric micelles are schematically illustrated in Figure 1.4.

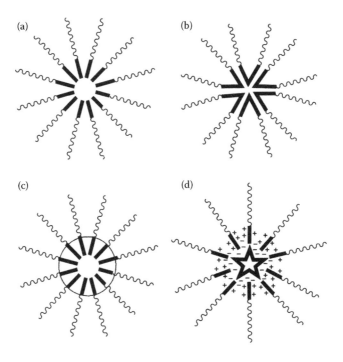

FIGURE 1.4 Schematic diagram of various polymeric micelles: (a) self-assembled micelle of diblock copolymers, (b) self-assembled micelle of triblock copolymers, (c) cross-linked micelle, and (d) PIC micelle.

Self-assembled micelles composed of amphophilic triblock copolymers of PEO hydrophilic blocks and PPO hydrophobic core, such as Pluronic P85 (P85), are an interesting example (Batrakova and Kabanov 2008). Above the CMC, this copolymer forms aggregated micelles of average size ~15 nm. The copolymer may utilize caveolae-mediated endocytosis for entry into cells bypassing early endosomes/lysosomes and eventually reaching the endoplasmic reticulum and mitochondria (Sahay et al. 2010b). Interestingly, the respective micelles internalize into cells through another mechanism—clathrin-dependent endocytosis (Sahay et al. 2008). Also, it has been suggested that Pluronics (poloxamers) could inhibit transmembrane transporter proteins (such as the P-glycoprotein), thus overcoming multidrug resistance in cancer cells (Batrakova et al. 2001; Kabanov et al. 2005). However, it should be noted that micellar formulations of a drug may have a different effect on regulation of some genes in comparison with the free drug (Nishiyama and Kataoka 2006b).

PEGylated micelles composed of poly(ethylene oxide)-b-poly(L-aspartic acid) [PEO-b-P(Asp)] are intensively investigated as a nanocarrier system for delivery of doxorubicin (Matsumura et al. 2004; Yokoyama et al. 1999). The carboxylic groups from the biodegradable P(asp) block have been covalently conjugated with doxorubicin and the conjugate spontaneously forms polymeric micelles with a diameter of 15–60 nm, which may incorporate an additional amount of non-conjugated drug in their cores.

It has been demonstrated that by targeting micelles to caveolae-mediated mechanism of endocytosis, one can achieve selective entry of these nanocarriers into cancer cells but not into normal cells. This has been demonstrated with core-cross-linked polymeric micelles (cl-micelles) of poly(ethylene oxide)-b-poly(methacrylic acid) (PEO-b-PMA) copolymer (Bronich et al. 2005). These cl-micelles enter predominantly in cancer epithelial cells because of their ability to target differences in endocytosis mechanisms between cancer and normal cells (Sahay et al. 2010a). The cl-micelles are endocytosed by cancer cells via the caveolae-mediated pathway, but in adjacent normal cells, this endocytic pathway is absent from the apical side of the cells.

Recent advances in polymer science have allowed the development of "smart" polymeric micelles possessing sensitivity to environmental changes, which can be targeted to specific cells and tissues (Nishiyama et al. 2005; Nishiyama and Kataoka 2006a). Many different pH-sensitive micelles have been described that can exploit the acidic environment at tumor tissue to unload their contents (Torchilin 2009). For example, this could be achieved by attaching functional groups that can be protonated/deprotonated (such as amines or carboxylic acids) to the block copolymers in such a way that the micelle formation is controlled by the protonation state of these groups (Torchilin 2006). Temperature sensitivity is another interesting characteristic in the development of stimuli-responsive polymeric nanocarriers (Torchilin 2009). Thermosensitive polymers display a lower critical solution temperature in aqueous solution, below which the polymers are water soluble and above which they become water insoluble as a result of change in polymer conformation and hydration of the polymer chains.

1.4.4 POLYMER NANOPARTICLES

Polymer nanoparticles for drug delivery applications are usually spherical colloidal structures of sizes much larger than polymeric micelles, being usually around 50–500 nm. Such nanoparticles are composed of low-toxic, biocompatible, and biodegradable synthetic polymers, such as various polyester copolymers with poly(lactic acid) (PLA) (Sahoo and Labhasetwar 2006), like PLGA, PEG-PLA, PEG-PLGA, poly(ε-caprolactone) (Dash and Konkimalla 2012), and homo- and copolymers of poly(alkyl cyanoacrylate) (PACA) (Murthy and Reddy 2006; Yordanov 2012).

Bioactive compounds can be encapsulated, covalently attached, or adsorbed on the surface of such nanocarriers. The extent of drug loading as well as the drug release profile from polymer nanoparticles may depend on the type of polymer and its physicochemical properties, the nature of polymer–drug interactions, particle size and morphology, particle porosity, the rate of bioerosion of the polymer, and so forth. Polymer nanoparticles can be classified into different types: hydrophobic nanoparticles with physically adsorbed surfactants on their surfaces, core–shell nanoparticles with hydrophilic polymers covalently attached on their surfaces, nanocapsules (with oily or aqueous core), hybrid organic–inorganic nanoparticles incorporating inorganic functional nanocrystals, and so on (Figure 1.5).

Polymer nanoparticles are usually prepared by two main approaches: (i) starting from initial monomers that are polymerized (e.g., by emulsion polymerization)

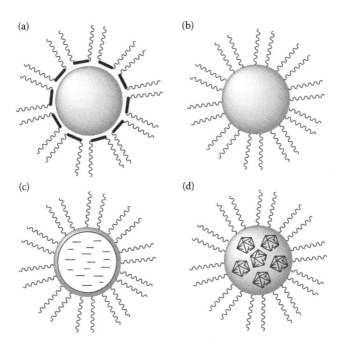

FIGURE 1.5 Schematic diagram of various polymer nanoparticles: (a) nanoparticle stabilized by physical adsorption of surfactants, (b) core–shell nanoparticle, (c) nanocapsule, and (d) hybrid nanoparticle containing inorganic nanocrystals.

and (ii) starting from presynthesized polymer (e.g., by nanoprecipitation and emulsification/solvent evaporation) (Rao and Geckele 2011). These nanocarriers are usually more stable in biological media than nanocarriers based on biopolymers. Also, like in the case of polymeric micelles, the physicochemical properties of nanoparticles can be controlled by modification of the synthetic polymer material used for their preparation. Drugs that are not covalently associated with the polymer can be released from nanoparticles by diffusion, which can be accelerated by bioerosion of the polymer in biological media.

There are various types of polymer nanoparticles: those that differ in their surface properties, their protein binding profiles, and their target locations in the body (Aggarwal et al. 2009). Nanoparticles of the first type are relatively hydrophobic, composed mainly of water-insoluble homopolymers and stabilized in aqueous dispersions by amphophilic compounds that are physically adsorbed on the nanoparticle surface. Such nanoparticles are easily opsonized by plasma opsonins and are rapidly captured by phagocytes, as discussed in Section 1.2.1. Such nanoparticles have a short lifetime in blood circulation and are fast accumulated in the liver, spleen, lungs, and other tissues where macrophages are found. These nanoparticles can be utilized as delivery systems to increase the accumulation of antibiotics or antiviral agents in infected phagocytes (see Section 1.3.1). In vivo experiments with mice have demonstrated the high efficiency of antibiotic-loaded polymer nanoparticles for the treatment of intracellular infections (Briones et al. 2008). The second type of nanoparticles are nanoparticles of the core–shell type, where the core is composed of a hydrophobic polymer capable of entrapping hydrophobic drugs, and the shell is composed of a hydrophilic polymer (usually PEG) to ensure particle stability in aqueous media and decrease opsonization of the nanocarriers. Therefore, such nanocarriers adsorb small amount of opsonins from the blood plasma, which decreases their uptake by phagocytes and prolongs their lifetime in the blood circulation. Such nanoparticles could be used to target anticancer drugs to solid tumors via the EPR effect (see Section 1.3.2). For example, cyanoacrylate nanoparticles with methoxy-PEG onto their surface (PEG-PHDCA) carrying a recombinant tumor necrosis factor alpha (TNF-α) therapeutic load are less recognized by macrophages, which results in increased circulating half-life of TNF-α from 28 min to 11 h (Fang et al. 2006). However, most of the preclinical and clinical studies for cancer treatment with cyanoacrylate nanoparticles have been performed with not such long-circulating carriers, but with homopolymer nanoparticles with physically adsorbed surfactants on their surface. Some surfactants, like polysorbate 80, can enhance adsorption of apolipoproteins from blood plasma and may enhance nanoparticle transport through the BBB (Andrieux and Couvreur 2009; Wohlfart et al. 2012) and increase the drug uptake in various cancer cells (Yordanov et al. 2012). Indeed, using poly(isohexyl cyanoacrylate) nanoparticles as carriers for anticancer drugs has resulted in improved treatment of glioblastoma models (Wohlfart et al. 2011). However, it is rather possible that the EPR effect may also contribute for the enhanced localization of nanoparticles in the tumor, especially in advanced cancers. In phase II clinical trials, mitoxantrone-loaded poly(butyl cyanoacrylate) nanoparticles improved slightly the survival rates in patients with hepatic cancer (Zhou et al. 2009). Similarly to other nanomedicines, polymer nanoparticles can reduce some of the side effects of loaded drugs, but new side effects may appear. For example, it has been shown that entrapment of doxorubicin in poly(isohexyl cyanoacrylate) nanoparticles may reduce its cardiotoxicity but may increase the bone marrow toxicity (Gibaud et al. 1994). Other examples of unexpected side effects are fever, bone pain, and allergic reactions that have been noted in a phase I trial with doxorubicin-loaded poly(isohexyl cyanoacrylate) nanoparticles, although these effects have been well tolerated and were all rapidly reversible (Kattan et al. 1992). The nanoparticles of the third type possess hydrophilic shell and are modified with targeting ligands, such as folate, transferrin, and antibodies. For example, an interesting approach for surface modification of PACA-based copolymer nanoparticles has been developed, which allows the decoration of the nanoparticle surface with various desirable targeting ligands via azide-alkyne "click" chemistry (Nicolas et al. 2008). The therapeutic efficacy of such nanoparticles, however, needs further in vivo evaluations with tumor models.

1.5 CONCLUSIONS

It can be concluded that many different colloidal systems and targeting concepts are currently investigated for their potential applications for targeted drug delivery. However, only few formulations are approved for clinical use in cancer chemotherapy and few other applications (such as liposomal vaccines). Although these formulations proved to be successful at least in decreasing some undesirable side effects, new side effects may appear. Therefore, the emphasis in future development of nanomedicines should be focused on nanocarrier systems composed of nontoxic, non-immunogenic, biocompatible, and biodegradable materials. It seems that there are some anticancer nanomedicines that hold a real promise to be effective, but many preclinical and clinical evaluations still need to be performed. Nanomedicines may have many advantages, such as prolonged plasma lifetime of drugs, decreased clearance, increased drug stability, increased accumulation in pathological locations, decreased side effects of drugs, improved therapeutic index, and capability for stimuli-responsive targeting and drug release. However, many different problems with nanomedicines need further evaluation, such as the in vivo stability of colloidal formulations, adverse effects and toxicity, efficacy of drug targeting, achieving adequate and controlled drug release, and stability upon storage and sterilization. Nanomedicines intended for targeting of tumors need to be tested on more realistic tumor models in vivo (such as chemically induced cancers in animals), where the heterogeneity of antigen presentation in cancer cells may compromise the active targeting concept and the pathoanatomical heterogeneity of the tumor could limit the EPR-based accumulation of nanomedicines. In vitro tests on cell lines may be important for preliminary evaluation of cytotoxicity, drug uptake, and nanocarrier–cell interactions, but these experiments cannot provide any indication about the potential of nanomedicines to accumulate in the pathological location and adequately release the drug under real in vivo conditions. Despite the difficulties in the development of nanomedicines, there are few reports that demonstrate highly efficient formulations capable of achieving complete tumor eradication in tested animals, but such formulations need to be further clinically evaluated on humans. These and many other examples hold promise that successful drug targeting and delivery to pathological locations is possible and may really improve the chemotherapy of severe diseases in the near future.

ACKNOWLEDGMENTS

This work was supported by the Bulgarian National Science Fund (project DMU 03/111, 2011) and COST Action D43 of the European Community.

REFERENCES

Acharya, S., F. Dilnawaz, and S. Sahoo. 2009. Targeted epidermal growth factor receptor nanoparticle bioconjugates for breast cancer therapy. *Biomaterials* 30:5737–5750.

Aggarwal, P., J. Hall, C. McLeland, M. Dobrovolskaia, and S. McNeil. 2009. Nanoparticle interaction with plasma proteins as it relates to particle biodistribution, biocompatibility and therapeutic efficacy. *Adv. Drug Deliv. Rev.* 61:428–437.

Allen, T., C. Hansen, and D. Lopez de Menezes. 1995. Pharmacokinetics of long-circulating liposomes. *Adv. Drug Deliv. Rev.* 16:267–284.

Altin, J., and C. Parish. 2006. Liposomal vaccines—targeting the delivery of antigen. *Methods* 40:39–52.

Andresen, T., S. Jensen, and K. Jørgensen. 2005. Advanced strategies in liposomal cancer therapy: problems and prospects of active and tumor specific drug release. *Progr. Lipid Res.* 44:68–97.

Andrieux, K., and P. Couvreur. 2009. Polyalkylcyanoacrylate nanoparticles for delivery of drugs across the blood-brain barrier. *WIREs Nanomed. Nanobiotechnol.* 1:463–474.

Arima, Y., M. Toda, and H. Iwata. 2008. Complement activation on surfaces modified with ethylene glycol units. *Biomaterials* 29:551–560.

Bae, S., K. Ma, T. Kim, E. Lee, K. Oh, E. Park, K. Lee, and Y. Youn. 2012. Doxorubicin-loaded human serum albumin nanoparticles surface-modified with TNF-related apoptosis-inducing ligand and transferrin for targeting multiple tumor types. *Biomaterials* 33:1536–1546.

Bae, Y., N. Nishiyama, S. Fukushima, H. Koyama, Y. Matsumura, and K. Kataoka. 2005. Preparation and biological characterization of polymeric micelle drug carriers with intracellular pH-triggered drug release property: tumor permeability, controlled subcellular drug distribution, and enhanced in vivo antitumor efficacy. *Bioconjug. Chem.* 16:122–130.

Bareford, L., and P. Swaan. 2007. Endocytic mechanisms for targeted drug delivery. *Adv. Drug Deliv. Rev.* 59:748–758.

Barenholz, Y. 2001. Liposome application: problems and prospects. *Curr. Opin. Colloid Interf. Sci.* 6:66–77.

Barenholz, Y. 2012. Doxil®—the first FDA-approved nano-drug: lessons learned. 2012. *J. Control. Release* 160(2):117–134.

Batrakova, E., S. Li, W. Elmquist, D. Miller, V. Alakhov, and A. Kabanov. 2001. Mechanism of sensitization of MDR cancer cells by Pluronic block copolymers: selective energy depletion. *Br. J. Cancer* 85:1987–1997.

Batrakova, E. and A. Kabanov. 2008. Pluronic block copolymers: evolution of drug delivery concept from inert nanocarriers to biological response modifiers. *J. Control. Release* 130:98–106.

Bender, A., H. von Briesen, J. Kreuter, I. Duncan, and H. Rübsamen-Waigmann. 1996. Efficiency of nanoparticles as a carrier system for antiviral agents in human immunodeficiency virus-infected human monocytes/macrophages in vitro. *Antimicrob. Agents Chemother.* 40:1467–1471.

Bertrand, N., and J.-C. Leroux. 2012. The journey of a drug-carrier in the body: an anatomo-physiological perspective. *J. Control. Release* 161:152–163.

Bousquet, Y., P. Swart, N. Schmitt-Colin, F. Velge-Roussel, M. Kuipers, D. Meijer, N. Bru, J. Hoebeke, and P. Breton. 1999. Molecular mechanisms of the adsorption of a model protein (human serum albumin) on poly(methylidene malonate 2.1.2) nanoparticles. *Pharm. Res.* 16:141–147.

Brannon-Peppas, L., and J. Blanchette. 2004. Nanoparticle and targeted systems for cancer therapy. *Adv. Drug Deliv. Rev.* 56:1649–1659.

Briesen, H., P. Ramge, and J. Kreuter. 2000. Controlled release of antiretroviral drugs. *AIDS Rev.* 2:31–38.

Brigger, I., C. Dubernet, and P. Couvreur. 2002. Nanoparticles in cancer therapy and diagnosis. *Adv. Drug Deliv. Rev.* 54:631–651.

Briones, E., C. Colino, and J. Lanao. 2008. Delivery systems to increase the selectivity of antibiotics in phagocytic cells. *J. Control. Release* 125:210–227.

Bronich, T., P. Keifer, L. Shlyakhtenko, and A. Kabanov. 2005. Polymer micelle with cross-linked ionic core. *J. Am. Chem. Soc.* 127:8236–8237.

Byrne, J., T. Betancourt, and L. Brannon-Peppas. 2008. Active targeting schemes for nanoparticle systems in cancer therapeutics. *Adv. Drug Deliv. Rev.* 60:1615–1626.

Cai, S., Q. Yang, T. Bagby, and M. Forrest. 2011. Lymphatic drug delivery using engineered liposomes and solid lipid nanoparticles. *Adv. Drug Deliv. Rev.* 63:901–908.

Calvo, P., C. Remunan-López, J.L. Vila-Jato, and M.J. Alonso. 1997. Chitosan and chitosan/ethylene oxide-propylene oxide block copolymer nanoparticles as novel carriers for proteins and vaccines. *Pharm. Res.* 14:1431–1436.

Camner, P., M. Lundborg, L. Lastbom, P. Gerde, N. Gross, and C. Jarstrand. 2002. Experimental and calculated parameters on particle phagocytosis by alveolar macrophages. *J. Appl. Physiol.* 92:2608–2616.

Carmeliet, P. 2005. VEGF as a key mediator of angiogenesis in cancer. *Oncology* 69:4–10.

Carstensen, H., R. Müller, and B. Müller. 1992. Particle size, surface hydrophobicity and interaction with serum of parenteral fat emulsions and model drug carriers as parameters related to RES uptake. *Clin. Nutr.* 11:289–297.

Caruthers, S., S. Wickline, and G. Lanza. 2007. Nanotechnological applications in medicine. *Curr. Opin. Biotechnol.* 18:26–30.

Cavalli, R., M. Donalisio, A. Civra, P. Ferruti, E. Ranucci, F. Trotta, and D. Lembo. 2009. Enhanced antiviral activity of Acyclovir loaded into β-cyclodextrin-poly(4-acryloylmorpholine) conjugate nanoparticles. *J. Control. Release* 137:116–122.

Cedervall, T., I. Lynch, M. Foy, T. Berggard, S.C. Donnelly, G. Cagney, S. Linse, and K. Dawson. 2007. Detailed identification of plasma proteins adsorbed on copolymer nanoparticles. *Angew. Chem. Int. Ed. Engl.* 46:5754–5756.

Champion, J., and S. Mitragotri. 2009. Shape induced inhibition of phagocytosis of polymer particles. *Pharm. Res.* 26:244–249.

Chithrani, B., A. Ghazani, and W. Chan. 2006. Determining the size and shape dependence of gold nanoparticle uptake into mammalian cells. *Nano Lett.* 6:662–668.

Collins, D., E. Unanue, and C. Harding. 1991. Reduction of disulfide bonds within lysosomes is a key step in antigen processing. *J. Immunol.* 147:4054–4059.

Danhier, F., O. Feron, and V. Préat. 2010. To exploit the tumor microenvironment: passive and active tumor targeting of nanocarriers for anti-cancer drug delivery. *J. Control. Release* 148:135–146.

Daniels, T., T. Delgado, G. Helguera, and M. Penichet. 2006. The transferrin receptor part II: targeted delivery of therapeutic agents into cancer cells. *Clin. Immunol.* 121:159–176.

Dash, T., and V.B. Konkimalla. 2012. Poly-ε-caprolactone based formulations for drug delivery and tissue engineering: a review. *J. Control. Release* 158:15–33.

Datta, R., and S. Jaitawat. 2006. Nanotechnology—the new frontier of medicine. *Med. J. Armed Forces India* 62:263–268.

Desai, N., V. Trieu, Z. Yao, L. Louie, S. Ci, A. Yang et al. 2006. Increased antitumor activity, intratumor paclitaxel concentrations, and endothelial cell transport of cremophor-free, albumin-bound paclitaxel, ABI-007, compared with cremophor-based paclitaxel. *Clin. Cancer Res.* 12:1317–1324.

Desgrosellier, J., and D. Cheresh. 2010. Integrins in cancer: biological implications and therapeutic opportunities. *Nat. Rev. Cancer* 10:9–22.

Dienst, A., A. Grunow, M. Unruh, B. Rabausch, J. Nor, J. Fries et al. 2005. Specific occlusion of murine and human tumor vasculature by VCAM-1-targeted recombinant fusion proteins. *J. Natl. Cancer Inst.* 97:733–747.

Dobrovolskaia, M., P. Aggarwal, J. Hall, and S. McNeil. 2008. Preclinical studies to understand nanoparticle interaction with the immune system and its potential effects on nanoparticle biodistribution. *Mol. Pharmaceutics* 5:487–495.

Doherty, G., and H. McMahon. 2009. Mechanisms of endocytosis. *Annu. Rev. Biochem.* 78:857.

Dutta, D., S. Sundaram, J. Teeguarden, B. Riley, L. Fifield, J. Jacobs et al. 2007. Adsorbed proteins influence the biological activity and molecular targeting of nanomaterials. *Toxicol. Sci.* 100:303–315.

Ellens, H., J. Bentz, and F. Szoka. 1984. pH-induced destabilization of phosphatidylethanolamine-containing liposomes: role of bilayer contact. *Biochemistry* 23:1532–1538.

Elsaesser, A., and C. Howard. 2012. Toxicology of nanoparticles. *Adv. Drug Deliv. Rev.* 64:129–137.

Ernst, J.D., and O. Stendahl. 2006. *Phagocytosis of Bacteria and Bacterial Pathogenicity.* Cambridge: Cambridge Univ. Press.

Fang, C., B. Shi, Y. Pei, M. Hong, J. Wu, and H. Chen. 2006. In vivo tumor targeting of tumor necrosis factor-alpha-loaded stealth nanoparticles: effect of MePEG molecular weight and particle size. *Eur. J. Pharm. Sci.* 27:27–36.

Fang, J., H. Nakamura, and H. Maeda. 2011. The EPR effect: unique features of tumor blood vessels for drug delivery, factors involved, and limitations and augmentation of the effect. *Adv. Drug Deliv. Rev.* 63:136–151.

Farokhzad, O., and R. Langer. 2006. Nanomedicine: developing smarter therapeutic and diagnostic modalities. *Adv. Drug Deliv. Rev.* 58:1456–1459.

Freitas, R. Jr. 2005. What is nanomedicine?. *Nanomedicine: NBM* 1:2–9.

Genis, L., B. Galvez, P. Gonzalo, and A. Arroyo. 2006. MT1-MMP: universal or particular player in angiogenesis? *Cancer Metastasis Rev.* 25:77–86.

Gensini, G., A. Conti, and D. Lippi. 2007. The contributions of Paul Ehrlich to infectious disease. *J. Infect.* 54:221–224.

Gerweck, L., and K. Seetharaman. 1996. Cellular pH gradient in tumor versus normal tissue: potential exploitation for the treatment of cancer. *Cancer Res.* 56:1194–1198.

Gessner, A., A. Lieske, B. Paulke, and R. Muller. 2002. Influence of surface charge density on protein adsorption on polymeric nanoparticles: analysis by two-dimensional electrophoresis. *Eur. J. Pharm. Biopharm.* 54:165–170.

Gessner, A., A. Lieske, B. Paulke, and R. Muller. 2003. Functional groups on polystyrene model nanoparticles: influence on protein adsorption. *J. Biomed. Mater. Res. A* 65:319–326.

Gibaud, S., J. Andreux, C. Weingarten, M. Renard, and P. Couvreur. 1994. Increased bone marrow toxicity of doxorubicin bound to nanoparticles. *Eur. J. Cancer* 30:820–826.

Gibbs, W., R. Drew, and J. Perfect. 2005. Liposomal amphotericin B: clinical experience and perspectives. *Expert Rev. Anti Infect. Ther.* 3:167–181.

Gilbert, P., P. Collier, and M. Brown. 1990. Influence of growth rate on susceptibility to antimicrobial agents: biofilms, cell cycle, dormancy and stringent response. *Antimicrob. Agents Chemother.* 34:1865–1868.

Gladue, R., G. Bright, R. Isaacson, and M. Newborg. 1989. In vitro and in vivo uptake of azithromycin (CP-62,993) by phagocytic cells: possible mechanism of delivery and release at sites of infection. *Antimicrob. Agents Chemother.* 33:277–282.

Göppert, T., and R. Müller. 2005a. Protein adsorption patterns on poloxamer- and poloxamine-stabilized solid lipid nanoparticles (SLN). *Eur. J. Pharm. Biopharm.* 60:361–372.

Göppert, T., and R. Müller. 2005b. Polysorbate-stabilized solid lipid nanoparticles as colloidal carriers for intravenous targeting of drugs to the brain: comparison of plasma protein adsorption patterns. *J. Drug Target.* 13:179–187.

Göppert, T. and R. Müller. 2005c. Adsorption kinetics of plasma proteins on solid lipid nanoparticles for drug targeting. *Int. J. Pharm.* 302:172–186.

Gradishar, W., S. Tjulandin, N. Davidson, H. Shaw, N. Desai, P. Bhar et al. 2005. Phase III trial of nanoparticle albumin-bound paclitaxel compared with polyethylated castor oil-based paclitaxel in women with breast cancer. *J. Clin. Oncol.* 23:7794–7803.

Gref, R., M. Luck, P. Quellec, M. Marchand, E. Dellacherie, S. Harnisch et al. 2000. 'Stealth' corona-core nanoparticles surface modified by polyethylene glycol (PEG): influences of the corona (PEG chain length and surface density) and of the core composition on phagocytic uptake and plasma protein adsorption. *Colloids Surf. B* 18:301–313.

Gregoriadis, G., A. Bacon, B. McCormack, and P. Laing. 2006. Genetic vaccines: a role for liposomes. In *Nanoparticulates as Drug Carriers*, ed. V. Torchilin, 43–55, London: Imperial College Press.

Greish, K., T. Sawa, J. Fang, T. Akaike, and H. Maeda. 2004. SMA–doxorubicin, a new polymeric micellar drug for effective targeting to solid tumours. *J. Control. Release* 97:219–230.

Greish, K., A. Nagamitsu, J. Fang, and H. Maeda. 2005. Copoly(styrene-maleic acid)-pirarubicin micelles: high tumor-targeting efficiency with little toxicity. *Bioconjugate Chem.* 16:230–236.

Grüll, H., and S. Langereis. 2012. Hyperthermia-triggered drug delivery from temperature-sensitive liposomes using MRI-guided high intensity focused ultrasound. *J. Control. Release* 161(2):317–327.

Gulyaev, A., S. Gelperina, I. Skidan, A. Antropov, G. Kivman, and J. Kreuter. 1999. Significant transport of doxorubicin into the brain with polysorbate 80-coated nanoparticles. *Pharm. Res.* 16:1564–1569.

Labarre, D., C. Vauthier, C. Chauvierre, B. Petri, R. Muller, and M. Chehimi. 2005. Interactions of blood proteins with poly(isobutylcyanoacrylate) nanoparticles decorated with a poly-saccharidic brush. *Biomaterials* 26:5075–5084.

Gupta, A., R. Naregalkar, V. Vaidya, and M. Gupta. 2007. Recent advances on surface engineering of magnetic iron oxide nanoparticles and their biomedical applications. *Nanomed.* 2:23–39.

Gupta, A. 2011. Nanomedicine approaches in vascular disease: a review. *Nanomedicine: NBM* 7:763–779.

Harada, A., and K. Kataoka. 1999. Chain length recognition: core–shell supermolecular assembly from oppositely charged block copolymers. *Science* 283:65–67.

Harush-Frenkel, O., N. Debotton, S. Benita, and Y. Altschuler. 2007. Targeting of nanoparticles to the clathrin-mediated endocytic pathway. *Biochem. Biophys. Res. Commun.* 353:26–32.

Harush-Frenkel, O., E. Rozentur, S. Benita, and Y. Altschuler. 2008. Surface charge of nanoparticles determines their endocytic and transcytotic pathway in polarized MDCK cells. *Biomacromolecules* 9:435–443.

Hawley, A., S. Davis, and L. Illum. 1995. Targeting of colloids to lymph nodes: influence of lymphatic physiology and colloidal characteristics. *Adv. Drug Deliv. Rev.* 17:129–148.

Hillaireau, H., and P. Couvreur. 2009. Nanocarriers' entry into the cell: relevance to drug delivery. *Cell. Mol. Life Sci.* 66:2873–2896.

Hu, J., H. Liu, and L. Wang. 2000. Enhanced delivery of AZT to macrophages via acetylated LDL. *J. Control. Release* 69:327–335.

Hu, Y., T. Litwin, A. Nagaraja, B. Kwong, J. Katz, N. Watson, and D. Irvine. 2007. Cytosolic delivery of membrane-impermeable molecules in dendritic cells using pH-responsive core–shell nanoparticles. *Nano Lett.* 7:3056–3064.

Huang, M., Z. Ma, E. Khor, and L.-Y. Lim. 2002. Uptake of FITC-chitosan nanoparticles by A549 cells. *Pharm. Res.* 19:1488–1494.

Iversen, T.-G., T. Skotland, and K. Sandvig. 2011. Endocytosis and intracellular transport of nanoparticles: present knowledge and need for future studies. *Nano Today* 6:176–185.

Jahn, B., A. Rampp, C. Dick, A. Jahn, M. Palmer, and S. Bhakdi. 1998. Accumulation of amphotericin B in human macrophages enhances activity against *Aspergillus fumigatus conidia*: quantification of conidial kill at the single-cell level. *Antimicrob. Agents Chemother.* 42:2569–2575.

Jeon, S., J. Lee, J. Andrade, and P. Degennes. 1991. Protein–surface interactions in the presence of polyethylene oxide: I. Simplified theory. *J. Colloid Interface Sci.* 142:149–158.

Kabanov, A., E. Batrakova, S. Sriadibhatla, Z. Yang, D. Kelly, and V. Alakhov. 2005. Polymer genomics: shifting the gene and drug delivery paradigm. *J. Control. Release* 101:259–271.

Kabanov, A., and H. Gendelman. 2007. Nanomedicine in the diagnosis and therapy of neurodegenerative disorders. *Prog. Polym. Sci.* 32:1054–1082.

Kagan, V., H. Bayir, and A. Shvedova. 2005. Nanomedicine and nanotoxicology: two sides of the same coin. *Nanomedicine: NBM* 1:313–316.

Kaprelyants, A., J. Gottschal, and D. Kell. 1993. Dormancy in nonsporulating bacteria. *FEMS Microbiol. Rev.* 104:271–286.

Kaplan, I., J. Wadia, and S. Dowdy. 2005. Cationic TAT peptide transduction domain enters cells by macropinocytosis. *J. Control. Release* 102:247–253.

Kataoka, K., A. Harada, and Y. Nagasaki. 2001. Block copolymer micelles for drug delivery: design, characterization and biological significance. *Adv. Drug Deliv. Rev.* 47:113–131.

Kattan, J., P. Droz, P. Couvreur, P. Marino, A. Boutan-Laroze, P. Rougier et al. 1992. Phase I clinical trial and pharmacokinetic evaluation of doxorubicin carried by polyisohexylcyanoacrylate nanoparticles. *Invest. New Drugs* 10:191–199.

Kaufmann, S. 2008. Elie Metchnikoff's and Paul Ehrlich's impact on infection biology. *Microbes Infect.* 10:1417–1419.

Kawasaki, E., and T. Player. 2005. Nanotechnology, nanomedicine, and the development of new, effective therapies for cancer. *Nanomedicine: NBM* 1:101–109.

Kevin, R., and S. Otto. 2005. Reversible covalent chemistry in drug delivery. *Curr. Drug Discov. Tech.* 2:123–160.

Kirpotin, D., K. Hong, N. Mullah, D. Papahadjopoulos, and S. Zalipsky. 1996. Liposomes with detachable polymer coating: destabilization and fusion of dioleoyl phosphatidylethanolamine vesicles triggered by cleavage of surface-grafted poly(ethylene glycol). *FEBS Lett.* 388:115.

Kirpotin, D., D. Drummond, Y. Shao, M. Shalaby, K. Hong, U. Nielsen et al. 2006. Antibody targeting of long-circulating lipidic nanoparticles does not increase tumor localization but does increase internalization in animal models. *Cancer Res.* 66:6732–6740.

Kitchens, K., A. Foraker, R. Kolhatkar, P. Swaan, and H. Ghandehari. 2007. Endocytosis and interaction of poly(amidoamine) dendrimers with Caco-2 cells. *Pharm. Res.* 24:2138–2145.

Kono, K. 2001. Thermosensitive polymer-modified liposomes. *Adv. Drug Deliv. Rev.* 53:307–319.

Kratz, F., A. Warnecke, K. Scheuermann, C. Stockmar, J. Schwab, P. Lazar et al. 2002. Probing the cysteine-34 position of endogenous serum albumin with thiol-binding doxorubicin derivatives: improved efficacy of an acid-sensitive doxorubicin derivative with specific albumin-binding properties compared to that of the parent compound. *J. Med. Chem.* 45:5523–5533.

Kratz, F., G. Ehling, H. Kauffmann, and C. Unger. 2007. Acute and repeat-dose toxicity studies of the (6-maleimidocaproyl)hydrazone derivative of doxorubicin (DOXO-EMCH), an albumin-binding prodrug of the anticancer agent doxorubicin. *Hum. Exp. Toxicol.* 26:19–35.

Kratz, F. 2008. Albumin as a drug carrier: design of prodrugs, drug conjugates and nanoparticles. *J. Control. Release* 132:171–183.

Krause, R. 1999. Paul Ehrlich and O.T. Avery: pathfinders in the search for immunity. *Vaccine* 17:S64–S67.

Kreuter, J. 2001. Nanoparticulate systems for brain delivery of drugs. *Adv. Drug Deliv. Rev.* 47:65–81.

Kreuter, J., T. Hekmatara, S. Dreis, T. Vogel, S. Gelperina, and K. Langer. 2007. Covalent attachment of apolipoprotein A-I and apolipoprotein B-100 to albumin nanoparticles enables drug transport into the brain. *J. Control. Release* 118:54–58.

Lahtinen, U., M. Honsho, R.G. Parton, K. Simons, and P. Verkade. 2003. Involvement of caveolin-2 in caveolar biogenesis in MDCK cells. *FEBS Lett.* 538:85–88.

Lai, S., K. Hida, C. Chen, and J. Hanes. 2008. Characterization of the intracellular dynamics of a non-degradative pathway accessed by polymer nanoparticles. *J. Control. Release* 125:107–111.

Lammers, T., W. Hennink, and G. Storm. 2008. Tumour-targeted nanomedicines: principles and practice. *Br. J. Cancer* 99:392–397.

Lanao, J., E. Briones, and C. Colino. 2007. Recent advances in delivery systems for anti-HIV1 therapy. *J. Drug Target.* 15:21–36.

Leroux, J., F. De Jaeghere, B. Anner, E. Doelker, and R. Gurny. 1995. An investigation on the role of plasma and serum opsonins on the internalization of biodegradable poly(D,L-lactic acid) nanoparticles by human monocytes. *Life Sci.* 57:695–703.

Leroux, J., E. Roux, D. Le Garrec, K. Hong, and D. Drummond. 2001. *N*-isopropylacrylamide copolymers for the preparation of pH-sensitive liposomes and polymeric micelles. *J. Control. Release* 72:71–84.

Liu, H., and Y. Wang. 1995. The sorption of lysozyme and ribonuclease onto ferromagnetic nickel powder 2. Desorption and competitive adsorption. *Colloids Surf. B* 5:35–42.

Lisziewicz, J., and E. Töke. 2013. Nanomedicine applications towards the cure of HIV. Nanomedicine 9(1):28–38.

Litzinger, D., and L. Huang. 1992. Phosphatidylethanolamine liposomes: drug delivery, gene transfer and immunodiagnostic applications. *Biochim. Biophys. Acta* 1113:201–227.

Löbenberg, R., L. Araujo, H. von Briesen, E. Rodgers, and J. Kreuter. 1998. Body distribution of azidothymidine bound to hexyl-cyanoacrylate nanoparticles after i.v. injection to rats. *J. Control. Release* 50:21–30.

Low, P., and S. Kularatne. 2009. Folate-targeted therapeutic and imaging agents for cancer. *Curr. Opin. Chem. Biol.* 13:256–262.

Lu, Y., and P. Low. 2002. Folate-mediated delivery of macromolecular anticancer therapeutic agents. *Adv. Drug Deliv. Rev.* 54:675–693.

Lurje, G., and H. Lenz. 2009. EGFR signaling and drug discovery. *Oncology* 77:400–410.

Lynch, I., T. Cedervall, M. Lundqvist, C. Cabaleiro-Lago, S. Linse, and K. Dawson. 2007. The nanoparticle–protein complex as a biological entity; a complex fluids and surface science challenge for the 21st century. *Adv. Colloid Interface Sci.* 134–135:167–174.

Lynch, I. and K. Dawson. 2008. Protein–nanoparticle interactions. *Nano Today* 3:40–47.

Ma, Z., and L.-Y. Lim. 2003. Uptake of chitosan and associated insulin in caco-2 cell monolayers: a comparison between chitosan molecules and chitosan nanoparticles. *Pharm. Res.* 20:1812–1819.

Maeda, H. 2001. The enhanced permeability and retention (EPR) effect in tumor vasculature: the key role of tumor-selective macromolecular drug targeting. *Adv. Enzyme Regul.* 41:189–207.

Maeda, H. 2010. Tumor-selective delivery of macromolecular drugs via the EPR effect: background and future prospects. *Bioconjugate Chem.* 21:797–802.

Maeda, H. 2012. Macromolecular therapeutics in cancer treatment: the EPR effect and beyond. *J. Control. Release* 164(2):138–144.

Maruyama, K. 2011. Intracellular targeting delivery of liposomal drugs to solid tumors based on EPR effects. *Adv. Drug Deliv. Rev.* 63:161–169.

Matsumura, Y., and H. Maeda. 1986. A new concept for macromolecular therapeutics in cancer chemotherapy: mechanism of tumoritropic accumulation of proteins and the antitumor agent SMANCS. *Cancer Res.* 46:6387–6392.

Matsumura, Y., T. Hamaguchi, T. Ura, K. Muro, Y. Yamada, Y. Shimada et al. 2004. Phase I clinical trial and pharmacokinetic evaluation of NK911, micelle-encapsulated doxorubicin. *Br. J. Cancer* 91:1775–1781.

Minko, T. 2004. Drug targeting to the colon with lectins and neoglycoconjugates. *Adv. Drug Deliv. Rev.* 56:491–509.

Moghimi, S., and S. Davis. 1994. Innovations in avoiding particle clearance from blood by Kupffer cells: cause for reflection. *Crit. Rev. Ther. Drug Carr. Syst.* 11:31–59.

Moghimi, S., and A. Hunter. 2001. Capture of stealth nanoparticles by the body's defences. *Crit. Rev. Ther. Drug Carr. Syst.* 18:527–550.

Moghimi, S., A. Hunter, and J. Murray. 2001. Long-circulating and target-specific nanoparticles: theory to practice. *Pharmacol. Rev.* 53:283–318.

Moghimi, S., A. Hunter, and J. Murray. 2005. Nanomedicine: current status and future prospects. *FASEB J.* 19:311–330.

Murthy, R., and L.H. Reddy. 2006. Poly(alkyl cyanoacrylate) nanoparticles for delivery of anti-cancer drugs. In *Nanotechnology for Cancer Therapy*, ed. M. Amiji, 243–250, NW: CRC Press.

Nagamitsu, A., K. Greish, and H. Maeda. 2009. Elevating blood pressure as a strategy to increase tumor targeted delivery of macromolecular drug SMANCS: cases of advanced solid tumors. *Jpn. J. Clin. Oncol.* 39:756–766.

Nagayama, S., K. Ogawara, Y. Fukuoka, K. Higaki, and T. Kimura. 2007. Time-dependent changes in opsonin amount associated on nanoparticles alter their hepatic uptake characteristics. *Int. J. Pharm.* 342:215–221.

Nicolas, J., F. Bensaid, D. Desmaele, M. Grogna, C. Detrembleur, K. Andrieux et al. 2008. Synthesis of highly functionalized poly(alkyl cyanoacrylate) nanoparticles by means of click chemistry. *Macromolecules* 41:8418–8428.

Nishioka, Y., and H. Yoshino. 2001. Lymphatic targeting with nanoparticulate system. *Adv. Drug Deliv. Rev.* 47:55–64.

Nishiyama, N., Y. Bae, K. Miyata, S. Fukushima, and K. Kataoka. 2005. Smart polymeric micelles for gene and drug delivery. *Drug Discov. Today Technol.* 2:21–26.

Nishiyama, N., and K. Kataoka. 2006a. Nanostructured devices based on block copolymer assemblies for drug delivery: designing structures for enhanced drug function. *Adv. Polym. Sci.* 193:67–101.

Nishiyama, N., and K. Kataoka. 2006b. Current state, achievements, and future prospects of polymeric micelles as nanocarriers for drug and gene delivery. *Pharmacol. Ther.* 112:630–648.

Nyström, A., and B. Fadeel. 2012. Safety assessment of nanomaterials: implications for nanomedicine. *J. Control. Release* 161(2):403–408.

Ogawara, K., K. Furumoto, S. Nagayama, K. Minato, K. Higaki, T. Kai et al. 2004. Pre-coating with serum albumin reduces receptor-mediated hepatic disposition of polystyrene nanosphere: implications for rational design of nanoparticles. *J. Control. Release* 100:451–455.

Olivier, J. 2005. Drug transport to brain with targeted nanoparticles. *NeuroRx* 2:108–119.

Oussoren, C., and G. Storm. 1998. Targeting to lymph nodes by subcutaneous administration of liposomes. *Int. J. Pharm.* 162:39–44.

Owens, D., and N. Peppas. 2006. Opsonization, biodistribution, and pharmacokinetics of polymeric nanoparticles. *Int. J. Pharm.* 307:93–102.

Panyam, J., and V. Labhasetwar. 2003. Dynamics of endocytosis and exocytosis of poly(D,L lactide-co-glycolide) nanoparticles in vascular smooth muscle cells. *Pharm. Res.* 20:212–220.

Park, K. 2010. Nano is better than micro for targeted vaccine delivery. *J. Control. Release* 144:117.

Parton, R., and K. Simons. 2007. The multiple faces of caveolae. *Nat. Rev. Mol. Cell Biol.* 8:185–194.

Patel, H. 1992. Serum opsonins and liposomes: their interaction and opsonophagocytosis. *Crit. Rev. Ther. Drug Carr. Syst.* 9:39–90.

Patil, S., A. Sandberg, E. Heckert, W. Self, and S. Seal. 2007. Protein adsorption and cellular uptake of cerium oxide nanoparticles as a function of zeta potential. *Biomaterials* 28:4600–4607.

Peek, L., C. Middaugh, and C. Berkland. 2008. Nanotechnology in vaccine delivery. *Adv. Drug Deliv. Rev.* 60:915–928.

Peters, T. 1985. Serum albumin. *Adv. Protein Chem.* 37:161–245.

Pinto-Alphandary, H., A. Andremont, and P. Couvreur. 2000. Targeted delivery of antibiotics using liposomes and nanoparticles: research and applications. *Int. J. Antimicrob. Agents* 13:155–168.

Prior, S., B. Gander, N. Blarer, H. Merkle, M. Subira, J. Irache et al. 2002. In vitro phagocytosis and monocytes–macrophage activation with poly(lactide) and poly(lactide-co-glycolide) microspheres. *Eur. J. Pharm. Sci.* 15:197–207.

Ponce, A., Z. Vujaskovic, F. Yuan, D. Needham, and M. Dewhirst. 2006. Hyperthermia mediated liposomal drug delivery. *Int. J. Hyperthon.* 22:205–213.

Poznansky, M., and R. Juliano. 1984. Biological approaches to the controlled delivery of drugs: a critical review. *Pharmacol. Rev.* 36:277–336.

Pucadyil, T., and S. Schmid. 2009. Conserved functions of membrane active GTPases in coated vesicle formation. *Science* 325:1217–1220.

Qaddoumi, M., H. Ueda, J. Yang, J. Davda, V. Labhasetwar, and V. Lee. 2004. The characteristics and mechanisms of uptake of PLGA nanoparticles in rabbit conjunctival epithelial cell layers. *Pharm. Res.* 21:641–648.

Rannard, S., and A. Owen. 2009. Nanomedicine: not a case of "one size fits all." *Nano Today* 4:382–384.

Rao, J., and K. Geckele. 2011. Polymernanoparticles: preparation techniques and size-control parameters. *Prog. Polym. Sci.* 36:887–913.

Rappoport, J. 2008. Focusing on clathrin-mediated endocytosis. *Biochem. J.* 412:415–423.

Rejman, J., M. Conese, and D. Hoekstra. 2006. Gene transfer by means of lipo- and polyplexes: role of clathrin and caveolae-mediated endocytosis. *J. Liposome Res.* 16:237–247.

Sant, S., S. Poulin, and P. Hildgen. 2008. Effect of polymer architecture on surface properties, plasma protein adsorption, and cellular interactions of pegylated nanoparticles. *J. Biomed. Mater. Res. A* 87:885–895.

Sahay, G., E. Batrakova, and A. Kabanov. 2008. Different internalization pathways of polymeric micelles and unimers and their effects on vesicular transport. *Bioconjugate Chem.* 19:2023–2029.

Sahay, G., J. Kim, A. Kabanov, and T. Bronich. 2010a. The exploitation of differential endocytic pathways in normal and tumor cells in the selective targeting of nanoparticulate chemotherapeutic agents. *Biomaterials* 31:923–933.

Sahay, G., V. Gautam, R. Luxenhofer, A. Kabanov. 2010b. The utilization of pathogen-like cellular trafficking by single chain block copolymer. *Biomaterials* 31:1757–1764.

Sahay, G., D. Alakhova, and A. Kabanov. 2010c. Endocytosis of nanomedicines. *J. Control. Release* 145:182–195.

Sahoo, S., and V. Labhasetwar. 2006. Biodegradable PLGA/PLA nanoparticles for anti-cancer therapy. In *Nanotechnology for Cancer Therapy*, ed. M. Amiji, 243–250, NW: CRC Press.

Sawant, R., J. Hurley, S. Salmaso, A. Kale, E. Tolcheva, T. Levchenko et al. 2006. "SMART" drug delivery systems: double-targeted pH-responsive pharmaceutical nanocarriers. *Bioconjugate Chem.* 17:943–949.

Scaltriti, M., and J. Baselga. 2006. The epidermal growth factor receptor pathway: a model for targeted therapy. *Clin. Cancer Res.* 12:5268–5272.

Schafer, V., H. von Briesen, R. Andreesen, A. Steffan, C. Royer, S. Troster et al. 1992. Phagocytosis of nanoparticles by human immunodeficiency virus (HIV)-infected macrophages: a possibility for antiviral drug targeting. *Pharm. Res.* 9:541–546.

Schafer, F., and G. Buettner. 2001. Redox environment of the cell as viewed through the redox state of the glutathione disulfide/glutathione couple. *Free Radic. Biol. Med.* 30:1191–1212.

Schmaljohann, D. 2006. Thermo- and pH-responsive polymers in drug delivery. *Adv. Drug Deliv. Rev.* 58:1655–1670.

Schnitzer, J. 1992. gp60 is an albumin-binding glycoprotein expressed by continuous endothelium involved in albumin transcytosis. *Am. J. Physiol. Heart Circ. Physiol.* 262:H246–H254.

Shapira, A., Y. Livney, H. Broxterman, and Y. Assaraf. 2011. Nanomedicine for targeted cancer therapy: towards the overcoming of drug resistance. *Drug Resist. Updates* 14:150–163.

Silverstein, S., and C. Kabbash. 1994. Penetration, retention, intracellular localization, and antimicrobial activity of antibiotics within phagocytes. *Curr. Opin. Hematol.* 1:85–91.

Simoes, S., J. Moreira, C. Fonseca, N. Duzgunes, and M. Lima. 2004. On the formulation of pH-sensitive liposomes with long circulation times. *Adv. Drug Deliv. Rev.* 56:947–965.

Simons, K., and D. Toomre. 2000. Lipid rafts and signal transduction. *Nat. Rev. Mol. Cell Biol.* 1:31–39.

Slingerland, M., H.-J. Guchelaar, and H. Gelderblom. 2012. Liposomal drug formulations in cancer therapy: 15 years along the road. *Drug Discov. Today* 17:160–166.

Solomon, R., and A. Gabizon. 2008. Clinical pharmacology of liposomal anthracyclines: focus on pegylated liposomal doxorubicin. *Clin. Lymphoma Myeloma* 8:21–32.

Szebeni, J., F. Muggia, A. Gabizon, and Y. Barenholz. 2011. Activation of complement by therapeutic liposomes and other lipid excipient-based therapeutic products: prediction and prevention. *Adv. Drug Deliv. Rev.* 63:1020–1030.

Tan, S., and S. Grimes. 2010. Paul Ehrlich (1854–1915): man with the magic bullet. *Singapore Med. J.* 51:842–843.

Tekle, C., B. Deurs, K. Sandvig, and T.-G. Iversen. 2008. Cellular trafficking of quantum dot–ligand bioconjugates and their induction of changes in normal routing of unconjugated ligands. *Nano Lett.* 8:1858–1865.

Torchilin, V. 2006. Recent approaches to intracellular delivery of drugs and DNA and organelle targeting. *Annu. Rev. Biomed. Eng.* 8:343–375.

Torchilin, V. 2009. Multifunctional and stimuli-sensitive pharmaceutical nanocarriers. *Eur. J. Pharm. Biopharm.* 71:431–444.

Torchilin, V. 2011. Tumor delivery of macromolecular drugs based on the EPR effect. *Adv. Drug Deliv. Rev.* 63:131–135.

Turk, J. 1994. Paul Ehrlich—the dawn of immunology. *J. R. Soc. Med.* 87:314–315.

Ulbrich, K., M. Michaelis, F. Rothweiler, T. Knobloch, P. Sithisarn, J. Cinatl et al. 2011. Interaction of folate-conjugated human serum albumin (HSA) nanoparticles with tumour cells. *Int. J. Pharm.* 406:128–134.

Vasir, J., M. Reddy, and V. Labhasetwar. 2005. Nanosystems in drug targeting: opportunities and challenges. *Curr. Nanosci.* 1:47–64.

Vasir, J., and V. Labhasetwar. 2008. Quantification of the force of nanoparticle–cell membrane interactions and its influence on intracellular trafficking of nanoparticles. *Biomaterials* 29:4244–4252.

Vizirianakis, I. 2011. Nanomedicine and personalized medicine toward the application of pharmacotyping in clinical practice to improve drug-delivery outcomes. *Nanomedicine: NBM* 7:11–17.

Vyas, S., Y. Katare, V. Mishra, and V. Sihorkar. 2000. Ligand directed macrophage targeting of amphotericin B loaded liposomes. *Int. J. Pharm.* 210:1–14.

Waeckerle-Men, Y., and M. Groettrup. 2005. PLGA microspheres for improved antigen delivery to dendritic cells as cellular vaccines. *Adv. Drug Deliv. Rev.* 57:475–482.

Wagner, S., F. Rothweiler, M. Anhorn, D. Sauer, I. Riemann, E.C. Weiss et al. 2010. Enhanced drug targeting by attachment of an anti av integrin antibody to doxorubicin loaded human serum albumin nanoparticles. *Biomaterials* 31:2388–2398.

Wike-Hooley, J., J. Haveman, and H. Reinhold. 1984. The relevance of tumour pH to the treatment of malignant disease. *Radiother. Oncol.* 2:343–366.

Winau, F., O. Westphal, and R. Winau. 2004. Paul Ehrlich—in search of the magic bullet. *Microbes Infect.* 6:786–789.

Wohlfart, S., A. Khalansky, C. Bernreuther, M. Michaelis, J. Cinatl Jr., M. Glatzel et al. 2011. Treatment of glioblastoma with poly(isohexyl cyanoacrylate) nanoparticles. *Int. J. Pharm.* 415:244–251.

Wohlfart, S., S. Gelperina, and J. Kreuter. 2012. Transport of drugs across the blood–brain barrier by nanoparticles. *J. Control. Release* 161:264–273.

Yang, F., C. Jin, Y. Jiang, J. Li, Y. Di, Q. Ni et al. 2011. Liposome based delivery systems in pancreatic cancer treatment: from bench to bedside. *Cancer Treatment Rev.* 37:633–642.

Yatvin, M., J. Weinstein, W. Dennis, and R. Blumenthal. 1978. Design of liposomes for enhanced local release of drugs by hyperthermia. *Science* 202:1290–1293.

Yokoyama, M., T. Okano, Y. Sakurai, S. Fukushima, K. Okamoto, and K. Kataoka. 1999. Selective delivery of adriamycin to a solid tumor using a polymeric micelle carrier system. *J. Drug Target.* 7:171–186.

Yordanov, G. 2012. Poly(alkyl cyanoacrylate) nanoparticles as drug carriers: 33 years later. *Bulg. J. Chem.* 1:61–73.

Yordanov, G., R. Skrobanska, and A. Evangelatov. 2012. Entrapment of epirubicin in poly(butyl cyanoacrylate) colloidal nanospheres by nanoprecipitation: formulation development and in vitro studies on cancer cell lines. *Colloids Surf. B* 92:98–105.

Zahr, A., C. Davis, and M. Pishko. 2006. Macrophage uptake of core–shell nanoparticles surface modified with poly(ethylene glycol). *Langmuir* 22:8178–8185.

Zhang, Z.-R., and H. Qin. 1999. Study on liver targeting and hepatocytes permeable valaciclovir polybutylcyanoacrylate nanoparticles. *World J. Gastroenterol.* 5:330–333.

Zhou, Q., X. Sun, L. Zeng, J. Liu, and Z. Zhang. 2009. A randomized multicenter phase II clinical trial of mitoxantrone-loaded nanoparticles in the treatment of 108 patients with unresected hepatocellular carcinoma. *Nanomedicine: NBM* 5:419–423.

2 Environmental Impact of Nanomaterials

Gospodinka Gicheva and Georgi Yordanov

CONTENTS

2.1 INTRODUCTION

Nanotechnology has emerged as a quite interdisciplinary field of modern science and technology. Nanomaterials have found a wide range of different applications in many aspects of human life. What makes nanomaterials interesting for fundamental and applied research is their small size (usually less than 100 nm), which is the reason for their remarkable physical and chemical properties, different from those of their larger counterparts of the same chemical composition. Nanoparticles have a large fraction of their atoms on the surface, which is among the reasons for their unique physicochemical characteristics, high surface energy and tendency for aggregation. Nanoparticles prepared by wet chemical methods are usually coated with a capping layer of organic molecules, which prevents aggregation and determines their surface chemistry, zeta potential, hydrophilicity/hydrophobicity, stability (stabilization via electrostatic or steric repulsion between nanoparticles), and surface functionalities. The unique characteristics of nanomaterials include specific electromagnetic, optical, catalytic, mechanical, thermal, pharmacokinetic, and targeting properties, which make them attractive for commercial, technological, and therapeutic applications (Edelstein and Cammaratra 1998). Currently, the list of consumer goods in the nano-field is constantly increasing, involving many different products containing nanomaterials, such as sunscreens, toothbrushes, dental bonding, paints, textiles, plastic wrap, waterless car wash, corrosion resistance, golf clubs, tennis rackets, solar batteries, catalysts, and microelectronic devices (Gajewicz et al. 2012; Savolainen et al. 2010). Engineered nanoparticles are also investigated for in situ applications in environmental remediation (Narayan 2010; Sanchez et al. 2011). However, recent studies have shown the evident toxicity of nanoparticles to living organisms, as well as its potentially negative impact on environmental ecosystems (ecotoxicity); various aspects of this topic are actively discussed in scientific literature (Elsaesser and Howard 2012; Gajewicz et al. 2012; Ju-Nam and Lead 2008; Sanchez et al. 2011; Savolainen et al. 2010; Shinde et al. 2012). A survey of industrial organizations that either manufacture, handle, research, or use nanomaterials has demonstrated that most of the organizations do not have nano-specific health and safety practices at the workplace (Conti et al. 2008). Nanotoxicology emerged as a new subspecialty of particle toxicology, which addresses the toxicology of nanoparticles that appear to have unusual toxicity profiles in comparison with larger particles (Shinde et al. 2012). Because of their small sizes, nanoparticles can easily cross biological barriers and enter into cells causing various toxic effects resulting in increased oxidative stress (with consequent damage to proteins, membranes, and DNA), inflammation, and cell death. Nanoparticles can get into contact with and penetrate a human body via the skin, eyes, respiratory system, and gastrointestinal (GI) system. Once inside a human organism, nanoparticles can be recognized by the innate immunity and be internalized by phagocytic cells, accumulate in the body (if not biodegradable), and may interfere with body regulatory mechanisms (Shinde et al. 2012), even causing changes in protein structure (Linse et al. 2007). Therefore, concerns are raised because toxicity and environmental impact of the newly developed nanomaterials are sometimes unknown and difficult to predict. Many different factors, such as the elemental composition of the nanoparticles, their surface area, surface characteristics, stability and tendency to aggregate, the shape of particles and their surface charge, and the properties of their degradation products, all play roles in their distribution through the environment and ecosystems and particularly through live organisms, including the human body (Gajewicz et al. 2012; Nel 2005; Nel et al. 2006). The large number of factors that influence the toxicity of these materials makes it difficult to generalize and predict the health risks of exposure to nanoparticles; therefore, the toxicity of each new nanoparticle should be tested separately, taking into account all material properties (Shinde et al. 2012).

In this chapter, we describe the basic properties of the most popular engineered nanomaterials (carbon nanostructures, oxide nanoparticles, metal nanoparticles, and quantum dots [QDs]) and summarize their toxic effects. The possible fate of nanomaterials when released in the environment is also discussed. Despite the great research in nanotoxicology that has clearly demonstrated the toxicity of many nanomaterials, there is very little information on treatment of nanowastes, which may be a serious problem in the very near future. Therefore, it is highly required to develop methods for the treatment of nanowaste and purification of air and water from manufactured nanosized pollutants. As discussed in this chapter, some of the classical methods for removal of colloidal impurities (like coagulation and filtration) and use of sorbents can be properly modified and adapted for this purpose, but more effort is needed to obtain effective technologies for treatment of nanowastes. Complete oxidation of carbon nanomaterials by combustion may be useful to prevent their release into the atmosphere. On the other hand, it has been shown that some types of nanomaterials could be used for purification of water and air (Kim and Van der Bruggen 2010; Narayan 2010). Some of these materials, such as nanoporous membranes and sorbents, could be also good candidates for application in the treatment of nanowastes. Regulatory actions to prevent the environmental release of nanowastes are also needed.

2.2 NANOMATERIALS AS ENVIRONMENTAL HAZARDS

2.2.1 MANUFACTURED NANOMATERIALS AND NANOWASTES

Large companies are increasingly interested in introducing nanotechnologies and specialty nanoparticles to their product portfolios. The increasing number of different types and quantities of commercial products containing manufactured nanomaterials (nanoproducts) will certainly create new types of waste (nanowastes) containing nanosized pollutants (Bystrzejewska-Piotrowska et al. 2009; Mraz 2005). The nanowastes can be released into the environment during the production of the commercial nanoproducts. The release of the nanoproducts in the environment after the end of their use also creates health and environmental warnings and nanowastes are recognized as the next big threat (Mraz 2005). Manufactured nanomaterials, which are released in the environment, are most likely to get accidentally in contact with a human organism by aspiration (in the cases of aerosols and fine solid nanoparticles), by direct contact with the skin and eyes or via the GI tract (by consuming contaminated water or food). Inhaled nanoparticles are usually captured by alveolar macrophages, but also may cross the alveolar–blood barrier and can be found to be distributed to the liver, heart, spleen, and brain (Shinde et al. 2012). Some nanoparticles may cross the GI–blood barrier and travel to distant organs (liver, kidneys, brain, lung, and spleen) via blood circulation (Choi et al. 2010; Shinde et al. 2012). Mechanisms of nanoparticle toxicity are discussed in more detail elsewhere (Elsaesser and Howard 2012). Nanoparticles also may affect microorganisms, plants, or animals, if released in the environment in significant quantities. Thin coatings, immobilized nanoparticles on surfaces, or nanomaterials embedded in solid matrices (nanocomposites) and nanoporous membranes may not pose a great health risk, but free nanostructures (of various shapes—spheres, polygonal, tubes, rods, prisms, cubes, etc.) could be toxic (Bystrzejewska-Piotrowska et al. 2009). Among them, carbon nanostructures, oxide, and metal nanoparticles make the majority of nanoparticles in use. The potential toxic effects of these nanomaterials, as well as nanomaterials containing toxic heavy metals (such as Cd-based semiconductor nanoparticles), are briefly considered in this chapter. It should be taken into account that chemically inert nanoparticles may be capable of entering living cells by endocytosis and accumulate there, thus disturbing the intracellular functions and causing toxicity. Nanoparticles that adhere to the cell membrane may disturb its structure and function, while nanotubes and other nanostructures with long aspect ratio may pierce cell membranes. A scheme of possible toxic effects of nanoparticles from

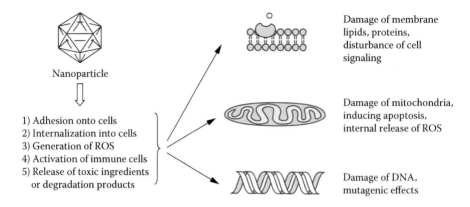

FIGURE 2.1 Scheme of possible toxic effects of nanoparticles. The main mechanism of nanoparticle toxicity is via oxidative stress and increase in ROS levels.

nanowastes is shown in Figure 2.1. The main mechanism of nanoparticle toxicity is via oxidative stress and increase in reactive oxygen species (ROS) levels, which can have various damaging effects on different cell structures, as discussed in the next sections.

Currently, little information is available on how to handle and treat discarded nanomaterials. It emphasizes the need for monitoring the fate of discarded nanomaterials and products containing nanoparticles. Methods for nanowaste management need to be developed before the first nanoparticle-containing products start to be disposed of. Nanowastes may contain nanoparticles made of not only toxic but expensive materials (such as gold and silver), which is another reason for development of effective methods for recovery of such nanomaterials. Importantly, legal regulations should be developed to hold, manage, and treat the toxic nanowastes as hazardous wastes and prevent their release into the environment. One of the major problems toward application of such regulations is the fact that, currently, there is no information on background concentrations of nanoparticles in the environment because of limitations in separation and analytical methodologies. Importantly, even though some nanomaterials may not be toxic, if mixed with other conventional waste containing toxic ingredients, the former may adsorb the toxic compounds and act as a Trojan horse to transport them into the cells (Musee 2011). Therefore, even nanowastes that are considered as non-toxic actually can be potentially dangerous.

2.2.2 CARBON NANOSTRUCTURES

In the last few decades, various carbon nanostructures such as fullerenes, carbon nanotubes (CNTs), and graphene have been given great attention in the field of nanoscience and nanotechnology, mainly because of their unique properties and several possible applications (Dinadayalane and Leszczynski 2010). Importantly, carbon nanoparticles have been proved to be formed during combustion processes and can be considered as nanosized pollutants in industrialized areas (Donaldson et al. 2005;

Murr and Soto 2005). Fullerenes are among the most investigated engineered carbon nanostructures, which are single spherical molecules with 60 atoms of carbon positioned at the vertices of a regular truncated icosahedron structure, commonly denoted as C_{60} (Kroto et al. 1985). There are also higher-mass fullerenes, but the C_{60} molecule and its derivatives are the most widely studied. Fullerenes are insoluble in water because of their hydrophobic character but can be dispersed in aqueous media by sonication (Brant et al. 2005) or using organic co-solvents (Deguchi et al. 2001). Water-soluble fullerenes can be obtained also by chemical modification with hydrophilic functional groups (Chiang et al. 1996). Numerous applications of fullerenes in personal care products, textile industry, and microelectronics, as well as several potential uses in medicine, have been proposed (Benn et al. 2011; Partha and Conyers 2009). CNTs can be viewed as rolled-up cylinders of graphene. There are two structural forms of CNTs: multi-walled CNTs and single-walled CNTs. A wide range of applications of CNTs was proposed and few of these have been already realized, mainly using CNTs to obtain composite materials with improved mechanical properties (Baughman et al. 2002).

Carbon nanomaterials may have many applications, but it has been found that these materials could be toxic and may have adverse effects to living systems (Hurt et al. 2006). For example, fullerenes dispersed in water as well as CNTs have been found to exhibit cytotoxicity to human cell lines (Magrez et al. 2006; Sayes et al. 2004; Tian et al. 2006). At the cellular level, mitochondria appear to be a major target for fullerenes (Foley et al. 2002) and CNTs (Zhu et al. 2006). It has been found that fullerene derivatives can induce ROS-mediated membrane damage to cells (Kamat et al. 2000). Studies that have compared the cytotoxicity of carbon nanomaterials to alveolar macrophages have shown that fullerene C_{60} is less toxic than CNTs and graphite (Uo et al. 2011). It should be taken into account that functionalization, and the production process, especially the presence of metals (that are used as catalysts during production of carbon nanostructures), may influence the toxicity of these nanomaterials. The aspect ratio (length/diameter) of CNTs could also determine their toxicological profile. Most nanoparticles after entering the body are usually engulfed by phagocytic cells via phagocytosis (Doherty and McMahon 2009). However, "frustrated phagocytosis," a process where a cell tries but fails to totally engulf a particle due to its size, has been observed with multi-walled CNTs (Poland et al. 2008). CNTs that have a high ratio between length and diameter are expected to have asbestos-like toxicity because of their shape similarity, which might induce lung cancer and mesothelioma in a similar manner as asbestos (Donaldson et al. 2010; Muller et al. 2006) (the toxicological risk of manufactured CNTs has been reviewed by Lam et al. 2006). Exposure to carbon nanomaterials is especially important to be evaluated for workers in industries that use such carbon nanomaterials, because CNTs can enter the human body via the respiratory tract with air inhalation and can distribute in the nervous system, lymph, blood, and internal organs (heart, spleen, kidney, bone marrow, and liver) (Kayat et al. 2011). Skin, eyes, and lung come first in contact with these nanostructures and have the highest risk of exposure that may cause irritation and inflammation. Both CNTs and graphite nanoparticles have been found to be genotoxic to human bronchial cells, causing dose- and time-dependent DNA damage (Lindberg et al. 2009). There are very little data on the effects of nanomaterials on anaerobic organisms. Assessment of the impacts of fullerene (C_{60}) on anaerobic organisms has demonstrated a neutral effect for C_{60} even at extremely high levels (50,000 mg C_{60} per kilogram of sludge) (Nyberg et al. 2008). In these studies, no effect on methanogenesis of bacteria or on the structure of the population has been observed. These and many other reports clearly indicate the need for more detailed evaluation of toxicity of carbon nanostructures, development of tools for personal protection from exposure to these nanostructures, and purification systems to prevent disposal of carbon nanowastes into the environment.

2.2.3 OXIDE NANOPARTICLES

The applications of manufactured oxide nanoparticles are rapidly expanding. Oxide nanostructures of titanium dioxide (TiO_2) and zinc oxide (ZnO) are used in paints, cements, sunscreens, catalysts, and UV protection, and can be released in wastewaters (Schmid and Riediker 2008). TiO_2 nanoparticles are used to protect glazing owing to their biocidal and anti-fouling properties. Upon illumination with light, such nanoparticles catalyze decomposition processes, facilitating breakdown of organic pollutants and preventing bacterial growth (Nakata and Fujishima 2012). Nanosized and amorphous metal oxides/hydroxides, like stannic dioxide (SnO_2), are usually the major components in electroplating sludge, large amounts of which are disposed by industry (Zhuang et al. 2012). Iron-containing nanoparticles of Fe_2O_3 are used as concrete additive and can be released to the atmosphere as a result of dismantling (Bystrzejewska-Piotrowska et al. 2009). Nanosized silicon dioxide (SiO_2) is used in varnish, UV protection, ceramics, electronics, pharmaceutical products, and so forth (Schmid and Riediker 2008). Commercial applications of ZnO nanoparticles involve their use in sunscreens and cosmetics, owing to their property of blocking broad UV-A and UV-B rays (Huang et al. 2008). Large amounts of manufactured oxide nanoparticles can be released in natural waters and therefore the possible ecotoxicity of such nanoparticles needs a careful evaluation (Blinova et al. 2010).

The toxicity of oxide nanoparticles is largely investigated mainly on microorganisms and various aquatic organisms (Aruoja et al. 2009; Bai et al. 2010; Federici et al. 2007; Franklin et al. 2007; Heinlaan et al. 2008; Kasemets et al. 2009; Li et al. 2011; Mortimer et al. 2010; Simon-Deckers et al. 2009). Nanosized oxides, like SiO_2, anatase (TiO_2), and ZnO, are known to be toxic against lung cells and can induce pulmonary inflammation and emphysema-like lung injury in mammals (Chen et al. 2004, 2006; Gordon et al. 1992; Limbach et al. 2005; Rehn et al. 2003). The mechanism of toxicity of oxide nanoparticles on the cellular level mainly involves increase of the level of ROS, oxidation of biomolecules, and enhancement of antioxidative systems in cells (Bai et al. 2010; Federici et al. 2007; Horie et al. 2010, 2012). Generation of ROS in brain cells (Long et al. 2006) and brain accumulation of oxide nanoparticles (Wu et al. 2011) raise warnings about nanoparticle neurotoxicity (Win-Shwe and Fujimaki 2011). Oxidative stress-related DNA damage has also been observed in cells treated with metal oxide nanoparticles; severe toxicity, such as toxicity of CuO nanoparticles, has been explained with release of free metal ions (Karlsson et al. 2008). The high surface-to-volume ratio in nanoparticles favors the release of ions. DNA-damaging and genotoxic effects have been reported for ZnO nanoparticles (Kumari et al. 2011; Sharma et al. 2011). It has been supposed that metal ion release is the most important factor; however, the adhesion of nanoparticles on cell membranes and cellular uptake of nanoparticles should also be considered (Horie et al. 2012). Nanoparticles dissolve more easily than microparticles, which may further increase the release of toxic metal ions (Franklin et al. 2007).

Nanoparticles of TiO_2 and ZnO from sunscreen can contribute to the formation of free radicals in skin cells and damage DNA (Dunford et al. 1997; McHugh and Knowland 1997), which demonstrates one of the harmful effects of nanoparticles for humans. This can result in mutations favoring cancer development. Nanoparticles of TiO_2 have been found to also penetrate the skin of mice and pigs, producing toxic effects (Wu et al. 2009). However, depending on the conditions, different mechanisms of oxidative stress could be observed even for the same metal oxide nanoparticles. For example, DNA microarray analysis has been performed to determine the gene expression profiles of human keratinocytes exposed to anatase (TiO_2) nanoparticles of different average sizes (7–200 nm) without illumination (Fujita et al. 2009). This analysis has shown that only genes involved in the inflammatory response and cell adhesion have been over-represented, but not the genes implicated in oxidative stress and apoptosis. These results suggest that without illumination, the TiO_2 nanoparticles have no significant impact on ROS-associated oxidative damage but affect the cell–matrix adhesion. The crystal structure of nanoparticles also can determine the mode of cell death induced by nanosized TiO_2 (Braydich-Stolle et al. 2009). Oxide nanoparticles (ZnO, TiO_2, CeO_2) can enter human cells by endocytosis. Evidence of dissolved ZnO nanoparticles in the culture media as well as inside endosomes was found, while TiO_2 (11 nm) and cerium(IV) oxide (CeO_2) were found undissolved in the endosomal cavities (Xia et al. 2008). This is probably due to the amphoteric character of ZnO, which is highly soluble in acidic (as inside endosomes, where pH ~4–5) and alkaline media. The toxicity of oxide nanoparticles (ZnO, Al_2O_3, TiO_2) has been tested on nematodes (Wang et al. 2009). It has been shown that both the bulk and the nanoparticle forms of these oxides affect the reproduction and growth capacity of these organisms. The bulk forms of Al_2O_3 and TiO_2 have been found to be about twice less toxic than the respective nanoparticles, while both (nanoparticle and bulk) forms of ZnO possessed similar toxicity (LC_{50} ~2.3 mg/L). It should be noted that computational and QSAR-based methods for statistical analysis could be quite useful for development of nanoparticle-toxicity models (Gajewicz et al. 2012; Puzyn et al. 2011).

Release of nanoparticles into natural waters and soil raises questions about possible toxic effects on plant development. Cells of plants, algae, and fungi possess cell walls that represent a barrier for the entrance of nanoparticles and prevent direct interaction between nanoparticles and the cell membrane. Generally, the mechanisms of interaction between nanoparticles and plant cells are poorly understood (Navarro et al. 2008). It has been observed that Zn- and Al-containing nanoparticles exert toxic effects on germination and growth of roots in agriculturally relevant plant species (Doshi et al. 2008; Lin and Xing 2007). ZnO nanoparticles can adhere onto the root surface, causing vacuolation and collapse of root epidermal and cortical cells, which has been probably the reason for the observed reduction of plant biomass (Lin and Xing 2008). Nanoparticles of TiO_2 (10–20 nm) at 5 mg/L concentrations have been found to damage cell membranes of planktonic microorganisms from surface water (Battin et al. 2009). Toxicity against plankton, which is the base of the marine food chain, may have quite disturbing effects on marine ecosystems.

2.2.4 Metal Nanoparticles

Most metals are highly reactive reducing agents and nanoparticles of such metals are therefore unstable and easily oxidized in the presence of oxygen and water. Metal nanoparticles that have found large applications in practice are composed of noble metals, mostly silver (Ag), gold (Au), platinum (Pt), and so forth. Such nanoparticles are usually easily prepared by chemical reduction from their salts and can be stabilized by coating with various ligands forming capping layers around the nanoparticles. Representative transmission electron microscopy (TEM) images of silver and gold nanoparticles prepared in our laboratory are shown in Figure 2.2. These ligands coordinate strongly with the nanoparticle surface (Daniel and Astruc 2004). It is essential to bear in mind the potential role of these capping ligands in toxicity and evaluation of the environmental impact of nanoparticles.

Cars may emit not only carbon-based aerosols of nanoparticles as a result of incomplete combustion but also nanoparticles of Pt, and Pd can be released as aerosols from automotive exhaust converters and other catalysts (Artelt et al. 1999; Stafford 2007). It has been announced that Pt nanoparticles, with sizes in the range 0.8–10 nm, are released from car catalysts during their lifetime (Artelt et al. 1999). Metal nanoparticles can be released into the environment mainly from products containing nanosilver. Silver nanoparticles represent well-known antibacterial agents (Lubick 2008; Parashar et al. 2011; Weir et al. 2008) that have found applications in domestic appliances to avoid growth of saprophytic bacteria and fungi (in refrigerators, air conditioning, vacuum cleaners, and washing machines, as well as in textiles, paints, plastics, etc.). The biocidal properties of silver have been utilized for many centuries, as for example in fabrication of cutlery and crockery, which prevented

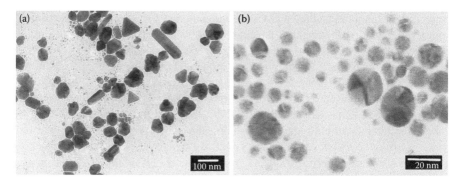

FIGURE 2.2 Representative TEM images of metal nanoparticles (NPs) prepared by reduction of aqueous solutions of metal salts with sodium citrate: (a) silver NPs (scale bar: 100 nm); (b) gold NPs (scale bar: 20 nm).

growth of bacteria and molds. Composite materials containing nanosilver may release nanoparticles and Ag^+ ions to the environment. The antibacterial mechanism of silver-containing products has been explained by a long-term release of silver ions (Ag^+) as a result of oxidation of metallic silver in contact with water (Kumar et al. 2005). It has been demonstrated that silver ions inhibit various enzymes in bacterial cells (Ratte 1999) and can block DNA transcription, bacterial respiration, and adenosine triphosphate production and can inactivate proteins by reacting with the thiol (SH) groups (Jeon et al. 2003). Taking into account their antimicrobial effect (Choi et al. 2008) and their toxic effects on nematodes (Ellegaard-Jensen et al. 2012), zebrafish embryos (Asharani et al. 2011), and so on, it can be expected that nanosilver might significantly affect ecological balance by cumulative aquatic exposure (Blaser et al. 2008). It has been found that the LC_{50} (lethal concentration 50%) of silver nanoparticles (~30 nm) toward the aquatic organism *Oryzias latipes* is ~1 mg/L for a treatment time of 48 h (Wu et al. 2010). Developmental, morphological, and histopathological changes in this organism, including edema and abnormalities in the spine, brain, and eyes, have been reported. Also, nanosilver can be internalized and cause growth inhibition in nematodes *Caenorhabditis elegans* (Meyer et al. 2010).

2.2.5 SEMICONDUCTOR NANOPARTICLES

Semiconductor nanoparticles, also known as QDs, are nanocrystals composed of inorganic semiconductor materials with crystal sizes within the range 1–10 nm (Wang et al. 2007). These nanoparticles are prepared by wet chemical methods and are usually composed of $A^{II}B^{VI}$ (CdS, CdSe, CdTe, ZnS, etc.), $A^{III}B^V$ (GaAs, InAs, InP, etc.), or Pb-based (PbS, PbSe) semiconductors. Such nanoparticles could be highly toxic because of the content of toxic heavy metals and because of their very small size. The sulfides and selenides of Cd and Pb have a very low solubility in water but can be slowly oxidized to form soluble salts. QDs are currently produced in relatively small amounts and can find applications mainly as fluorescent biological markers (Yordanov and Dushkin 2011), light-emitting diodes (Coe-Sullivan et al. 2003), and components of solar cells (Emin et al. 2012). Semiconductor nanocrystals can be dispersible in organic solvents, as well as in aqueous media depending on their surface coatings. The cytotoxicity of QDs has been found to depend on a number of factors including size, capping materials, color, dose of QDs, surface chemistry, processing parameters, and the type of surface coating ligands (usually various thiol compounds) used for modification and stabilization of QDs (Hardman 2006). Different mechanisms have been proposed to explain the cytotoxicity of QDs (Kirchner et al. 2005), including release of free heavy metal ions (appropriate coating of the QDs may reduce their cytotoxicity and allows their utilization as fluorescent biolabels), formation of ROS and oxidative stress (resulting in damage of membranes, organelles, and DNA), affecting biochemical pathways, or toxicity caused by the coating ligand. Semiconductor nanoparticles can induce cell death also by lipid peroxidation (Choi et al. 2007). The ecotoxicological effects of CdTe nanoparticles to freshwater mussels include immunotoxicity and oxidative stress in gills that can cause DNA damage (Gagné et al. 2008).

2.3　FATE OF DISPOSED NANOMATERIALS IN ECOSYSTEMS

Investigations of the nanoparticle's fate and impact in the environment is becoming more important owing to nanoparticle release already occurring to the environment, the known toxicity of nanoparticles, and the gaps in our knowledge leading to difficulties in risk assessment and management (Handy et al. 2008; Ju-Nam and Lead 2008). There are naturally occurring nanoparticles, such as iron and manganese oxyhydroxides, metal sulfides, carbonates, amorphous silica, aluminosilicates, macromolecular aggregates of humic and fulvic substances, polysaccharides and peptidoglycans (Lead and Wilkinson 2006; Nowack and Bucheli 2007), and various submicron structures from volcanic ashes. Knowledge of their chemistry and environmental impact would be highly useful for a better understanding of the fate, environmental behavior, and toxicity of manufactured nanoparticles (Ju-Nam and Lead 2008), although the physicochemical properties of the latter may be quite different from those of natural colloids.

Nanoparticles released into the environment may interact with air, water, and soil in various ways (Peralta-Videa et al. 2011). A scheme of possible pathways of engineered nanoparticles in ecosystems is shown in Figure 2.3. Nanoparticles released into the atmosphere may reach soil or different aquatic systems (surface water, seawater, and underground water). Nanoparticles or products of their degradation and chemical transformation may accumulate in living organisms (plants, aquatic organisms, terrestrial animals). Nanoparticles (including their degradation products) can come into contact with human organism by direct contact with contaminated air, water, and soil, or indirectly, via contaminated food (including plant and animals). The actual nanoparticle distribution and fate in natural ecosystems are, however, largely unknown and need to be studied.

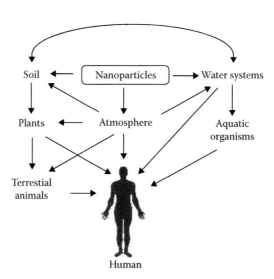

FIGURE 2.3　Scheme of possible pathways of engineered nanoparticles in ecosystems.

Nanoparticles can undergo various chemical transformations after their release into the environment, depending on their properties and the local environmental conditions. In the presence of oxygen (O_2) and sunlight, some nanoparticles can undergo abiotic photochemical degradation. For example, exposure to light can oxidize fullerene to fullerol, decrease the size of the C_{60} fullerenes, and alter the surface chemistry of the material forming many more water-soluble degradation products (Lee et al. 2009; Turco et al. 2011). Interestingly, it has been demonstrated that a small amount of fullerol carbon could become incorporated into fungal biomass (Schreiner et al. 2009). This work actually provides the first evidence for microbial conversion of a manufactured nanomaterial.

Nanowastes can also be released into natural waters (Brar et al. 2010), including marine environment (Matranga and Corsi 2012). Their stability in the aqueous environment is determined by the nanoparticle characteristics (hydrophilicity, surface chemistry, zeta potential, nanoparticle concentration) and the physicochemical parameters of the aqueous environment (ionic strength, hardness, pH, biochemical oxygen demand, alkalinity, and presence of organic matter). Hydrophobic and larger particles tend to aggregate and form sediments in water systems; however, small and hydrophilic nanoparticles are likely to be stable for long periods (Brar et al. 2010). Changes in the parameters of the aqueous environment can influence the nanoparticles' stability. The lower ionic strength and the presence of organic matter (mainly humic substances) in freshwater favor nanoparticle stability. Nanoparticle transfer from freshwater to seawater can decrease their zeta potential (because of the higher ionic strength of seawater due to salinity), thus causing aggregation and precipitation. Therefore, studies of nanoparticle stability, aggregation, and sedimentation at various conditions will help to better predict the fate of nanoparticles in aqueous systems. It should be noted that accumulation of nanoparticles or their degradation products (such as toxic ions of heavy metals) into aquatic planktonic organisms may lead to a biomagnification effect and increasing accumulation of these toxic materials in food chains (plankton organisms are located at the bottom of the aquatic food chain). Aquatic invertebrates are important objects for testing in aquatic nanotoxicology, because they are the ultimate recipients of most contaminants released into the environment (Baun et al. 2008). Nanoparticles can also undergo chemical changes in aqueous media. Taking into account the chemical properties of materials and known environmental conditions, some of these transformations could be suggested. For example, metal nanoparticles of iron can be easily oxidized to iron oxide, ZnO can be dissolved in alkaline or acidic solutions, while inert materials such as TiO_2 are expected to be quite stable and remain unchanged for a long time. Chemical transformation of silver nanoparticles into silver sulfide is expected to be the most probable fate of nanosilver (Choi et al. 2009; Kim et al. 2010), especially in aqueous environments rich in hydrogen sulfide (usually obtained from rotting organic materials or found in some natural waters). The main sources of nanosized pollutants in seawater involve industrial sewage and wastewater sludge, personal care products (sunscreens, cosmetics), and anti-fouling paints on vessel hulls (Matranga and Corsi 2012). Nanoparticles released into seawater may tend to aggregate and may be incorporated into sediments and come into contact with benthic organisms and be ingested by them. Stable nanoparticles may remain in suspension for a long time and may interact with planktonic and other organisms.

The fate of nanoparticles released to soil may vary depending on the physical and chemical characteristics of the nanoparticles, type of soil, and weather conditions—soil may act as a sorbent that strongly binds nanoparticles or nanoparticles can go free through the soil reaching underground waters—the actual situation may be different with different nanoparticles and may depend on many factors. The interaction of nanoparticles with various soil components may also change their properties (Handy et al. 2008). Humic and other organic substances in soil may affect the stability and the formation of aggregates of manufactured nanoparticles by interaction with their surfaces. Therefore, the type of soil and its condition (water content, type of organic materials, electrolytes, pH, mineral composition, oxygen content, etc.) is expected to determine nanoparticle stability and transport (aggregation, sorption, desorption, chemical transformation) (Darlington et al. 2009). For example, the influence of humic and fulvic acids, other organic substances, pH, and electrolytes on the stability of fullerenes has been studied (Chen and Elimelech 2006). It has been shown that humic substances may act as stabilizers, but also may induce aggregation, depending on their concentration and other conditions. Increased stability of gold nanoparticles has been observed by addition of humic substances, which can replace the surface-capping agent (Diegoli et al. 2008). Nanoparticles of Al_2O_3, coated with humic acid, have been found to be highly stable (Ghosh et al. 2010). Nanoparticles are usually small enough (if not aggregated) to pass through the spaces between soil particles and might travel further than the larger bulk particles; however, various interactions (electrostatic, hydrophobic) with soil materials (organics, minerals, bacteria) might alter their transport in soil environment (Darlington et al. 2009; Peralta-Videa et al. 2011). Most soil organics are acidic and negatively charged, which suggests that solid organic matter will possibly adsorb positively charged nanoparticles by electrostatic attraction. Soil solutions contain various natural colloids with high specific surface areas, which can be of great environmental importance as carriers of many substances, including pollutants and other materials. Such natural colloids could possibly also affect the nanoparticle transport through soil. Currently, there are only a few reports in the scientific literature about the nanomaterial transport in real soil systems, including TiO_2 nanoparticles (Fang et al. 2009) and CNTs (Jaisi and Elimelech 2009). In these studies, it has been found that TiO_2 nanoparticles can travel to large distances (to 3.7 m) in soil, thus probably reaching underground waters, while CNTs do not exhibit substantial transport in soils (up to 5–6 cm) because of effective retention by the soil matrix.

2.4 MANAGEMENT OF NANOWASTES

The fate of manufactured nanoparticles in wastewater treatment plants is largely unknown. Nanowastes should be safely handled with caution and treated as hazardous waste materials until special regulations and treatment methods are developed (Hallock et al. 2009). Recycling of nanoproducts will require separation of the used nanostructures from other components and possible reuse and elimination from wastes. Current methods of storage may not be sufficient in the case of nanowaste.

Nanowaste classification is an important prerequisite when planning nanowaste management; it is broadly discussed elsewhere (Musee 2011) and will not be considered here.

Currently, it may be assumed that the existing waste management technologies could remove nanoparticles from the waste streams; however, there are no hard data to support such an assumption. It has been illustrated that wastewater treatment systems have low removal efficiency (0%–40%) for nanoparticles (Westerhoff et al. 2008). Experiments with CeO_2 nanoparticles have shown that nanoparticles could not be completely removed by a wastewater clearing system (Limbach et al. 2008). Since the research on removing nanoparticles from wastewaters has just begun, there are insufficient data to make general conclusions about the efficacy of current wastewater treatment systems. Most probably, current methods used for wastewater treatment may require modifications in order to become more efficient in treatment of nanowastes. Another example of nanowaste is nanosized silica used for polishing in the semiconductor industry. Various treatment processes have been developed for treatment of this nanowaste, including coagulation, electrocoagulation, flotation, membrane filtration, and magnetic seeding aggregation (Chin et al. 2006; Golden et al. 2000; Huang et al. 2004; Lai and Lin 2004). Electroplating sludge containing nano-SnO_2 waste has been treated via fast crystal growth of amorphous Sn compound into acid-insoluble SnO_2 nanowires (Zhuang et al. 2012). Removal of silver nanoparticles in simulated wastewater treatment processes via aeration and utilization of sequence batch reactors has been demonstrated (Hou et al. 2012). Another method to remove nanosilver from wastewater has been proposed: sorption of the nanoparticles into biomass (Benn and Westerhoff 2008). Although the removal of nanoparticles by this method proved to be successful, these findings suggest that the high content of silver in the resulting biomass may limit its utilization for agricultural applications. This is so, because nanosilver from the utilized biomass can be transferred to soil, where it would act as an inhibitor of bacterial growth (this can cause adverse effects on useful microbial populations in the soil such as nitrogen-fixing bacteria) (Musee 2011).

Treatments for nanowastes may also involve other conventional methods, such as combustion (applicable in the case of carbon-rich and organic nanomaterials). Coagulation methods are largely applied for purification of water from various natural colloids (Joseph et al. 2012; Matilainen et al. 2010). For example, coagulation using polyaluminum chloride has been used to remove nanoparticles from industrial wastewaters (Chang et al. 2007). The possible fate of nanoparticles in each of the unit operations in a typical wastewater treatment plant is discussed in detail elsewhere (Brar et al. 2010). Novel nanoporous materials and membranes (Kim and Van der Bruggen 2010; Narayan 2010) need to be tested for separation of nanomaterials from contaminated waters. The authors of this chapter are currently involved in research on purification of water from nanosized pollutants, from which it has been found that nanosilver and other water-dispersible nanoparticles can be adsorbed onto porous activated charcoal and other specially designed sorbents (manuscripts in preparation). On the other hand, the different properties of the different types of nanomaterials make it unlikely to develop a universal technology capable of removing different types of nanoparticles from waste streams. It is rather possible that purification systems for nanowastes will be confined only to a particular type of nanoparticle or may be composed of multiple units, each of which are being designed to remove a different type of nanomaterial. Therefore, it would be easier to develop a purification system for each particular nanotechnology-related production process, which generates a known type and quantity of nanowaste.

2.5 CONCLUSIONS

Currently, nanotoxicological investigations reveal various aspects of toxicity of different nanomaterials. The main public health concerns with nanomaterials, however, will be with chronic low-dose exposures over long periods and the effects from such exposures are difficult to predict. Therefore, nanotoxicological research should be focused also on investigation of the effects from chronic exposure to nanomaterials. It is also critical to characterize very well the nanomaterial under examination in each particular case of nanotoxicological investigation, because toxicity of nanomaterials has been found to depend on a large number of parameters. The real impact of nanotechnological wastes needs to be addressed before nanowastes appear in the environment in large quantities, as well as before introducing new nanoproducts to the market. Methods for purification of water and air from nanosized pollutants are highly required. Conventional methods for removing colloidal impurities, such as coagulation, filtration, and electrocoagulation, can be utilized, but these methods might have low efficiency and most probably will require modifications in order to be adapted to remove smaller and highly stable nanoparticles. New methods for treatment of nanowaste, which take advantage of the specific properties of nanomaterials, are also needed.

ACKNOWLEDGMENTS

This work was supported by the Bulgarian National Science Fund (project DMU 03/86) and COST Action D43 of the European Community. TEM images were obtained with the technical help of Dr. Daniela Karashanova from the Institute of Optical Materials and Technology, Bulgarian Academy of Sciences.

REFERENCES

Artelt, S., O. Creutzenberg, H. Kock, K. Levsen, D. Nachtigall, U. Heinrich et al. 1999. Bioavailability of fine dispersed platinum as emitted from automotive catalytic converters: a model study. *Sci. Total Environ.* 228:219–242.

Aruoja, V., H. Dubourguier, K. Kasemets, and A. Kahru. 2009. Toxicity of nanoparticles of CuO, ZnO and TiO$_2$ to microalgae *Pseudokirchneriella subcapitata*. *Sci. Total. Environ.* 407:1461–1468.

Asharani, P., Y. Lian Wu, Z. Gong, and S. Valiyaveettil. 2011. Comparison of the toxicity of silver, gold and platinum nanoparticles in developing zebrafish embryos. *Nanotoxicology* 5:43–54.

Bai, W., Z. Zhang, W. Tian, X. He, Y. Ma, Y. Zhao et al. 2010. Toxicity of zinc oxide nanoparticles to zebrafish embryo: a physicochemical study of toxicity mechanism. *J. Nanopart. Res.* 12:1645–1654.

Baughman, R., A. Zakhidov, and W. de Heer. 2002. Carbon nanotubes—the route toward applications. *Science* 297:787–792.

Baun, A., N. Hartmann, K. Grieger, and K. Kusk. 2008. Ecotoxicity of engineered nanoparticles to aquatic invertebrates: a brief review and recommendations for future toxicity testing. *Ecotoxicology* 17:387–395.

Battin, T., F. Kammer, A. Weilhartner, S. Ottofuelling, and T. Hofmann. 2009. Nanostructured TiO2: transport, behavior and effects on aquatic microbial communities under environmental conditions. *Environ. Sci. Technol.* 43:8098–8104.

Benn, T.M., and P. Westerhoff. 2008. Nanoparticle silver released into water from commercially available socks fabrics. *Environ. Sci. Technol.* 42:4133–4139.

Benn, T., P. Westerhoff, and P. Herckes. 2011. Detection of fullerenes (C60 and C70) in commercial cosmetics. *Environ. Pollut.* 159:1334–1342.

Blaser, S., M. Scheringer, M. MacLeod, and K. Hungerbühler. 2008. Estimation of cumulative aquatic exposure and risk due to silver: contribution of nanofunctionalized plastics and textiles. *Sci. Total Environ.* 390:396–409.

Blinova, I., A. Ivask, M. Heinlaan, M. Mortimer, and A. Kahru. 2010. Ecotoxicity of nanoparticles of CuO and ZnO in natural water. *Environ. Pollut.* 158:41–47.

Brant, J., H. Lecoanet, M. Hotze, and M. Wiesner. 2005. Surface charge acquisition and characteristics of fullerene aggregates (n-C60) in aqueous suspensions. *Environ Sci Technol.* 39:6343–6351.

Brar, S., M. Verma, R. Tyagi, and R. Surampalli. 2010. Engineered nanoparticles in wastewater and wastewater sludge—evidence and impacts. *Waste Manag.* 30:504–520.

Braydich-Stolle, L., N. Schaeublin, R. Murdock, J. Jiang, P. Biswas, J. Schlager et al. 2009. Crystal structure mediates mode of cell death in TiO$_2$ nanotoxicity. *J. Nanopart. Res.* 11:1361–1374.

Bystrzejewska-Piotrowska, G., J. Golimowski, and P. Urban. 2009. Nanoparticles: their potential toxicity, waste and environmental management. *Waste Manage.* 29:2587–2595.

Chang, M., D. Lee, and J. Lai. 2007. Nanoparticles in wastewater from a science-based industrial park—coagulation using polyaluminum chloride. *J. Environ. Manage.* 85:1009–1014.

Chen, H.-W., S.-F. Su, C.-T. Chien, W.-H. Lin, S.-L. Yu, C.-C. Chou et al. 2006. Titanium dioxide nanoparticles induce emphysema-like lung injury in mice. *FASEB* 20:2393–2395.

Chen, K., and M. Elimelech. 2006. Aggregation and deposition kinetics of fullerene (C60) nanoparticles. *Langmuir* 22:10994–11001.

Chen, Y., J. Chen, J. Dong, and Y. Jin. 2004. Comparing study of the effect of nanosized silicon dioxide and microsized silicon dioxide on fibrogenesis in rats. *Toxicol. Ind. Health* 20:21–27.

Chiang, L., J. Bhonsle, L. Wang, S. Shu, T. Chang, and J. Hwu. 1996. Efficient one-flask synthesis of water-soluble (60) Fullerenols. *Tetrahedron* 52:4963–4972.

Chin, M.C.-J., P.-W. Chen, and L.-J. Wang. 2006. Removal of nanoparticles from CMP wastewater by magnetic seeding aggregation. *Chemosphere* 63:1809–1813.

Choi, A., S. Cho, J. Desbarats, J. Lovric, and D. Maysinger. 2007. Quantum dot-induced cell death involves Fas upregulation and lipid peroxidation in human neuroblastoma cells. *J. Nanobiotechnol.* 5:1, doi:10.1186/1477-3155-5-1.

Choi, H., Y. Ashitate, J. Lee, S. Kim, A. Matsui, N. Insin et al. 2010. Rapid translocation of nanoparticles from the lung airspaces to the body. *Nat. Biotechnol.* 28:1300–1303.

Choi, O., K. Deng, N. Kim, L. Ross Jr., R. Surampalli, and Z. Hu. 2008. The inhibitory effects of silver nanoparticles, silver ions, and silver chloride colloids on microbial growth. *Water Res.* 42:3066–3074.

Choi, O., T. Clevenger, B. Deng, R. Surampalli, L. Ross, and Z. Hu. 2009. Role of sulfide and ligand strength in controlling nanosilver toxicity. *Water Res.* 43:1879–1886.

Coe-Sullivan, S., W.-K. Woo, J. Steckel, M. Bawendi, and V. Bulovic. 2003. Tuning the performance of hybrid organic/inorganic quantum dot light-emitting devices. *Organic Electronics* 4:123–130.

Conti, J., K. Killpack, G. Gerritzen, L. Huang, M. Mircheva, M. Delmas et al. 2008. Health and safety practices in the nanomaterials workplace: results from an international survey. *Environ. Sci. Technol.* 42:3155–3162.

Daniel, M., and D. Astruc. 2004. Gold nanoparticles: assembly, supramolecular chemistry, quantum-size-related properties, and applications toward biology, catalysis, and nanotechnology. *Chem. Rev.* 104:293–346.

Darlington, T., A. Eeigh, M. Spencer, O. Nguyen, and S. Oldenburg. 2009. Nanoparticle characteristics affecting environmental fate and transport through soil. *Environ. Toxicol. Chem.* 28:1191–1199.

Deguchi, S., R. Alargova, and K. Tsujji. 2001. Stable dispersions of fullerenes, C60 and C70, in water. Preparation and characterisation. *Langmuir* 17:6013–6017.

Diegoli, S., A. Manciulea, Begum, I. Jones, J. Lead, and J. Preece. 2008. Interactions of charge stabilised gold nanoparticles with organic macromolecules. *Sci. Total Environ.* 402:51–61.

Dinadayalane, T., and J. Leszczynski. 2010. Unique diversity of carbon–carbon bonds: structures and properties of fullerenes, carbon nanotubes and graphene. *Struct. Chem.* 21:1155–1169.

Doherty, G., and H. McMahon. 2009. Mechanisms of Endocytosis. *Annu. Rev. Biochem.* 78:857–902.

Donaldson, K., L. Tran, L. Jimenez, R. Duffin, D.E. Newby, N. Mills et al. 2005. Combustion-derived nanoparticles: a review of their toxicology following inhalation exposure. *Part. Fibre Toxicol.* 2:10.

Donaldson, K., F. Murphy, R. Duffin, and C. Poland. 2010. Asbestos, carbon nanotubes and the pleural mesothelium: a review of the hypothesis regarding the role of long fibre retention in the parietal pleura, inflammation and mesothelioma. *Part. Fibre Toxicol.* 7:5.

Doshi, R., W. Braida, C. Christodoulatos, M. Wazne, and G. O'Connor. 2008. Nanoaluminum: transport through sand columns and environmental effects on plants and soil communities. *Environ. Res.* 106:296–303.

Dunford, R., A. Salinaro, L. Cai, N. Serpone, S. Horikoshi, H. Hidaka et al. 1997. Chemical oxidation of DNA damage catalysed by inorganic sunscreen ingredients. *FEBS Lett.* 418:87–90.

Edelstein, A., and R. Cammaratra (Eds.). 1998. *Nanomaterials: Synthesis, Properties and Applications (Second Edition).* Boca Raton: Taylor & Francis.

Ellegaard-Jensen, L., K. Jensen, and A. Johansen. 2012. Nanosilver induces dose-response effects on the nematode *Caenorhabditis elegans. Ecotoxicol. Environ. Saf.* 80:216–223.

Elsaesser, A., and C. Howard. 2012. Toxicology of nanoparticles. *Adv. Drug Deliv. Rev.* 64:129–137.

Emin, S., M. Yanagida, W. Peng, and L. Han. 2012. Evaluation of carrier transport and recombinations in cadmium selenide quantum-dot-sensitized solar cells. *Sol. Energy Mater. Solar Cells* 101:5–10.

Fang, J., X. Shan, and B. Wen. 2009. Stability of titania nanoparticles in soil suspensions and transport in saturated homogeneous soil columns. *Environ. Pollut.* 157:1101–1109.

Federici, G., B. Shaw, and R. Handy. 2007. Toxicity of titanium dioxide nanoparticles to rainbow trout (*Oncorhynchus mykiss*): gill injury, oxidative stress, and other physiological effects. *Aquat. Toxicol.* 84:415–430.

Foley, S., C. Crowley, M. Smaihi, C. Bonfils, B. Erlanger, P. Seta et al. 2002. Cellular localisation of a water-soluble fullerene derivative. *Biochem. Biophys. Res. Commun.* 294:116–119.

Franklin, N., N. Rogers, S. Apte, G. Batley, G. Gadd, and P. Casey. 2007. Comparative toxicity of nanoparticulate ZnO, bulk ZnO, and ZnCl₂ to a freshwater microalga (*Pseudokirchneriella subcapitata*): the importance of particle solubility. *Environ. Sci. Technol.* 41:8484–8490.

Fujita, K., M. Horie, H. Kato, S. Endoh, M. Suzuki, A. Nakamura et al. 2009. Effects of ultrafine TiO₂ particles on gene expression profile in human keratinocytes without illumination: involvement of extracellular matrix and cell adhesion. *Toxicol. Lett.* 191:109–117.

Gagné, F., J. Auclair, P. Turcotte, M. Fournier, C. Gagnon, S. Sauvé et al. 2008. Ecotoxicity of CdTe quantum dots to freshwater mussels: impacts on immune system, oxidative stress and genotoxicity. *Aquat. Toxicol.* 86:333–340.

Gajewicz, A., B. Rasulev, T. Dinadayalane, P. Urbaszek, T. Puzyn, D. Leszczynska et al. 2012. Advancing risk assessment of engineered nanomaterials: application of computational approaches. *Adv. Drug Deliv. Rev.* 64(15):1663–1693, doi:10.1016/j.addr.2012.05.014.

Ghosh, S., H. Mashayekhi, P. Bhowmik, and B. Xing. 2010. Colloidal stability of Al₂O₃ nanoparticles as effected by coating of structurally different humic acids. *Langmuir* 26:873–879.

Golden, J., R. Small, L. Pagan, C. Shang, and S. Raghavan. 2000. Evaluating and treating CMP wastewater. *Semicond. Int.* 23:92–103.

Gordon, T., L. Chen, J. Fine, R. Schlesinger, W. Su, T. Kimmel et al. 1992. Pulmonary effects of inhaled zinc oxide in human subjects, guinea pigs, rats, and rabbits. *Am. Ind. Hyg. Assoc. J.* 53:503–509.

Hallock, M., P. Greenley, L. DiBerardinis, and D. Kallin. 2009. Potential risks of nanomaterials and how to safely handle materials of uncertain toxicity. *J. Chem. Health Saf.* Jan/Feb:16–23.

Handy, R., F. von der Kammer, J. Lead, M. Hassellöv, R. Owen, and M. Crane. 2008. The ecotoxicology and chemistry of manufactured nanoparticles. *Ecotoxicology* 17:287–314.

Hardman, R. 2006. A toxicologic review of quantum dots: toxicity depends on physicochemical and environmental factors. *Environ. Health Perspect.* 114:165–172.

Heinlaan, M., A. Ivask, I. Blinova, H. Dubourguier, and A. Kahru. 2008. Toxicity of nanosized and bulk ZnO, CuO and TiO₂ to bacteria *Vibrio fischeri* and crustaceans *Daphnia magna* and *Thamnocephalus platyurus*. *Chemosphere* 71:1308–1316.

Horie, M., K. Nishio, K. Fujita, H. Kato, S. Endoh, M. Suzuki et al. 2010. Cellular responses by stable and uniform ultrafine titanium dioxide particles in culture-medium dispersions when secondary particle size was 100 nm or less. *Toxicol. In Vitro* 24:1629–1638.

Horie, M., H. Kato, K. Fujita, S. Endoh, and H. Iwahashi. 2012. In vitro evaluation of cellular response induced by manufactured nanoparticles. *Chem. Res. Toxicol.* 25:605–619.

Hou, L., K. Li, Y. Ding, Y. Li, J. Chen, X. Wub et al. 2012. Removal of silver nanoparticles in simulated wastewater treatment processes and its impact on COD and NH4 reduction. *Chemosphere* 87:248–252.

Huang, C., W. Jiang, and C. Chen. 2004. Nano silica removal from IC wastewater by precoagulation and microfiltration. *Water Sci. Technol.* 50:133–138.

Huang, Z., X. Zheng, D. Yan, G. Yin, X. Liao, Y. Kang et al. 2008. Toxicological effect of ZnO nanoparticles based on bacteria. *Langmuir* 24:4140–4144.

Hurt, R., M. Monthioux, and A. Kane. 2006. Toxicology of carbon nanomaterials: status, trends, and perspectives on the special issue. *Carbon* 44:1028–1033.

Jaisi, D., and M. Elimelech. 2009. Single-walled carbon nanotubes exhibit limited transport in soil columns. *Environ. Sci. Technol.* 43:9161–9166.

Jeon, H., S. Yi, and S. Oh. 2003. Preparation and antibacterial effects of Ag-SiO₂ thin films by sol-gel method. *Biomaterials* 24:4921–4928.

Joseph, L., J. Flora, Y.-G. Park, M. Badawy, H. Saleh, and Y. Yoon. 2012. Removal of natural organic matter from potential drinking water sources by combined coagulation and adsorption using carbon nanomaterials. *Sep. Purif. Technol.* 95:64–72.

Ju-Nam, Y., and J. Lead. 2008. Manufactured nanoparticles: an overview of their chemistry, interactions and potential environmental implications. *Sci. Total Environ.* 400:396–414.

Kamat, J., T. Devasagayam, K. Priyadarsini, and H. Mohan. 2000. Reactive oxygen species mediated membrane damage induced by fullerene derivatives and its possible biological implications. *Toxicology* 155:55–61.

Karlsson, H., P. Cronholm, J. Gustafsson, and L. Möller. 2008. Copper oxide nanoparticles are highly toxic: a comparison between metal oxide nanoparticles and carbon nanotubes. *Chem. Res. Toxicol.* 21:1726–1732.

Kasemets, K., A. Ivask, H.C. Dubourguier, and A. Kahru. 2009. Toxicity of nanoparticles of ZnO, CuO and TiO2 to yeast *Saccharomyces cerevisiae. Toxicol. In Vitro* 23: 1116–1122.

Kayat, J., V. Gajbhiye, R. Tekade, and N. Jain. 2011. Pulmonary toxicity of carbon nanotubes: a systematic report. *Nanomed. Nanotechnol. Biol. Med.* 7:40–49.

Kim, B., C.-S. Park, M. Murayama, and M. Hochella. 2010. Discovery and characterization of silver sulfide nanoparticles in final sewage sludge products. *Environ. Sci. Technol.* 44:7509–7514.

Kim, J., and B. Van der Bruggen. 2010. The use of nanoparticles in polymeric and ceramic membrane structures: review of manufacturing procedures and performance improvement for water treatment. *Environ. Pollut.* 158:2335–2349.

Kirchner, C., T. Liedl, S. Kudera, T. Pellegrino, A. Javier, H. Gaub et al. 2005. Cytotoxicity of colloidal CdSe and CdSe/ZnS nanoparticles. *Nano Lett.* 5:331–338.

Kroto, H., J. Heath, S. O'Brien, R. Curl, and R. Smalley. 1985. C60: Buckminsterfullerene. *Nature* 318(6042):162–163.

Kumar, R., S. Howdle, and H. Münstedt. 2005. Polyamide/silver antimicrobials: effect of filler types on the silver ion release. *J. Biomed. Mater. Res. Part B: Appl. Biomater.* 75B:311–319.

Kumari, M., S. Khan, S. Pakrashi, A. Mukherjee, and N. Chandrasekaran. 2011. Cytogenetic and genotoxic effects of zinc oxide nanoparticles on root cells of *Allium cepa, J. Hazard. Mater.* 190:613–621.

Lai, C., and S. Lin. 2004. Treatment of chemical mechanical polishing wastewater by electrocoagulation: system performances and sludge settling characteristics. *Chemosphere* 54:235–242.

Lam, C., J. James, R. McCluskey, S. Arepalli, and R. Hunter. 2006. A review of carbon nanotube toxicity and assessment of potential occupational and environmental health risks. *Crit. Rev. Toxicol.* 36:189–217.

Lead, J., and K. Wilkinson. 2006. Aquatic colloids and nanoparticles: current knowledge and future trends. *Environ. Chem.* 3:159–171.

Lee, J., M. Cho, J. Fortner, J. Hughes, and J. Kim. 2009. Transformation of aggregate C-60 in the aqueous phase by UV irradiation. *Environ. Sci. Technol.* 43:4878–4883.

Li, M., L. Zhu, and D. Lin. 2011. Toxicity of ZnO nanoparticles to *Escherichia coli*: mechanism and the influence of medium components. *Environ. Sci. Technol.* 45:1977–1983.

Limbach, L., Y. Li, R. Grass, T. Brunner, M. Hintermann, M. Muller et al. 2005. Oxide nanoparticle uptake in human lung fibroblasts: effects of particle size, agglomeration, and diffusion at low concentrations. *Environ. Sci. Technol.* 39:9370–9376.

Limbach, L., R. Bereiter, E. Müller, R. Krebs, R. Gälli, and W. Stark. 2008. Removal of oxide nanoparticles in a model wastewater treatment plant: influence of agglomeration and surfactants on clearing efficiency. *Environ. Sci. Technol.* 42:5828–5833.

Lin, D., and B. Xing. 2007. Phytotoxicity of nanoparticles: inhibition of seed germination and root growth. *Environ. Pollut.* 150:243–250.

Lin, D., and B. Xing. 2008. Roof uptake and phytotoxicity of ZnO nanoparticles. *Environ. Sci. Technol.* 42:5580–5585.

Lindberg, H., G. Falck, S. Suhonen, M. Vippola, E. Vanhala, J. Catalán et al. 2009. Genotoxicity of nanomaterials: DNA damage and micronuclei induced by carbon nanotubes and graphite nanofibres in human bronchial epithelial cells in vitro. *Toxicol. Lett.* 186:166–173.

Linse, S., C. Cabaleiro-Lago, W. Xue, I. Lynch, S. Lindman, E. Thulin et al. 2007. Nucleation of protein fibrillation by nanoparticles. *Proc. Natl. Acad. Sci. USA* 104:8691–8696.

Long, T., N. Saleh, R. Tilton, G. Lowry, and B. Veronesi. 2006. Titanium dioxide (P25) produces reactive oxygen species in immortalized brain microglia (BV2): implications for nanoparticle neurotoxicity. *Environ. Sci. Technol.* 40:4346–4352.

Lubick, N. 2008. Nanosilver toxicity: ions, nanoparticles—or both? *Environ. Sci. Technol.* 42:8617.

Magrez, A., S. Kasas, V. Salicio, N. Pasquier, J. Seo, M. Celio et al. 2006. Cellular toxicity of carbon-based nanomaterials. *Nano Lett.* 6:1121–1125.

Matilainen, A., M. Vepsäläinen, and M. Sillanpää. 2010. Natural organic matter removal by coagulation during drinking water treatment: a review. *Adv. Colloid Interf. Sci.* 159:189–197.

Matranga, V., and I. Corsi. 2012. Toxic effects of engineered nanoparticles in the marine environment: model organisms and molecular approaches. *Marine Environ. Res.* 76:32–40.

McHugh, P., and J. Knowland. 1997. Characterization of DNA damage inflicted by free radicals from a mutagenic sunscreen ingredient and its location using an in vitro genetic reversion assay. *Photochem. Photobiol.* 66:276–281.

Meyer, J., C. Lord, X. Yang, E. Turner, A. Badireddy, S. Marinakos et al. 2010. Intracellular uptake and associated toxicity of silver nanoparticles in *Caenorhabditis elegans. Aquat. Toxicol.* 100–102:140–150.

Mortimer, M., K. Kasemets, and A. Kahru. 2010. Toxicity of ZnO and CuO nanoparticles to ciliated protozoa *Tetrahymena thermophila. Toxicology* 269:182–189.

Mraz, S. 2005. Nanowaste: the next big threat? *Machine Design* 77:46–53.

Muller, J., F. Huaux, and D. Lison. 2006. Respiratory toxicity of carbon nanotubes: how worried should we be? *Carbon* 44:1048–1056.

Murr, L., and K. Soto. 2005. A TEM study of soot, carbon nanotubes, and related fullerene nanopolyhedra in common fuel-gas combustion sources. *Mater. Charact.* 55:50–65.

Musee, N. 2011. Nanowastes and the environment: potential new waste management paradigm. *Environ. Int.* 37:112–128.

Nakata, K., and A. Fujishima. 2012. TiO2 photocatalysis: design and applications. *J. Photochem. Photobiol. C: Photochem. Rev.*, 13:169–189.

Narayan, R. 2010. Use of nanomaterials in water purification. *Mater. Today* 13:44–46.

Navarro, E., A. Baun, R. Behra, N. Hartmann, J. Filser, A. Miao et al. 2008. Environmental behavior and ecotoxicity of engineered nanoparticles to algae, plants and fungi. *Ecotoxicology* 17:372–387.

Nel, A. 2005. Air pollution-related illness: effects of particles. *Science* 308:804–806.

Nel, A., T. Xia, L. Madler, and N. Li. 2006. Toxic potential of materials at the nanolevel. *Science* 311:622–627.

Nowack, B., and T. Bucheli. 2007. Occurrence, behaviour and effects of nanoparticles in the environment. *Environ. Pollut.* 150:5–22.

Nyberg, L., R. Turco, and L. Nies. 2008. Assessing the impact of nanomaterials on anaerobic microbial communities. *Environ. Sci. Technol.* 42:1938–1943.

Parashar, U., V. Kumar, T. Bera, P. Saxena, G. Nath, S. Srivastava et al. 2011. Study of mechanism of enhanced antibacterial activity by green synthesis of silver nanoparticles. *Nanotechnology* 22:415104.

Partha, R., and J. Conyers. 2009. Biomedical applications of functionalized fullerene-based nanomaterials. *Int. J. Nanomed.* 4:261–275.

Peralta-Videa, J., L. Zhao, M. Lopez-Moreno, G. de la Rosa, J. Hong, and J. Gardea-Torresdey. 2011. Nanomaterials and the environment: a review for the biennium 2008–2010. *J. Hazard. Mater.* 186:1–15.

Poland, C., R. Duffin, I. Kinloch, A. Maynard, W. Wallace, A. Seaton et al. 2008. Carbon nanotubes introduced into the abdominal cavity of mice show asbestos-like pathogenicity in a pilot study. *Nat. Nanotechnol.* 3:423–428.

Puzyn, T., B. Rasulev, A. Gajewicz, X. Hu, T. Dasari, A. Michalkova et al. 2011. Using nano-QSAR to predict the cytotoxicity of metal oxide nanoparticles. *Nat. Nanotechnol.* 6:175–178.

Ratte, H. 1999. Bioaccumulation and toxicity of silver compounds: a review. *Environ. Toxicol. Chem.* 18:89–108.

Rehn, B., F. Seiler, S. Rehn, J. Bruch, and M. Maier. 2003. Investigations on the inflammatory and genotoxic lung effects of two types of titanium dioxide: untreated and surface treated. *Toxicol. Appl. Pharmacol.* 189:84–95.

Sanchez, A., S. Recillas, X. Font, E. Casals, E. Gonzalez, and V. Puntes. 2011. *Trends Anal. Chem.* 30:507–516.

Sayes, C., J. Fortner, W. Guo, D. Lyon, A. Boyd, K. Ausman et al. 2004. The differential cytotoxicity of water-soluble fullerenes. *Nano Lett.* 4:1881–1887.

Savolainen, K., H. Alenius, H. Norppa, L. Pylkkänen, T. Tuomi, and G. Kasper. 2010. Risk assessment of engineered nanomaterials and nanotechnologies—a review. *Toxicology* 269:92–104.

Schmid, K., and M. Riediker. 2008. Use of nanoparticles in Swiss industry: a targeted survey. *Environ. Sci. Technology* 42:2253–2260.

Schreiner, K., T. Filley, R. Blanchette, B. Bowen, R. Bolskar, W. Hockaday et al. 2009. White-Rot basidiomycete-mediated decomposition of C-60 fullerol. *Environ. Sci. Technol.* 43:3162–3168.

Sharma, V., S. Singh, D. Anderson, D. Tobin, and A. Dhawan. 2011. Zinc oxide nanoparticle induced genotoxicity in primary human epidermal keratinocytes. *J. Nanosci. Nanotechnol.* 11:3782–3788.

Shinde, S., N. Grampurohit, D. Gaikwad, S. Jadhav, M. Gadhave, and P. Shelke. 2012. Toxicity induced by nanoparticles. *Asian Pac. J. Trop. Dis.*, 2(4):331–334.

Simon-Deckers, A., S. Loo, M. Mayne-Lhermite, N. Herlin-Boime, C. Menguy, C. Reynaud et al. 2009. Size-, composition- and shape-dependent toxicological impact of metal oxide nanoparticles and carbon nanotubes toward bacteria. *Environ. Sci. Technol.* 43:8423–8429.

Stafford, N. 2007. Catalytic converters go nano. *Chem. World* 4:16.

Tian, F., D. Cui, H. Schwarz, G. Estrada, and H. Kobayashi. 2006. Cytotoxicity of single-wall carbon nanotubes on human fibroblasts. *Toxicol. In Vitro* 20:1202–1212.

Turco, R., M. Bischoff, Z. Tong, and L. Nies. 2011. Environmental implications of nanomaterials: are we studying the right thing? *Curr. Opin. Biotechnol.* 22:527–532.

Uo, M., T. Akasaka, F. Watari, Y. Sato, and K. Tohji. 2011. Toxicity evaluations of various carbon nanomaterials. *Dent. Mater. J.* 30:245–263.

Wang, H., R. Wick, and B. Xing. 2009. Toxicity of nanoparticulate and bulk ZnO, Al_2O_3 and TiO_2 to the nematode *Caenorhabditis elegans*. *Environ. Pollut.* 157:1171–1177.

Wang, X., M. Ruedas-Rama, and E. Hall. 2007. The emerging use of quantum dots in analysis. *Anal. Lett.* 40:1497–1520.

Weir, E., A. Lawlor, A. Whelan, and F. Regan. 2008. The use of nanoparticles in antimicrobial materials and their characterization. *Analyst* 133:835–845.

Westerhoff, P., Y. Zhang, J. Crittenden, and Y. Chen. 2008. Properties of commercial nanoparticles that affect their removal during water treatment. In: Grassian V. (Ed.). *Nanoscience and Nanotechnology: Environmental and Health Impacts*. NJ: John Wiley and Sons; pp. 71–90.

Win-Shwe, T.-T., and H. Fujimaki. 2011. Nanoparticles and neurotoxicity. *Int. J. Mol. Sci.* 12:6267–6280.

Wu, J., W. Liu, C. Xue, S. Zhou, F. Lan, L. Bi et al. 2009. Toxicity and penetration of TiO_2 nanoparticles in hairless mice and porcine skin after subchronic dermal exposure. *Toxicol. Lett.* 191:1–8.

Wu, J., C. Wang, J. Sun, and Y. Xue. 2011. Neurotoxicity of silica nanoparticles: brain localization and dopaminergic neurons damage pathways. *ACS Nano* 5:4476–4489.

Wu, Y., Q. Zhoua, H. Li, W. Liua, T. Wanga, and G. Jianga. 2010. Effects of silver nanoparticles on the development and histopathology biomarkers of Japanese medaka (*Oryzias latipes*) using the partial-life test. *Aquat. Toxicol.* 100–102:160–167.

Xia, T., M. Kovochich, M. Liong, L. Mädler, B. Gilbert, H. Shi et al. 2008. Comparison of the mechanism of toxicity of zinc oxide and cerium oxide nanoparticles based on dissolution and oxidative stress properties. *ACS Nano* 2–10:2121–2134.

Yordanov, G., and C. Dushkin. 2011. Quantum dots for biomedical applications: present status and prospects. *Ann. Univ. Sofia Fac. Chim.* 102–103:81–102.

Zhu, Y., Q. Zhao, Y. Li, X. Cai, and W. Li. 2006. The interaction and toxicity of multi-walled carbon nanotubes with *Stylonychia mytilus*. *J. Nanosci. Nanotechnol.* 6:1357–1364.

Zhuang, Z., X. Xu, Y. Wang, Y. Wang, F. Huang, and Z. Lin. 2012. Treatment of nanowaste via fast crystal growth: with recycling of nano-SnO_2 from electroplating sludge as a study case. *J. Hazardous Mater.* 211–212:414–419.

3 Magnetic Field Directed Self-Assembly of Magnetic Nanoparticles into Higher-Order Structures

Tugce Ozdemir, Deniz Sandal,
Romina Cuculayef, Mustafa Culha, Amitav Sanyal,
Naz Zeynep Atay, and Seyda Bucak

CONTENTS

3.1 INTRODUCTION

In recent years, there has been growing interest in the utilization of metal nanoparticles (NPs) in materials and biomedical sciences due to the characteristic intrinsic properties of these nano-objects that provide novel solutions to many challenges in these areas (Saha et al. 2012; Shenhar and Rotello 2003). In particular, metal NPs play a vital role in biomedical sciences in the area ranging from early detection to cure of diseases. Metal NPs are very versatile objects since their functional attributes arise from the metal core, and their surface coatings can be easily tailored toward specific applications. A range of functionalities can be included within a NP complex using surface chemistry that allows attachment of cell-specific ligands for targeted delivery, surface coatings to increase circulation times and enhance bioavailability, specific materials on the surface or in the NP core that enable storage of a therapeutic cargo until the target site is reached, and materials sensitive to local or remote actuation cues that allow controlled delivery of therapeutics to the target cells (Galvin et al. 2012).

Among various metal NPs, magnetic metal oxide NPs are attractive in various applications in biomedical sciences since they can be imaged using magnetic resonance imaging (MRI) and actuated by external magnetic fields (Ho et al. 2011). Magnetic NPs (MNPs), because of their sizes comparable to those of biomolecules such as proteins, antibodies, DNA, or RNA, can interact with or bind to them. Their magnetic properties have led to fabrication of magnetic carriers for biomedical applications such as DNA and RNA purification, cell separation, and drug delivery. They have also been exploited for hyperthermia and MRI contrast enhancement (Gupta and Gupta 2005; Hafeli and Pauer 1999; Mornet et al. 2000; Shinkai 2002).

Superparamagnetism is a form of magnetism, which is observed with small ferromagnetic or ferrimagnetic NPs. Superparamagnetism occurs in NPs that is, composed of a single magnetic domain. Under this condition, it is considered that the magnetization of the NPs is a single giant magnetic moment, the sum of all the individual magnetic moments carried by the atoms of the NP. When an external magnetic field is applied to the superparamagnetic NPs, they tend to align along the magnetic field, leading to a net magnetization. In the absence of an external magnetic field, however, the dipoles are randomly oriented and there is no net magnetization (Awschalom and Divencenzo 1995). The size dependence of magnetic properties of Fe_3O_4 NPs synthesized from non-aqueous homogeneous solutions of polyols has been recently investigated (Caruntu et al. 2007).

Because their biocompatibility and toxicity are well investigated and better understood, MNPs, especially iron oxide NPs, are better-suited materials as contrast agents for MRI and for image-directed delivery of therapeutics. Given tunable magnetic properties and various surface chemistries from the coating materials, most applications of engineered MNPs take advantage of their superb MRI contrast enhancing capability as well as surface functionalities (Huang et al. 2012).

Use of MNPs in magnetic drug targeting is especially preferred in tumor treatment, since the NPs are concentrated within the target region because of the magnetic field influence, with considerably reduced undesirable side effects (Gitter and Odenbach 2011). Magnetic targeting promises to improve the efficacy and safety of different classes of therapeutic agents by enabling their active guidance to the site of disease and minimizing dissemination to non-target tissues (Chorny et al. 2011). Recent results demonstrate that Fe_3O_4 NPs could inhibit the function of P-glycoprotein so as to increase the relevant drug accumulation in target cancer cells and thus enhance the cytotoxicity suppression of As_2O_3. This may open doors for potential application of Fe_3O_4 NPs in the cancer chemotherapy of leukemia in the clinical area (Guo et al. 2009).

A recent study by Cheng et al., using an in vivo dual-modal optical/MRI of mice, shows that by placing a magnet near the tumor, MNPs tend to migrate toward the tumor after intravenous injection and show high tumor accumulation, eightfold higher than that without magnetic targeting (Cheng et al. 2011).

MNPs also have extensive industrial applications in nanobiotechnology. Studies of defect chemistry of oxide NPs for creating new functionalities pertinent to energy applications including dilute magnetic semiconductors, giant dielectrics, or white light generation are well understood (Li et al. 2011). Magnetic bioseparations of proteins have been carried out using magnetophoresis (Bucak et al. 2003). Magnetic particles have also been used for cell clarification using a countercurrent magnetic separation device (Bucak et al. 2011).

Another area of growing interest is the creation of magnetic microstructures by assembly of NPs since hierarchical organization modulates the magnetic property of such structures. For example, the magnetic properties of the magnetoferritin particles are directly affected by the hierarchical organization (Kostiainen et al. 2011). Magnetoferritin NPs dispersed in water exhibit typical magnetism of single-domain noninteracting NPs; however, the same NPs organized into fcc superstructures show clearly the effects of the altered magnetostatic (e.g., dipole–dipole) interactions by exhibiting different hysteresis of the field-dependent magnetization. In another recent study, monodisperse iron oxide NPs with a diameter of 20 nm were synthesized by thermal decomposition, and the structural and magnetic characterization of these self-assembled NP arrays have been reported (Benitez et al. 2011).

Recent developments indicate that the application of a magnetic field could be an elegant way to orient and self-assemble MNPs into nano- and microscale structures, in which dipole–dipole interactions between adjacent MNPs force them to form reversible anisotropic structures (Furst et al. 1998; Niu et al. 2004; Sheparovych et al. 2006; Vuppu et al. 2003).

Patterning of MNPs has been of interest in several applications and to date mostly accomplished by employment of techniques such as microcontact printing (Palacin et al. 1996), taking advantage of layer-by-layer self-assembly (Xue et al. 2005), electron beam treatment (Erokhina et al. 2004), and focused ion beams (Anders et al. 2002). Despite all these developments, fabrication of three-dimensional (3D) microstructures from nanoscopic building blocks still remains challenging. Undoubtedly, development of novel strategies to fabricate discrete 3D microstructures that are simple and cost-effective will be beneficial.

In this chapter, we explain how the application of a magnetic field accentuated by the presence of magnetic beads creates a field gradient that can be an elegant way to orient and self-assemble magnetic particles into microscale structures (Ozdemir et al. 2010). We also demonstrate that the self-assembly of MNPs can be easily tuned to generate a plethora of different microstructures by simple variation of both the strength and direction of the magnetic field, as well as the underlying template formed from magnetic microparticles (Cuculayev et al. 2012). The strategy utilized benefits from magnetic accentuation of microbeads by the application of external magnetic field, coupled with the superparamagnetism of MNPs and solvent evaporation, to furnish well-defined hierarchical structures composed of MNPs.

The strategy utilizes a template that consists of preexisting channels of a commercial compact disc (CD) that are roughly filled with the micrometer-size beads composed of MNPs embedded in a polymer matrix (Ademtech beads with an average size of 500 nm). Thereafter, the CD channels filled with microbeads are coated with a thin layer of polydimethylsiloxane (PDMS). The surface is exposed to a suspension containing superparamagnetic NPs. Magnetic accentuation of microbeads by application of external magnetic field, coupled with superparamagnetism of MNPs and solvent evaporation, leads to well-defined hierarchical structures of densely packed standing cylinders composed of MNPs. A schematic representation of the process is given in Figure 3.1.

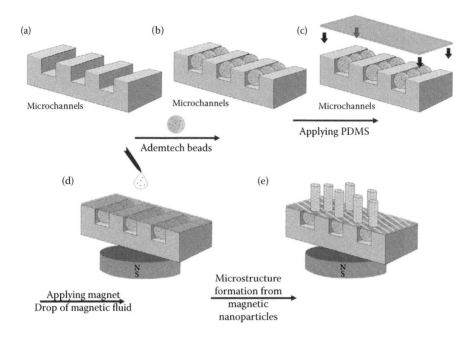

FIGURE 3.1 (a) Schematic representation of the preexisting channels of a commercial CD. (b) Ademtech beads are then added to fill these channels. (c) PDMS layer is applied on top to fix the Ademtech beads in place. (d) A magnet is placed underneath the surface where a drop of magnetic nanofluids is applied. (e) Upon drying of the magnetic nanofluid, micrometer-size magnetic structures are obtained. (From *Nanotechnology*, IOP Publishing. With permission.)

3.2 ENHANCEMENT OF THE MAGNETIC FORCE

The magnetic field in our experiments was created by a disc-shaped neodymium handheld magnet, generating a magnetic field of approximately 20 to 100 T on the solid surface depending on the distance of the magnet from the surface. The force acting on each particle depends on the volume of the magnetic particle, magnetic field strength, and the gradient of the magnetic field as shown in Equation 3.1 (Hafeli et al. 2005), where F_m is the magnetic force acting on the individual particle, χ is the magnetic susceptibility, V_{core} refers to the volume of the magnetic core, μ_0 is the magnetic constant (permeability), H is the magnetic field intensity, and ∇H is the magnetic field gradient.

$$F_m = \chi V_{core} \mu_0 H \nabla H \tag{3.1}$$

It is straightforward to deduce that the force acting on small particles is less. However, magnetite particles below a certain size (<30 nm) become superparamagnetic (as opposed to paramagnetic) and their magnetic susceptibility increases.

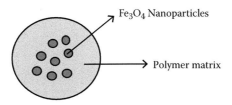

FIGURE 3.2 Schematic representation of magnetic beads. (From *Nanotechnology*, IOP Publishing. With permission.)

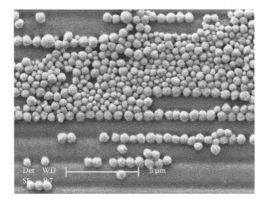

FIGURE 3.3 SEM image of the purchased Ademtech magnetic beads. (From *Nanotechnology*, IOP Publishing. With permission.)

Therefore, there is a discontinuity in the force acting on the particle as the size drops below 30 nm. As the size of the synthesized particles is rather small (<10 nm), although the particles are superparamagnetic (Fe_3O_4), when a magnetic field of 0.1 T is applied, the magnetic force acting on each particle is not very high. In order to enhance the force on the particles, instead of trying to increase the magnetic field, as an attractive alternative, the magnetic field gradient may be enhanced by disturbing the uniformity of the field. In our experiment, this is achieved by introducing polymer-coated magnetic beads of size 500 nm between the magnet and the MNPs. A schematic representation of these beads and the SEM image of these MNP-embedded polymeric particles are shown in Figures 3.2 and 3.3, respectively.

3.3 TEMPLATE PREPARATION FOR PATTERNING AND CHARACTERIZATION

A commercial CD purchased from Princo with approximately 0.860 ± 0.030 μm size channels (channel size determined by atomic force microscopy, Figure 3.4) was used as a template for the assembly of the 500 nm magnetic beads (500 nm –COOH functionalization, Ademtech).

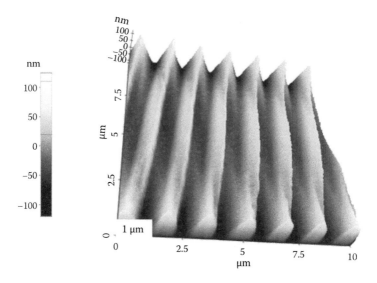

FIGURE 3.4 **(See color insert.)** Atomic force microscopy image showing the channels of the commercial CD. (From *Nanotechnology*, IOP Publishing. With permission.)

The magnetic beads were assembled into the channels on the CD surfaces with the "convective assembly" method that was based on the assembly of micrometer- and nanometer-size particles during evaporation of the solvent from a moving droplet owing to the capillary effect (Denkov et al. 1993; Yamaki et al. 1995). The experimental setup was explained elsewhere (Kahraman et al. 2008). Briefly, a piece of a CD, whose aluminum foil part protecting the micropatterns was peeled off, was located on a μ-stage (PI, Germany). A glass slide with an angle of 30° was placed on the CD with the help of a fixed clamp. A 20 μL bead solution (0.1% concentration) was placed at the cross section of the glass slide and the CD. Then, the stage was moved at a certain velocity.

The magnetic beads are placed on the template in three different ways: one is by filling the microchannels of the CD surface using a convective self-assembly setup. Another was by randomly dispersing the beads on the CD surface, and the last method was to place the beads as a complete monolayer on the surface, again using convective self-assembly.

Figure 3.5a and b shows a commercial CD with the magnetic beads on its surface. In Figure 3.5a, the CD microchannels are filled with the magnetic beads, and in Figure 3.5b, these beads form a monolayer on the CD surface.

When an external magnetic field is applied, these magnetic beads (regardless of their initial distribution) with MNPs on them aggregate as shown in Figure 3.6.

In order to immobilize the intended distribution of the magnetic beads on the CD surface and separate them from the magnetic particles that are later applied, a thin layer (about 1 mm) of PDMS polymer (Sylgard 184 Silicone Elastomer Kit, Dow Corning) was applied on the surface after distributing the magnetic beads. Figure 3.7

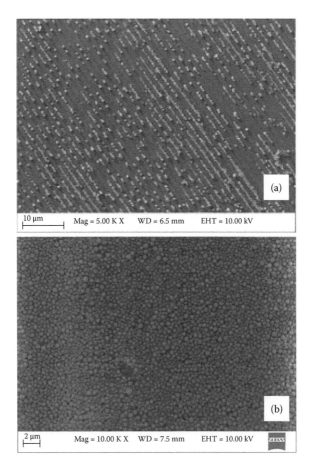

FIGURE 3.5 (a) Magnetic beads in CD. (b) Beads form a monolayer on the CD surface. (From *Nanotechnology*, IOP Publishing. With permission.)

shows the surface once the PDMS is applied, prior to the application of the magnetic particles.

Alternatively, instead of using a PDMS layer to keep the magnetic nano beads (MNBs) in place, thin glass slides (microscope cover slides with 26 × 76 mm dimensions and 0.8 to 1 mm thickness, Pearl) are used. The advantage of this system is its ability to easily remove the glass slide and reuse the patterned MNB surface underneath over and over, significantly reducing the cost of each experiment.

Synthesized magnetic particles of diameter about 6 nm suspended in organic solvents were placed on surfaces coated with PDMS (or glass) with or without magnetic beads, and patterning of NPs upon drying was observed in the presence and absence of a magnetic field. SEM (XL30 ESEM-FEG/EDAX system) was used to observe the pattern formation on flat surfaces.

FIGURE 3.6 SEM image of MNPs with magnetic beads under magnetic field. (From *Nanotechnology*, IOP Publishing. With permission.)

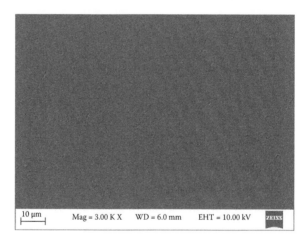

FIGURE 3.7 Image of the CD surface with magnetic beads after coating with a thin layer of PDMS. (From *Nanotechnology*, IOP Publishing. With permission.)

3.4 SYNTHESIS AND CHARACTERIZATION OF MNPs

The organic-phase synthesis of magnetite was described previously (Sun et al. 2004), and this procedure was used with minor modifications. In a typical synthesis, 0.706 g of iron(III) acetylacetonate (Fluka, 97%), 2.3039 g of 1,2-tetradecanediol (Aldrich, 90%), 1.9 mL of oleic acid (Riedel-de Häen, 99%), 1.69 mL of oleylamine (Fluka, ≥70%), and 20 mL of dibenzyl ether (Merck, synthesis grade) were mixed under magnetic stirring in an N_2 environment. The solution was heated to 300°C at a rate of 2.5°C/min in order to obtain 6 nm particles. Particles are precipitated upon addition of methanol and separated by centrifugation followed by drying in a vacuum oven. NPs are then suspended in the desired organic solvent.

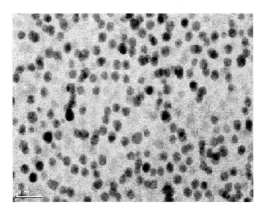

FIGURE 3.8 TEM image of the synthesized nanoparticles. (From *Nanotechnology*, IOP Publishing. With permission.)

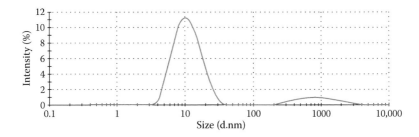

FIGURE 3.9 DLS data of the synthesized nanoparticles. (From *Nanotechnology*, IOP Publishing. With permission.)

To obtain the size and morphology of the synthesized particles, both transmission electron microscopy (TEM) (Model: TecnaiG2F30) and dynamic light scattering (DLS) (Malvern) techniques were used. A typical TEM image is given in Figure 3.8. Based on the TEM images of synthesized magnetite, it can be concluded that the particle cores are spherical with approximately 6.1 ± 2.1 nm in radius with a fairly narrow size distribution. DLS results show (Figure 3.9) that the average hydrodynamic radius is measured to be 10.1 ± 3.1, which is in perfect agreement with the TEM results where the stabilizing surfactant layer contributes to about 2 nm of thickness around the core. Crystal structure information was obtained by using XRD (Rigaku D/MAX-Ultima+/PC). The XRD results exhibit the characteristic diffraction pattern of Fe_3O_4 (Figure 3.10).

3.5 FORMATION OF STRUCTURES ON THE SURFACES WITH MNPs

The MNPs were suspended in hexane (Riedel-de Häen, 95%), heptane (Lab-scan, 95%), and decane (Riedel-de Häen, 95%). Each suspension was dropped on the PDMS-coated CD, with and without beads underneath the PDMS layer, in the presence and absence of an external magnetic field. When an external magnetic field

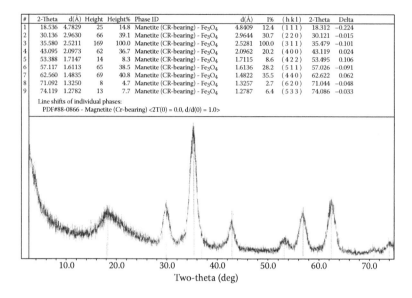

#	2-Theta	d(Å)	Height	Height%	Phase ID	d(Å)	I%	(h k l)	2-Theta	Delta
1	18.536	4.7829	25	14.8	Manetite (CR-bearing) - Fe_3O_4	4.8409	12.4	(1 1 1)	18.312	−0.224
2	30.136	2.9630	66	39.1	Manetite (CR-bearing) - Fe_3O_4	2.9644	30.7	(2 2 0)	30.121	−0.015
3	35.580	2.5211	169	100.0	Manetite (CR-bearing) - Fe_3O_4	2.5281	100.0	(3 1 1)	35.479	−0.101
4	43.095	2.0973	62	36.7	Manetite (CR-bearing) - Fe_3O_4	2.0962	20.2	(4 0 0)	43.119	0.024
5	53.388	1.7147	14	8.3	Manetite (CR-bearing) - Fe_3O_4	1.7115	8.6	(4 2 2)	53.495	0.106
6	57.117	1.6113	65	38.5	Manetite (CR-bearing) - Fe_3O_4	1.6136	28.2	(5 1 1)	57.026	−0.091
7	62.560	1.4835	69	40.8	Manetite (CR-bearing) - Fe_3O_4	1.4822	35.5	(4 4 0)	62.622	0.062
8	71.092	1.3250	8	4.7	Manetite (CR-bearing) - Fe_3O_4	1.3257	2.7	(6 2 0)	71.044	−0.048
9	74.119	1.2782	13	7.7	Manetite (CR-bearing) - Fe_3O_4	1.2787	6.4	(5 3 3)	74.086	−0.033

Line shifts of individual phases:
PDF#88-0866 - Magnetite (Cr-bearing) <2T(0) = 0.0, d/d(0) = 1.0>

Two-theta (deg)

FIGURE 3.10 XRD data for the synthesized Fe_3O_4 NPs. (From *Nanotechnology*, IOP Publishing. With permission.)

was applied, microstructures only from the heptane and decane suspensions in the presence of magnetic beads were observed. We believe that this is due to the fast evaporation rate of hexane (i.e., 9 times faster compared to normal butyl acetate standard). Because of this fast evaporation rate, there is not sufficient time for the NPs to organize themselves into such high structures.

In all cases, no particular structure was observed in the absence of the applied magnetic field. Figure 3.11a–f shows the effect of external field and magnetic beads on the structure formation with Fe_3O_4 NPs upon solvent evaporation.

The presence of the external magnetic field (100 mT) has a clear effect on the behavior of MNPs during evaporation of the solvent from the droplet. Figure 3.11a shows that on a plain surface, MNPs simply clump together during solvent evaporation, leading to no particular structure formation. In the absence of an external magnetic field, when there is a layer of magnetic beads (500 nm) underneath the PDMS surface (0.04 mT), the MNPs dry in a slightly different manner. Their drying pattern suggests that the beads are acting like small permanent magnets and induce a weak dipole in the NPs, resulting in a slight change in the structure of the dried MNPs (Figure 3.11b). When an external magnetic field is applied, solid cylindrical structures of micrometer sizes start to form from the MNPs (Figure 3.11c). When the field gradient is enhanced upon addition of magnetic beads, the cylinders can be observed to align up, in the direction of the field, showing clearly the combined effect of the external field and the magnetic beads under the PDMS surfaces (Figure 3.11d). When decane is used, where the solvent evaporation is slower than that of heptane (<1 compared to 4.3 times *n*-butyl acetate), in the presence of magnetic beads and the applied external field, long tubular structures of a few micrometers can

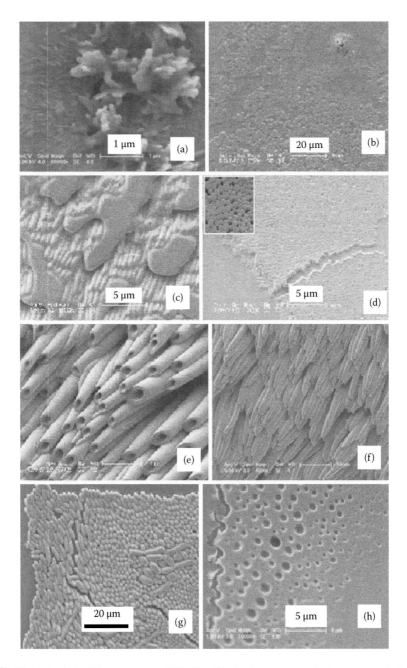

FIGURE 3.11 (a) MNPs in heptane on PDMS surface in the presence of external magnetic field. (b) MNPs in heptane on magnetic beads coated with PDMS. (c) MNPs in heptane on PDMS surface in the presence of external magnetic field. (d) MNPs in heptane on magnetic beads coated with PDMS in the presence of external magnetic field. (e–g) MNPs in decane on magnetic beads coated with PDMS in the presence of external magnetic field. (f) MNPs in decane on magnetic beads coated with PDMS. (From *Nanotechnology*, IOP Publishing. With permission.)

be obtained (Figure 3.11e, f, and g). It is also noteworthy that these structures are very regular and monodisperse. As it can be seen from the figures, solid as well as hollow cylinders can be obtained. Both of these structures can be found in proximity of each other of the same sample. This suggests that the presence of the beads results in a particular alignment of the magnetic field lines leading to the formation of solid or hollow cylinders within the same sample.

When hydrophobic glass slides were used instead of PDMS, even in the absence of magnetic beads between the MNPs, some organization is observed upon application of an external magnetic field. However, it is clear that this organization is accentuated upon placing MNBs on the template, increasing the magnetic force acting on the NPs, resulting in organization at a higher level, leading to the formation of much larger structures, somewhat similar to those obtained when PDMS was used. These structures, although comparable in size, have less smooth surfaces and appear to have a more porous structure (Figure 3.12).

To investigate the effect of the strength and direction of the applied field, each magnetic particle in decane suspension was dropped on the PDMS-coated CD, and the strength of the external magnetic field is varied by placing the magnet at varying distances from the CD surface (25, 47, and 100 mT) and applying the field from different directions (top or bottom of the CD).

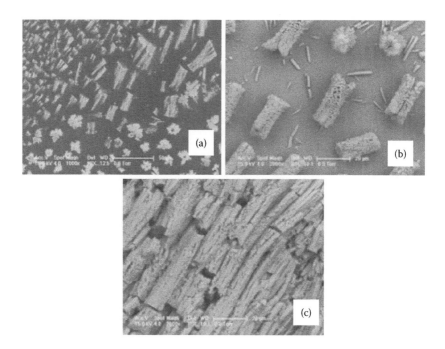

FIGURE 3.12   MNPs on various glass slide surfaces under 100 mT magnetic field. (a) Glass surface only. (b and c) Glass surface over MNBs in microchannels of CD. (From *Journal of Nanoscience and Nanotechnology*, American Scientific Publishers.)

When an external magnetic field is applied, microstructures from all suspensions are observed. In all cases, no particular structure is observed in the absence of the applied magnetic field. Figures 3.13 through 3.15 show the effect of varying the strength and direction of an external field on the structure formation with Fe_3O_4 NPs upon solvent evaporation.

Whether the magnetic beads are in the microchannels, spread as a monolayer, or placed randomly on the CD surface, the magnetic force acting on the particles is clearly sufficient to form organized structures that are elongated in the direction

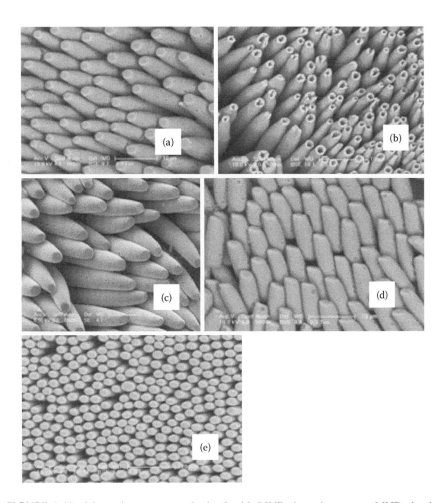

FIGURE 3.13 Magnetic structures obtained with MNPs in *n*-decane, on MNBs in the microchannels of the CD, coated with PDMS, upon application of a magnetic field from different directions and at different strengths: (a) 25 mT/bottom, (b) 47 mT/bottom, (c) 100 mT/bottom, (d) 25 mT/top, and (e) 47 mT/top. (From *Journal of Nanoscience and Nanotechnology*, American Scientific Publishers.)

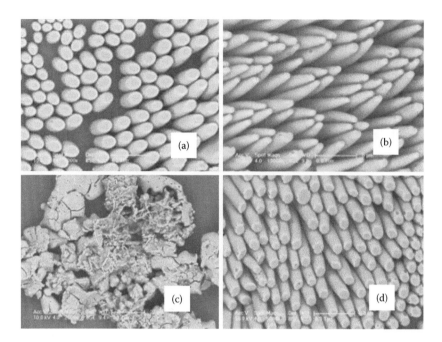

FIGURE 3.14 Magnetic structures obtained with MNPs in *n*-decane, on a monolayer of MNBs over the CD surface, coated with PDMS, upon application of a magnetic field from different directions and at different strengths: (a) 25 mT/bottom, (b) 47 mT/bottom, (c) 25 mT/top, and (d) 47 mT/top. (From *Journal of Nanoscience and Nanotechnology*, American Scientific Publishers.)

away from the template surface even at a magnetic field strength of 25 mT when the field is applied from the bottom. It should be noted that when a monolayer of magnetic beads is present, relatively shorter structures are obtained upon application of 25 mT.

On the other hand, when the field is applied from the top, generating 25 or 47 mT, depending on the organization of the template, organized structures are found to be either elongated or flat on the surface or not able to form at all. The results can be summarized as follows: When a monolayer of magnetic beads is present as a template, no structure can form upon application of 25 mT from the top. When the field strength is increased to 47 mT, the elongated cylindrical structures are observed. When magnetic beads are randomly placed on the template, short cylindrical structures that are upright are formed, and upon increase of the field strength to 47 mT, longer structures that lay flat on the surface is obtained. When the magnetic beads are placed in the microchannels of the template, application of 25 mT from the top results in the formation of structures that lay flat on the surface, whereas application of 47 mT results in the formation of structures that are short but upright.

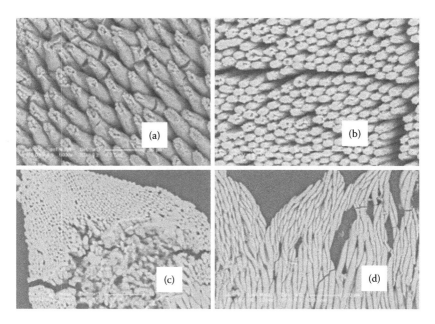

FIGURE 3.15 Magnetic structures obtained with MNPs in *n*-decane, on randomly distributed MNBs over the CD surface, coated with PDMS, upon application of a magnetic field from different directions and at different strengths: (a) 25 mT/bottom, (b) 47 mT/bottom, (c) 25 mT/top, and (d) 47 mT/top. (From *Journal of Nanoscience and Nanotechnology*, American Scientific Publishers.)

3.6 CONCLUSIONS

The influence of the presence of patterned magnetic beads on the behavior of the MNPs from drying droplets under the influence of magnetic field was demonstrated. In the presence of a magnetic field, the formation of micrometer-size cylindrical structures composed of MNPs was observed on the PDMS surfaces prepared as a thin layer on submicrometer-size magnetic beads assembled into channels of a commercially available CD. Under the same experimental conditions, the absence of submicrometer-size magnetic beads yielded arbitrary MNP clusters. It was also observed that the solvent evaporation rate from the droplet spotted on surfaces influenced the formation of unique structures. Decane, with a lower evaporation rate (<1, where butyl acetate = 1) than hexane (9, where butyl acetate = 1), and heptane (4.3, where butyl acetate = 1), led to the formation of better-defined structures even up to 20 μm in length. The influence of the different ways of patterning magnetic beads on the behavior of the MNPs from drying droplets under the influence of varying magnetic field and direction was demonstrated. At a constant magnetic field, applied from a specific direction, it is also shown that changing the drying surface from PDMS to regular glass leads to the formation of slightly different magnetic structures that appear to be more porous. This study shows that by carefully designing the template and changing the magnetic field and direction of the field, the structures that can be obtained from MNPs can be controlled, which lends this method to have great potential for forming large magnetic structures.

ACKNOWLEDGMENTS

We would like to thank Bilge Gedik from the SEM facility at Bogazici University for the images and useful discussions. This work was supported by TUBITAK Grant No. 107M580 and the Bogazici University Research Fund (Project No. 5569). We also acknowledge the support of Yeditepe University during the course of this study.

REFERENCES

Anders, S., S. Sun, C.B. Murray et al. 2002. Lithography and self-assembly for nanometer scale magnetism. *Microelectron. Eng.* 61:569–75.

Awschalom, D.D., and D.P. Divencenzo. 1995. Complex dynamics of mesoscopic magnets. *Phys. Today* 48:43–8.

Benitez, M.J., D. Mishra, P. Szary et al. 2011. Structural and magnetic characterization of self-assembled iron oxide nanoparticle arrays. *J. Phys.: Condens. Matter* 23:126003.

Bucak, S., D.A. Jones, P.E. Laibinis et al. 2003. Protein separations using colloidal magnetic nanoparticles. *Biotechnol. Prog.* 19:477–84.

Bucak, S., S. Sharpe, S. Kuhn et al. 2011. Cell clarification and size separation using continuous countercurrent magnetophoresis. *Biotechnol. Prog.* 27:744–50.

Caruntu, D., G. Caruntu, and C.J. O'Connor. 2007. Magnetic properties of variable-sized Fe_3O_4 nanoparticles synthesized from non-aqueous homogeneous solutions of polyols. *J. Phys. D: Appl. Phys.* 40:5801–9.

Cheng, L., K. Yang, Y. Li et al. 2011. Multifunctional nanoparticles for upconversion luminescence/MR multimodal imaging and magnetically targeted photothermal therapy. *Biomaterials* 33:2215–22.

Chorny, M., I. Fishbein, S. Forbes et al. 2011. Magnetic nanoparticles for targeted vascular delivery. *IUBMB Life* 63:613–20.

Cuculayev, R., A. Sanyal, N.Z. Atay et al. 2012. The effect of the strength and direction of magnetic field on the assembly of magnetic nanoparticles into higher structures *J. Nanosci. Nanotechnol.* 12:2761–6.

Denkov, N.D., O.D. Velev, P.A. Kralchevsky et al. 1993. Two-dimensional crystallization. *Nature* 361:26.

Erokhina, S., T. Berzina, L. Cristofolini et al. 2004. Patterned arrays of magnetic nano-engineered capsules on solid supports. *J. Magn. Magn. Mater.* 272:1353–4.

Furst, E.M., C. Suzuki, M. Fermigier et al. 1998. Permanently-linked monodisperse paramagnetic chains. *Langmuir* 14:7334–6.

Galvin, P., D. Thompson, K. Ryan et al. 2012. Nanoparticle-based drug delivery: case studies for cancer and cardiovascular applications. *Cell. Mol. Life Sci.* 69(3):389–404.

Gitter, K., and S. Odenbach. 2011. Quantitative targeting maps based on experimental investigations for a branched tube model in magnetic drug targeting. *J. Magn. Magn. Mater.* 323(23):3038–42.

Guo, D.D., W.F. Song, and B.A. Chen. 2009. Enhanced cytotoxicity suppression of arsenic trioxide to leukemia cancer cells by using magnetic nanoparticles. In: *Proceedings of the 2009 2nd International Conference on Biomedical Engineering and Informatics*, eds. R. Shi, W.J. Fu, Y.Q. Wang et al. vols. 1–4:1348–52, USA: IEEE eXpress Conference Publishing.

Gupta, A.K., and M. Gupta. 2005. Cytotoxicity suppression and cellular uptake enhancement of surface modified magnetic nanoparticles. *Biomaterials* 26:1565–73.

Hafeli, U.O., and G.J. Pauer. 1999. In vitro and in vivo toxicity of magnetic microspheres. *J. Magn. Magn. Mater.* 194:76–82.

Hafeli, U.O., M.A. Lobedann, J. Steingroewer et al. 2005. Optical method for measurement of magnetophoretic mobility of individual magnetic microspheres in defined magnetic field. *J. Magn. Magn. Mater.* 293:224–39.

Ho, D., X.L. Sun, and S.H. Sun. 2011. Monodisperse magnetic nanoparticles for theranostic applications. *Acc. Chem. Res.* 44:875–82.

Huang, J., X.D. Zhong, L.Y. Wang et al. 2012. Improving the magnetic resonance imaging contrast and detection methods with engineered magnetic nanoparticles. *Theranostics* 2(1):86–102.

Kahraman, M., N. Tokman, and M. Culha. 2008. Silver nanoparticle thin films with nanocavities for surface-enhanced Raman scattering. *ChemPhysChem* 9:902–10.

Kostiainen, M.A., P. Ceci, M. Fornara et al. 2011. Hierarchical self-assembly and optical disassembly for controlled switching of magnetoferritin nanoparticle magnetism. *ACS Nano* 5(8):6394–402.

Li, G.S., L.P. Li, and J. Zheng. 2011. Understanding the defect chemistry of oxide nanoparticles for creating new functionalities: a critical review. *Sci. China Chem.* 54(6):876–86.

Mornet, S., A. Vekris, J. Bonnet et al. 2000. DNA–magnetite nanocomposite materials. *Mater. Lett.* 42:183–8.

Niu, H., Q. Chen, M. Ning et al. 2004. Synthesis and one-dimensional self-assembly of acicular nickel nanocrystallites under magnetic fields. *J. Phys. Chem. B* 108:3996–9.

Ozdemir, T., D. Sandal, M. Culha et al. 2010. Assembly of magnetic nanoparticles into higher structures on patterned magnetic beads under the influence of magnetic field. *Nanotechnology* 21:125603.

Palacin, S., P.C. Hidber, J.-P. Bourgoin et al. 1996. Patterning with magnetic materials at the micron scale. *Chem. Mater.* 8:1316–25.

Saha, K., S.S. Agasti, C. Kim et al. 2012. Gold nanoparticles in chemical and biological sensing. *Chem. Rev.* 112:2739–79.

Shenhar, R., and V.M. Rotello. 2003. Nanoparticles: scaffolds and building blocks. *Acc. Chem. Res.* 36:549–61.

Sheparovych, R., Y. Sahoo, M. Motornov et al. 2006. Polyelectrolyte stabilized nanowires from Fe_3O_4 nanoparticles via magnetic field induced self-assembly. *Chem. Mater.* 18:591–3.

Shinkai, M. 2002. Functional magnetic particles for medical applications. *Biosci. Bioeng.* 94:606–13.

Sun, S., H. Zeng, D.B. Robinson et al. 2004. Monodisperse MFe_2O_4 (M = Fe, Co, Mn) nanoparticles. *J. Am. Chem. Soc.* 126:273–9.

Vuppu, A.K., A.A. Garcia, and M.A. Hayes. 2003. Video microscopy of dynamically aggregated paramagnetic particle chains in an applied rotating magnetic field. *Langmuir* 19: 8646–53.

Xue, W., T.H. Cui, X.J. Lei et al. 2005. In situ deposition/positioning of magnetic nanoparticles with ferroelectric nanolithography. *J. Mater. Res.* 20:712–8.

Yamaki, M., J. Higo, and K. Nagayama. 1995. Size-dependent separation of colloidal particles in two-dimensional convective self-assembly. *Langmuir* 11:2975–8.

4 Particle–Surfactant Interaction at Liquid Interfaces

Eduardo Guzmán, Eva Santini,
Libero Liggieri, Francesca Ravera,
Giuseppe Loglio, Armando Maestro,
Ramón G. Rubio, Jürgen Krägel,
Dmitry Grigoriev, and Reinhard Miller

CONTENTS

4.1 INTRODUCTION

In recent years, the study of the effect of solid particles of micro/nanometric dimensions on the properties of liquid interfaces has undergone an intense development (Binks 2002; Pugh 1996). Under specific conditions, especially if interacting with surfactants, particles are in fact able to segregate at the interface and stabilize it, opposing droplet/bubble coalescence. This makes these systems particularly important for the stabilization of bubbles, liquid films, and disperse systems such as emulsions and foams (Aveyard et al. 2003; Binks et al. 2008; Grigoriev et al. 2007a;

Limage et al. 2010a; Stocco et al. 2011). The ability of nanoparticles to stabilize disperse systems has been known for more than a century, as it is with the case of the Pickering emulsions (Pickering 1907). Nowadays, these systems receive special attention from scientists and engineers because of their increasing applicability in the field of nanomaterial tailoring and fabrication and in other technological fields such as oil recovering, food processing, cosmetics, and membrane-based separation and purification techniques (Even and Gregory 1994). Additionally, the increasing interest in nanostructured materials raises questions related to nanoparticle risks and hazards for health and environment such as the pollution of water environments (Santini et al. 2010, 2012a) and the impact on lung surfactant physiology (Guzmán et al. 2011, 2012). These issues are usually related to the interaction with liquid interfaces.

The affinity of the particles for the liquid/air or liquid/liquid interfaces is essentially determined by their hydrophilic/hydrophobic character (Binks and Horozov 2006). The free-energy change, ΔE_p, associated with the transfer of one spherical particle of radius R from a liquid phase to a planar interface, against air or another liquid, can be expressed in terms of the contact angle, Θ, of the particles with the first liquid, the interfacial tension γ, and the particle radius R (Levine et al. 1989):

$$\Delta E_p = -\pi R^2 \gamma (1 - \cos\Theta)^2. \tag{4.1}$$

Equation 4.1 evidences the strong dependence of ΔE_p on particle wettability, which makes Θ a key parameter driving the attachment at the interface. Generally, partially wetted particles ($0 < \Theta < 180°$) are irreversibly absorbed at the interface with an attachment energy that may exceed by several orders of magnitude the thermal energy kT (Binks 2002). That happens for example for nanometric particles at $\Theta = 90°$ and common values of interfacial tension. This situation is schematically represented in Figure 4.1 where it is clear that the total hydrophilicity and hydrophobicity of particles correspond to $\Theta = 0°$ and $180°$, respectively. The particle-laden interfacial layer should be considered as a multiphase region with three interfaces: one of them between two fluid interfaces, that is, air/oil–water, and two solid–fluid interfaces. It is noteworthy that the behavior of particles transferring to interfaces is substantially different from that of surfactants. These are subjected to exchange processes between the bulk phases and the fluid interface in time scales that can be several orders of magnitude smaller than those involving nanoparticles (Binks and Horozov 2006; Levine et al. 1989).

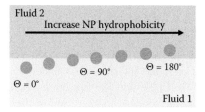

FIGURE 4.1 Position of a particle in an arbitrary fluid–fluid interface depending on the free energy of their attachment ΔE_p accounted for by the phase contact angle Θ. From left to right, the increase in the hydrophobicity of the particle is shown.

The wettabililty properties of the particles can be tuned by the chemical modification of the nanoparticle surface (Zang et al. 2009) or by the adsorption at the
particle surface of surface-active modifiers, such as long-chain surfactants (Maestro
et al. 2012; Santini et al. 2011) or even short-chain alcohols (Maestro et al. 2010).
This latter approach is particularly interesting since it allows a fine control on the
particle attachment to the interface by the simple addition to the particle dispersion of surface-active molecules in proper concentration. The interaction between
particles and surfactants introduces synergistic effects at the interfacial layers that are
strongly dependent on the size, shape, and chemical nature of the particle, as well as on
the nature and concentration of the surfactant (Ahualli et al. 2011; Iglesias et al. 2011).
The formation of NP–surfactant composite layers at a fluid interface modifies the
stability and physicochemical behavior of the system. However, the microscopically
non-homogeneous character of the interfacial layers confers a critical character to
the definition of quantities that describe the mechanical properties of the systems at
the macroscopic level. In these systems, the interfacial tension should be considered
as an effective thermodynamic variable entering the mechanical equilibrium conditions of the macroscopic interface, instead of the ordinary interfacial tension. For the
sake of simplicity, the terms *interfacial tension* and *surface tension* are used below,
meaning the corresponding effective magnitudes. It is worth of mentioning that the
definition of the effective quantities makes possible the application of the standard
thermodynamic equations and methods to the study of these composite systems
(Levine et al. 1989; Menon and Wasan 1988; Menon et al. 1988).

The interfacial properties of the composite layers are strongly correlated to the
synergetic interactions between particles and surfactants that add significant complexity to the interfacial structure and dynamics of these systems, influencing interfacial tension and interfacial rheology (Liggieri et al. 2011; Ravera et al. 2006a,
2008; Whittby et al. 2012). The latter is intrinsically related to the particle-induced
interfacial stabilization in disperse systems (Limage et al. 2010a). In fact, both the
dilational (Liggieri et al. 2011) and the shear (Zang et al. 2010a,b) complex viscoelastic modulus are related to the ability of the systems to dampen external disturbances of the interfacial layer.

This chapter reviews the current trends and future perspectives in the study of
composite particle–surfactant interfacial layers, with special attention to the effect
of nanoparticles on the mechanical and structural properties of these systems and to
how the synergic interactions, occurring at the interface, contribute to the stabilization of dispersed systems with potential technological applications.

4.2 EQUILIBRIUM PROPERTIES

4.2.1 THERMODYNAMIC MODELS

The thermodynamic properties of particle-laden interfaces are related to the particle
partitioning between the continuous phases and the fluid interface (Levine et al.
1989). The classical equations of state, like Langmuir, Frumkin, and so forth (Prosser
and Franses 2001), providing for common surfactants equilibrium relationships
between interfacial tension, bulk surfactant concentration, and surface coverage, are

not appropriate for these composite systems, for which new thermodynamic models have been developed (Fainerman et al. 2003, 2006a; Miller et al. 2006).

The initial attempts to the theoretical description of particle-laden interfaces were based on the Volmer and van der Waals equations, assuming an area per molecule in the monolayer identical to the area of the particles (Binks 2002). However, this approach fails because it leads to unrealistic dependences of surface pressure, Π, on particle sizes that are ascribable to the different interfacial behavior of common surfactants and particles. A correct modeling of the interfacial thermodynamics of particle-laden interfaces must consider, necessarily, the significant difference between the sizes of particles, surfactant, and solvent molecules. According to Fainerman et al. (2006b), it is possible to obtain an equation of state for particle-laden interfaces, or composite particle–surfactant interfacial layers, following a theoretical approach similar to that previously derived for the characterization of protein and mixed proteins/surfactant interfacial layers (Fainerman et al. 2003). This approach is based on the same assumptions of the classical approach (Defay and Prigogine 1966) describing the surface thermodynamics of solutions containing molecules with different sizes.

The approach by Fainerman et al. (2006b) was developed on the basis of the classical Butler equation (Butler 1932). That requires, as a first step, the definition of the chemical potentials of the different species present in the surface layer and the bulk phases. The chemical potential is a thermodynamic quantity describing the state of the system that depends on the composition, temperature, T, pressure, P, and, for systems involving interfacial phases, surface tension. A general expression for the chemical potential of each component of the system including solvent and adsorbed species, such as surfactant or nanoparticles, in the bulk or the interfacial layer can be written as

$$\mu_i = \mu_i^0 + RT \ln f_i x_i - \gamma \Omega_i. \tag{4.2}$$

Here, $\mu_i^0(T,P) = \mu_i^0$ is the standard chemical potential of component i, $x_i = m_i / \sum m_i$ is the corresponding molar fraction, m_i is the number of moles, f_i is the activity coefficient, and Ω_i is the molar area. For the chemical potential of components in solution, it is necessary to consider zero the term associated to the interface; that is, $\gamma \Omega_i = 0$. For dilute solutions, it is possible to assume the same chemical potential of the solvent ($i = 0$) in the bulk and in the surface layer. The latter is essential to define an equation of state for surface layers formed by any number of components characterized by its own geometry,

$$\Pi = -\frac{kT}{\omega_0}(\ln x_0 + \ln f_0), \tag{4.3}$$

with $\Pi = \gamma_0 - \gamma$ being the surface pressure, k being the Boltzmann constant, and ω_0 being the molecular area of a solvent molecule. Equation 4.3 is a very general equation of state because it has been derived only considering the properties of the solvent. However, a more precise formulation of the equation of state requires considering the composition of the interfacial layer. This can be done by introducing the surface excesses, Γ_i, of all species including the solvent, directly related to the partial surface fractions θ_j of the jth components.

In order to derive an expression for the thermodynamic properties of these mixed systems, it is necessary to introduce the activity coefficient f_i of the different species of the system, taking into account additivity of enthalpy (H) and entropy (E) contributions,

$$\ln f_i^{S} = \ln f_i^{H} + \ln f_i^{E}. \tag{4.4}$$

The enthalpic contribution can be expressed as (Fainerman et al. 2003)

$$\ln f_0^{H} = a\theta^2, \ \ln f_j^{H} = an_j\theta_0^2, \tag{4.5}$$

where θ is the fraction of surface covered by molecules or solid particles with an average area ω, a is the Frumkin interaction parameter for non-ideality, and $n_i = \omega_i/\omega_0$. The entropic contribution must be derived considering a first-order, Flory-type model for mixtures of any number of molecules of different molar area (Lucassen-Reynders 1981)

$$\ln f_i^{E} = 1 - n_j \sum_{i \ge 0} (\theta_i/n_i) + \ln\left[n_j \sum_{i \ge 0} (\theta_i/n_i) \right]. \tag{4.6}$$

Taking into account the above considerations, it is possible to define the equation of state for an interfacial layer composed of solvent and any other components, valid for $n = \omega/\omega_0 > 1$.

$$-\frac{\Pi\omega_0}{kT} = \ln(1-\theta) + \theta(1-\omega_0/\omega) + a\theta^2 \tag{4.7}$$

Considering the area of a surface layer A via the relation $\theta = \omega/A$, and replacing Equation 4.7, as the enthalpy contribution to the solvent activity coefficient does not depend on θ, and the condition $n = \omega/\omega_0 \gg 1$ being valid, the Π–A isotherm was obtained (Fainerman et al. 2006b) valid for particle layers,

$$\Pi = -\frac{kT}{\omega_0}\left[\ln\left(1-\frac{\omega}{A}\right) + \left(\frac{\omega}{A}\right) \right] - \Pi_{coh}, \tag{4.8}$$

where $\Pi_{coh} = \frac{kT}{\omega_0} a\left(\frac{\omega}{A}\right)^2$ is the cohesion pressure, which is a parameter that includes all the characteristics of long-range forces between all the components of the monolayer. As shown by Equation 4.8, the surface pressure does not depend on the particle size, but it is determined by the monolayer coverage ω/A and by the parameters ω_0

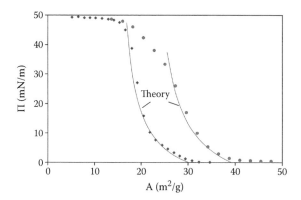

FIGURE 4.2 Dependence of surface pressure Π on the monolayer coverage for polymeric particles 113 nm in diameter without dispersant ($\blacklozenge$) and with dispersant ($\bullet$); solid lines are calculated using Equation 4.8 with the following parameters: $\omega = 15.5$ m^2/g, $\omega_0 = 0.12$ nm, and $\Pi_{coh} = 7$ mN/m in the absence of dispersant, and $\omega = 23$ m^2/g, $\omega_0 = 0.12$ nm, and $\Pi_{coh} = 10$ mN/m in the presence of dispersant. (Reprinted with permission from Fainerman et al. 2006b, *Langmuir* 22:1701–5. Copyright 2012 American Chemical Society.)

and Π_{coh}. For this reason, Equation 4.8 is valid for particles occupying the interface and also for aggregates of particle/molecules. The Π–A isotherms calculated according to Equation 4.8 are compared in Figure 4.2 with experimental results obtained for monolayers of polymeric particles in the absence and presence of a dispersant agent. The theoretical isotherm fits well the data in the absence of dispersant agent whereas its presence reduces the agreement between experimental and calculated results. This makes necessary to define a new model that takes into account the presence of additional soluble components in the monolayer. This can be done following a scheme similar to that applied for monolayers of proteins + surfactant (Fainerman et al. 2004).

In this case, Equation 4.8 is modified as

$$\Pi = -\frac{kT}{\omega_0}\left[\ln(1-\theta_P - \theta_S) + \theta_P + \theta_S\left(1-\frac{\omega_0}{\omega_S}\right) + a_S\theta_S^2\right] - \Pi_{coh}(c_s), \qquad (4.9)$$

where θ_S and θ_P are the surface coverages of surfactant and particles, respectively, a_S is the surfactant adsorption constant, and ω_S is the surfactant molecular area. The dependence on the surfactant concentration of Π_{coh} reflects the influence of the surfactant on the cohesion between particles. For hydrophilic nanoparticles, the interaction with surfactants induces their hydrophobization and this leads to the decrease in $\Pi_{coh}(c_S)$ (Fainerman et al. 2003). This, together with the increase of θ_S, leads to an increase in Π. Further increase in surfactant concentration can induce a new increase of $\Pi_{coh}(c_S)$ and the subsequent decrease of Π owing to the hydrophilization of the particles. This latter effect occurs directly for hydrophobic nanoparticles.

4.2.2 Thermodynamic Experimental Results

As shown above, the equilibrium properties of mixed particles–surfactant systems depend on the relation between the interfacial tension and the interfacial coverage. However, for most of these systems, this relation is not easily evaluable, as in the case of soluble or dispersible materials, owing to the uncertainty in the determination of the interfacial monolayer coverage. To avoid this problem in some studies, the equilibrium state is expressed by the equilibrium relation between the interfacial tension and either the surfactant or the particle bulk concentration (Maestro et al. 2010; Ravera et al. 2006a). While, for common surfactants, the adsorption equilibrium can be easily described with a classical model like the Frumkin isotherm, relating the interfacial tension, the bulk surfactant concentration, and the surface coverage, the scenario changes significantly when mixed systems are considered (see Figure 4.3 for an example).

A paradigmatic model system to investigate the properties of these mixed layers is offered by hydrophilic silica nanoparticles plus cationic surfactants (Liggieri et al. 2011; Maestro et al. 2012; Ravera et al. 2006a) like CTAB (hexadecyltrimethylammonium bromide) or DTAB (dodecyltrimethylammonium bromide). These systems show an almost negligible decrease in surface tension in the low surfactant concentration regime, owing to the low coverage of the interfacial layer. In fact, under these conditions, the very low adsorption of surfactant on the nanoparticles does not modify appreciably their hydrophobicity, limiting consequently their segregation at the interface.

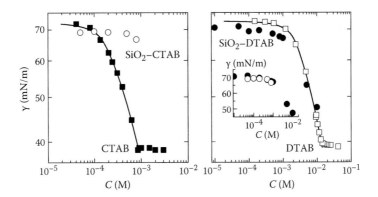

FIGURE 4.3 Equilibrium interfacial tension as a function of the surfactant concentration; symbols as follows: CTAB (■), SiO$_2$–CTAB (○), DTAB (●), and SiO$_2$–DTAB (□). Solid lines represent the Frumkin equation of state for the pure surfactant as it is defined in Ref. with $1/\Gamma_\infty = 0.439$ nm^2 per molecule, $a = 0.66$ mM, and $h/RT = 1.85$ for CTAB and $1/\Gamma_\infty = 0.468$ nm^2 per molecule, $a = 8.35$ mM, and $h/RT = 1.40$ for DTAB. The inset shows the comparative data γ–C for the SiO$_2$–C$_n$TAB complex; symbols as in the main figure. (Reprinted from Maestro et al. 2012, *Soft Matter* 8:837–43. Reproduced by permission of The Royal Society of Chemistry.)

Increasing the surfactant concentration, the increased hydrophobicity of complexes favors their attachment at the interface. Moreover, the hydrophobic interactions between the hydrophobic tails of the chains may lead to a larger decrease in surface tension.

The inset in Figure 4.3 shows that the length of the alkyl chain in the surfactant does not seem to play a significant role in the surface tension of the particle–surfactant suspensions, even though the γ versus C curves are different for the two pure surfactant solutions.

In addition to the direct interaction between nanoparticles and surfactant molecules, it is important to consider that the interactions between adjacent surfactant molecules play a key role in the adsorption of the complexes at the interface. The above considerations are supported by the existence of depletion of the surfactant solution owing to surfactant adsorption onto the nanoparticle surface. The latter leads to the existence of only residual amounts of free surfactant available in the bulk, making it possible to ascribe the decrease of the interfacial tension to the attachment of the surfactant–particle complexes to the interface and not to the adsorption of free surfactant molecules (Ravera et al. 2006a; Santini et al. 2012b). This intricate balance of interactions that govern the formation of interfacial layers formed by nanoparticle–surfactant mixtures contrasts with the scenario observed in the formation of layers by nanoparticles decorated with grafted polymeric chains where the properties of the polymers govern the characteristics of the layers (Stefaniu et al. 2010).

The above scenario concerns the spontaneous accumulation of nanoparticle–surfactant complexes at a fluid interface. Another way to investigate the equilibrium characteristics of particle-laden interfaces is by compression of these layers. When the segregation of nanoparticles can be considered irreversible, this kind of experiment allows studying overcompressed states of the particle layers (Liggieri et al. 2011). However, it exists a packing threshold for the nanoparticles at the interface, below which the surface pressure does not change owing to the fact that the nanoparticles are not close enough. This packing threshold is determined by the chemical nature and the hydrophobicity of the particle–surfactant complexes (Santini et al. 2012b). Figure 4.4 shows the compression isotherm of different monolayers formed by CTAB and hydrophilic silica nanoparticles. The spontaneous accumulation of nanoparticles does not lead to the formation of close-packed structures as evidenced by the increase of surface pressure obtained as a response to the compression of the layer. Such response is ascribable to the reorganization and packing of the complexes at the interface. The surface pressure increases until a kink is reached, corresponding to maximum packing of the monolayer. Further compressions provide the expulsion of the complexes from the monolayer with the organization of the nanoparticles as multilayers (Kundu 2011).

The observed behavior can be explained assuming the formation of insoluble aggregates owing to the layer compression, which cannot be obtained by spontaneous accumulation of material at the interface. It is also worth of mentioning that the increase of the packing degree at the collapse condition with the decrease of the surfactant concentration, which is ascribable to the different coating of the nanoparticle and, consequently, to its different hydrophobicity. This provokes the formation of

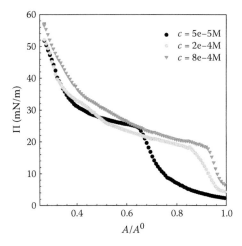

FIGURE 4.4 Compression isotherm, Π versus A/A^0, with A^0 being the initial area of the Langmuir trough. (Adapted from Liggieri et al. 2011, *Soft Matter* 7:7699–709. Reproduced by permission of The Royal Society of Chemistry.)

layers after spontaneous accumulation with different coverages. The dependence of the collapse pressure on the CTAB concentration has been explained, accounting for the different hydrophobic degrees of the complexes (Liggieri et al. 2011), which is related to the energy needed to expel the nanoparticles from the monolayer.

When the oil/water interfaces are concerned, beside the phenomena so far presented for mixed layers at the air/water interface, other transfer processes are often considered such as the possible transfer of naked particles, surfactant molecules, or particle–surfactant complexes (Santini et al. 2012b; Whittby et al. 2012). This can affect the synergetic character of the interaction between particles and surfactant molecules.

4.2.3 Morphology of the Composite Interfacial Layer

To investigate the structure of particle-laden interfaces, it is necessary to focus on two main aspects. These are the bi-dimensional distribution of the nanoparticles at the fluid interface, which defines the texture characteristics of the layer, and the position of the nanoparticles in the vertical direction with respect to the interface. The latter is related to the contact angle of the particles or to their wettability.

The organization of the adsorbed complexes at the interfacial layer is related to their packing. This means that, depending on the coverage, the interfacial layer may present a different morphology. Santini et al. (2011) pointed out by Brewster angle microscopy (BAM) that interfacial layers composed of CTAB + silica nanoparticle complexes present different structural textures depending on the surfactant concentration. As in previously discussed studies, the CTAB concentration determines the tendency of the complexes to segregate at the interface and the degree of interaction of such complexes once at the surface. For the lowest surfactant concentrations, there

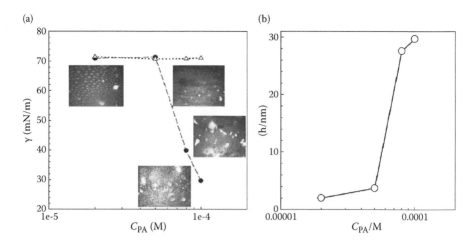

FIGURE 4.5   (a) Long-time surface tension of 1 wt% aqueous silica nanoparticles (diameter = 30 nm) plus palmitic acid (PA) dispersions for different PA concentrations in the dispersions and (●) of the supernatant obtained after their ultracentrifugation (△). The inserted images correspond to the BAM images obtained for the monolayers at the dispersion surface. (b) Ellipsometric thickness obtained for the monolayer at the dispersion surface (○). (Reprinted from Santini et al. 2012, *Phys. Chem. Chem. Phys.* 14:607–15. Reproduced by permission of the PCCP Owner Societies.)

is no evidence of formation of organized structures, whereas increasing the amount of surfactant molecules induces the formation of aggregates at the interfacial layers. This is explained by assuming that the interaction between the hydrophobic tails of CTAB adsorbed on particles becomes more important as the CTAB concentration increases. In fact, the number and size of these interfacial aggregates increase with CTAB concentration until the formation of a closer-packed structure.

Similar results are observed for other surfactant + particle systems such as palmitic acid (PA) + silica nanoparticles (Santini et al. 2012b) where, already at low surfactant concentration, isolated islands are formed, which, upon increasing the surfactant concentration, tend to coalesce until the formation of a continuous layer (see Figure 4.5a).

Additional information on interfacial organization can be obtained from ellipsometric data (see Figure 4.5b). The average thickness of the mixed PA–silica nanoparticle layer is found to be lower than the diameter of a single nanoparticle (30 nm) for the lowest surfactant concentration, while, when the PA concentration increases, it achieves values closer to the nanoparticle diameter. This is in accordance with the conclusions derived from the BAM images; that is, an increase in PA concentration leads to a higher degree of interfacial coverage and consequently to the formation of a close-packed layer.

It is important to notice that, as better focused on in the next section, particle wettability plays a key role in the self-organization of mixed particle–surfactant layers at fluid interfaces (Maestro et al. 2012).

4.2.4 WETTABILITY OF NANOPARTICLES AT THE INTERFACE

Partial wettability is the key factor driving the segregation of particles at the liquid interface. Among the methods for the determination of the contact angle of particles, there are those exploiting the monolayer compression, which may be based on the excluded area concept (Grigoriev et al. 2007b) or on the evaluation of the collapse pressure (Clint and Taylor 1992; Hórvölgyi et al. 1996) and others based on ellipsometric measurements (Hunter et al. 2009).

The determination of the contact angle, according to the excluded area principle, is applicable to insoluble surfactant molecules with known Π–A isotherms as a calibration curve to infer the area occupied by the nanoparticles, that is, the excluded area. To this aim, it is necessary to assume that the particles do not interact with the surfactant at least in the region of low to intermediate surface pressures (see Figure 4.6). It is thus possible to determine the area occupied by the particles as the difference between the total accessible area and the area covered by the surfactant. The latter can be obtained from the previously determined compression isotherm of the insoluble surfactant as the A value corresponding to the measured surface pressure.

The contact angle is determined from the area shift at a given surface pressure in the isotherms as a function of the number of particles in the monolayer, N_p. In fact, the difference, ΔA, between the total area A^0 and the area occupied by the surfactant A_{surf} is proportional to the number of particles. Assuming spherical particles, this allows the estimation of the contact angle from the slope of the excluded area as a function of surface pressure,

$$\frac{\mathrm{d}\Delta A}{\mathrm{d}\Pi} = -\frac{2N_p \pi \gamma_0^2 R^2 \cos^2 \Theta}{\gamma^3}, \qquad (4.10)$$

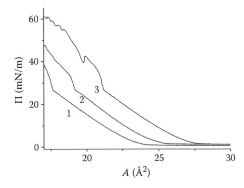

FIGURE 4.6 Π–A isotherms of an insoluble surfactant in the absence (1) and presence of increasing amounts of SiO_2 microparticles (2 and 3). (Reprinted from Grigoriev et al. 2007, *Phys. Chem. Chem. Phys.* 9:6447–54. Reproduced by permission of the PCCP Owner Societies.)

where γ_0 and γ are the surface tensions of liquid phase in the initial uncovered state and in a certain compressed state, respectively, and Θ is the contact angle corresponding to the initial uncovered state. In principle, the change of contact angle due to the decrease of the surface tension during compression should be considered and Young's equation should be used together with Equation 4.11. However, under measurement conditions where only very low Π variation is concerned, this effect is negligible.

The essential limitation of the excluded area approach to the determination of the contact angle in mixed particle–surfactant systems consists in the assumption of no interactions between the nanoparticles and the monolayer components. Any interaction modifies the area occupied by the different components and consequently the particles' contact angle.

Another method utilizing the compression isotherm exploits the monolayer collapse to indirectly evaluate the particle contact angle (Clint and Taylor 1992; Hórvölgyi et al. 1996). This method relies on the relationship between the hydrophobicity of the particles and the area, A_C, at which the system reaches the collapse point (Clint and Taylor 1992) where particles are expelled out of the layer. Such a relationship, which can be obtained by energetic considerations, reads

$$\Pi_C A_C' = \gamma \pi R^2 (1 \pm \cos \Theta)^2, \qquad (4.11)$$

where Π_C is the collapse pressure and A_C' is the ratio between the area at collapse and the number of particles at the surface (N_p). It is worth noticing that for many systems, such as water-dispersible nanoparticles, an accurate determination of the real number of particles at the surface is not possible. This problem may be overcome by passing through the evaluation of the area corresponding to the percolation point (or gelling point) A_g (Hórvölgyi et al. 1996). This state corresponds to the formation of a continuous particle layer whose area, as a first approximation, is equal to the cross-sectional area of one particle multiplied by N_p. The value of A_g can be estimated from the Π–A isotherm as the intersection with the A axis of the straight line fitting the isotherm just above the collapse area (see Figure 4.7). The value of A_g can be used to estimate the number of particles at the surface (N_p); that is,

$$N_p = \frac{A_g}{A_p}. \qquad (4.12)$$

Here, A_p is the area of a single particle.

The main problem for the application of this method to mixed surfactant–particle layers is that the modification of the contact angle due to monolayer compression is not considered. This means that the calculated contact angle may not represent the contact angle of the particle spontaneously adsorbed at the interface, but that in the close-packed state of the monolayer.

An alternative method that overcomes the difficulties discussed above relies on ellipsometric measurements (Hunter et al. 2009). For the application of this method, the particle-laden interface is modeled as a two-layer continuous system,

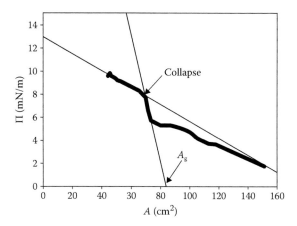

FIGURE 4.7 Example of the estimation of A_g from a compression isotherm Π versus A. The solid lines allow the determination of the collapse point as the intersection between them. (Reprinted from Santini et al. 2011, *Colloids Surf. A* 382:186–91. Copyright 2011, with permission from Elsevier.)

characterized by specific thicknesses and refractive indices (see Figure 4.8), in order to account for a partial wetting situation. The model assumes that the upper layer (Layer 1) is formed by a mixture of adsorbed nanoparticle–surfactant composites and air, while the lower layer (Layer 2) is formed by a mixture of the nanoparticle–surfactant composites and water. In Figure 4.8, h_{Si-air} and n_{Si-air} and h_{Si-H_2O} and h_{Si-H_2O} are the thickness and the complex refractive index of the upper silica–air and the lower

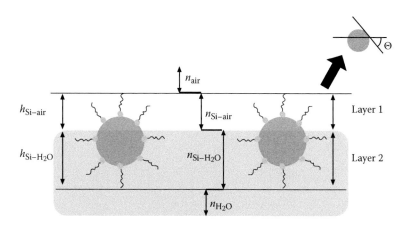

FIGURE 4.8 Scheme of the two-layer ellipsometric model used to analyze the nanocomposite interfacial system formed by silica and C_nTAB. The particles of radius r and refractive index n_{Si} are attached to the interface. h is the vertical distance between the center of the sphere and the air–water interface. h_{Si-air} and h_{Si-H_2O} represent the thickness of the two layers, respectively, that is, the part of the nanocomposite in air and water, respectively. Θ is the contact angle of the particle at the air–water interface measured in water.

silica–water layer, respectively. This allows a relation between the particle contact angle Θ and the upper layer thickness ($h_{\text{Si–air}}$) to be derived as (Hunter et al. 2009)

$$\cos\Theta = 1 - \frac{2h_{\text{Si-air}}}{d}, \tag{4.13}$$

with $d = h_{\text{Si–air}} + h_{\text{Si–H}_2\text{O}}$. To obtain reliable values of the two-layer thickness, as well as the phase contact angle, the ideal response is modeled by means of the effective medium approximation (Aspnes 1982). Further details can be found in the literature (Hunter et al. 2009; Zang et al. 2009).

Other techniques usually employed in the determination of the contact angle of nanoparticles at fluid interfaces like the nano-gel-trapping technique (Paunov 2003), the film-calliper method (Horozov et al. 2008), or the Washburn–Rideal method (Siebold et al. 1997) are difficult to apply in mixed nanoparticle–surfactant layers.

A paradigmatic example is found in the change of wettability of complexes formed by silica nanoparticles and C_nTAB (see Figure 4.9) (Maestro et al. 2012). The addition of a low concentration of C_nTAB to the dispersion of silica nanoparticles leads to an increase in the hydrophobicity of the complexes as a result of the favorable electrostatic interaction between the nanoparticles and the surfactant molecules, which favors their adsorption as individual ions onto the surface of the nanoparticles. Further increases of the surfactant concentration lead to the increase of the hydrophobic interactions between the hydrocarbon tails of surfactant molecules attached to the particle surface. This interaction could change the structure of the adsorbed surfactant layer onto the silica nanoparticles, inducing the formation of a more hydrophilic interfacial layer. Concretely, it is assumed that the hydrophobic

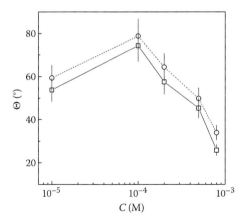

FIGURE 4.9   Contact angle, obtained from the analysis of the ellipsometric results by means of the two-layer model proposed, for the silica–surfactant nanocomposite at the air–water interface as a function of the surfactant concentration; symbols as follows: CTAB ($\square$) and DTAB ($\bigcirc$). (Reprinted from Maestro et al. 2012, *Soft Matter* 8:737–43. Reproduced by permission of The Royal Society of Chemistry.)

interaction between the tails of the adsorbed surfactant molecules can lead to the rearrangement of the surfactant molecules on the nanoparticle–water interface. This situation leads to the formation of adsorbed complexes in which a second layer of surfactant is adsorbed, even though these bilayers or hemimicelles were not presented onto nanoparticle surfaces in the initial bulk species. This fact renders the complexes more hydrophilic, thus decreasing the contact angle of the nanoparticles attached to the interface. It should be noted that, in accordance with Liggieri et al. (2011), the values of $\Gamma/\Gamma_{saturation}$ are on the order of 10^{-2} for all surfactant concentrations. Thus, to explain the formation of hemimicelles and bilayers of surfactant onto the nanoparticles (Kekicheff et al. 1989), one can consider that surfactant-modified nanoparticles act as carriers of cationic surfactants to the interface, and once there, the surfactants can be rearranged into the interfacial layer formed by the tuned nanoparticles. Indeed, once at the interface, the surfactants adsorbed at the particle surface should be favored to migrate to the particle surface exposed to air. In fact, the surfactant molecules do not expose their hydrophobic chains to the water phase. The particle surface in contact with water is therefore depleted and could adsorb new surfactant molecules (to maintain the adsorption equilibrium at the water–silica surface) through the interaction with particles (decorated with surfactant molecules) arriving from the aqueous bulk. These "carrier particles" are mostly rejected in the bulk since they become less hydrophobic. Such process has the net effect of enriching the surfactant fraction adsorbed on the particles at the liquid interface. The presence of this process is supported by the dilational rheological response of this type of systems (Liggieri et al. 2011). In addition to the "carrier particles" mechanism, one also has to consider the exchange of residual amounts of free surfactant molecules from the bulk to the interface.

4.3 DYNAMIC PROPERTIES

The equilibration process of the mixed particle–surfactant systems is a dynamic process essentially driven by the interaction between particles and surfactant molecules, the surfactant adsorption properties, and the hydrophilic/hydrophobic character of the solid components present in the suspension. For the transfer process to the interface, diffusion of particles and surfactant–nanoparticle complexes must be considered as well as, in some cases, the overcoming of interfacial energy barriers (Stocco et al. 2011).

Studies on the dynamic properties of mixed nanoparticle–surfactant layers, on the basis of the interfacial tension measurement during the equilibration of the systems (dynamic surfact tension measurements) have pointed out that the time scales of the involved processes are significantly larger—by three orders of magnitude—than those of surfactant adsorption (Ravera et al. 2006a; Whittby et al. 2012).

Different scenarios for the partition of the particles and surfactant between the two bulk phases and the interface may be taken into consideration, associated to different kinds of relaxation processes: (a) both components can be dispersed (or dissolved) in only one of the phases, (b) the components can be only dispersed (or dissolved) in two different phases, and (c) only one of the components can be dispersed (or dissolved) in both phases.

The first scenario is surely the most conventional and includes, for example, the case of partially hydrophobic complexes formed in the aqueous bulk phase and transferring to the fluid interfaces. If such complexes are obtained by water-soluble surfactant and hydrophilic particles, the transfer in the second fluid phase can be assumed negligible. Well-investigated systems of this type are the mixtures of cationic surfactants, like CTAB, and negative charged nanoparticles, such as hydrophilic silica nanoparticles (Liggieri et al. 2011; Maestro et al. 2012; Ravera et al. 2006a).

The composition of the dispersions investigated in those studies presented negligible amounts of free surfactant, which is instead totally adsorbed on the solid particles forming partially hydrophobic complexes. The adsorption kinetics of these systems shows a bimodal character with a first step corresponding to the transfer of nanoparticle–surfactant complexes to the interface. This transfer is diffusion controlled, as confirmed by the characteristic time comparable to that of diffusion in liquid of particles of the same size, and does not present any appreciable effect of the surfactant concentration (Ravera et al. 2006a). The second step corresponds to different rearrangement processes of the composite film, with a possible redistribution of the surfactant molecules between the particle/liquid and the liquid/fluid interface. It is important to notice that the feature of this second step depends on the surfactant concentration. The layer reorganization is in fact expected to be strongly affected by the amount of surfactant transferred to the surface and by the nanoparticle coverage reached after the first step of the process, which in turn depends on the surfactant adsorbed on particles.

The existence of a two-step adsorption process has been also observed for interfacial layers of silica nanoparticles and octadecylamine (ODA) at the water/hexane interface (see Figure 4.10) (Whittby et al. 2012). Here, the surfactant and the

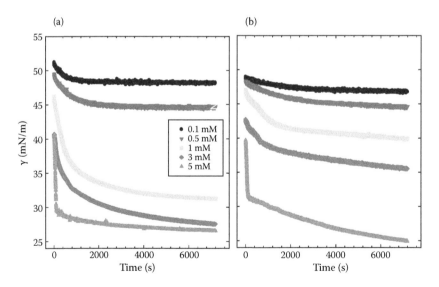

FIGURE 4.10 Dynamic interfacial tension during the adsorption process. (a) ODA adsorption at the solution/hexane interface. (b) Mixed system silica nanoparticles (concentration 1 wt%) plus ODA at the dispersion/hexane interface. (Reprinted with permission from Whittby et al. 2012, *J. Phys. Chem. C* 116:3350–8. Copyright 2012 American Chemical Society.)

particles are dissolved (or dispersed) in different phases (scenario b) and the interaction between them occurs only at the fluid interface. The first step, showing a characteristic time around 10^2 s, refers to the adsorption of the oil-soluble surfactant and is not appreciably modified by the presence of the particles in the other phase. The second step shows larger time scales in the presence of nanoparticles with an equilibration time around 10^4 s. The longer time observed in the presence of nanoparticles is related to the attachment of the nanoparticles at the preformed ODA layer and the possible reorganization processes of the material at the interface and in the two bulk phases induced by the synergetic interactions between ODA and silica that occurs at the fluid interface.

As a last example, it is interesting to mention an experimental study where the relaxation process of the mixed layer occurs also in the presence of transfer of surfactant or particles from one phase to another (scenario c). This may occur for silica nanoparticle and PA systems at water/hexane interfaces (Santini et al. 2012b). PA is not soluble alone in water beyond a bulk concentration of 2×10^{-5} M. However, in this system, silica–PA complexes are formed in the aqueous phase owing to the hydrogen bonds between the non-dissociated silanol groups at the surface of the silica nanoparticles and the carboxylic group of the PA. During the equilibration of the aqueous dispersion–hexane system, the complexes are transferred to the liquid–liquid interface where complexes break up and a part of PA, because of its strong hydrophobic character, is released into the oil phase. This transfer of PA in the second phase is evidenced from the feature of the dynamic interfacial tension (see Figure 4.11) that, after the initial decrease caused by the adsorption of complexes, passes through a minimum before increasing to values closer to that of the clean interface. This feature is typical of the adsorption processes in the presence of

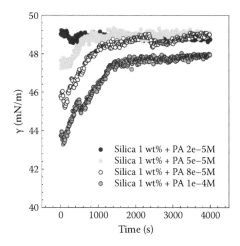

FIGURE 4.11 Dynamic interfacial tension of aqueous silica nanoparticles (1 wt%) plus PA dispersions versus pure hexane for different PA concentrations in the dispersion. (Reprinted from Santini et al. 2012b, *Phys. Chem. Chem. Phys.* 14:607–615. Reproduced by permission of the PCCP Owner Societies.)

surfactant transfer across the interface and it has been already pointed out for common surface-active species (Ravera et al. 2001).

The kinetic behavior of PA–silica complexes is ascribable to the relative higher diffusion coefficient of PA molecules in the hexane phase compared to that of complexes in the dispersion. For this reason, the flux of surface-active material to the interface becomes quickly not sufficient to balance the desorption flux toward the hexane.

In summary, from these examples, it is possible to conclude that the principal factor governing the dynamics of formation/equilibration of mixed nanoparticle–surfactant layers is the hydrophilic/hydrophobic character of the particles that, being determined by the interaction with surfactants, may change during the relaxation process and participate with the adsorption dynamic characteristics of surfactants.

4.4 SURFACE RHEOLOGY AND MECHANICAL PROPERTIES

Surface rheology is the response of interfacial layers to external perturbations induced by mechanical forces, like shear or dilational stresses. In composite layers, surface rheology is extremely important because, on one hand, it provides a description of the mechanical properties of the layer and, on the other hand, it is related to their structural characteristics and to the adsorption dynamic and, in general, to all the kinetic processes involving surfactant and nanoparticles. For example, the evaluation of the dependence of viscoelasticity on the frequency of the perturbation allows obtaining information on the existence of kinetics processes in the interfacial layer (Liggieri et al. 2005; Liggieri and Miller 2010; Miller and Liggieri 2010; Ravera et al. 2005).

Additionally, it is expected that the mechanical properties of the interfacial layers play an important role in the stability of films, foams, and emulsions. This correlation between mechanical properties and stability of dispersed systems has been previously observed for surfactant-stabilized films and emulsions (Santini et al. 2007a,b).

Nowadays, studies of the rheological response, both dilational (Liggieri et al. 2011; Ravera et al. 2008) and shear (Zang et al. 2010a,b), of particle-laden fluid interfaces are of increasing interest. Despite its recognized importance, there is not a comprehensive theoretical description of shear rheology, whereas some models are available for the dilational response of these complex layers (Miller et al. 2006; Ravera et al. 2008).

4.4.1 DILATIONAL RHEOLOGY

Dilational rheology concerns the response of interfacial layers to changes in the interfacial area or compression/expansion deformations. The stress associated to this kind of surface deformation is the variation of the interfacial tension (or pressure). This allows defining a dilational modulus (or dilational viscoelasticity) as

$$E = -\frac{\delta \Pi}{\delta \ln A}. \tag{4.14}$$

Under quasi equilibrium variation of the surface area, E can be assumed as a thermodynamic quantity (dilational elasticity) related to the equilibrium adsorption state of the interfacial layer.

Assuming that Equation 4.14 is calculated at quasi equilibrium variation of the surface area, and on the basis of some thermodynamic considerations, it is possible to obtain a formulation for the rheological behavior of composite layers. One of the first definitions of dilational modulus for composite layers was given by Lucassen (1992), which provided an expression of E in the case of non-interacting particles located in fixed positions and characterized by a certain internal compressibility; that is,

$$\frac{1}{E} = \sum_{1}^{n} \frac{X_i}{E_i},$$
(4.15)

where X_i is the surface fraction having the dilational modulus E_i. This equation was utilized later and applied to mixed nanoparticle–surfactant systems. In particular, introducing the equation of the state discussed above (Equation 4.7) for particle-laden interfaces (Fainerman et al. 2006b), an expression was formulated by Miller et al. (2006) for the surface elasticity of a monolayer formed exclusively by nanoparticles E_P^0:

$$E_P^0 = \frac{kT}{\omega_0} \left[\frac{\theta_P}{1 - \theta_P} - \theta_P \right] = E_0 \left[\frac{\theta_P^2}{1 - \theta_P} \right], \text{with } E_0 = kT/\omega_0.$$
(4.16)

Consider particles mixed with surfactants, the situation is slightly more complex and it is necessary to consider the surface pressure as a function of two variables, $\Pi(\theta_S, \theta_P)$, following the definition given in Equation 4.15,

$$E = -\frac{d\Pi}{d\ln A} = -\left(\frac{\partial\Pi}{\partial\ln\theta_S} \right)_{\theta_P} \frac{d\ln\theta_S}{d\ln A} - \left(\frac{\partial\Pi}{\partial\ln\theta_P} \right)_{\theta_S} \frac{d\ln\theta_P}{d\ln A}.$$
(4.17)

From the equations $\theta_S = \omega_S N_S/A$ and $\theta_P = \omega_S N_P/A$, N_S and N_P are the total numbers of surfactant molecules and particles at the interface, respectively. It is assumed that both particles and surfactants behave as insoluble and incompressible,

$$\frac{d\ln\theta_S}{d\ln A} = \frac{d\ln\theta_P}{d\ln A} = -1.$$
(4.18)

Hence, the elasticity of the layer can be defined in terms of the partial elasticities of the surfactant, E_S, and the particles, E_P, which can be calculated from Equation 4.19:

$$E = E_S + E_P$$
(4.19)

$$E_S = E_0 \left[\frac{\theta_S}{1-\theta_P-\theta_S} - \theta_S \left(1 - \frac{\omega_0}{\omega_S}\right) - 2a_S\theta_S^2 \right] - \frac{d\Pi_{coh}}{d\ln\theta_S} \tag{4.20}$$

$$E_P = E_0 \frac{\theta_P(\theta_P+\theta_S)}{1-\theta_P-\theta_S} \tag{4.21}$$

It is important to notice that the treatment leading to Equation 4.19 does not take into account the possibility for an exchange of surfactant between bulk and interface. In practice, the surfactant molecules are assumed as insoluble. However, a similar behavior is expected for soluble surfactants at sufficiently high frequencies of perturbation, that is, when the equilibrium between bulk and interface is not established, while this equilibrium occurs instantaneously under low-frequency conditions. The average surface coverage for the surfactant can be expressed under such conditions as $\theta_S = \theta_S^0(1-\theta_P)$, where $\theta_S^0(c_S)$ is the equilibrium surface coverage for the surfactant in the absence of particles. Therefore, at constant surfactant bulk concentration, instead of Equation 4.18, we will obtain

$$[d\ln\theta_S] = d\ln(1-\theta_P) = \frac{\theta_P}{1-\theta_P} d\ln\theta_P. \tag{4.22}$$

Accounting for $d\ln\theta_P = -d\ln A$, and with Equation 4.17, we obtain

$$E = E_P - \frac{\theta_P}{1-\theta_P} E_S. \tag{4.23}$$

Increasing the frequency of oscillations, the elasticity of the composite layer should change between two limiting cases, given by Equations 4.19 and 4.23. The validity of the expressions given above for the elasticity of composites is limited by the fact that they do not take into account possible phenomena involving particles and surfactant, which can become important under more dynamic conditions. For example, the area occupied by a particle or a surfactant molecule may change during the area oscillation owing to changes of the contact angle with surface tension (Lucassen-Reynders 1981). Also, the surfactant molecules are characterized by an intrinsic compressibility that can be related to a change in their tilt angle with changing surface pressure (Kovalchuk et al. 2005). This makes a more general expression for the elasticity modulus necessary.

The models reported above have been developed considering the equilibrium aspects of the dilational rheology of mixed layers and, for this reason, are related to the thermodynamics of the system. Dilational rheology is, however, very important for investigating the dynamic aspects of the interfacial layers, especially because the response of the interfacial tension to dynamic area changes. As better explained elsewhere (Miller and Liggieri 2010; Ravera et al. 2005, 2006b), for harmonic perturbation of the interfacial area, the dilational viscoelasticity defined in Equation 4.14 can be effectively expressed as a function of the perturbation frequency, ν. In the frequency domain, E is complex and frequency dependent; that is,

$$E = E_R + iE_I = E_R + 2\pi i\nu\eta. \tag{4.24}$$

Here, the real part E_R is the dilational elasticity, and the imaginary part E_I is related to the dilational viscosity $\eta = \nu E_I$. The dependence of E on the area perturbation triggers different processes involved in the re-equilibration of the interfacial layer. It is known (Noskov and Loglio 1998) that if a layer initially at equilibrium is perturbed by a small variation of its area, it behaves as a linear system. Thus, according to its definition and the properties of linear systems, $E(\nu)$ completely characterizes the dynamics of the interfacial layer, because it provides the response of the interfacial tension to any low-amplitude perturbation of the surface area (Loglio et al. 1979). Several studies about the dilational viscoelasticity of surfactant systems are available and, especially for soluble surfactant, the classic Lucassen–van der Tempel (Lucassen and van der Tempel 1972) model has been widely employed to describe diffusion-controlled adsorption. Together with these diffusive processes, several extensions have been proposed taking into account other reorganization processes inside the adsorbed layer, such as molecular reorientation, aggregation, or chemical reactions (Miller and Liggieri 2010; Ravera et al. 2006b). The approach proposed by Ravera et al. (2006b) has been effectively used for particles plus surfactant systems (Liggieri et al. 2011; Ravera et al. 2008). These studies pointed out that the mechanical response of such mixed systems is the result of a diffusive process together with other kinetic surface processes involving the adsorbed surfactant molecules and the nanoparticle–surfactant complexes. This approach assumes that additional variables are needed to describe the state of a complex interfacial layer out of equilibrium. Thus, a generic interfacial process, described by a generic variable, X, is considered to vary, during the area oscillation, according to a linear rate equation; that is, $dX/dt = -K(X - X^0)$.

Taking into account the dependence of the surface pressure on X as well as on the total adsorption Γ, that is, $\Pi = \Pi(\Gamma, X)$, according to Equation 4.14, the dilational viscoelasticity can be written as

$$E = -E_{0\Gamma}\frac{d\ln\Gamma}{d\ln A} - E_{0X}\frac{d\ln X}{d\ln A}, \tag{4.25}$$

where $E_{0\Gamma} = \partial\Pi/\partial\ln\Gamma$ and $E_{0X} = \partial\Pi/\partial\ln X$ are the quantities calculated from the surface equation of state. Solving the mass balance at the interface together with the kinetic equation for the relaxation process, it is possible to obtain an expression for E. For a single surface kinetic process besides diffusion, the following expression is obtained:

$$E = \frac{E_{0\Gamma} - i\lambda E_{0G}}{-(1+\xi-i\xi)i\lambda + 1 + G + \xi - i\xi/1 + G}, \tag{4.26}$$

where $\xi = \sqrt{v_D/2v}$, with v_D being the characteristic frequency of the diffusion transfer, and $\lambda = v_k/v$, with v_k being the characteristic frequency of the surface kinetic process, while G is a thermodynamic quantity, $G = 1 - \dfrac{dc_s/d\Gamma}{\partial c_s/\partial \Gamma}$.

The limit case of Equation 4.26 for very fast equilibration (λ very large) of the relaxation process leads to the Lucassen–van der Tempel equation,

$$E = E_{0G} \frac{1 + \xi + i\xi}{1 + 2\xi + 2\xi^2}. \tag{4.27}$$

Considering the diffusion as vanishing, the other limit case is obtained, which is the classical situation of an insoluble monolayer,

$$E = \frac{E_{0\Gamma} + E_{0G}\lambda^2}{1 + \lambda^2} + (E_{0\Gamma} - E_{0G})\frac{i\lambda}{1 + \lambda^2}. \tag{4.28}$$

It is noteworthy that the presence of the relaxation process confers to E particular characteristics at the corresponding characteristic frequency. These are a maximum in the imaginary part of E and an inflection point in the real part (Ravera et al. 2008). Equation 4.26 is quite general and can be applied to any relaxation process described by a linear rate equation.

The complex dynamics of the interfacial behavior of surfactant + particle layers can be effectively investigated by measuring the rheological response to harmonic perturbation at varying frequencies. This is of key importance for the potential applications in the field of foams and emulsion, being the mechanical properties of the layers at the bases of their stabilization.

An important requirement to maximize the information by dilational rheology experiments is to perform measurements in a wide frequency range. To this aim, a multi-technique approach can be effective for obtaining the wide-range rheological spectrum appropriate to detect the relaxation process occurring in the interfacial layers. Liggieri et al. (2011) pointed out that the combination of two oscillating drop/bubble techniques, working in a drop shape tensiometer (10^{-2} to 10^{-1} Hz) and a capillary pressure tensiometer (10^{-1} to 10^2 Hz), together with the electrocapillary waves technique (10 to 10^3 Hz), provided good performance in the investigation of nanoparticle–surfactant systems. In particular, the dilational rheology of spontaneously formed layers of silica nanoparticles and CTAB at the water/air interface was investigated by these techniques. Figure 4.12 shows some results obtained in this study. It is noteworthy to mention the almost perfect matching between the dilational viscoelasticity data obtained for the different techniques in the overlapping regions.

Interpreting the dilational viscoelasticity obtained with a multi-process theoretical model (Equation 4.26), it is possible to detect the different processes involved in the rheological response of the particle-laden interface. The different processes show different characteristic frequencies that can be evaluated by inflection points in the

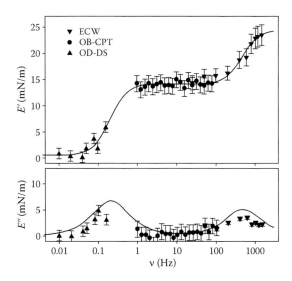

FIGURE 4.12 Real and imaginary parts of the dilational viscoelasticity of the 1 wt% silica dispersion plus CTAB for $c_{CTAB} = 1 \times 10^{-4}$ M. Notice that the results for different c_{CTAB} are similar to those shown in the figure. Measurements in the different frequency windows are obtained with different methodologies, as described in the text. The curves are obtained by the fitting to the rheological model. (Reprinted from Liggieri et al. 2011, *Soft Matter* 7:7699–709. Reproduced by permission of The Royal Society of Chemistry.)

real part of the viscoelasticity (E') and maximum in the imaginary part (E''). For low surfactant concentrations, the rheological spectra show a low-frequency rheological process ascribable to the diffusive exchange of material between the interface and the bulk and a second process with higher characteristic frequency owing to the kinetic reorganization of the interfacial layer. It is important to note that the characteristic frequency observed for the diffusion process, ν_D, is compatible with the characteristic time of the nanoparticle diffusion obtained from the adsorption studied, of the same dispersion to the fluid interface (Ravera et al. 2006a). A similar diffusive process was reported by Santini et al. (2012b) as a result of a low-frequency rheological study on silica nanoparticles + PA layers.

The increase of surfactant concentration induces the occurrence of two kinetic processes, while the diffusion is no more evident in the investigated frequency range. The diffusion exchange with the bulk dispersion becomes in fact negligible, increasing the surfactant concentration, and above a surfactant concentration threshold that tunes the attachment of the nanoparticles from reversible to irreversible, the complexes can be assumed to be almost insoluble.

For a deeper understanding of the kinetic processes occurring in these mixed interfacial layers, it is interesting to analyze the characteristic frequencies found out by this rheological study, as a function of the CTAB concentration (Figure 4.13). The low characteristic frequency reported in Figure 4.13 can be ascribed to processes involving particles whereas the high-frequency process instead likely involves the

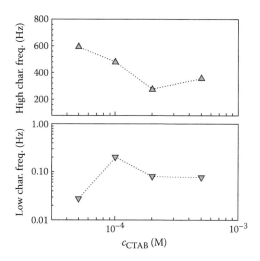

FIGURE 4.13 Characteristic frequencies of the surface tension relaxation processes versus the CTAB concentration in 1 wt% silica dispersions. (Reprinted from Liggieri et al. 2011, *Soft Matter* 7:7699–709. Reproduced by permission of The Royal Society of Chemistry.)

surfactant (Liggieri et al. 2011). The latter is confirmed by the fact that both E_0 and E_1 parameters were found, for this high-frequency process, to be similar to those typical in the processes of rearrangement of surfactants within the interfacial layer (Ravera et al. 2005), such as molecular reorientation or variation of the molar area. This process may be attributed to the redistribution of surfactant between the particles and the liquid surface or to some reorganization of the adsorption layer at the particle interface arising from the interaction with other particles during the area compression–expansion. The low characteristic frequency is ascribable to a diffusive process for the lowest surfactant concentrations, while it may correspond to some kind of time-dependent reorganization of the structure of the particle layer for the larger surfactant concentration.

4.4.2 SHEAR RHEOLOGY

The study of the response of particle + surfactant layers against shear deformation has been less developed than dilational rheology (Degen et al. 2011). Shear rheology analyzes the response of the interfacial layer to change in its shape without modification of the area (Oh and Slattery 1978). The response against shear is intimately correlated to the cohesive interactions on the interfacial layer. For this reason, it is well known that for fluid monolayers containing small surfactants, shear viscoelasticity is negligible (Fuller and Vearmant 2012), while highly packed layers—eventually solid or gel-like systems—present high values of shear viscoelasticity (Cicuta et al. 2003). Contrary to dilational rheology, shear rheology experiments of particle monolayers are much more scarce and, in most cases, deal with particle monolayers with no surfactant added. Safouane et al. (2007) carried out a systematic study of the shear rheology of planar silica particle monolayers at the air/water interface, varying the hydrophobicity of the particles. In the case of hydrophobic particles (36% or more silanol groups silanized), the monolayers are heterogeneous and show fractures upon compression. Moreover, at high compressions, the monolayers become thick; thus, they can no longer be considered as two-dimensional objects. The fractures self-heal upon releasing the strain. They found that, at a fixed frequency of 0.1 rad/s, both G' and G'' decrease with hydrophobicity, and G'' is two orders of magnitude larger than that for typical surfactant monolayers, which makes the contribution of the subphase negligible, thus making the measurements more reliable. Hydrophilic particles show almost negligible values of G' and G'', whereas for intermediate hydrophobicities (36% of silanol groups silanized), $G' \approx G''$. More hydrophobic particles show $G' > G''$, the monolayers being rigid, probably due to the formation of a network with strong hydrophobic interactions. Even though the authors reported the surface density dependence of the dilational elasticity, no shear modulus measurements were performed as a function of the monolayer density.

More recently, Zang et al. (2010b, 2011) have extended the previous work and have studied the shear modulus dependence on the strain amplitude. They found that the behavior of particle monolayers is similar to that of three-dimensional soft solids where the monolayer melts above a critical amplitude, and G'' shows a maximum. Above this yield amplitude, $G'' > G'$. It was also found that at constraint strain, $G'' > G'$ at low frequencies, whereas it is lower at high frequencies. They postulated that, as in three-dimensional soft solids, this behavior is associated with the decrease of the structural relaxation time with an increase in the strain-rate amplitude. One of the most important results of this work is that the authors demonstrated the validity of the strain-rate frequency superposition principle of measurement developed by Wyss et al. (2007), which allows calculating the structural relaxation time, even when it occurs at frequencies well below the ones accessible to standard experimental techniques. It must be remarked that the above principle assumes that the relaxation depends only on the amplitude of the deformation, not on its time dependence.

Vandebril et al. (2010) have carried out measurements of the shear modulus of a hematite nanoparticle monolayer at the interface between two fluid polymers. They have demonstrated that the particles stabilize the dispersion between two immiscible polymers by a mechanism similar to that of the Pickering emulsions. They found a strain dependence qualitatively similar to that of the silica particles at the air/water interface (Zang et al. 2010b). An interesting finding is that whereas hematite nanoparticles form a brittle interfacial gel at the air/water interface, they form a stronger gel-like behavior at the polymer–polymer interface.

Unfortunately, so far there are no systematic studies of the shear modulus as a function of the monolayer density that would allow relating the rheological behavior to the phase (fluid, hexatic, or solid) of the monolayer (Bonales et al. 2011).

Besides, to the best of our knowledge, only few studies deal with the evaluation of the response of particle + surfactant layers against shear deformation, contrary to the great number of studies concerning the dilational rheology of those systems. Perhaps this has happened because for phenomena such as Ostwald ripening in emulsions, dilational elasticity has a higher influence. Maas et al. (2010) have studied mixed monolayers of lipids and silica nanoparticles of two different hydrophobicities at the water/oil interface. The lipids were dissolved in the oil phase whereas the particles were suspended in the aqueous phase. The authors followed the kinetics of adsorption at the interface by monitoring the shear modulus and found two-stage kinetics. The first process (approximately 1 h) lasts until the film is cohesive and $G' > G''$. In the second period (much slower), lipids and particles are accumulated onto the interface at a constant rate until equilibrium is reached. As already found for monolayers of particles (Safouane et al. 2007), at the end of the adsorption, G' and G'' are much higher than that for the usual surfactant monolayers. For a given concentration of particles (5% in the case of Ludox HS-30 silica nanoparticles), above a concentration at which the lipid forms a saturated film, the shear modulus remains constant because no more lipid is adsorbed from the oil phase. Similarly, for a given lipid concentration (1 mM of stearic acid), the modulus remains constant above a particle concentration of 10%. The authors concluded that nanoparticle–lipid films were mainly elastic and had a rather low linear viscoelastic threshold. More recently, Degen et al. have evaluated via interfacial shear rheology the competition of maghemite nanoparticles (γ-Fe$_2$O$_3$) on the surface adsorption and stability of different surfactant films (SDS, CTAB, and Brij 35) (Degen et al. 2011). They conclude that the different forces acting between not only particles but also particle–surfactant molecules play an important role in the structure and the homogeneity of the resultant interfacial layer, with the behavior of the macroscopic shear elasticity of each evaluated system being a good descriptor of that conclusion.

4.5   PARTICLE–SURFACTANT SYSTEMS IN STABILIZATION OF FOAMS AND EMULSIONS

To the best of our knowledge, not only Pickering emulsions—emulsions stabilized solely by nano- or microparticles—but also foams stabilized by colloidal particles can be stable during long periods (Aveyard et al. 2003). In fact, foams and emulsions can be considered as thermodynamically metastable systems, and their unstable nature is a critical issue in all the applications where they are involved (from food and cosmetics to oil recovery among others) as well as their use as precursors of light and resistant nanomaterials. Their instability arises from the high energy associated with the respective fluid/fluid interface and thus constitutes a driving force for decreasing the total interfacial area of the foam and emulsion through coalescence—film rupture—and coarsening (Ostwald ripening or disproportionation) of the bubbles—exchange of gas between bubbles owing to differences in Laplace pressure.

The study of the stabilization of dispersed systems such as foams and emulsions has attracted much attention in the last two decades because of the recognized importance of this type of system as precursors of light and resistant nanomaterials with multiple applications. Generally, the stabilization of these dispersed systems is achieved by the segregation of surface-active materials to the fluid interface. This segregation creates an interfacial layer that can stabilize the interface, avoiding the coalescence and coarsening of bubbles/droplets. Traditionally, surfactants and polymers, or even proteins (Weaire and Hutzler 1999), have been used as a stabilizing agent of dispersed systems. However, the short stability time conferred for these relatively simple molecules to the dispersed systems makes it necessary for one to focus the research on other stabilizing agents like micro- and nanoparticles (Cervantes-Martinez et al. 2008). In this latter case, the bubbles/droplets coated with particles form a colloidal armor that prevents the destabilization mechanisms of dispersed systems (Abkarian et al. 2009). Yet, the wettability properties are essential in the stabilization mechanism of dispersed systems and the use of chemically hydrophobized particles shows only a good performance in a limited hydrophobicity range. This makes it necessary to look for alternative methods of hydrophobization, with the combination of particles and surfactant (Gonzenbach et al. 2006) being a good choice for the stabilization of foams and emulsions because of the synergism between them, which leads to enhanced foam/emulsion stability. Note that it is necessary to consider the existence of a threshold on the surfactant amount that can be added to modify the properties of the particles in order to obtain stable dispersed systems. Above that threshold, the particles become too hydrophilic and this induces the destabilization of foams and emulsions (Subramaniam et al. 2006). It is possible to consider as a general rule that the most stable dispersed systems are those formed from dispersion with intermediate surfactant concentrations (Binks et al. 2008).

In order to understand the stabilization mechanism of dispersed systems from a physical point of view, it is necessary to consider two key points: (i) the wettability properties of the stabilizing agent (hydrophilic–lipophilic balance) and (ii) the mechanical properties of the formed layers, mainly the dilational rheological behavior. In fact, this is due to the shape evolution of a ripening drop or bubble that gives rise to a purely compressional deformation of the interface as long as the drop remains spherical.

It is expected that those systems without an enhanced wettability do not allow the formation of good foams/emulsions. Studies on the stabilization of oil in water emulsion by a combination of hydrophilic silica nanoparticle and CTAB (Ravera et al. 2008) pointed out the importance of the accumulation of particles at the droplet interfaces to obtain stable emulsions with lifetimes longer than days. This accumulation of nanoparticles at the interface is governed by the wettability properties of the complexes, allowing one to control the size distribution and the shape of the droplet (Limage et al. 2010b).

A key physical parameter in the stabilization of dispersed systems is the dilational elasticity E of the coated surfaces. As pointed out by Cervantes-Martinez et al. (2008), the destabilization of dispersed systems occurs because the derivative of the bubble/drop capillary pressure P with respect to its radius is negative ($dP/dR = -2\gamma/R^2$). The irreversible attachment of material to the bubble/droplet surface implies high desorption energies, which allows considering constant coverage, meaning that the concentration of the surface-active species at the interface (and therefore the surface tension) varies only with the interfacial area A (and hence the bubble/droplet size). This provides an interfacial dilational elasticity $E = d\gamma/d\ln(A)$, and the derivative of the bubble/droplet capillary pressure can be written as $dP/dR = -2\gamma/R^2 + 4E/R^2$. Hence, a bubble/droplet becomes stable when $E > \gamma/2$, which is the so-called Gibbs stability criterion.

Studies using carbon nanoparticles and CTAB (Santini et al. 2010, 2012a) have shown that the formation of an interfacial layer is not enough to obtain stable foams/emulsions; it is necessary that the elasticity modulus overcomes a certain value to avoid the coalescence between the droplets/bubbles. Generally, an elasticity modulus several times higher than that obtained for surfactant layers is necessary to obtain dispersed systems with enhanced properties. The improvement in the stability of dispersed systems with the characteristic of the surfactant–nanoparticle complexes was pointed out by Binks et al. (2008). They show that the characteristics of the segregated complexes play a key role in the stabilization of the formed foams, with the interaction between the nanoparticles at the air/water interface being essential in the stabilization of the foams. This is in agreement with the effect of the cohesion between nanoparticles on the stability of nanoparticle-stabilized foams found by Stocco et al. (2009, 2011). They point out that the cohesion between nanoparticles in the interfacial layers caused by the hydrophobic interactions leads to higher values of the viscoelasticity moduli, allowing a better stabilization of foams.

The enhanced stability of dispersed systems owing to the synergism between particles and surfactant molecules arises from the adsorption of coated particles around bubbles/droplets forming elastic layers that prevent their coalescence and disproportionation and reduce the drainage between the bubbles/droplets as a consequence of the increase of the viscosity of the continuous phase (Binks et al. 2008; Limage et al. 2010a). However, it is important to consider that it is not possible to use the above criteria as an absolute one to determine the stability of emulsions/foams (Santini et al. 2012b) because other viscoelastic and structural parameters such as shear viscoelasticity, bending elasticity, and thickness of the interfacial layer can play a key role in the stabilization of dispersed systems.

4.6 CONCLUSIONS

In this chapter, a revision of the principal results obtained in the field of mixed particle–surfactant layers at the liquid interface is presented. The segregation of such materials to fluid interfaces modifies their mechanical properties and morphology, and in some cases, these effects can be exploited to control the interfacial properties of composite systems. A complete understanding of these types of systems implies a multi-focus approach based on the acquisition of the structural equilibrium and dynamic properties of the interfacial layer. The segregation of particles to the fluid interface is mainly controlled by their wettability, which is in turn determined by their interaction with surfactants. Under certain conditions, the particle attachment at the interface is energetically strongly favored so that it can be considered irreversible. This intricate balance of interactions (particle–surfactant, surfactant–surfactant, and particle–particle) in the interfacial layers confers a complex physicochemical behavior to these mixed systems. In particular, the coverage of the layers may induce particular mechanical properties to coated bubbles and droplets, and this is the reason for the strong relevance of these systems to the stabilization of dispersed systems such as foams, emulsions, and films.

Despite the important development of systems formed by particles and surfactant, knowledge on some aspects remains in the dark. Further developments from both experimental and theoretical points of view will be useful in the future to obtain a better understanding of the physical mechanisms governing the intrinsic equilibrium and dynamic properties of these mixed systems composed of nanoparticles and surfactants. A deeper knowledge of the structure and interfacial properties of these composite materials is expected, in fact, to favor the development of new ideas in a wide range of applications such as the production of light nanomaterials with tailored properties or in other fields such as oil recovering, food processing, cosmetic, membrane-based separation, and purification.

ACKNOWLEDGMENT

The research was financially supported by the European Space Agency (FASES, PASTA), the Deutsche Luft- und Raumfahrt (DLR 50WM1129), the Italian Space Agency (LIFT), the Deutsche Forschungsgemeinschaft SPP 1506 (Mi418/18-1), and COST actions D43, CM1101, and MP1106.

REFERENCES

Abkarian, M., Subramaniam, A.B., Kim, S.H., Larsen, R.J., Yang, S.M. and Stone, H.A. 2009. Microscopic mechanisms of the brittleness of viscoelastic fluids. *Phys. Rev. Lett.* 99:188301.

Ahualli, S., Iglesias, G.R., Wachter, W., Dulle, M., Minami, D. and Glatter, O. 2011. Adsorption of anionic and cationic surfactants on anionic colloids: supercharging and destabilization. *Langmuir* 27:9182–92.

Aspnes, D.E. 1982. Optical properties of thin films. *Thin Solid Films* 89:249–62.

Aveyard, R., Binks, B.P. and Clint, J.H. 2003. Emulsions stabilised solely by colloidal particles. *Adv. Colloid Interface Sci.* 100:503–46.

Binks, B.P., Kirkland, M. and Rodrigues, J.A. 2008. Origin of stabilisation of aqueous foams in nanoparticle–surfactant mixtures. *Soft Matter* 4:2373–82.

Binks, B.P. and Horozov, T.S. 2006. *Colloidal Particles at Liquid Interfaces*. Cambridge: Cambridge University Press.

Binks, B.P. 2002. Particles as surfactants: similarities and differences. *Curr. Opin. Colloid Interface Sci.* 7:21–41.

Bonales, L.J., Rubio, J.E.F., Ritacco, H., Vega, C., Rubio, R.G. and Ortega, F. 2011. Freezing transition and interaction potential in monolayers of microparticles at fluid interfaces. *Langmuir* 27:3391–400.

Butler, J.A.V. 1932. The thermodynamics of the surfaces of solutions. *Proc. Roy. Soc. Ser. A* 138:348–75.

Cervantes Martinez, A., Rio, E., Delon, G., Saint-Jalmes, A., Langevin, D. and Binks, B.P. 2008. On the origin of the remarkable stability of aqueous foams stabilised by nanoparticles: link with microscopic surface properties. *Soft Matter* 4:1531–5.

Cicuta, P., Stancik, E.J. and Fuller, G.G. 2003. Shearing or compressing a soft glass in 2D: time–concentration superposition. *Phys. Rev. Lett.* 90:236101.

Clint, J.H. and Taylor, S.E. 1992. Particle size and interparticle forces of overbased detergents: a Langmuir trough study. *Colloids Surf. A* 65:61–7.

Defay, R. and Prigogine, I. 1966. *Surface Tension and Adsorption*. London: Longmans.

Degen, P., Wieland, D.C.F., Leick, S., Paulus, M., Rehage, H. and Tolan, M. 2011. Effect of magnetic nanoparticles on the surface rheology of surfactant films at the water surface. *Soft Matter* 7:7655–62.

Even, W.R. and Gregory, D.P. 1994. Emulsion-derived foams—preparation, properties, and application. *MRS Bull.* 19:29–33.

Fainerman, V.B., Kovalchuk, V.I., Grigoriev, D.O., Leser, M.E. and Miller, R. 2006a. Theoretical analysis of surface pressure of monolayers formed by nano-particles. In *Surface Chemistry in Biomedical and Environmental Science*, 79–90. Heidelberg: Springer.

Fainerman, V.B., Kovalchuk, V.I., Lucassen-Reynders, E.H., Grigoriev, D.O., Ferri, J.K., Leser, M.E., Michel, M., Miller, R. and Möhwald, H. 2006b. Surface-pressure isotherms of monolayers formed by microsize and nanosize particles. *Langmuir* 22:1701–5.

Fainerman, V.B., Zholob, S.A., Leser, M.E., Michel, M. and Miller, R. 2004. Competitive adsorption from mixed nonionic surfactant/protein solutions. *J. Colloid Interface Sci.* 274:496–501.

Fainerman, V.B., Lucassen-Reynders, E.H. and Miller, R. 2003. Description of the adsorption behaviour of proteins at water/fluid interfaces in the framework of a two-dimensional solution model. *Adv. Colloid Interface Sci.* 106:237–59.

Fuller, G.G. and Vearmant, J. 2012. Complex fluid–fluid interfaces: rheology and structure. *Annu. Rev. Chem. Biomol. Eng.* 3:519–43.

Gonzenbach, U.T., Studart, A.R., Tervoort, E. and Gauckler, L.J. 2006. Ultrastable particle-stabilized foams. *Ang. Chem. Int. Ed.* 45:3526–30.

Grigoriev, D., Miller, R., Shchukin, D. and Möhwald, H. 2007a. Interfacial assembly of partially hydrophobic silica nanoparticles induced by ultrasonic treatment. *Small* 3:665–71.

Grigoriev, D.O., Krägel, J., Dutschk, V., Miller, R. and Möhwald, H. 2007b. Contact angle determination of micro- and nanoparticles at fluid/fluid interfaces: the excluded area concept. *Phys. Chem. Chem. Phys.* 9:6447–54.

Guzmán, E., Liggieri, L., Santini, E., Ferrari, M. and Ravera, F. 2012. Influence of silica nanoparticles on dilational rheology of DPPC–palmitic acid Langmuir monolayers. *Soft Matter* 8:3938–48.

Guzmán, E., Liggieri, L., Santini, E., Ferrari, M. and Ravera, F. 2011. Effect of hydrophilic and hydrophobic nanoparticles on the surface pressure response of DPPC monolayers. *J. Phys. Chem. C* 115:21715–22.

Horozov, T.S., Braz, D.A., Fletcher, P.D.I., Binks, B.P. and Clint, J.H. 2008. Novel film-calliper method of measuring the contact angle of colloidal particles at liquid interfaces. *Langmuir* 24:1678–81.

Hórvölgyi, Z., Fendler, J.H., Máté, M. and Zrínryi, M. 1996. An experimental approach to the determination of two-dimensional gel-point: a film balance study. *Prog. Colloid Polym. Sci.* 102:126–30.

Hunter, T., Jameson, G.J., Wanless, E.J., Dupin, D. and Armes, S.P. 2009. Adsorption of submicrometer-sized cationic sterically stabilized polystyrene latex at the air-water interface: contact angle determination by ellipsometry. *Langmuir* 25:3440–9.

Iglesias, G.R., Wachter, W., Ahualli, S. and Glatter, O. 2011. Interactions between large colloids and surfactants. *Soft Matter* 7:4619–22.

Kekicheff, P., Christenson, H.K. and Ninham, B.W. 1989. Adsorption of cetyltrimethylammonium bromide to mica surfaces below the critical micellar concentration. *Colloids Surf.* 40:31–43.

Kovalchuk, V.I., Miller, R., Fainerman, V.B. and Loglio, G. 2005. Dilational rheology of adsorbed surfactant layers—role of the intrinsic two-dimensional compressibility. *Adv. Colloid Interface Sci.* 114–115:303–12.

Kundu, S. 2011. Layer-by-layer assembly of thiol-capped Au nanoparticles on a water surface and their deposition on H-terminated Si(001) by the Langmuir–Blodgett method. *Langmuir* 27:3930–7.

Levine, S., Bown, B.D. and Partridge, S.J. 1989. Stabilization of emulsions by fine particles I. Partitioning of particles between continuous phase and oil/water interface. *Colloids Surf.* 38:325–43.

Liggieri, L., Santini, E., Guzmán, E., Maestro, A. and Ravera, F. 2011. Wide-frequency dilational rheology investigation of mixed silica nanoparticle-CTAB interfacial layers. *Soft Matter* 7:7699–709.

Liggieri, L. and Miller, R. 2010. Relaxation of surfactants adsorption layers at liquid interfaces. *Curr. Opin. Colloid Interface Sci.* 15:256–63.

Liggieri, L., Ferrari, M., Mondelli, D. and Ravera, F. 2005. Surface rheology as a tool for the investigation of processes internal to surfactant adsorption layers. *Faraday Discus.* 129:125–40.

Limage, S., Krägel, J., Schmitt, M., Dominici, C., Miller, R. and Antoni, M. 2010a. Rheology and structure formation in diluted mixed particle-surfactant systems. *Langmuir* 26:16754–61.

Limage, S., Schmitt, M., Vincent-Bonnieu, S., Dominici, C. and Antoni, M. 2010b. Characterization of solid-stabilized water/oil emulsions by scanning electron microscopy. *Colloids Surf. A* 365:154–61.

Loglio, G., Tesei, U. and Cini, R. 1979. Spectral data of surface viscoelastic modulus acquired via digital Fourier transformation. *J. Colloid Interface Sci.* 71:316–20.

Lucassen, J. 1992. Dynamic dilational properties of composite surfaces. *Colloids Surf.* 65:139–49.

Lucassen, J. and van der Tempel, M. 1972. Dynamic measurements of dilational properties of a liquid interface. *Chem. Eng. Sci.* 27:1283–91.

Lucassen-Reynders, E.H. 1981. *Anionic Surfactant: Physical Chemistry of Surfactant Action.* New York: Marcel Dekker.

Maas, M., Ooi, Ch.C. and Fuller, G.G. 2010. Thin film formation of silica nanoparticle/lipid composite films at the fluid–fluid interface. *Langmuir* 26:17867–73.

Maestro, A., Guzmán, E., Santini, E., Ravera, F., Liggieri, L., Ortega, F. and Rubio, R.G. 2012. Wettability of silica nanoparticle–surfactant nanocomposite interfacial layers. *Soft Matter* 8:837–43.

Maestro, A., Bonales, L.J., Ritacco, H., Rubio, R.G. and Ortega, F. 2010. Effect of the spreading solvent on the three-phase contact angle of microparticles attached at fluid interfaces. *Phys. Chem. Chem. Phys.* 12:14115–20.

Menon, V.B. and Wasan D.T. 1988. Characterization of oil–water interfaces containing finely divided solids with applications to the coalescence of water-in-oil emulsions: a review. *Colloids Surf. A* 29:7–27.

Menon, V.B., Nikolov, A.D. and Wasan, D.T. 1988. Interfacial effects in solids-stabilized emulsions: measurements of film tension and particle interaction energy. *J. Colloid Interface Sci.* 124:317–27.

Miller, R. and Liggieri, L. 2010. *Interfacial Rheology*. Leiden: Brill.

Miller, R., Fainerman, V.B., Kovalchuk, V.I., Grigoriev, D.O., Leser, M.E. and Michel, M. 2006. Composite interfacial layers containing micro-size and nano-size particles. *Adv. Colloid Interface Sci.* 128–130:17–26.

Noskov, B.A. and Loglio, G. 1998. Dynamic surface elasticity of surfactant solutions. *Colloids Surf. A* 143:167–83.

Oh, S.G. and Slattery, J.C. 1978. Disk and biconical interfacial viscometers. *J. Colloid Interface Sci.* 67:516–25.

Paunov, V.N. 2003. Novel method for determining the three-phase contact angle of colloid particles adsorbed at air–water and oil–water interfaces. *Langmuir* 19:7970–6.

Pickering, S.U. 1907. Emulsions. *J. Chem. Soc. Trans.* 91:2001–21.

Prosser, A.J. and Franses, E.I. 2001. Adsorption and surface tension of ionic surfactants at the air–water interface: review and evaluation of equilibrium models. *Colloids Surf. A* 178:1–40.

Pugh, R.J. 1996. Foaming, foam films, antifoaming and defoaming. *Adv. Colloid Interface Sci.* 64:67–142.

Ravera, F., Ferrari, M., Liggieri, L., Loglio, G., Santini, E. and Zanobini, A. 2008. Liquid–liquid interfacial properties of mixed nanoparticle–surfactant systems. *Colloids Surf. A* 323:99–108.

Ravera, F., Santini, E., Loglio, G., Ferrari, M. and Liggieri, L. 2006a. Effect of nanoparticles on the interfacial properties of liquid/liquid and liquid/air surface layers. *J. Phys. Chem. B* 110:19543–51.

Ravera, F., Ferrari, M. and Liggieri, L. 2006b. Modelling of dilational visco-elasticity of adsorbed layers with multiple kinetic processes. *Colloids Surf. A* 282–283:210–6.

Ravera, F., Ferrari, M., Santini, E. and Liggieri, L. 2005. Influence of surface processes on the dilational visco-elasticity of surfactant solutions. *Adv. Colloid Interface Sci.* 117:75–100.

Ravera, F., Ferrari, M. and Liggieri, L. 2001. Adsorption and partitioning of surfactants in liquid–liquid systems. *Adv. Colloid Interface Sci.* 88:129–77.

Safouane, M., Langevin, D. and Binks, B.P. 2007. Effect of particle hydrophobicity on the properties of silica particle layers at the air–water interface. *Langmuir* 23:11546–53.

Santini, E., Guzmán, E., Ravera, F., Ciajolo, A., Alfè, M., Liggieri, L. and Ferrari, M. 2012a. Soot particles at the aqueous interface and effects on foams stability. *Colloids Surf. A* 413:216–23.

Santini, E., Guzmán, E., Ravera, F., Ferrari, M. and Liggieri, L. 2012b. Properties and structure of interfacial layers formed by hydrophilic silica dispersions and palmitic acid. *Phys. Chem. Chem. Phys.* 14:607–15.

Santini, E., Krägel, J., Ravera, F., Liggieri, L. and Miller, R. 2011. Study of the monolayer structure and wettability properties of silica nanoparticles and CTAB using the Langmuir trough technique. *Colloids Surf. A* 382:186–91.

Santini, E., Ravera, F., Ferrari, M., Alfè, M., Ciajolo, A. and Liggieri, L. 2010. Interfacial properties of carbon particulate-laden liquid interfaces and stability of related foams and emulsions. *Colloids Surf. A* 365:189–98.

Santini, E., Ravera, F., Ferrari, M., Stubenrauch, C., Makievski, A. and Krägel, J. 2007a. A surface rheological study of non-ionic surfactants at the water–air interface and the stability of the corresponding thin foam films. *Colloids Surf. A* 298:12–21.

Santini, E., Liggieri, L., Sacca, M., Classe, D. and Ravera, F. 2007b. Interfacial rheology of Span 80 adsorbed layers at paraffin oil–water interface and correlation with the corresponding emulsion properties. *Colloids Surf. A* 309:270–9.

Siebold, A., Walliser, A., Nardin, M., Oppliger, M. and Schultz, J. 1997. Capillary rise for thermodynamic characterization of solid particle surface. *J. Colloid Interface Sci.* 186:60–70.

Stefaniu, C., Chanana, M., Wang, D., Novikov, D.V., Brezesinski, G. and Möhwald, H. 2010. Biocompatible magnetite nanoparticles trapped at the air/water interface. *ChemPhysChem* 11:3585–8.

Stocco, A., Rio, E., Binks, B.P. and Langevin, D. 2011. Aqueous foams stabilized solely by particles. *Soft Matter* 7:1260–7.

Stocco, A., Drenckhan, W., Rio, E., Langevin, D. and Binks, B.P. 2009. Particle-stabilised foams: an interfacial study. *Soft Matter* 5: 2215–22.

Subramaniam, A.B., Mejean, C., Abkarian, M. and Stone, H.A. 2006. Microstructure, morphology, and lifetime of armored bubbles exposed to surfactants. *Langmuir* 22:5986–90.

Vandebril, S., Vermant, J. and Moldenaers, P. 2010. Efficiently suppressing coalescence in polymer blends using nanoparticles: role of interfacial rheology. *Soft Matter* 6:3353–62.

Weaire, D. and Hutzler, S. 1999. *The Physics of Foams*. Oxford: Oxford University Press.

Whittby, C., Fornasiero, D., Ralston, J., Liggieri, L. and Ravera, F. 2012. Properties of fatty amine–silica nanoparticle interfacial layers at the water–hexane interface. *J. Phys. Chem. C* 116:3050–8.

Wyss, H.M., Miyazaki, K., Mattsson, J., Hu, Z., Reichman, D.R. and Weitz, D.A. 2007. Strain-rate frequency superposition: a rheological probe of structural relaxation in soft materials. *Phys. Rev. Lett.* 98:238303.

Zang, D.Y., Rio, E., Delon, G., Langevin, D., Wei, B. and Binks, B.P. 2011. Influence of the contact angle of silica nanoparticles at the air–water interface on the mechanical properties of the layers composed of these particles. *Mol. Phys.* 109:1057–66.

Zang, D.Y., Rio, E., Langevin, D., Wei, B. and Binks, B.P. 2010a. Viscoelastic properties of silica nanoparticle monolayers at the air-water interface. *Eur. Phys. J. E* 31:125–34.

Zang, D., Langevin, D., Binks, B.P. and Wei, B. 2010b. Shearing particle monolayers: strain-rate frequency superposition. *Phys. Rev. E* 81:011604.

Zang, D.Y., Stocco, A., Langevin, D., Wei, B.B. and Binks, B.P. 2009. An ellipsometry study of silica nanoparticle layers at the water surface. *Phys. Chem. Chem. Phys.* 11: 9522–9.

5 Magnetic-Core Microgels

Rafael Contreras-Caceres, Marco Laurenti,
Jorge Perez-Juste, Jorge Rubio- Retama,
Enrique Lopez-Cabarcos, and
Antonio Fernandez-Barbero

CONTENTS

5.1   INTRODUCTION

The preparation and characterization of thermosensitive poly(*N*-isopropylacrylamide) (pNIPAM) microgels produced via surfactant-free emulsion polymerization (Pelton 1986) have attracted extensive interest because of their potential applications in many fields, such as templates for nanoparticles (Schmidt et al. 2008), contrast agents (Rowe et al. 2009), sensors (Yin et al. 2010), catalyst support (Carregal-Romero et al. 2010; Fernandez-Barbero et al. 2009), or biomedical applications (Madeiros et al. 2011).

In many of these works, the thermosensitive microgels are combined with inorganic components such as quantum dots (Jaczewski et al. 2009), silver (Xu et al. 2006), gold (Kawano et al. 2009) or magnetic nanoparticles (Luo et al. 2010) to yield nanostructured and multifunctional hybrid material. The combination between the organic and the inorganic components establishes a symbiotic relation in which the microgels give colloidal stability as well as stimuli-responsive features, while the inorganic counterparts provide quantum properties such as photoluminescence (Agrawal et al. 2008; Bai et al. 2010), surface plasmon resonance (Karg et al. 2009), or magnetism (Schachschal et al. 2010).

In some works, the localization of the inorganic material is on the outer part of the microgels and this sort of decoration can be produced either by exploiting charge interactions between the nanoparticles and the microgels (Karg et al. 2009; Sauzedde et al. 1999) or by covalent bonds between the microgels and the magnetic material (Zhang and Wang 2009). However, one of the major drawbacks concomitant with this strategy is a reduction of the colloidal stability, which leads to system coagulation. Another approach used to synthesize hybrid magnetic nanoparticles is based on the preparation of magnetic nanoparticles in a bulk solution, where microgels have been previously dispersed. This approach usually yields nanoparticles homogeneously distributed within the microgels (Pich et al. 2004). Nevertheless, this sort of decoration presents two major disadvantages: First, not all the inorganic nanoparticles can be done in the presence of microgels, limiting this strategy only to syntheses performed under mild conditions. Second, the high specific surface of the nanoparticles can favor interactions between the nanoparticles and the microgel matrix, which normally affect the thermal response of the microgels being possible to block the low critical solubility temperature (LCST). This is the reason why only microgels with a limited amount of inorganic material have been produced using this method (Rubio-Retama et al. 2007, 2010).

To overcome the previous disadvantages, many efforts have been done to produce hierarchical microgels in which the inorganic nanoparticles are located in the core of the system (Contreras-Caceres et al. 2009, 2010; Dagallier et al. 2010; Sanchez-Iglesias et al. 2009). However, these methodologies have not been successfully applied to Fe_3O_4 nanoparticles since its surface inhibits the polymerization reaction. This impediment has been related to their capacity to mediate in redox processes (Rebodos and Vikesland 2010) and to transfer an electron to the radical, which subsequently stops the polymerization. To overcome this problem, some authors have incorporated pNIPAM on the surface of magnetic nanoparticles using a layer-by-layer deposition technique (Yamamoto et al. 2008). However, this procedure requires many steps when one wants to grow a thick polymer layer around the inorganic core. Other authors cover the surface of the magnetic nanoparticles with SiO_2 to overcome the above problems (Dagallier et al. 2010; Karg et al. 2006; Luo et al. 2010). However, this method presents two major disadvantages: first, the process turns into a tedious multistep procedure, and second, the incorporation of a second inorganic material diminishes the maximum magnetic moment of the final material. These problems have been recently solved (Contreras-Caceres et al. 2011) by covering the nanoparticles with 3-butenoic acid (3-bt), which hinders the nanoparticles' surface oxidation. With this molecule, it is possible to create a simple method that yields thermosensitive microgels, which present a magnetic core formed by a cluster of Fe_3O_4 nanoparticles. In this work, we perform a parametric study about the role of this molecule as well as the hydrophobic–hydrophilic character of the nanoparticles during the synthesis of the microgels and how they influence the architecture of the final hybrid material. One can take advantage of the porosity of the previously synthesized Fe_3O_4@pNIPAM, to diffuse gold salt across the polymer mesh, in the presence of a weak salt across the polymer mesh. It allows gold salt to reach the central part of the bicomponent particle, starting nucleation just there. Three-component core–shell Fe_3O_4@Au@pNIPAM nanoparticles are then obtained. The final structure presents three different properties: trapping ability due to the microgel shell, optical response coming from the gold nanoparticles at the center, and magnetic features due to the magnetic core.

As was previously reported, core–shell metal@pNIPAM composite colloids are able to mechanically trap surface-enhanced Raman scattering (SERS) (Alvarez-Puebla et al. 2009). SERS is a unique ultrasensitive technique that allows for the unequivocal identification of analytes in a wide variety of matrices, with minimal processing, if any, prior to the analysis (Moskovits 2005). Notwithstanding, this technique still encompasses several limitations, one of the most restrictive ones is that only analytes with suitable functional groups (i.e., thiol, nitrile, amine, and carboxylic) provide sufficiently good signal for ultrasensitive analytical purposes in a convenient time. This is usually not a drawback when dealing with bio-related problems, as most of the bio-relevant structures contain at least one of such groups (Kneipp et al. 1999; Rosi and Mirkin 2005). However, organic pollutants and other hazardous materials in environmental problems are characterized by an extra-ordinary diversity of chemical structures including nonfunctionalized aliphatic and aromatic compounds or molecules (alcohols, ethers, ketones, halides, etc.) with no affinity for gold and silver surfaces, the most common plasmonic materials (Murray et al. 2010). These molecular systems are *a priori* impossible to detect directly, via SERS, and thus, indirect approaches are required. All these approaches are based on the surface functionalization of the plasmonic nanoparticles with different receptors, in an attempt to increase the local concentration of the molecule to be determined near the electric field generated by the particle. Examples of these strategies include electrostatic attraction by the counterions (Alvarez-Puebla et al. 2009), highly selective molecules such as aptamers (Kim et al. 2010; Neumann et al. 2009), antibodies (Porter et al. 2008; Sanles-Sobrido et al. 2009) or calixarenes (Guerrini et al. 2006, 2009), and thiolated aliphatic monolayers (Bantz and Haymes 2009; Jones et al. 2009). Unfortunately, all these methods are highly selective, so that only certain molecules are effectively adsorbed to the plasmonic surface. A new family of hybrid materials that are capable of mechanically trapping molecules from aqueous solution have recently been developed (Contreras-Caceres et al. 2010). These materials comprise a metallic core surrounded by a thermoresponsive pNIPAM shell. However, and while the trapping properties are clearly efficient, since each single core nanoparticles is isolated by the surrounding polymer shell (Contreras-Caceres et al. 2009; Rubio-Retama et al. 2010), the formation of hot spots is completely inhibited, thereby restricting the detection limits of the system. Additionally, the concept of dynamic hot spots has recently been demonstrated by incorporation of silver or gold nanoparticles within a macroscopic gel matrix (Contreras-Caceres et al. 2010; Sanchez-Iglesias et al. 2009), resulting in the generation of hot spots when the gel was dehydrated and the particles became closer to each other. This idea can be applied in a similar way to microgel spheres through the incorporation of multiple nanoparticles, with the additional advantage that the microgel collapse can be externally triggered by simply increasing the temperature. Another strategy that has been proven successful to lower the actual detection limits in SERS is the addition of a magnetic functionality to the SERS colloidal platform (Rebodos and Vikesland 2010), which permits the rapid concentration of the plasmonic hybrid material within a small region prior to SERS analysis; thus, a small amount of the sensing platform would be required, decreasing the concentration of analyte needed to obtain a meaningful SERS signal (Contreras-Caceres et al. 2011).

In the second part of this chapter, we engineered and fabricated a SERS substrate comprising magnetite and silver particles encapsulated within a pNIPAM thermo-responsive shell. Additionally, proof of concept for the sequestration of uncommon molecular systems is also included, through the first SERS analysis of pentachloro-phenol (PCP), a ubiquitous environmental pollutant extensively used as herbicide, insecticide, fungicide, algaecide, and disinfectant and also as a preserver of wooden materials. Rapid and sensitive identification of PCP is extremely important as it is highly soluble (it easily contaminates tap water) and bioaccumulates in fatty tissues, so even small exposures may eventually reach dangerous levels. In fact, severe expo-sure results in harmful effects on the liver, kidneys, blood, lungs, nervous system, immune system, and gastrointestinal tract. Chronic effects after exposure to low lev-els include damage to the liver, kidneys, blood, and nervous system, as well as can-cer (classified as B2 carcinogen by the Environmental Protection Agency) (United States, Environmental Protection Agency 2007).

5.2 SYNTHESIS OF THE MAGNETIC NANOPARTICLES AND SURFACE MODIFICATION

Two methods have been used to produce magnetic nanoparticles: thermal decom-position of iron(III) acetylacetonate and a coprecipitation of $FeCl_3$ and $FeCl_2$ in a stoichiometric ratio 2:1.

Method I: Thermal Decomposition. In this case, the synthesis of the magnetic nanoparticles was a modification of a previously described method (Guardia et al. 2010a,b). For the synthesis of 13-nm-edge-length particles, 0.353 g (1 mmol) of iron(III) acetylacetonate was mixed with 0.688 g (4 mmol) of 10-undecenoic acid in 25 mL of dibenzyl ether. After 1 h under vacuum in a Schlenk line, the solution was heated up to 200°C with a constant heating rate of 6–7°C/min under an argon blanket flow and vigorous stirring. After 2 h at 200°C, the solution was heated up to reflux with a constant heating rate of 5.2°C/min up to 300°C and kept at this temperature for 1 h. After cooling to room temperature, a mixture of toluene and acetone was added to the solution and then centrifuged to precipitate. The precipitate was washed several times with a mixture of toluene and acetone. Finally, the particles were stored in ethanol.

Method II: Coprecipitation of Fe^{3+} and Fe^{2+}. The particles are synthesized using a modified coprecipitation recipe reported previously (Massart 1981). Tetra-methylammonium hydroxide (TMAOH) is used as an alkaline base instead of ammo-nium hydroxide. A total of 25 mL of TMAOH 1 M and 22 mL of H_2O are mixed under mechanic stirring at 70°C. After that, a solution containing 3 mL of H_2O, 650 mg of $FeCl_3$, and 250 mg of $FeCl_2$ is added drop-wise. The stirring stands for 30 min. After this time, the nanoparticles are recovered by magnetic decantation and washed several times with milli-Q water.

Surface modification of the magnetic nanoparticles. The modification of the magnetic nanoparticles was carried out as follows. In a vial of 15 mL, 350 μL of magnetic nanoparticles was diluted in 10 mL of water and sonicated for 15 min to reduce particle aggregation. Subsequently, the vial was immersed in a water bath at 70°C, and different amounts of 3-bt were added. This solution stood for 1 h at these conditions and then the nanoparticles were centrifuged for 30 min to remove the 3-bt excess. With the aim of reducing the aggregation effects just before the centrifugation, 200 μL of 0.2 M CTAB was added before starting the centrifugation. After that, the supernatant was discarded and the precipitate was diluted to 10 mL with water and redispersed by sonication for 15 min.

Figure 5.1 shows a representative transmission electron microscopy (TEM) image of the synthesized magnetic nanoparticles. From the TEM images, one can observe that the size of the nanoparticles obtained by thermal decomposition was 13 ± 2 nm while the size for those nanoparticles produced by coprecipitation was around 6 ± 1 nm. These two different syntheses permitted us to obtain magnetic nanoparticles with different hydrophobic–hydrophilic character that was exploited to control the magnetic-core nature of the hybrid microgels.

The x-ray analysis showed that the crystalline structure of the magnetic nanoparticles is in both cases Fe_3O_4. The dispersion in water of the magnetic nanoparticles synthesized by thermal decomposition produced nanoparticle aggregates with a mean hydrodynamic diameter of 188 ± 27 nm (Qiu et al. 2010). With the aim of getting magnetic clusters with smaller sizes, we treated these magnetic nanoparticles with different amounts (μL) of 3-bt. In this way and after incubating these nanoparticles with 3-bt during 1 h at 70°C, it was possible to reduce the mean hydrodynamic diameter of the clusters. Table 5.1 reports the hydrodynamic diameter measured by

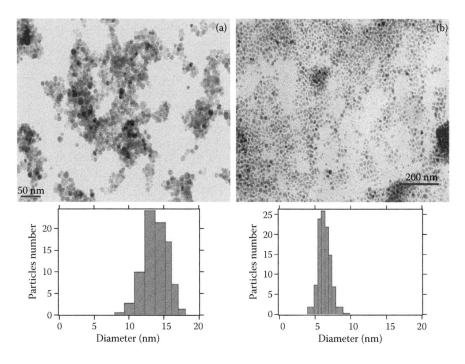

FIGURE 5.1   TEM image of (a) 13 ± 2 nm edge length iron oxide nanoparticles synthesized by thermal decomposition and (b) 6 ± 1 nm edge length iron oxide nanoparticles synthesized by coprecipitation method. (Adapted from Laurenti, M. et al., *Langmuir*, 27, 10484–10491, 2011. With permission.)

TABLE 5.1

Amount of 3-bt Added for Surface Treatment, Electrophoretic Mobility (μ), Hydrodynamic Diameter (D_h) of Magnetic Cluster Treated with 3-bt but without CTAB, D_h of Magnetic Clusters Treated with 3-bt and CTAB, D_h of pNIPAM@Fe$_3$O$_4$, and Colloidal Stability

Sample	3-bt (μL)	μ (10^{-8} m^2·V^{-1}·s^{-1}) (Before CTAB Addition)	D_h (nm) Magn. Cluster (Before CTAB Addition)	D_h (nm) Magn. Cluster (After CTAB Addition)	D_h (nm) Fe$_3$O$_4$@ pNIPAM	Colloidal Stability
S0I	0	3.8 ± 0.1	188 ± 21	188 ± 21	Not measured	<2 h
S10I	10	4.2 ± 0.1	120 ± 11	134 ± 12	600 ± 21	>2 weeks
S20I	20	4.4 ± 0.1	105 ± 16	110 ± 21	630 ± 23	>2 weeks
S40I	40	5.2 ± 0.1	67 ± 12	78 ± 15	674 ± 28	>2 weeks
S100I	100	5.3 ± 0.1	48 ± 9	55 ± 15	724 ± 32	>2 weeks
S160II	160	–	12 ± 7	12 ± 7	4800 ± 120	>2 weeks

Source: Adapted from Laurenti, M. et al., *Langmuir*, 27, 10484–10491, 2011. With permission.

Note: Samples from S0I to S100I are synthesized by method I and those from S160II are synthesized by method II.

dynamic light scattering (DLS) of the clusters treated with different amounts of 3-bt. From this study, we can infer that 3-bt was able to disaggregate the magnetic clusters and to reduce their sizes. The driving force of the size reduction would be the increment of the superficial charge of the magnetic nanoparticles that occurs when 3-bt is added. Figure 5.2 summarizes the process that is involved in the nanoparticle disaggregation. To document that disaggregation as a consequence of the extra surface charge provided to the particles by the 3-bt, we have measured the electrophoretic mobility and the hydrodynamic diameter of the cores after their treatment with 3-bt and before the CTAB addition. From these measurements, we observed an increment of the electrophoretic mobility from 3.8×10^{-8} m^2·V^{-1}·s^{-1} for untreated particles (sample S0I) up to 5.3×10^{-8} m^2·V^{-1}·s^{-1} for sample S100I.

The reason why samples S40I and S100I present similar electrophoretic mobility would be attributed to the fact that electrophoretical mobility is the result of two competitive processes: charge density and friction coefficient. Sample S100I has a higher surface charge than sample S40I, since it was prepared with a higher concentration of 3-bt, which would lead to a greater mobility. However, the increment of superficial charge would provoke an increment of the water adsorbed on the cluster surface, as is described by Lopez-Leon et al. (2005), increasing the friction coefficient of the cluster and thus reducing its electrophoretic mobility. It is the balance between terms, charge and friction, that explains the observed result. In the case of using 200 µL of 3-bt for treating the magnetic nanoparticles synthesized by thermal decomposition, we observe that the sizes of the magnetic clusters were similar to those obtained after adding 100 µL. This effect would indicate that above 100 µL of 3-bt, the charge density of the particles is maximum and cluster segregation stops,

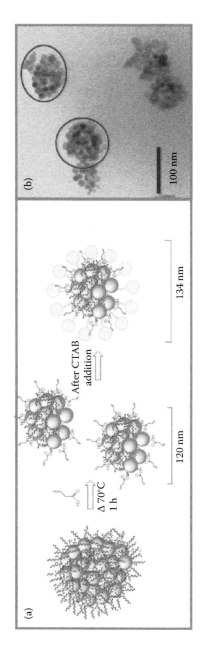

FIGURE 5.2 (a) Representation of the iron nanoparticles' surface modification and the effect on the hydrodynamic diameter. (b) TEM micrograph of a nanoparticle cluster obtained at the end of the treatment with 3-bt. (Adapted from Laurenti, M. et al., *Langmuir*, 27, 10484–10491, 2011. With permission.)

resulting in single particles being impossible to obtain. In order to get the smallest magnetic cluster, nanoparticles produced by coprecipitation were synthesized and modified with 3-bt. Before the treatment with 3-bt, they showed a hydrodynamic diameter of 20 ± 10 nm, which falls down to 12 ± 7 nm after the addition of 160 µL of 3-bt.

5.3 SYNTHESIS OF THERMORESPONSIVE MICROGELS@MAGNETIC CORES

The previous solution was put into a three-neck round-bottom flask and heated up to 70°C under mechanical stirring (350 rpm) and N_2 atmosphere. Subsequently, 0.226 g of N-isopropylacrylamide and 0.031 g of N,N-methylenebisacrylamide were added to the magnetic nanoparticles' dispersion. The mixture was stirred for 15 min, the polymerization was initiated by adding 100 µL of 2,2′-azobis-(2-methylpropionamidine) dihydrochloride 0.1 M, the N_2 flow was removed, and the polymerization was kept for 2 h. Finally, the milk-like dispersion was cooled down and the magnetic microgels were collected by placing the reaction flask on a neodymium magnet of 0.6 T for 24 h. A brown precipitate appeared on the bottom of the flask, which was collected and redispersed in 10 mL of water. The cleaning process was repeated at least 10 times. When NIPAM is polymerized in the presence of the magnetic clusters, we are able to obtain hybrid microgels. Figure 5.3 shows the shape of the hybrid microgels obtained after using different magnetic clusters.

The microgel synthesis without 3-bt (sample S0I) leads to large aggregates of the magnetic material, resulting in irregular clusters covered by pNIPAM. These hybrid particles show poor stability, provoking colloidal aggregation and strong sedimentation. The average particle size for pNIPAM@Fe_3O_4 in sample S0I is around 500 nm, with the magnetic cores of these microgels being close to 400 nm, as determined by TEM. Such an amount of inorganic counterpart decreases the colloidal stability of these microgels, making it impossible to measure their size by DLS. The use of magnetic clusters modified with 3-bt and polymerized with NIPAM leads to stable hybrid microgels with size, measured by DLS, between 320 and 370 nm (in the collapsed state) and between 724 and 600 nm (in the swollen state) for samples S10I–S100I. In the TEM pictures, it can be observed that the pNIPAM reaction carried out in the presence of magnetic clusters with sizes above 55 nm (samples from S10I to S100I) produced hybrid microgels with the inorganic material located in the core, which maintains the magnetic cluster formed before the microgels synthesis. Furthermore, the shape of the clusters evolved from irregular to pseudo-spherical when the amount of 3-bt was increased. By contrast, when the smallest magnetic clusters were used (S160II), hybrid particles with a mean hydrodynamic diameter of 1000 nm (in the collapsed state) and 4800 nm (in the swollen state) were obtained. In this case, the hybrid microgels present the magnetic nanoparticles randomly distributed in the inner part of the microgels with multiple magnetic clusters per microgel subunit. These results indicate that there is a strong influence of the aggregation state of the magnetic nanoparticles on the final hybrid microgel architecture.

Despite the localization of the magnetic nanoparticles, all the samples present volume phase transition. The particles shrink continuously with increasing temperature

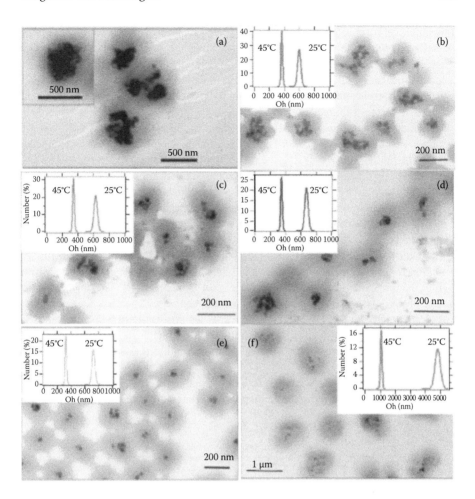

FIGURE 5.3 TEM pictures showing the core–shell structure of the pNIPAM@Fe$_3$O$_4$ microgels. The magnetic-core size reduces increasing the 3-bt concentration: (a) 0 μL (S0I), (b) 10 μL (S10I), (c) 20 μL (S20I), (d) 40 μL (S40I), (e) 100 μL (S100I), and (f) 160 μL (S160II). In the inset, the size distribution of the different hybrid microgels at 25°C and 45°C is reported. (Adapted from Laurenti, M. et al., *Langmuir*, 27, 10484–10491, 2011. With permission.)

as a consequence of the temperature-dependent Flory parameter (Hirotsu 1994). Furthermore, the shrinking–swelling cycles are reversible in all cases without any significant hysteresis. For microgels with a well-defined magnetic core, only the pNIPAM shell manifests swelling capacity, while the inorganic core keeps constant. That has been proved by comparing the relative shell swelling for pNIPAM@Fe$_3$O$_4$ with different core sizes. All curves overlap, indicating that the swelling of the organic shell is not influenced by the presence of the inorganic core. It is worth pointing out the big difference in size exhibited by the microgels prepared with the smallest magnetic clusters, which have a hydrodynamic diameter of 4800 nm in the swollen state. This result shows the tremendous influence the cluster size has on the microgel formation and their architecture. As can be inferred from the previous results, the variation of the amount of 3-bt influences the aggregation state of the magnetic clusters used during the synthesis and permits controlling the size of the magnetic clusters, allowing the entrapment of different amounts of magnetic material. The different particle decoration obtained from the synthesis could be attributed to the different action mechanism involved in the synthesis of the hybrid microgels. Several authors have described the mechanism of the neat pNIPAM microgel production as a nucleation process of colloidally unstable pNIPAM polymer chains and nanogels that are formed during the first step of the polymerization reaction, which are colloidally unstable at the reaction temperature. Because of their colloidal instability, they coagulate, forming a stable particle, which results in the final microgel. Using small-angle neutron scattering (Fernandez-Barbero et al. 2002), these nanogels have been detected showing a size around 22 nm above the LCST. However, when we introduce the functionalized magnetic cores, the vinyl groups provided by 3-bt could form covalent bonds with the pNIPAM growing radicals. As a result, the polymerization process could create a polymer shell around the cores, which would grow thanks to the excess pNIPAM oligomers or nanogels formed during the first stage of the polymerization, in a process named "seed-feed" that has been previously described by Neetu and Lyon (2007). This growing process stops when the microgels reach a size that provides enough colloidal stability, which is the same for hybrid and neat microgels produced under the same reaction conditions (370 nm at the collapsed state, see Figure 5.4).

This scenario changes when the size of the magnetic clusters is reduced and the specific surface of the inorganic nanoparticle is much higher than in the previous

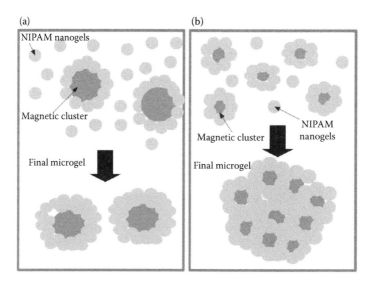

FIGURE 5.4 Microgel formation mechanism (a) for magnetic cluster bigger than the nanogels and (b) for magnetic cluster smaller than the nanogels. (Adapted from Laurenti, M. et al., *Langmuir*, 27, 10484–10491, 2011. With permission.)

case. Under this condition, more pNIPAM oligomers and nanogels would be required to cover the smallest magnetic clusters. This could provoke a rapid consumption of the pNIPAM oligomers and nanogels, yielding to small hybrid nanogels, which are unstable. In order to gain colloidal stability, neighbor hybrid nanogels would aggregate with each other, creating the final hybrid microgel. The big difference in size of S160II regarding the other hybrid microgels could arise from the smaller gain of colloidal stability that the aggregation of hybrid nanogels produces in comparison with the aggregation of neat nanogels.

With the aim of studying the magnetic behavior of the hybrid microgels, we have measured the magnetization moment of two different microgel dispersions below and above the LCST as a function of the magnetic field using a superconducting quantum interference device. From these experiments, we can observe that pNIPAM@Fe$_3$O$_4$ microgels show superparamagnetic behavior with very low hysteresis.

In Figure 5.5a, one can observe that the superparamagnetic behavior is independent of the particle swelling for sample S100I since the magnetization curves are coincident above and below the LCST. That would indicate that the distances between nanoparticles within the magnetic cluster are very stable and the saturation magnetic moment remains constant independently of the polymer swelling. However, for sample S160II, in Figure 5.5b, the swelling determines the distance between the magnetic nanoparticles and influences the saturation magnetic moment of the hybrid microgels, which is stronger at high temperatures because of the proximity, in the

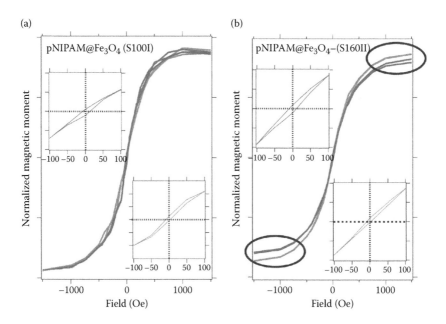

FIGURE 5.5 Magnetization for S100I (a) and S160II (b). Black and gray lines correspond to experiments performed at temperatures below and above the LCST, respectively. The insets show the hysteresis found for both temperatures. (Adapted from Laurenti, M. et al., *Langmuir*, 27, 10484–10491, 2011. With permission.)

collapsed state, of the magnetic clusters that are spread within the microgel (Gupta and Gupta 2005; Pankhurst et al. 2003; Tepper et al. 2003). Superparamagnetism is an important property of single-domain magnetic nanoparticles originating from the fast flipping of the magnetic moments owing to thermal energy. Without an external magnetic field, the magnetization of the nanoparticles appears to be zero, but in the presence of a magnetic field, their magnetic moments align along the field direction, leading to a net magnetization. This property is very interesting since it ensures the disappearance of the nanoparticles' magnetization after removing the magnetic field, allowing in this way the redispersion of the colloidal magnetic material. However, in the presence of a magnetic field, the magnetic material is aligned and can be collected or concentrated (Gupta and Gupta 2005; Pankhurst et al. 2003).

5.4 SILVER NUCLEATION AND GROWTH WITHIN FE₃O₄@pNIPAM MICROGELS

With the aim of fabricating a multicomponent system capable of including SERS capabilities to the global structure, silver nanoparticles were grown into the pNIPAM network. The procedure was performed through a nucleation and growth process. $AgNO_3$ (100 µL, 25 mM) was added under mild magnetic stirring to 10 mL of Fe_3O_4@pNIPAM hybrid particle. The mixture stood for 30 min at 25°C to allow a homogeneous diffusion of Ag^+ into the gel network. Then, 300 µL of $NaBH_4$ 10 mM was added into the sample under vigorous magnetic stirring to promote the nucleation of silver nanoparticles into the pNIPAM microgel network (Lu et al. 2006) (Ag-system 1) (Figure 5.6). Silver overgrowth (Yang et al. 2005) was achieved by adding a mixture composed of CTAB (2.5 mL, 0.2 M), glycine, pH 9.5 (2.5 mL, 0.4 mM), and 800 µL of $AgNO_3$ 0.25 mM to 5 mL of the previously prepared Fe_3O_4@Ag@pNIPAM. Then, 600 µL of ascorbic acid 100 mM was added to the system under vigorous magnetic stirring. The solution was left to growth in a water bath at 27°C for 30 min. After that, the solution was centrifuged at 3500 rpm for 30 min to remove the excess ascorbic acid, the supernatant was discarded, and the precipitate was diluted in 5 mL of water (Ag-system 2) (Figure 5.6).

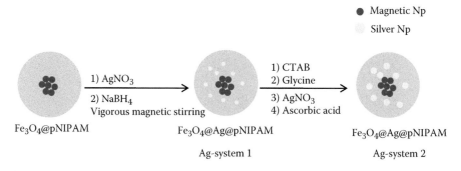

FIGURE 5.6 Schematic representation for the growth of silver dots and subsequent silver growth process. (Adapted from Contreras-Caceres, R. et al., *Langmuir*, 27, 4520–4525, 2011. With permission.)

Figure 5.7Aa shows TEM images of the $Fe_3O_4@Ag@pNIPAM$ particles, and as can be observed, silver seeds were then generated inside the pores of the polymer shell by adsorption of silver ions followed by fast in situ reduction using sodium borohydride (Figure 5.7Aa). The resulting system shows a well-dispersed collection of small silver nanoparticles inside the entire pNIPAM matrix. We found, however, that these particles are not particularly efficient as SERS platforms owing to the small Ag particle size and the low level of plasmon coupling. Thus, an additional growth step was carried out by adding silver ions from solution and using a mixture of CTAB and glycine to promote the epitaxial deposition of Ag on the preformed seeds (Contreras-Caceres et al. 2011). The morphology of the final SERS platform can be seen in Figure 5.7Ab, where clearly the magnetic particles are concentrated in the center, whereas the grown silver nanoparticles are preferentially close to the shell surface.

UV-vis spectroscopy shows a localized surface plasmon resonance (LSPR) band in both systems, with a maximum centered at approximately 421 nm, when the polymer is expanded (below 32°C), which becomes broader and red-shifts upon increasing the temperature, owing to the collapse of the pNIPAM shell and a subsequent increase in the plasmon coupling when more Ag nanoparticles get closer to each other. This effect has been found to be reversible as previously shown for Au

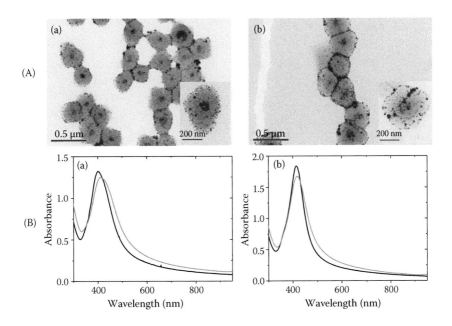

FIGURE 5.7 Representative TEM images of (Aa) the $Fe_3O_4@pNIPAM$ nanohybrid materials containing silver seeds and (Ab) the final $Fe_3O_4@Ag@pNIPAM$ composite microgel. UV-vis spectra of (Ba) Ag-system 1 and (Bb) Ag-system 2 below (black line) and above (gray line) the LCST. (Adapted from Contreras-Caceres, R. et al., *Langmuir*, 27, 4520–4525, 2011. With permission.)

nanorods (Alvarez-Puebla et al. 2012). The shift is not remarkable here because of the small silver particle size.

5.5 SERS CHARACTERIZATION

An initial SERS characterization proving the optical enhancing properties of the material was initially carried out using 1-naphthalenethiol (1NAT) as molecular probe and three excitation laser lines, from the visible to the NIR. For all the laser lines, the SERS spectra (Figure 5.8Ba) show well-defined bands with high intensity, which are characteristic of 1NAT: ring stretching (1553, 1503, and 1368 cm^{-1}), CH bending (1197 cm^{-1}), ring breathing (968 and 822 cm^{-1}), ring deformation (792, 664, 539, and 517 cm^{-1}), and CS stretching (389 cm^{-1}), allowing ultrasensitive detection in a wide spectral window of excitation wavelengths. However, the SERS intensity was found to be temperature dependent and in agreement with the observed changes in the LSPR band. For all the laser lines, the intensity of the SERS signal consistently increases as the gel collapses (at high temperature). The volume reduction of the pNIPAM shell produced when the temperature is increased drives the embedded

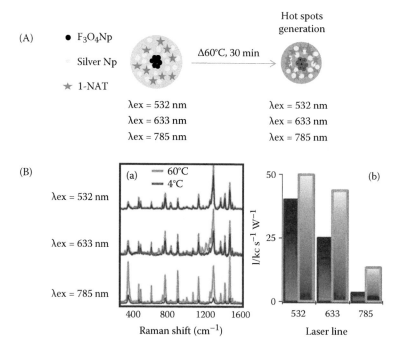

FIGURE 5.8 Schematic representation of the swelling–deswelling process used for SERS experiments (A). (Ba) SERS spectra of 1 NAT in Fe_3O_4@Ag@pNIPAM in the swollen (4°C black) and the collapsed (60°C, gray) states, for excitation laser lines (532, 633, and 785 nm). (Bb) Comparison of intensities of the band at 1368 cm^{-1} at low (black) and high (gray) temperature. (Adapted from Contreras-Caceres, R. et al., *Langmuir*, 27, 4520–4525, 2011. With permission.)

silver nanoparticles closer to each other, thus promoting the interaction between their respective electromagnetic fields and therefore further increasing the enhanced Raman signal owing to the formation of hot spots. This statement is also supported by the significant increase of the signal in all cases, but it is more pronounced when the laser energy is decreased (toward the IR), ranging from barely 1.2-fold in the case of the green line (532 nm) to over 4-fold in the case of the NIR line (785 nm) (see Figure 5.8Bb), which can be easily explained considering that the electromagnetic coupling between two or more metallic nanoparticles has been consistently reported to red-shift the corresponding LSPR, so that the overlap between LSPR and laser line is improved for longer wavelength excitations (Aldeanueva-Potel et al. 2009; Jain et al. 2007).

5.6 SERS ULTRADETECTION

For testing the ultrasensitive power of the material, two different experiments were devised. In the first experiment, aliquots of the composite colloid with concentrations of 1NAT ranging from 10^{-5} to 10^{-8} M were prepared at 4°C. After 2 h, 10 µL of each sample was cast onto a glass slide and air-dried (leading to microgel collapse, similar to the effect of temperature increase) (Figure 5.9a). In the second experiment, we diluted 20-fold the Fe_3O_4@Ag@pNIPAM colloid and then mixed 1-mL aliquots of the dilute colloid with 1NAT concentrations ranging from 10^{-8} to 10^{-13} M. After 2 h, the magnetic particles were collected at the wall of the vial with a permanent magnet (110 mT) and an iron nail so that the magnetic particles are concentrated in a small spot. Carefully, 10 µL of the concentrated spot was cast and air-dried prior to SERS measurements (Figure 5.9b).

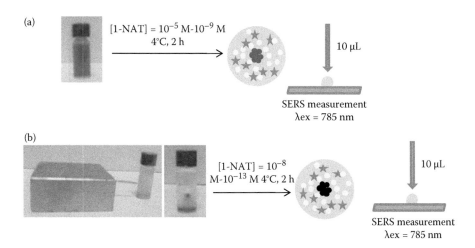

FIGURE 5.9 Schematic representation of the protocol used in SERS ultradetection in both cases (a) without the application of an external magnetic field and (b) applying an external magnetic field to concentrate the nanohybrid system. (Adapted from Contreras-Caceres, R. et al., *Langmuir*, 27, 4520–4525, 2011. With permission.)

Whereas the detection limit determined for 1NAT with the original composite microgel was around 10^{-8} M, dilution of the Fe_3O_4@Ag@pNIPAM dispersion and subsequent concentration into a small spot with a permanent magnet readily allowed us to decrease the detection limit down to 10^{-12} M, that is, by four orders of magnitude (Figure 5.10a). Such an improvement is mainly due to a more efficient use of the sensing material; a decrease in the amount of adsorbent (plasmonic material) allows in turn a decrease in the amount of absorbate (analyte), thereby reaching a sufficient level to be observed in SERS (Yang et al. 2005). Additionally, the magnetic accumulation of the composite microgel particles effectively increases the amount of material that is actually sampled by the laser beam, thus increasing the amount of scattering centers and the signal reaching the detector.

A final experiment was carried out to test the applicability of the prepared material for real applications, involving the first reported SERS ultrasensitive detection of

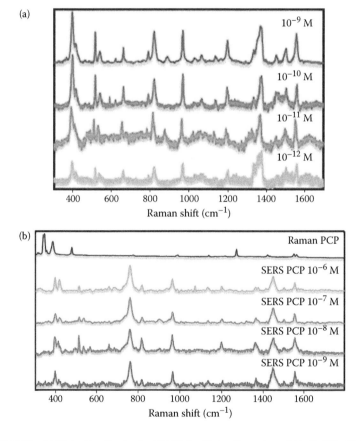

FIGURE 5.10 (a) Detection limits for 1NAT in dilute dispersion of Fe_3O_4@Ag@pNIPAM after concentration of the material using a permanent magnet. (b) SERS ultradetection of PCP in dilute dispersions of Fe_3O_4@Ag@pNIPAM after concentration of the material in a spot using a permanent magnet. Excitation laser line: 785 nm. (Adapted from Contreras-Caceres, R. et al., *Langmuir*, 27, 4520–4525, 2011. With permission.)

PCP. On the basis of the second experiment, we decided to use dilute dispersions of Fe_3O_4@Ag@pNIPAM, which are added to a certain volume of PCP molecules with concentrations ranging from 10^{-5} down to 10^{-12} M, at low temperature (4°C). The temperature was then raised to 60°C to collapse the microgel, inducing molecular trapping, and finally concentrated with a magnet in a small spot, where it was analyzed with the Raman spectrometer. The SERS spectrum recorded matched band to band with the Raman spectrum of pure PCP, but with different relative intensities. The enhanced vibrational spectrum is dominated by strong peaks at 1154 and 1448 cm^{-1} (ring stretching), 963 cm^{-1} (ring breathing and C-Cl stretching), 761 cm^{-1} (C-Cl stretching), 419 cm^{-1} (OH out-of-plane deformation), and 396 cm^{-1} (out-of-plane ring deformation) (Pawlukoj et al. 2001), which are perfectly recognized down to the nanomolar regime (Figure 5.10b). This sensitivity illustrates the capability of SERS to achieve the low detection limits required for PCP (1 ppb; 1 mg L^{-1}), as mandated by the Environmental Protection Agency (United States, Environmental Protection Agency 2007), and demonstrates that SERS can be used as a general ultrasensitive analytical technique, through careful design and implementation of composite-enhancing substrates.

ACKNOWLEDGMENTS

This work has been funded by the Spanish Ministerio de Economía y Competitividad/ FEDER (project MAT2011-28385), Andalusian Government/FEDER (Project P010-FQM 06104), and EU-COST-Action D43.

REFERENCES

Agrawal, M., Rubio-Retama, J., Zafeiropoulos, N. E., Gaponik, N., Gupta, S., Cimrova, V., Lesnyak, V., Lopez-Cabarcos, E., Tzavalas, S., Rojas-Reyna, R., Eychmueller, A., Stamm, M. 2008. Switchable photoluminescence of CdTe nanocrystals by temperature-responsive microgels. *Langmuir* 24:9820–4.

Aldeanueva-Potel, P., Faoucher, E., Alvarez-Puebla, R. A., Liz-Marzan, L. M., Brust, M. 2009. Recyclable molecular trapping and SERS detection in silver-loaded agarose gels with dynamic hot spots. *Anal. Chem.* 81:9233–8.

Alvarez-Puebla, R. A., Zubarev, E. R., Kotov, N. A., Liz-Marzan, L. M. 2012. Self-assembled nanorod supercrystals for the ultrasensitive SERS diagnostics. *Nano Today* 7:6–9.

Álvarez-Puebla, R. A., Contreras-Caceres, R., Pastoriza-Santos, I., Perez-Juste, J., Liz-Marzan, L. M. 2009. Au@pNIPAM colloids as molecular traps for surface-enhanced, spectroscopic, ultra-sensitive analysis. *Angew. Chem. Int. Ed.* 48:138–43.

Alvarez-Puebla, R. A., Aroca, R. F. 2009. Synthesis of silver nanoparticles with controllable surface charge and their application to surface-enhanced Raman scattering. *Anal. Chem.* 81:2280–5.

Bai, S., Wu, C., Gawlitza, K., Von Klitzing, R., Ansorge-Schumacher, M., Wang, D. 2010. Using hydrogel microparticles to transfer hydrophilic nanoparticles and enzymes to organic media via stepwise solvent exchange. *Langmuir* 26:12980–7.

Bantz, K. C., Haymes, C. L. 2009. Surface-enhanced Raman scattering detection and discrimination of polychlorinated biphenyls. *Vib. Spectrosc.* 50:29–35.

Carregal-Romero, S., Buurma, N. B., Perez-Juste, J., Liz-Marzan, L. M., Herves, P. 2010. Catalysis by Au@pNIPAM nanocomposites: effect of the cross-linking density. *Chem. Mater.* 22:3051–9.

Contreras-Caceres, R., Abalde-Cela, S., Guardia-Giros, P., Fernandez-Barbero, A., Perez-Juste, J., Alvarez-Puebla, R. A., Liz-Marzan, L. M. 2011. Multifunctional microgel magnetic/optical traps for SERS ultradetection. *Langmuir* 27:4520–5.

Contreras-Caceres, R., Pacifico, J., Pastoriza-Santos, I., Perez-Juste, J., Fernandez-Barbero, A., Liz-Marzan, L. M. 2009. Au@pNIPAM thermosensitive nanostructures: control over shell cross-linking, overall dimensions, and core growth. *Adv. Funct. Mater.* 19:3070–6.

Contreras-Caceres, R., Pastoriza-Santos I., Alvarez-Puebla, R. A., Perez-Juste, J., Fernandez-Barbero, A., Liz-Marzan, L. M. 2010. Growing Au/Ag nanoparticles within microgel colloids for improved surface-enhanced Raman scattering detection. *Chem. A Eur. J.* 16:9462–7.

Dagallier, C., Dietsch, H., Schurtenberger, P., Scheffold, F. 2010. Thermoresponsive hybrid microgel particles with intrinsic optical and magnetic anisotropy. *Soft Matter* 6:2174–7.

Fernandez-Barbero, A., Fernandez-Nieves, A., Grillo, I., Lopez-Cabarcos, E. 2002. Structural modifications in the swelling of inhomogeneous microgels by light and neutron scattering. *Phys. Rev. E* 66:51803–10.

Fernandez-Barbero, A., Suarez, I., Sierra-Martin, B., Fernandez-Nieves, A., De las Nieves, F. J., Marquez, M., Rubio-Retama, J., Lopez-Cabarcos, E. 2009. Gels and microgels for nanotechnological applications. *J. Adv. Colloid Interface Sci.* 147:88–108.

Guardia, P., Perez, N., Labarta, A., Batlle, X. 2010a. Controlled synthesis of iron oxide nanoparticles over a wide size range. *Langmuir* 26:5843–7.

Guardia, P., Perez-Juste, J., Labarta, A., Batlle, X., Liz-Marzan, L. M. 2010b. Heating rate influence on the synthesis of iron oxide nanoparticles: the case of decanoic acid. *Chem. Commun.* 46:6108–10.

Guerrini, L., Garcia-Ramos, J. V., Domingo, C., Sanchez-Cortes, S. 2006. Functionalization of Ag nanoparticles with dithiocarbamate calix[4]arene as an effective supramolecular host for the surface-enhanced Raman scattering detection of polycyclic aromatic hydrocarbons. *Langmuir* 22:10924–6.

Guerrini, L., Garcia-Ramos, J. V., Domingo, C., Sanchez-Cortes, S. 2009. Sensing polycyclic aromatic hydrocarbons with dithiocarbamate-functionalized Ag nanoparticles by surface-enhanced Raman scattering. *Anal. Chem.* 81:953–60.

Gupta, A. K., Gupta, M. 2005. Synthesis and surface engineering of iron oxide nanoparticles for biomedical applications. *Biomaterials* 26:3995–4021.

Hirotsu, S. 1994. Static and time-dependent properties of polymer gels around the volume phase transitions. *Phase Transitions* 47:183–240.

Jain, P. K., Huang, W., El-Sayed, M. A. 2007. On the universal scaling behavior of the distance decay of plasmon coupling in metal nanoparticle pairs: a plasmon ruler equation. *Nano Lett.* 7:2080–8.

Jaczewski, D., Tomczak, N., Han, M. Y., Vancso, G. 2009. Stimulus responsive PNIPAM/QD hybrid microspheres by copolymerization with surface engineered QDs. *Macromolecules* 42:1801–4.

Jones, C. L., Bantz, K. C., Haynes, C. L. 2009. Partition layer-modified substrates for reversible surface-enhanced Raman scattering detection of polycyclic aromatic hydrocarbons. *Anal. Bioanal. Chem.* 394:303–11.

Karg, M., Lu, Y., Carbo-Argibay, E., Pastoriza-Santos, I., Perez-Juste, J., Liz-Marzan, L. M. 2009. Multiresponsive hybrid colloids based on gold nanorods and poly(NIPAM-co-allylacetic acid) microgels: temperature- and pH-tunable plasmon resonance. *Langmuir* 25:3163–7.

Karg, M., Pastoriza-Santos, I., Liz-Marzan, L. M., Hellweg, T. 2006. A versatile approach for the preparation of thermosensitive PNIPAM core–shell microgels with nanoparticle cores. *ChemPhysChem* 7:2298–301.

Kawano, T., Niidome, Y., Mori, T., Katayama, Y., Niidome, T. 2009. PNIPAM gel-coated gold nanorods for targeted delivery responding to a near-infrared laser. *Bioconjugate Chem.* 20:209–12.

Kim, N. H., Lee, S. J., Moskovits, M. 2010. Aptamer-mediated surface-enhanced Raman spectroscopy intensity amplification. *Nano Lett.* 10:4181–5.

Kneipp, K. E., Kneipp, H., Itzkan, I., Dasari, R. R., Feld, M. S. 1999. Ultrasensitive chemical analysis by Raman spectroscopy. *Chem. Rev.* 99:2957–76.

Laurenti, M., Guardia, P., Contreras-Caceres, R., Perez-Juste, J., Fernandez-Barbero, A., Lopez-Cabarcos, E., Rubio-Retama, J. 2011. Synthesis of thermosensitive microgels with a tunable magnetic core. *Langmuir* 27:10484–91.

Lopez-Leon, T., Carvalho, E. L. S., Sijo, B., Ortega-Vinuesa, J. L., Bastos-Gonzalez, D. 2005. Physicochemical characterization of chitosan nanoparticles: electrokinetic and stability behavior. *J. Colloid Interface Sci.* 283:344–51.

Lu, Y., Mei, Y., Dreschsler, M., Ballauff, M. 2006. Thermosensitive core–shell particles as carriers for Ag nanoparticles: modulating the catalytic activity by a phase transition in networks. *Angew. Chem. Int. Ed.* 45:813–6.

Luo, B., Song, X. J., Zhang, F., Xia, A., Yang, W. L., Hu, J. H., Wang, C. C. 2010. Multifunctional thermosensitive composite microspheres with high magnetic susceptibility based on magnetite colloidal nanoparticle clusters. *Langmuir* 26:1674–9.

Madeiros, S. F., Santos, A. M., Fessi, M., Elaissari, A. 2011. Stimuli-responsive magnetic particles for biomedical applications. *Int. J. Pharm.* 403:139–61.

Massart, R. 1981. Preparation of aqueous magnetic liquids in alkaline and acidic media. *IEEE Trans. Magn.* 17:1247–8.

Moskovits, M. 2005. Surface-enhanced Raman spectroscopy: a brief retrospective. *J. Raman Spectrosc.* 36:485–96.

Murray, K. E., Thomas, S. M., Bodour, A. A. 2010. Prioritizing research for trace pollutants and emerging contaminants in the freshwater environment. *Environ. Pollut.* 158:3462–71.

Neetu, S., Lyon, A. L. 2007. Au nanoparticle templated synthesis of pNIPAm nanogels. *Chem. Mater.* 19:719–26.

Neumann, O., Zhang, D. M., Tam, F., Lal, S., Wittung-Statshede, P., Halas, N. J. 2009. Direct optical detection of aptamer conformational changes induced by target molecules. *Anal. Chem.* 81:10002–6.

Pankhurst, Q. A., Connolly, J., Jones, S. K., Dobson, J. 2003. Applications of magnetic nanoparticles in biomedicine. *J. Phys. D: Appl. Phys.* 36:R167–81.

Pawlukoj, A., Natkaniec, I., Majerz, I., Sobczyk, L. 2001. Inelastic neutron scattering studies on low frequency vibrations of pentachlorophenol. *Spectrochim. Acta* 57:2775–9.

Pelton, R. H., Chivante, P. 1986. Preparation of aqueous lattices with *N*-isopropylacrylamide. *Colloids Surf.* 20:247–56.

Pich, A., Bhattacharya, S., Lu, Y., Boyko, V., Adler, H. J. P. 2004. Temperature-sensitive hybrid microgels with magnetic properties. *Langmuir* 20:10706–11.

Porter, M. D., Lipert, R. J., Siperko, L. M., Wang, G., Narayanana, R. 2008. SERS as a bioassay platform: fundamentals, design, and applications. *Chem. Soc. Rev.* 37:1001–11.

Qiu, P., Jensen, C., Charity, N., Towner, R., Mao, C. 2010. Oil phase evaporation-induced self-assembly of hydrophobic nanoparticles into spherical clusters with controlled surface chemistry in an oil-in-water dispersion and comparison of behaviors of individual and clustered iron oxide nanoparticles. *J. Am. Chem. Soc.* 132:17724–32.

Rebodos, R. L., Vikesland, P. J. 2010. Effects of oxidation on the magnetization of nanoparticulate magnetite. *Langmuir* 26:16745–53.

Rosi, L. N., Mirkin, C. A. 2005. Nanostructures in biodiagnostics. *Chem. Rev.* 105:1547–62.

Rowe, M. D., Chang, C., Thamm, D. H., Kraft, S. L., Harmon, J. L., Vogt, P. A., Sumerlin, B. S., Boye, S. G. 2009. Tuning the magnetic resonance imaging properties of positive contrast agent nanoparticles by surface modification with RAFT polymers. *Langmuir* 25:9487–99.

Rubio-Retama, J., Zafeiropoulos, N. E., Frick, B., Seydel, T., Lopez-Cabarcos, E. 2010. Investigation of the relationship between hydrogen bonds and macroscopic properties in hybrid core-shell gamma-Fe_2O_3-P(NIPAM-AAS) microgels. *Langmuir* 26:7101–6.

Rubio-Retama, J., Zafeiropoulos, N. E., Serafinelli, C., Rojas-Reyna, R., Voit, B., Lopez-Cabarcos, E., Stamm, M. 2007. Synthesis and characterization of thermosensitive PNIPAM microgels covered with superparamagnetic gamma-Fe$_2$O$_3$ nanoparticles. *Langmuir* 23:10280–5.

Sanchez-Iglesias, A., Grzelczak, M., Rodriguez-Gonzalez, B., Guardia-Giros, P., Pastoriza-Santos, I., Perez-Juste, J., Prato, M., Liz-Marzan, L. M. 2009. Synthesis of multifunctional composite microgels via in situ Ni growth on pNIPAM-coated Au nanoparticles. *ACS Nano* 3:3184–90.

Sanles-Sobrido, M., Rodriguez-Lorenzo, L., Lorenzo-Abalde, S., Gonzalez-Fernandez, A., Correa-Duarte, M. A., Alvarez-Puebla, R. A., Liz-Marzan, L. M. 2009. Label-free SERS detection of relevant bioanalytes on silver-coated carbon nanotubes: the case of cocaine. *Nanoscale* 1:153–8.

Sauzedde, F., Elaissari, A., Pichot, A. 1999. Hydrophilic magnetic polymer latexes. 1. Adsorption of magnetic iron oxide nanoparticles onto various cationic latexes. *Colloid Polym. Sci.* 277:846–55.

Schachschal, S., Balaceanu, A., Melian, C., Demco, D., Eckert, T., Richtering, W., Pich, A. 2010. Polyampholyte microgels with anionic core and cationic shell. *Macromolecules* 43:4331–9.

Schmidt, S., Hellweg, T., Von Klitzing, R. 2008. Packing density control in P(NIPAM-co-AAc) microgel monolayers: effect of surface charge, pH, and preparation technique. *Langmuir* 24:12595–602.

Tepper, T., Ilievski, F., Ross, C. A., Zaman, T. R., Ram, R. J., Sung, S. Y., Stadler, B. J. H. 2003. Magneto-optical properties of iron oxide films. *J. Appl. Phys.* 93:6948–51.

United States Environmental Protection Agency. Available at http://www.epa.gov/ttn/atw/hlthef/pentachl.html. Last modified November 6, 2007.

Xu, H., Xu, J., Zhu, Z., Liu, H. 2006. In-situ formation of silver nanoparticles with tunable spatial distribution at the poly(*N*-isopropylacrylamide) corona of unimolecular micelles. *Macromolecules* 39:8451–5.

Yamamoto, K., Matsukuma, D., Nanasetani, K., Aoyagi, T. 2008. Effective surface modification by stimuli-responsive polymers onto the magnetite nanoparticles by layer-by-layer method. *Appl. Surf. Sci.* 255:384–7.

Yang, Z., Tseng, W.-L., Lin, Y.-W., Chang, H.-T. 2005. Impacts that pH and metal ion concentration have on the synthesis of bimetallic and trimetallic nanorods from gold seeds. *J. Mater. Chem.* 15:2450–4.

Yin, J., Li, C., Wang, D., Liu, S. 2010. FRET-derived ratiometric fluorescent K$^+$ sensors fabricated from thermoresponsive poly(*N*-isopropylacrylamide) microgels labeled with crown ether moieties. *J. Phys. Chem. B* 114:12213–20.

Zhang, F., Wang, C. C. 2009. Preparation of P(NIPAM-co-AA) microcontainers surface-anchored with magnetic nanoparticles. *Langmuir* 25:8255–62.

6 The Central Role of Interparticle Forces in Colloidal Processing of Ceramics

Davide Gardini, Carlo Baldisserri, and Carmen Galassi

CONTENTS

6.1 INTRODUCTION

Interparticle forces acting among colloidal particles suspended in a liquid medium determine the physical behavior of dispersions. Colloidal stability, rheology, electrokinetic phenomena, and sedimentation are all affected by the nature and extent of such forces. Interparticle interactions determine the physical behavior of paints, foods, pharmaceuticals, adhesives, printing inks, cosmetics, detergents, ceramics, and many other systems of industrial relevance. However, a comprehensive quantitative understanding of the observed behavior of all colloidal systems is at present hardly obtainable.

Currently available theoretical treatment is limited, on the one hand, to model systems in which the colloidal particles are spherical, monodispersed, and weakly interacting (diluted suspensions), while interparticle interaction by direct contact is usually dealt with using the *hard-sphere* model. The influence of surfactants on

131

interparticle interactions can be taken into account, but quantitative treatment is limited to particles coated by a single layer of a known surfactant.

Unfortunately, most available models are still short of being able to describe many colloidal systems of industrial interest. Therefore, further theoretical advances are crucial to the progress of colloidal science, and some effort should be aimed at the development of improved models. Referring to suspensions of ceramic particles (as well as other aforementioned systems), deviations from model predictions are due to several factors. The solid fraction in industrial colloidal systems is often quite high, and the particle-size distribution can be broad (polydispersion). In addition, particles are usually nonspherical. In order to obtain ceramic composites with specific properties or to satisfy processing requirements, two or more powders are often mixed together. Moreover, industrial suspensions can contain a wide spectrum of additives (inorganic salts, surfactants, neutral polymers, polyelectrolytes, etc.), added with the aim of imparting particular properties to the systems. Even though the currently available theoretical knowledge is hardly able to quantitatively describe all systems, it can provide, in some cases, valuable qualitative suggestions as to the direction to follow to solve the scientific and technological problems encountered in such complex systems. Researchers and scientists involved in the colloidal processing of ceramics are well aware that the control of interparticle forces is crucial to the correct design of materials and processes, and a lot of effort has been put in place to fully exploit the existing theoretical background. In particular, it is worth mentioning here two extensions of the well-known DLVO theory to include the cases of sterically polymer-stabilized systems (Napper 1983) and heterogeneous systems containing mutually attracting dissimilar inorganic particles (Hogg et al. 1966).

In this chapter, some literature examples will be considered to illustrate the application of such concepts to the control of interparticle interactions in ceramic suspensions. In most applications, the main reason for such a control is to avoid the formation of hard agglomerates, because these cause defects to be present in the sintered bodies, with consequent risk of cracking.

Both water-based suspensions and suspensions in organic solvents will be considered in the following paragraphs. Although literature studies are available on both types of suspensions, full understanding of the basic mechanisms of colloidal stabilization is still lacking. Cases in point are the electrosteric stabilization of aqueous suspensions by polyelectrolytes and the stabilization of organic solvent-based suspensions by surfactants. In advanced shaping techniques like direct coagulation casting (DCC) or temperature-induced gelation (TIG), the creation of gels is a required processing step, in which case interparticle interactions are specifically driven toward the formation of three-dimensional structures by controlling either pH or temperature, respectively. One of the commonly used techniques to obtain indirect information about the degree of structuring of these complex systems is rotational rheometry, whose main features will be briefly described.

6.2 COLLOIDAL PROCESSING OF CERAMICS

The world's oldest ceramic object ever found (the *Venus of Dolni Věstonice*, in modern-day Czech Republic) dates back to 26,000 years ago (Vandiver et al. 1989).

As such, it is at least 10,000 years older than the ancient pottery found in China (Boaretto et al. 2009) and Japan (Rice 1999).

All ancient ceramic items were made by dispersing clay in water and hand forming the resulting plastic bodies. Firing was carried out at relatively high temperatures (600°C–900°C). In our days, traditional ceramics like tableware, whiteware, and sanitaryware, as well as most advanced ceramics, are still made by processing ceramic particles in a similar way. The three-step process (dispersion, shaping, and firing) constitute the classic *ceramic processing*. In the dispersion stage, the ceramic powders are dispersed in a liquid medium, leading to the creation of a colloidal suspension due to ceramic powders being characterized by a particle-size distribution that includes significant submicrometric fractions. In the subsequent shaping stage, the suspensions are poured in molds and—after drying to eliminate the liquid medium—the ceramic powder ends up as a compact solid object of the desired shape, called *green body* (cold consolidation).

The final stage, that is, firing at high temperatures (from 800°C up to 2000°C, depending on material), is often preceded by a preheating step at intermediate temperatures to eliminate the organic additives present in the material. Because of the creation of strong bonds between the particles (*sintering*), the firing step densifies the cold consolidated body into the so-called *sintered body*. In the latter, ceramic particles are arranged in microstructures that impart the object a higher mechanical resistance and the desired functional properties. For many applications, the density of the green bodies obtained after the shaping stage and the subsequent drying to eliminate the liquid phase must be as high as possible, in such a way that, after sintering, nearly full density objects are obtained. For the purposes of our discussion, the dispersion stage is the most interesting one, because it is during the preparation of the fluid system that proper tailoring of the properties of the suspension is possible in view of the subsequent shaping and firing stages. In particular, it is in that stage that it is possible to control the interparticle interactions to avoid or minimize the formation of hard agglomerates and to eliminate large agglomerates by sedimentation or filtration techniques (Lange 1989).

The dispersion stage, along with all treatments performed to obtain homogeneous suspensions free from large and hard agglomerates, constitutes the *colloidal processing of ceramics* (Sigmund et al. 2000). This point is of particular relevance for ceramics, because the shaping methods used to form such materials are not based on the casting of molten materials like in the case of metallic and polymeric objects. Because of this fact, the final ceramic products "remember" the size, shape, and orientation of the starting particles, which, as a consequence, significantly affect the properties of the sintered bodies.

The possibility of controlling interparticle interactions is the main advantage of colloidal processing with respect to dry powders processing. Although in ancient time such awareness was obviously neither present nor required, the mere fact that formulation adjustments (like the addition of water or clay) could be done to get the desired object allows placing the birth of colloidal processing of ceramics to thousands of years ago.

Very likely, in ancient times, the control of the process to obtain reproducible and good quality products was virtually nonexistent. In the middle of the 19th century, the advent of the modern industrial era and the need to produce an ever-increasing number of pieces displaying constant performance made the reliability of the manufacturing process paramount. In order to satisfy such need, ceramic process engineers and technicians began to correlate the characteristic of the final products (as to mechanical, thermal, chemical, and electrical properties) to either the properties of the raw starting materials or the operative conditions of the manufacturing process.

In this way, and with reference to specific systems, some empirical process/property correlations were found. Such correlations, however, were not extendible to different systems and did not allow the prediction of the final properties of the product should some variables be changed, even within the same system. That is, such empirical correlations did not allow the controlled design of the properties of ceramics. The empirical approach lasted essentially unchanged up to the middle of the 20th century (Reed 1995). When one takes into account the complexity of the ceramic systems and the large number of variables involved in the process, this state of affairs is hardly surprising. Moreover, in past years, available characterization techniques for raw materials were undeveloped, as well as effective means of controlling process variables. Even more critical, there was no shared awareness of how materials properties and process parameters so greatly affect the final properties of ceramics.

Nowadays, such awareness is widespread among operators in the ceramic field. Notwithstanding, in ceramic factories, the trial-and-error approach remains in use in most situations to these days. While the control of the whole ceramic process on the basis of empirical correlations is not easy to achieve, such control would not be easily obtained even with our advanced knowledge on interparticle and additives/ceramic particles interactions. For example, our present-day knowledge of the interactions between ceramic particles and polyelectrolytes—the most common type of polymeric deflocculant used for the stabilization of aqueous suspensions—is yet to be completed, even though some early models were proposed starting from the 1970s (Hesselink 1977).

Despite difficulties, significant progress in the fundamental understanding of colloidal suspensions was made in the last 50 years in many directions. The increased availability of monodispersed colloidal dispersions to be tested as model systems and the possibility of measuring particle size and forces between particles at close range (down to a few nanometers), together with advances in the formal description of the fluid dynamics around the particles, have all been contributing factors toward a comprehensive understanding of the behavior of "simple" systems (Russel et al. 1989).

The fast development of computer simulation strategies allowed the numerical solution of complex fluid dynamics problems, thus providing graphical representations of the agglomeration mechanisms. Along with such scientific and technological advances, another key factor has contributed to improve the understanding and control of the ceramic process, that is, the consideration of the latter from an engineering standpoint. Such approach tries to describe the ceramic process as a sequence of rationalized unit operations whereby the matter undergoes physicochemical transformations. In particular, the dispersion of the ceramic powders in a liquid should not to be considered simply as a physical mixing of two different phases, but rather the stage in which an engineered two-phase system is created by adding inorganic or organic additives in proper amounts, so as to modify the interactions among the particles and to produce suspensions with particular properties.

6.3   PARTICLES AGGLOMERATION

In general, the main goal of the dispersion step is to obtain homogeneous, agglomerate-free, and low-viscosity suspensions. The latter requirement is to make the flow of the suspension easier and facilitate the shaping stage, while homogeneity allows having the same properties at each point of the sample. The presence of agglomerates, which tend to form spontaneously because of the ever-present van der Waals attractive forces, must be avoided for several reasons. The formation of the so-called *hard agglomerates*, that is, clusters of particles that cannot be removed by usual means (e.g., by sonication), should especially be avoided, because they give some undesired effects. The first effect is that they increase the viscosity of the suspension, thus complicating the shaping step when the suspension must fill molds having complex geometries and narrow features. The second effect is that they decrease the density of both green and sintered bodies, thus worsening the final properties. For example, the presence of agglomerates in a powder of yttria-stabilized zirconia decreases the density of the sintered bodies. The agglomerate-free powders can be sintered with high density at a much lower temperatures; that is, in the presence of agglomerates, the sintering temperature must be increased to obtain the same sintered density (Barsoum 2003; Rhodes 1981).

The third and more critical disadvantage is that the different thermal expansion coefficients of agglomerates with respect to the surrounding ceramic matrix could generate microcracks during the sintering step, with the risk of rupturing or deforming the pieces. While microstructural heterogeneities like inclusions, large grains, porosity, or surface cracks from machining can be eliminated using purer raw materials or by controlling more carefully the subsequent powder treatments (for inclusions), by changing the sintering conditions (for large grains and porosity), or by producing near-net shape objects that do not require strong surface machining (for machining surface cracks), the presence of agglomerates cannot be avoided by any of the previous solutions (Lange 1989). For such reason, controlling the agglomeration phenomena becomes important.

Complete elimination of agglomeration is very difficult to obtain, yet agglomeration phenomena should be minimized as much as possible. In particular, the formation of hard agglomerates must be prevented. The colloidal stabilization of industrial ceramic suspensions seldom yields full-fledged hard-sphere systems (no significant interparticle interaction unless particles come into contact) or soft-sphere systems (repulsive interparticle interactions at any distance, well before approaching the deep attractive minimum at close range). However, it is possible to obtain weakly flocculated systems in which attractive interactions slightly dominate up to some interparticle distance, at which repulsive and attractive interactions become comparable and a repulsive energetic barrier is created. Although this energetic barrier prevents the particles from falling in the deep attractive primary minimum, it allows the creation of *soft agglomerates*, that is, agglomerates that can be fragmented by applying low-intensity viscous forces (Lewis 2000). If the suspension is electrostatically stabilized, the microstructural state made of soft agglomerates corresponds to the shallow secondary minimum of the potential energy of the well-known DLVO theory.

The van der Waals attractive energy potential between two spheres of radius a depends on the distance h between the two particles' surfaces according to the following equation:

$$V_{vdW}(h) = -\frac{A}{6}\left(\frac{2a^2}{h^2 + 4ah} + \frac{2a^2}{h^2 + 4ah + 4a^2} + \ln\frac{h^2 + 4ah}{h^2 + 4ah + 4a^2} \right). \tag{6.1}$$

In Equation 6.1, A is the Hamaker constant that gives the overall magnitude of the interaction and depends on the physicochemical nature (dielectric constants or refractive indexes) of both the interacting particles and the liquid between them. Since the distance between the particles can be expressed in other ways (e.g., as the distance D between the centers of the spheres, so that $D = 2a + h$), slightly different but equivalent forms of Equation 6.1 can be found in literature. The same caveat applies to all the equations in the following in which the interparticle distance appears. At small interparticle separation ($h \ll 2a$) Equation 6.1 can be approximated by the following one:

$$V_{vdW}(h) = -\frac{Aa}{12h}. \tag{6.2}$$

The control of the state of agglomeration is made more critical by the fact that, in most cases, industrial ceramic suspensions are concentrated; that is, the solid fraction is high. This is due to the fact that, in many applications, fully dense ceramics are required. The high solid content brings about the first important complication, because in such conditions, the magnitude of the van der Waals attractive force becomes much larger, owing to smaller interparticle distance. Moreover, for many years, in the ceramic as well as in other industrial sectors, an increasing trend is seen toward using ever-smaller particles (down to a few nanometers). While using fine powders results in increased product performance and/or lower sintering temperature, it makes it more difficult to control the interparticle interactions. In suspensions of nanopowders, at the same volumetric fraction of solid particles in the suspension, the interparticle distances are reduced, and the surface area is increased with respect to suspensions of coarser powders (Figure 6.1).

With small particles (below 1 μm), surface phenomena become largely predominant on volumetric ones. The number of reactive surface sites increases, and it becomes increasingly more difficult to saturate them in order to control the colloidal interactions.

The colloidal size of the particles also significantly affects the viscosity of the suspension and, in turn, the easy processing of suspension during shaping or application.

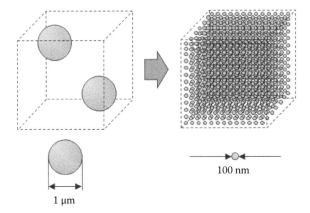

FIGURE 6.1   By decreasing the size of a particle by one order of magnitude (e.g., from 1 μm to 100 nm), a two-orders-of-magnitude decrease of the particles' surface area is obtained. However, since the number of the so-obtained finer particles increases by three orders of magnitude, the total surface area increases by one order of magnitude. A sharp decrease of the interparticle distance also occurs.

The viscosity can increase up to a hundred times with respect to the viscosity of the dispersing liquid, even for diluted suspension containing a solid volumetric fraction lower than 0.05, as was shown for carbon black in mineral oil (Macosko 1994; Mewis and Spaull 1976). Therefore, a suspension that should be diluted in terms of solid content can behave like a concentrated suspension, because the size of the particle is in the colloidal range. In this view, the "concentration" of a suspension is determined by both the nominal volume fraction and the particle size of the solids.

When the particles become sufficiently small, the effect of the sedimentation is reduced, while the Brownian motion is enhanced. For example, if the particle size decreases by an order of magnitude, the velocity of sedimentation decreases by two orders of magnitude, as predicted by Stokes' law for the sedimentation of an isolated spherical particle of radius a (Equation 6.3):

$$v_s = \frac{2}{9} \frac{(\rho_p - \rho_L) g a^2}{\eta_L}.$$
(6.3)

On the other hand, according to the Stokes–Einstein equation for the translational diffusion coefficient D_T (Equation 6.4), the Brownian motion (due to the collisions among the liquid molecules in thermal agitation and the particles) increases by an order of magnitude:

$$D_T = \frac{kT}{6\pi \eta_L a}.$$
(6.4)

In Equations 6.3 and 6.4, ρ_P and ρ_L are the densities of the particle and the liquid, respectively, η_L is the viscosity of the liquid, g is the gravity acceleration, k is the Boltzmann constant, and T is the absolute temperature. Taking into account the previous considerations, the forces that play the main role in a colloidal suspension are the attractive ones (in that they tend to agglomerate the particles) and the forces due to the Brownian motion (in that they increase the probability of collision among the particles), while the gravitational force (partially compensated by the buoyancy force) is less important. If the suspension is flowing, particles and agglomerates also experience viscous forces due to the friction with the liquid, which tend to drag the particles and break down the agglomerates in smaller units. Viscous forces are also present in suspensions at rest, owing to the relative motion between solid and liquid phases from gravity and Brownian motion, but these are usually too weak to break down the agglomerates.

At the solid/liquid interface between the ceramic particles and the surrounding liquid, a charge separation spontaneously occurs because of the absorption or desorption of ions. Such a phenomenon creates an electrical double layer (EDL) at the interface, made by the charges on the solid surface and by the electrostatically attracted ions of opposite charge (*counterions*) in the solution, which form an ionic cloud around the particle. The counterions screen the electrostatic surface potential, which therefore becomes vanishingly small at some distance from the solid surface. The value of the electrostatic potential at the (somewhat undefined) shear plane is defined as the *zeta potential*, which is a measurable quantity.

The higher the zeta potential is, the stronger is the electrostatic repulsion between the approaching particles, when their ionic clouds start to overlap. Unfortunately, the electrostatic repulsion, which is naturally present in aqueous suspensions, is not usually strong enough to overcome the attractive forces and to avoid the formation of hard agglomerates. Therefore, in order to avoid coagulation, it is necessary to artificially introduce in the system some repulsive forces that are strong enough to avoid such effect. Typically, the methods used to stabilize a colloidal suspension are either by increasing the electrostatic repulsion between the particles or creating a steric repulsion by adsorbing organic molecules on the particles' surface. The two approaches can be applied to both aqueous suspensions and suspensions in organic solvents, even though the nature of the additives (referred to as *deflocculants* or *dispersants*) and the mechanisms of stabilization are generally different. Adding a deflocculant to stabilize a suspension is also done in order to approach the hard-sphere or soft-sphere systems used as paradigms. However, unlike model systems, in suspensions of industrial interest, other additives are often present. This fact, together with the high content of solids and the submicrometer size of particles, complicates the control of interparticle interactions. Additives are added to impart useful properties to the suspensions. The liquid medium is chosen according to its ability to dissolve the additives and, for this reason, it is sometimes referred to as *solvent*.

In addition to *deflocculants*, added to colloidally stabilize the suspensions, *binders* and other additives are often present. Binders are added to link the solid particles by creating networks, and so improve the mechanical strength of the green bodies for subsequent handling. *Plasticizers* are added to impart the green bodies' flexibility, which they do by lowering the glass transition temperature of the binders. *Wetting agents* increase the wetting of particles by the liquid medium, while *antifoaming agents* prevent foam formation during the preparation of suspensions. Depending on the application, other substances (*biocides, flocculants, porous agents, lubricants*, etc.) may be added.

Even if each additive is added with a specific aim, once present in the system, it will interact with all the other components, that is, solvent, solid particles, and any other additive. In some cases, competitive interaction causes detrimental effects. For example, polyvinyl butyral (PVB), which is usually employed as a binder in the preparation of tape casting suspensions, can be adsorbed on the particle surface, thus interfering with the dispersant. Depending on the concentration and molecular weight of the polymer, PVB can act as a dispersant rather than as a binder (Bhattacharjee et al. 1993). In this view, a proper selection of the additives is fundamental, as is the order in which additions are made. For example, it was reported that in the preparation of lead zirconate titanate (PZT) screen printing inks, it is better to dissolve the deflocculant (phosphate ester) in a fugitive agent (methyl ethyl ketone, MEK) before adding it to the powder with respect to other sequences (Thiele et al. 2000). Moreover, although such additives are present in relatively small amounts, a slight change of their concentration can have dramatic effects on the macroscopic behavior of the system. A right choice of the nature, amount, and number of additives is fundamental for controlling interparticle interactions and tailor the properties of suspensions.

As an example, we list here some typical additives used for the preparation of suspensions to be used in tape casting, a technology created in the 1940s to manufacture large-area ceramic sheets, typically between 100 and 400 μm thick. Historically, such suspensions were prepared in organic solvents or, more frequently, in binary mixtures of organic solvents for enhanced dissolution of different additives. Commonly used solvent mixtures were MEK/ethanol, MEK/toluene, trichloroethylene (TCE)/ethanol, or TCE/acetone. Azeotropic mixtures were employed to avoid differential volatilization of the solvents.

For colloidal stabilization in such nonaqueous media, surfactants like fatty acids (oleic, stearic, or natural mixtures like the Menhaden fish oil), phosphate ester, glycerol trioleate, or sulfosuccinates have been used as deflocculants. PVB and polymethyl methacrylate (PMMA) were extensively used as binders. Finally, polyethylene glycol (PEG) or benzyl butyl phthalate (BBP) found application as plasticizers (Mistler and Twiname 2000; Moreno 1992a,b).

Heeding increasing health and environment concerns, eco-friendly water-based suspensions for tape casting were studied, starting from the 1990s. For such suspensions, typical deflocculants are sodium carboxymethyl cellulose or ammonium salts of polyacrylic acids (PAAs). Suitable binders for water-based formulations are cellulose ethers, polyvinyl alcohols, or polyvinyl acetate, while glycerol, BBP, PEG, or polypropylene glycol can be used as plasticizers (Hotza and Greil 1995). Addition of organic molecules requires a preliminary screening of their mutual interactions and of their interaction with the solid particle. Proper amounts of each additive must be determined so that each additive can perform its own function without interfering with the others.

Other interparticle forces, the so-called *non-DLVO forces*, may play a role in colloidal stabilization (Israelachvili 1992). For example, when the interparticle distance becomes small, the finite size of the molecules of the dispersing liquid could play a role (*structural forces*). In that case, the interaction energy can show an oscillating behavior, where the minima correspond to distances that are multiple of the molecular diameter. A good arrangement of the liquid molecules can be obtained at these minima. If the surface orients the solvated molecules of the liquid in a preferred direction, the energy of interaction could be affected (*solvation forces*). Forces acting between two hydrophobic surfaces may also give a significant contribution to the interaction energy (*hydrophobic forces*) (Ducker et al. 1994; Horn 1990).

6.4 COLLOIDAL STABILIZATION

As emphasized in the previous paragraphs, the colloidal stabilization of ceramic suspensions is a key step for obtaining flawless ceramics, which can be reached by creating proper combinations of electrostatic and steric repulsive interactions by introducing appropriate amounts of selected additives.

Electrostatic stabilization. In the case of spherical particles of radius a, if the inverse $1/\kappa$ of the Debye–Hückel parameter κ is used as a measure of the thickness of the EDL, the physical situation of thin double layer is defined by the condition $\kappa a > 10$. In this case, the electrostatic repulsion between spherical particles is described by the Derjaguin equation (Equation 6.5):

$$V_{electr}(h) = 2\varepsilon_r \varepsilon_0 a \psi_0^2 \ln[1 + \exp(-\kappa h)], \qquad (6.5)$$

in which V_{electr} is the electrostatic repulsive energy potential, ε_r is the dielectric constant (or relative permittivity) of the liquid, ε_0 is the permittivity of vacuum, and ψ_0 is the surface potential. The Debye–Hückel parameter κ is given by

$$\kappa = \sqrt{\frac{2F^2 I}{\varepsilon_r \varepsilon_0 RT}}, \qquad (6.6)$$

with

$$I = \frac{1}{2} \sum_i c_{i,\infty} z_i^2, \qquad (6.7)$$

in which F is the Faraday constant, R is the universal constant of gases, T is the absolute temperature, I is the ionic strength of the liquid solution, and $c_{i,\infty}$ and z_i are the bulk molar concentration and valence of the ith ion in the solution, respectively.

According to this model, in order to increase the electrostatic repulsion, one can increase the surface charge on the particles (i.e., the surface potential ψ_0) or decrease the ionic strength I of the liquid solution. The surface potential is often approximated with the zeta potential, as the latter is a measurable property. Metal oxide particles in water have hydroxyl groups on the surface, so that their surface charge is very sensitive to pH changes. A transition from acidic to basic conditions causes metal oxides to reverse the sign of their surface charge from positive to negative. The pH at which this inversion occurs is called *point of zero charge* (p.z.c.), which coincides with the *isoelectric point* (i.e.p.), that is, the pH at which the zeta potential is zero, and no electrostatic repulsion is present unless specific ion adsorption occurs. Therefore, in order to maximize the electrostatic repulsion between the particles, the operational pH should be far from the i.e.p., and this is the main technique used in the aqueous ceramic suspensions to electrostatically stabilize them.

In the case of clay particles, the charge separation at the solid/liquid interface can occur also by isomorphic substitution in the crystal lattice; that is, lower valence cations (Mg^{2+} or K^+) can replace lattice cations (Si^{4+} or Al^{3+}). The facets of the clay platelets are always negatively charged, whereas the sign of the charge on the edges can change with pH. Edges are negatively charged in basic environments (pH > 9) and positively charged in acidic environments. Because of the lamellar shape of the clay particles, agglomerates (the so-called *house of cards*) can be formed in acidic environments owing to the electrostatic attraction between faces and edges, while electrostatic repulsion can be obtained at high pH values, as the charges on both edges and surfaces have the same sign. In order to decrease the ionic strength, it is possible to replace the interlayer bivalent cations (Ca^{2+}, Mg^{2+}) with monovalent cations as Na^+, by adding inorganic additives like sodium carbonate, sodium silicate, or a mixture of the two.

The DLVO theory for lyophobic colloids (Derjaguin and Landau 1941; Verwey and Overbeek 1948) describes the interaction potential V_T as a function of the distance h between two solid particles dispersed in a liquid when the attractive van der Waals forces are hindered by the electrostatic repulsion owing to the interaction between the EDLs:

$$V_T(h) = V_{vdW}(h) + V_{electr}(h). \tag{6.8}$$

The potential–distance curve typically displays a deep primary minimum at very small distances, and a shallow secondary minimum at higher distance, separated by a repulsive energy barrier whose height can be increased by decreasing the ionic strength (Figure 6.2).

Steric stabilization. Another very effective way for stabilizing a colloidal suspension consists in the addition of organic molecules that can be physically adsorbed or grafted onto the surface of the solid particles. A thick layer is thus created around the particles, which prevents the particles from coming in close contact with each other. Organic molecules suited to the purpose can be surfactants (especially for solvent-based suspensions), neutral polymers, or polyelectrolytes (aqueous-based suspensions). In order to be effective, the polymer–solvent interaction should be energetically favored (*good solvent* condition, Flory–Huggins parameter χ lower than 0.5). The polymer should form thick and dense layers and cover the whole

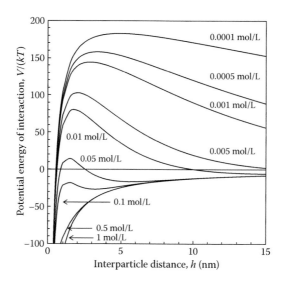

FIGURE 6.2 Potential energy of interaction for electrostatically stabilized suspensions as a function of electrolyte concentration. Equations 6.2 and 6.5 were used for van der Waals attraction and electrostatic repulsion, respectively. Parameters: $a = 0.1$ μm, $\psi_0 = 100$ mV, $A = \left(\sqrt{A_P} - \sqrt{A_L}\right)^2$ with $A_P = 2 \times 10^{-19}$ J (for particle) and $A_L = 4 \times 10^{-20}$ J (for liquid), $\varepsilon_r = 78.5$, $T = 25°C$, and $z = 1$.

surface of the particles. In addition, it should be sufficiently pinned to the surfaces to avoid desorption when collisions occur. Assuming a constant radial density for the polymer in the adsorbed layer of thickness δ—which is quite a good approximation for thin layers with respect to the particle size, $\delta \ll a$—the steric potential in the interpenetrating region ($\delta < h < 2\delta$) is due to a mixing (or osmotic) contribution given by Equation 6.9 (Napper 1983):

$$V_{\text{steric,mix}}(h) = \frac{4\pi akT}{\bar{V}_{\text{sol}}} \phi_{\text{pol}}^2 \left(\frac{1}{2} - \chi\right) \delta^2 \left(1 - \frac{h}{2\delta}\right)^2. \tag{6.9}$$

In Equation 6.9, $\bar{V}_{\text{sol}}$ is the partial molecular volume of the solvent and ϕ_{pol} is the volume fraction of the polymer in the adsorbed layer. When the distance between the surfaces of the particles becomes smaller than the thickness of the layer ($h < \delta$), the mixing contribution assumes a different form, and an elastic (or volume restriction) contribution arises (Fritz et al. 2002):

$$V_{\text{steric,mix}}(h) = \frac{4\pi akT}{\bar{V}_{\text{sol}}} \phi_{\text{pol}}^2 \left(\frac{1}{2} - \chi\right) \delta^2 \left(\frac{h}{2\delta} - \frac{1}{4} - \ln\frac{h}{\delta}\right), \tag{6.10}$$

$$V_{\text{steric,el}}(h) = \frac{2\pi akT}{M_{\text{pol}}} \phi_{\text{pol}} \rho_{\text{pol}} \delta^2 \left\{ \frac{h}{\delta} \ln\left[\frac{h}{\delta}\left(\frac{3 - h/\delta}{2} \right)^2 \right] \right.$$

$$\left. -6\ln\left(\frac{3 - h/\delta}{2} \right) + 3\left(1 + \frac{h}{\delta} \right) \right\}, \tag{6.11}$$

in which M_{pol} and ρ_{pol} are the molecular weight and the density of the polymer, respectively. In the case of purely steric stabilization, the total energy potential is given, in an analogous way to the DLVO theory, by the sum of the attractive (van der Waals) and the repulsive (steric) contributions:

$$V_T(h) = V_{\text{vdW}}(h) + V_{\text{steric}}(h). \tag{6.12}$$

Electrosteric stabilization. For the stabilization of aqueous ceramic suspensions, polyelectrolytes, that is, polymers with charges distributed along their backbone, are often used. Ammonium or sodium polyacrylates are the more common ones. In this case, the stabilization is due to a combination of electrostatic and steric contributions, and accordingly, it is referred to as *electrosteric stabilization*. In such cases, the total energy of interaction includes three contributions:

$$V_T(h) = V_{\text{vdW}}(h) + V_{\text{electr}}(h) + V_{\text{steric}}(h). \tag{6.13}$$

The arrangement of polyelectrolyte molecules at the solid/liquid interface is one of the critical factors that determine the effectiveness of the electrosteric stabilization, and it depends on both pH and ionic strength. If the polyelectrolyte adsorbed at the interface has a flat conformation, that is, with loop and tails coiled near the surface, the steric contribution is low. Conversely, when the polyelectrolyte chains extend into the solution, the steric contribution is high (Figure 6.3).

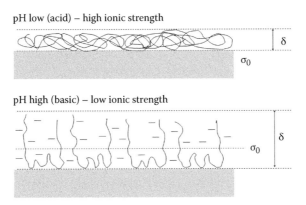

FIGURE 6.3 Dependency of the conformation of adsorbed anionic polyelectrolyte on a ceramic surface on pH and ionic strength. δ is the thickness of the adsorbed polymer, and σ_0 is the plane of charge.

6.5 RHEOLOGY OF CONCENTRATED SUSPENSIONS

In order to obtain information about the microstructural state of concentrated suspensions and their degree of stability, many researchers use—rather than direct, sophisticated, but expensive techniques like small-angle neutron scattering (Tadros 1990)—a phenomenological and cheap technique like rotational rheometry. By rotational rheometry, it is only possible to gain indirect information about the microstructural state of the system through the measurement of a macroscopic variable (the viscosity). However, viscosity is very sensitive to changes in the microstructural state, making it a good parameter for detecting changes in the microscopic state. From another standpoint, viscosity is the property that best describes the flow of suspensions, which is an important aspect in many stages of the processing, especially in the shaping stage. Therefore, gathering information on the flow behavior of a suspension is twice as important, both to check the degree of flocculation and to improve the design of the process.

The mechanisms of breakdown and buildup of the soft agglomerates in response to a change in the flow conditions determine the shear-thinning (decrease of shear viscosity with increasing of shear rate) and thixotropic (decreasing of shear viscosity with time after a sudden increase of shear rate) behavior of the suspension. Typically, shear-thinning is sought after because it allows easy flow of the suspension during the application, when the shear rates are relatively high, but hinders the flow (by increasing the viscosity) when the shear rate is zero.

To obtain information about the microstructural state of the suspension, the mechanical moduli (G' and G'') are specifically addressed. These can be measured by applying small oscillating strains/stresses to the samples at different frequencies, so that the material response is not dependent on the applied load (*linear viscoelastic region*). Liquid-like behavior is observed at low frequency, because the structures have sufficient time to relax after the application of the stress. At high frequency, when the structure has not enough time to relax because of quick variation of the stress, solid-like behavior is expected. By changing the frequency of oscillation, it is possible to detect the relaxation times corresponding to the different structural units present in the suspension (*flocs, aggregates, networks*). Unlike viscosity measurements, in which the microstructures are usually broken down owing to the application of relatively high shear stresses, the tests in oscillation mode apply low shear strains so that the system is only slightly perturbed and the correlation between material response and its microstructure is more direct.

Besides the dependence of the viscosity on the shear rate (*flow curve*), and that of the mechanical moduli on the oscillation frequency (*mechanical spectra*), the rheological measurements are commonly used to obtain information on the dependence of the viscosity η on the volume fraction of the solid phase ϕ, the latter being the main parameter affecting the rheological behavior of suspensions. The experimental data are often fitted to the Quemada (Equation 6.14) or Krieger–Dougherty (Equation 6.15) models (Quemada 1977; Krieger and Dougherty 1959):

$$\eta_r = \left(1 - \frac{\phi}{\phi_M}\right)^{-2}, \tag{6.14}$$

$$\eta_r = \left(1 - \frac{\phi}{\phi_M}\right)^{-[\eta]\phi_M}, \qquad (6.15)$$

in which η_r is the relative viscosity of the suspension, that is, the ratio between the viscosity of the suspension and the viscosity of the liquid medium, ϕ_M is the maximum packing factor (0.74 for monodisperse spheres), and $[\eta]$ is the intrinsic viscosity (2.5 for spherical particles). Such relationships accurately model the rapid increase of viscosity observed when the solid content approaches the maximum packing factor.

For colloidally stable suspensions, it should be taken into account that the effective volume occupied by the particles is higher than the nominal one, because the particles are surrounded by the electric double layers or polymeric layers that increase their hydrodynamic radius. Therefore, Equations 6.14 and 6.15 are often modified by substituting the volume fraction of solids ϕ with an effective volume fraction ϕ_{eff}, defined as

$$\phi_{eff} = \phi\left(1 + \frac{\delta}{a}\right)^3. \qquad (6.16)$$

In Equation 6.16, δ is either the thickness of the electric double layer (for electrostatically stabilized systems) or the thickness of the adsorbed polymer (for sterically stabilized systems). If the suspensions are flocculated, part of the liquid is trapped inside the agglomerates, so that the effective volume fraction of the solid increases in a manner that depends on the density of agglomerates:

$$\phi_{eff} = \phi C_{FP}, \qquad (6.17)$$

in which C_{FP} is the corrective factor. C_{FP} is close to 1 for dense agglomerates and higher than 1 for fractal agglomerates. C_{FP} can be related to the fractal dimension of agglomerates (Gardini et al. 2005).

The volume fraction of solids also affects the mechanical spectra. In particular, in a log–log plot, the slope of the elastic modulus G' ranges from 2 (typical of liquid-like behavior) to 0 (typical of solid-like behavior) as shown by PMMA particles (84 nm) dispersed in decalin when the volumetric fraction of solids range from 0.638 to 0.783 (Frith et al. 1990).

Another use of the rheological characterization is to compare the effectiveness of deflocculants and to determine their optimal concentration. Typically, the viscosity of suspensions with variable concentration of a given deflocculant is measured keeping the solid loading constant. By comparing the so-obtained flow curves, it is possible to find the concentration at which the flow curve, or the viscosity at a given shear rate, is the lowest. In the same way, by comparing the flow curves of suspensions with different deflocculants at their optimal concentration, and finding the one with the lowest viscosity, the best deflocculant can be chosen. This approach has proved rather useful in practical applications, because it gives direct information about the effectiveness of deflocculants. However, since the procedure yields the macroscopic result of many microscopic phenomena, it cannot provide information on the stabilization mechanism.

For electrostatically stabilized systems, some attempts have been made to correlate the rheological behavior with parameters directly related with the colloidal stability. For example, the static yield stress, that is, the threshold stress under which—at least at short time—no flow is observed, has been correlated with the squared zeta potential. For a given volume fraction of solids, the maximum yield stress is observed at the i.e.p. ($\zeta = 0$), and it decreases far from the i.e.p. Such dependence was modeled taking into account van der Waals attraction and electrostatic repulsion. As the volume fraction of solids increases, the yield stress increases following a power law with an exponent of about 4–5 (Johnson et al. 2000).

6.6   WATER-BASED SUSPENSIONS

For water-based ceramic suspensions, the most common and effective means of colloidal stabilization is by adding polyelectrolytes that adsorb on the particle surfaces and stabilize the system through a combination of electrostatic and steric repulsions. Polyelectrolytes are polymers with ionizable groups (carboxylic, sulfonic, amino groups) distributed along their polymeric chain. However, it should be stressed that the addition of polyelectrolytes can also cause flocculation, either by *charge neutralization* of the particle with polyelectrolytes of opposite charge or by *bridging flocculation*, which occurs when the same high-molecular-weight polymer chains are adsorbed on different particles, or by means of *depletion flocculation*, which occurs in concentrated suspensions when the nonadsorbed polymeric chains are squeezed out from the space between the particles, forcing the latter to approach each other. Therefore, to get an effective stabilization using polyelectrolytes, operative conditions have to be chosen in a proper way. More specifically, pH and ionic strength strongly affect the surface charge on the ceramic particles, the degree of ionization and the conformation of the polyelectrolyte in the solution, and, consequently, its adsorption on the ceramic surface.

In order to obtain a strong repulsive interaction, it is necessary that the loops and tails of the adsorbed polyelectrolyte extend into the solution (Figure 6.3). This condition is obtained when the ionizable groups along the chain are fully dissociated and their screening by counterions dissolved in the solution does not occur (low ionic strength). In this way, the ionic charges repel each other, and the chains are elongated.

Anionic polyelectrolytes, like polyacrylates, increase their degree of ionization with increasing pH, while cationic polyelectrolytes, for example, polyethylene imine (PEI), are fully ionized in acidic environments. Depending on pH, the number of molecules adsorbed on the particles' surface per unit area changes. Adsorption isotherms measured at different pH values provide information on the behavior of the polymer/particle interface. For example, the adsorption isotherms of PAA (molecular weight, 5000) on silicon nitride (Si_3N_4) (specific surface area, 10.1 m²/g) show that at pH 9, the adsorption is very weak, but it can be increased by lowering the pH to 3 (Hackley 1997).

The polymer added to the system is fully adsorbed only at low concentrations. When the concentration is increased, an equilibrium between the adsorbed and free polymer molecules in the solution is established, so that deviations from the 100%

adsorption curve are observed. This can be explained by considering that, at high pH, silicon nitride is negatively charged and a mutual repulsion with the strongly ionized PAA can occur. Electroacoustic measurements of dynamic mobility (or zeta potential) of Si_3N_4 particles show that the i.e.p. can be shifted from pH 6.3 for uncoated particles toward lower pH values by increasing the PAA concentration. At high pH, the curves at different concentrations of PAA overlap with the curve without polymer, indicating lack of polymer absorption. Moreover, at sufficiently high concentration of polymer, the curves overlap over the entire range of pH, indicating adsorption saturation. The opposite situation (shift of the i.e.p. toward higher pH and no polymer adsorption at low pH) was observed for the cationic polyelectrolyte PEI on silicon nitride (Zhu et al. 2007).

On the basis of the information gathered by adsorption isotherms and viscosity measurements at different polyelectrolyte concentrations, stability maps, that is, diagrams, can be constructed in which curves separating stability regions from regions where flocculation occurs are displayed as functions of pH and adsorbed polyelectrolyte. This was done, for example, for electrosterically stabilized aqueous alumina suspensions with sodium salt of poly(methacrylic acid) (Cesarano and Aksay 1988).

6.7   ORGANIC SUSPENSIONS

In the production of many advanced ceramics, the liquid medium is not water but an organic solvent. The mechanisms of interaction between additives and ceramic particles in organic solvents have been investigated for many years. In solvents with a lower dielectric constant than water, the surface charge is high and the DLVO theory still applies. However, it is not clear what mechanism promotes the electrical charging of the surface. For many years, it was assumed that, in organic media, only steric stabilization was possible, even when flocculation occurred at the secondary minimum. However, since the pinning of the deflocculant to the surface requires acid–base interactions, electrostatic repulsion is also present. The latter may give an important contribution to stabilization and, in some cases, altogether eliminate the secondary minimum (Fowkes 1987).

The hypothesis that electrosteric rather than steric repulsion occurs in organic solvents is supported, for example, by the good dispersion of barium titanate ($BaTiO_3$) in organic solvents obtained using (ethoxylated) phosphate ester (Mikeska and Cannon 1984). $BaTiO_3$ is a high-dielectric-constant material, widely used for the production of multilayer capacitors, whose layers are manufactured by tape casting of organic suspensions. In particular, suspensions containing an acrylic binder and an azeotropic mixture of ethanol–MEK as solvent were studied.

The need to use an organic solvent arises from the high solubility of $BaTiO_3$ in water, which would alter the Ba/Ti molar ratio in the sintered bodies and therefore worsen their dielectric properties. For these systems, the higher effectiveness of the phosphate ester was explained by the ionization of P–OH groups. The hydrogen ions resulting from the ionization are adsorbed on the $BaTiO_3$ surface, strongly increasing its surface charge. This generates an electrostatic repulsion at long distance and aids the sticking of the negatively charged ends of the phosphate ester. The hydrophobic tails of the phosphate ester extend in the nonpolar media, creating a steric repulsion at short distance. The phosphate ester was chosen, among several candidates, mainly on the basis of the results of viscosity measurements, as described in Section 6.5.

In other cases, the action of deflocculants on the interparticle interactions in an organic medium was studied by considering only steric stabilization. For example, the interparticle interactions in the presence of fatty acids (oleic and octanoic) and poly(12-hydroxy stearic acid) (PHS) for the dispersion of $BaTiO_3$ powder in decane were described by considering the van der Waals attraction given by Equation 6.2 together with the steric repulsion given by Equation 6.9 in the interpenetrative region (Bergström et al. 1997). Calculations suggest that PHS provides a sufficient steric repulsion to colloidally stabilize the system, while fatty acids cause a deeper secondary minimum at shorter distances to appear, so that flocculated systems are formed. Such predictions were confirmed by sedimentation tests that show an ill-defined sediment/supernatant interface and a very long time of settling for PHS, whereas a sharp interface and short time of settling are observed for fatty acids. In a similar way, the interaction of fatty acids (propionic acid, pentanoic acid, heptanoic acid, and oleic acid) on alumina (diameter, 0.4 μm; specific surface area, 6.9 m^2/g) in decalin was described by considering only the steric repulsive potential given by Equation 6.9. Results were contrasted with the consolidation behavior promoted by centrifugation (Bergström et al. 1992). Despite the approximation affecting the calculations, the energy estimates provide useful information and correlate well with the observed results. The most weakly flocculated suspension, that is, the one stabilized with oleic acid, has an incompressible behavior, with a constant density profile. Conversely, the most strongly flocculated suspension, that is, the one with propionic acid, is compressible and along the tube a volumetric fraction gradient is established, which has been found to depend on the centrifugation speed.

6.8   HETEROCOAGULATION

To increase fracture toughness or other properties like corrosion resistance, a secondary powder (e.g., $MoSi_2$ or TiN) is sometimes added to the primary one (e.g., Si_3N_4) to create ceramic composites suitable for many advanced applications. Moreover, secondary powders like Al_2O_3 or Y_2O_3 are often added in small amounts as sintering aids. When such multiphase systems undergo colloidal processing, the dissimilar inorganic particles dispersed in the liquid often carry on their surface electrical charges of opposite sign. In this situation, there is a tendency to form agglomerates (*heterocoagulation*). For such complex systems, the proper conditions should be determined in order to keep the different particles apart.

The van der Waals attractive energy potential between two spherical particles of radii a_1 and a_2 is given by Equation 6.18 as a function of the interparticle distance h:

$$V_{vdW}(h) = -\frac{A}{6}\left[\frac{2a_1a_2}{h^2 + 2h(a_1 + a_2)}\right.$$
$$\left. + \frac{2a_1a_2}{h^2 + 2h(a_1 + a_2) + 4a_1a_2} + \ln\left(\frac{h^2 + 2h(a_1 + a_2)}{h^2 + 2h(a_1 + a_2) + 4a_1a_2}\right)\right]. \quad (6.18)$$

When the distance h is small ($h \ll a_1 + a_2$), Equation 6.18 can be approximated by Equation 6.19:

$$V_{vdW}(h) = -\frac{A}{6}\frac{a_1 a_2}{a_1 + a_2}\frac{1}{h}. \tag{6.19}$$

The electrostatic interaction between two spherical particles with surface potentials ψ_{01} and ψ_{02} is given by Equation 6.20 (Hogg et al. 1966):

$$V_{electr}(h) = \frac{\varepsilon_r \varepsilon_0}{4}\left(\psi_{01}^2 + \psi_{02}^2\right)\frac{a_1 a_2}{a_1 + a_2}\left[2\frac{\psi_{01}\psi_{02}}{\psi_{01}^2 + \psi_{02}^2}\ln\frac{1 + \exp(-\kappa h)}{1 - \exp(-\kappa h)}\right.$$
$$\left. + \ln(1 - \exp(-2\kappa h))\right]. \tag{6.20}$$

When the surface potentials are of opposite sign, Equation 6.20 gives an attractive interaction. This phenomenon can be exploited to create agglomerates between dissimilar particles, in particular to create core–shell-structured particles. Equations 6.18 and 6.20 were used to describe the heterocoagulation of nanometric silica on micrometric alumina particles in aqueous systems (Cerbelaud et al. 2008) and the heterocoagulation of small alumina or magnesia particles on large, monodispersed silica particles in ethanol (Wang and Nicholson 2001).

6.9 ADVANCED SHAPING TECHNIQUES

The control of the interparticle interactions can also be exploited, rather than to keep the particles separated, to obtain flocculated suspensions or three-dimensional networks in a controlled way. Since the 1990s, new shaping techniques were proposed on the basis of such possibility, which provided an alternative to the classic shaping techniques (like dry pressing and subsequent machining, slip casting, pressure casting, and injection molding) for mass production of ceramics with complex geometries (Sigmund et al. 2000). With these novel techniques, the suspensions are consolidated directly inside the molds by forming physical or chemical gels that block the particles in their positions. Such techniques require concentrated and well-stabilized suspensions as starting materials to obtain three-dimensional networks with adjacent and uniformly distributed particles. For our discussion, techniques where physical gels are formed are relevant, as the creation of chemical gels occur by *in situ* polymerization of monomers or by networking of polymers already present in the systems (*gel casting, starch consolidation*), with no changes of interparticle interactions involved.

The so-called DCC (Graule et al. 1994) is based on the destabilization of electrostatically stable suspensions by changing the pH toward the p.z.c. or by increasing the ionic strength. In this way, the van der Waals forces overcome the electrostatic repulsion, and rigid bodies are formed. In order to have homogeneous gelations, avoiding the creation of localized agglomerates, the changes of pH and ionic strength are promoted without addition of acid, base, or salts, but rather through *in situ* enzyme-catalyzed chemical reactions that occur directly in the suspension. The two most common reactions are the hydrolysis of urea catalyzed by urease and the hydrolysis of amides catalyzed by amidase to shift the pH from acid to basic, and the hydrolysis of esters by esterase and oxidation of glucose by glucoseoxidase to obtain the opposite changes of pH. By changing the pH and ionic strength in the proper way, it is possible to coagulate the ceramic particles. When the pH is near the i.e.p., the system passes from a liquid-like state to a solid-like state, as well as when the salt concentration is sufficiently high to compress the electric double layer and reduce the energetic barrier. Taking into account these effects, a stability map for alumina particles in a concentrated aqueous suspension was drawn (Figure 6.4) (Graule et al. 1994).

Another technique is *temperature-induced forming* (Bell et al. 1999), which is based on the destabilization of an electrostatically stabilized suspension by exploiting the increase of solubility of the ceramic powders with increasing temperature and the subsequent bridging flocculation induced by the adsorption of a polymer present in the system (e.g., PAA). Upon heating, the ionic strength increases owing to the release of ions from the ceramic powder, and the electrostatic repulsion decreases. Moreover, the deflocculant (e.g., citric acid) used to stabilize the suspension at room temperature is desorbed. In such conditions, the particles are flocculated by bridging flocculation.

Yet another forming technique that exploits the control of interparticle forces is the so-called TIG. In this case, the gelation of the suspension is obtained by changing the polymer–solvent interaction in sterically stabilized suspensions by varying the temperature. A decrease of the temperature causes a decrease of the solubility of the polymer, which means that the polymeric chains dangling into the solution are

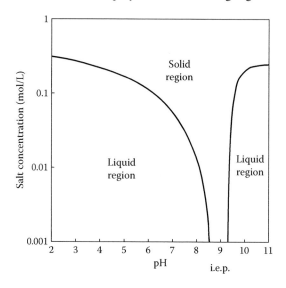

FIGURE 6.4 Stability diagram for an aqueous alumina suspension in relation with pH and salt concentration.

compressed back on the particle surface, thus reducing the thickness of the polymer layer and causing the gelation. The process is reversible.

6.10   CONCLUSIONS

The key role of the interparticle forces in the control of the colloidal stability of ceramic suspensions was highlighted. The main aim is to avoid the formation of hard agglomerates that strongly affect the properties of the final products. However, the possibility of controlling interparticle forces also allows the creation of core–shell particles by controlling the heterocoagulation between different particles. Advanced forming techniques can be used in which the suspensions are flocculated by decreasing the electrostatic or steric repulsion. Some examples of the application of such concepts in aqueous- and solvent-based suspensions were given.

REFERENCES

Barsoum, M.W. 2003. *Fundamentals of Ceramics*. Institute of Physics Publishing, Bristol and Philadelphia.

Bell, N.S., L.W. Wang, W.M. Sigmund, F. Aldinger. 1999. Temperature Induced Forming: Application of Bridging Flocculation to Near-Net Shape Production of Ceramic Parts. *Z. Metallkd.* 90:388–92.

Bergström, L., C.H. Schilling, I.A. Aksay. 1992. Consolidation Behavior of Flocculated Alumina Suspensions. *J. Am. Ceram. Soc.* 75:3305–14.

Bergström, L., K. Shinozaki, H. Tomiyama, N. Mizutani. 1997. Colloidal Processing of a Very Fine BaTiO$_3$ Powder—Effect of Particle Interactions on the Suspension Properties, Consolidation, and Sintering Behavior. *J. Am. Ceram. Soc.* 80:291–300.

Bhattacharjee, S., M.K. Paria, H.S. Maiti. 1993. Polyvinyl Butyral as a Dispersant for Barium Titanate in a Non-aqueous Suspension. *J. Mater. Sci.* 28:6490–5.

Boaretto, E., X.H. Wu, J.R. Yuan et al. 2009. Radiocarbon Dating of Charcoal and Bone Collagen Associated with Early Pottery at Yuchanyan Cave, Hunan Province, China. *Proc. Natl. Acad. Sci. U.S.A.* 106:9595–600.

Cerbelaud, M., A. Videcoq, P. Abélard, C. Pagnoux, F. Rossignol, R. Ferrando. 2008. Heteroaggregation between Al$_2$O$_3$ Submicrometer Particles and SiO$_2$ Nanoparticles: Experiment and Simulation. *Langmuir* 24:3001–8.

Cesarano III, J., I.A. Aksay. 1988. Stability of Aqueous a-Al$_2$O$_3$ Suspensions with Poly(Methacrylic Acid) Polyelectrolyte. *J. Am. Ceram. Soc.* 71:250–55.

Derjaguin, B.V., L.D. Landau. 1941. Theory of Stability of Highly Charged Lyophobic Sols and Adhesion of Highly Charged Particles in Solutions of Electrolytes. *Acta Physicochim. URSS* 14:633–52.

Ducker, W.A., Z. Xu, D.R. Clarke, J.N. Israelachvili. 1994. Forces between Alumina Surfaces in Salt Solutions: Non-DLVO Forces and the Implications for Colloidal Processing. *J. Am. Ceram. Soc.* 77:437–43.

Fowkes, F.M. 1987. Dispersions of Ceramic Powders in Organic Media. In *Advances in Ceramics, Volume 21: Ceramic Powder Science*, 411–421. The American Ceramic Society, Inc., Westerville, OH, USA.

Frith, W.J., T.A. Strivens, J. Mewis. 1990. Dynamic Mechanical Properties of Polymerically Stabilized Dispersions. *J. Coll. Interface Sci.* 139:55–62.

Fritz, G., V. Schädler, N. Willenbacher, N.J. Wagner. 2002. Electrosteric Stabilization of Colloidal Dispersions. *Langmuir* 18:6381–90.

Gardini, D., C. Galassi, R. Lapasin. 2005. Rheology of Hydroxyapatite Dispersions. *J. Am. Ceram. Soc.* 88:271–76.

Graule, T.J., F.H. Baader, L.J. Gauckler. 1994. Shaping of Ceramic Green Compact Direct from Suspensions by Enzyme Catalyzed Reactions. *Cfi/Ber. DKG.* 71:317–22.

Hackley, V.A. 1997. Colloidal Processing of Silicon Nitride with Poly(Acrylic Acid): I, Adsorption and Electrostatic Interactions. *J. Am. Ceram. Soc.* 80:2315–25.

Hesselink, F.Th. 1977. On the Theory of Polyelectrolyte Adsorption. *J. Colloid Interface Sci.* 60:448–66.

Hogg, R., T.W. Healy, D.W. Fuerstenau. 1966. Mutual Coagulation of Colloidal Dispersions. *Trans. Faraday Soc.* 62:1638–51.

Horn, R.G. 1990. Surface Forces and Their Action in Ceramic Materials. *J. Am. Ceram. Soc.* 73:1117–35.

Hotza D., P. Greil. 1995. Review: Aqueous Tape Casting of Ceramic Powders. *Mater. Sci. Eng.* A202:206–17.

Israelachvili, J.N. 1992. *Intermolecular and Surface Forces*, 2nd ed. Academic Press, London.

Johnson, S.B., G.V. Franks, P.J. Scales, D.V. Boger, T.W. Healy. 2000. Surface Chemistry–Rheology Relationships in Concentrated Mineral Suspensions. *Int. J. Miner. Process.* 58:167–304.

Krieger, I.M., T.J. Dougherty. 1959. A Mechanism for Non-Newtonian Flow in Suspensions of Rigid Spheres. *T. Soc. Rheol.* 3:137–52.

Lange, F.F. 1989. Powder Processing Science and Technology for Increased Reliability. *J. Am. Ceram. Soc.* 72:3–15.

Lewis, J.A. 2000. Colloidal Processing of Ceramics. *J. Am. Ceram. Soc.* 83:2341–59.

Macosko, C.W. 1994. *Rheology. Principles, Measurements, and Applications*. John Wiley & Sons, Inc., New York, USA.

Mewis, J., A.J.B. Spaull. 1976. Rheology of Concentrated Dispersions. *Adv. Colloid Interface Sci.* 6:173–200.

Mikeska, K., W.R. Cannon. 1984. Dispersants for Tape Casting Pure Barium Titanate. In *Forming of Ceramics, Volume 9*, eds. J.A. Mangels and G.L. Messing, 164–183. The American Ceramic Society, Inc., Columbus, OH, USA.

Mistler, R.E., E.R. Twiname. 2000. *Tape Casting. Theory and Practice*. The American Ceramic Society, Westerville, OH, USA.

Moreno, R. 1992a. The Role of Slip Additives in Tape-Casting Technology: Part I—Solvents and Dispersants. *Am. Ceram. Soc. Bull.* 71:1521–31.

Moreno, R. 1992b. The Role of Slip Additives in Tape-Casting Technology: Part II—Binders and Plasticizers. *Am. Ceram. Soc. Bull.* 71:1647–57.

Napper, D.H. 1983. *Polymeric Stabilization of Colloidal Dispersions*. Academic Press, London.

Quemada, D. 1977. Rheology of Concentrated Disperse Systems and Minimum Energy Dissipation Principle. I. Viscosity-Concentration Relationship. *Rheol. Acta* 16:82–94.

Reed, J.S. 1995. *Principles of Ceramics Processing*. John Wiley & Sons, Inc., New York, USA.

Rhodes, W.H. 1981. Agglomerate and Particle Size Effects on Sintering Yttria-Stabilized Zirconia. *J. Am. Ceram. Soc.* 64:19–22.

Rice, P.M. 1999. On the Origins of Pottery. *J. Archaeol. Method Theory* 6:1–54.

Russel, W.B., D.A. Saville, W.R. Schowalter. 1989. *Colloidal Dispersions*. Cambridge University Press, Cambridge, United Kingdom.

Sigmund, W.M., N.S. Bell, L. Bergström. 2000. Novel Powder-Processing Methods for Advanced Ceramics. *J. Am. Ceram. Soc.* 83:1557–74.

Tadros, Th.F. 1990. Use of Viscoelastic Measurements in Studying Interactions in Concentrated Dispersions. *Langmuir* 6:28–35.

Thiele, E.S., N. Setter. 2000. Lead Zirconate Titanate Particle Dispersion in Thick-Film Ink Formulations. *J. Am. Ceram. Soc.* 6:1407–12.

Vandiver, P.B., O. Soffer, B. Klima, J. Svoboda. 1989. The Origins of Ceramic Technology at Dolni Vestonice, Czechoslovakia. *Science* 246:1002–8.

Verwey, E.J.W., J.Th.G. Overbeek. 1948. *Theory of Stability of Lyophobic Colloids.* Elsevier, Amsterdam, Netherlands.

Wang, G., P.S. Nicholson. 2001. Heterocoagulation in Ionically Stabilized Mixed-Oxide Colloidal Dispersions in Ethanol. *J. Am. Ceram. Soc.* 84:1250–6.

Zhu, X., T. Uchikoshi, T.S. Suzuki, Y. Sakka. 2007. Effect of Polyethylenimine on Hydrolysis and Dispersion Properties of Aqueous Si_3N_4 Suspensions. *J. Am. Ceram. Soc.* 90:797–804.

7 Synthesis of Anisotropic Gold Nanocrystals Mediated by Water-Soluble Conjugated Polymers and Lead and Cadmium Salts

Marco Laurenti, Jorge Rubio-Retama,
Kyriacos C. Kyriacou, Epameinondas Leontidis,
and Enrique López-Cabarcos

CONTENTS

7.1 INTRODUCTION

The controlled synthesis of gold or silver nanostructures has attracted considerable attention because of their widespread use in catalysis, photonics, electronics, optoelectronics, biological labeling, imaging, sensing, and surface-enhanced Raman scattering (SERS) (Chen et al. 2005b; Maier et al. 2001; Taton et al. 2000; Tkachenko et al. 2003; Zhang et al. 2005). Gold and silver nanostructures can be also encapsulated into microgels to produce hybrid materials with thermoresponsive properties or magnetic and optical properties for SERS ultradetection (Contreras-Caceres et al. 2008, 2011). Besides size, shape is an important factor that controls the electronic and optical properties of metal nanoparticles. In particular, the morphological control of asymmetric Au or Ag nanostructures allows tuning the localized surface plasmon resonance (SPR) in the visible and near-infrared regions (Hu et al.

155

2004; Liz-Marzán 2006; Mayer and Hafner 2011; Pastoriza-Santos and Liz-Marzán 2002). Therefore, synthetic processes allowing good shape control are highly desirable (Grzelczak et al. 2008). Despite the strong research interest in this area, the fabrication of anisotropic metal nanocrystals is still a challenge nowadays. For instance, the reduction of $HAuCl_4$ can be done with a wide range of substances and a variety of methods. For colloidal gold nanorods and nanoprisms, the currently preferred synthetic route is a seed-mediated growth process (Chen et al. 2009; Ha et al. 2007; Huang et al. 2007; Jana et al. 2001; Nikoobakht and El-Sayed 2003; Turkevich et al. 1951). The role of all the chemicals involved in this synthetic process, such as ascorbic acid, cetyl trimethylammonium bromide (CTAB), and chloroaurate ion, has been extensively studied (Daniel and Astruc 2004). In recent years, considerable research attention has been focused on the role of simple additives such as theoretically "inert" electrolytes. Additives, such as halide ions, have been found to be major factors for the shape control of metal nanostructures by directing crystal growth (Filankembo and Pileni 2000; Leontidis et al. 2002; Magnussen 2002; Mayer and Hafner 2011; Millstone et al. 2008).

Heavy metal ions have also been found to control the morphology of gold nanostructures. Chen et al. have reported a modified seed-mediated technique for fabricating gold nanorods/wires wherein the shape of the gold nanomaterials evolved from fusiform into one-dimensional rods and other geometries (Chen et al. 2005a, 2009). The presence of silver ions strongly influences the formation of fusiform nanoparticles, changing the mechanism of particle formation from template domination (CTAB) to silver ion domination, depending on the concentration of silver nitrate used. Sun et al. used the replacement reaction between silver nanostructures and an aqueous $HAuCl_4$ solution to generate metal nanostructures with hollow interiors (Sun et al. 2004). In 2008, Sun et al. have demonstrated the role played by Cu^{2+} on the selective synthesis of gold cuboids and decahedral nanoparticles (Sun et al. 2008).

A different approach to control the morphology of the gold nanocrystals is to use polymers as templates or reducing agents. Zhai and McCullough reported in 2004 the stabilization of gold nanoparticles mediated by poly(3-hexylthiophene) in organic solvent (Zhai and McCullough 2004). The synthesis of regioregular poly(3-hexylthiophene)-stabilized gold nanoparticles was accomplished using a room temperature, two-phase, one-pot reaction involving the reduction of tetrachloroauric acid by sodium borohydride in the presence of regioregular poly(3-hexylthiophene). They prevent the oxidation of the polymer by the tetrachloroauric acid solution using a large amount of the surfactant tetraoctylammonium bromide. The stabilization of the gold nanoparticles was possible since the thiophene rings attach the gold surface and the lateral polymer chains prevent aggregation. More recently, Tang et al. reported the synthesis of dendritic gold in a one-step hydrothermal reduction of $HAuCl_4$ using ammonium formate as a reducing agent in the presence of poly(vinylpyrrolidone) (PVP) (Tang et al. 2008). They found a relationship between an excess of ammonium formate and the dendritic Au particle shape. In the same process, PVP acts as a stabilizer and may serve not only as a reductant but also as a capping reagent. This work suggests that the use of two or more capping reagents with different adsorption abilities could be beneficial for the formation of hyperbranched

Au nanoparticles. In their recent work, Lim et al. described a simple approach to synthesize anisotropic Au nanostructures with various shapes by reducing HAuCl$_4$ with PVP in aqueous solutions (Lim et al. 2008). The morphology of Au nanostructures evolved from nanotadpoles to nanokites and finally to triangular and hexagonal microplates. These authors concluded that the slow reduction rate associated with the mild reducing power of PVP plays a critical role in the formation of nanoplates during nucleation, as well as in their growth into anisotropic nanostructures.

All the above studies illustrate that the control of the mechanism of crystal growth achieved by various additives plays an important role in the size and shape obtained, and therefore must be of primary concern in the effort to obtain tailored nanostructures. We have studied the synthesis of Au nanostructures using the water-soluble conjugated polymer poly[2-(3-thienyl)-ethoxy-4-butylsulfonate] (PTEBS) for the reduction of HAuCl$_4$ and the stabilization of the Au nanocrystals. We have focused on PTEBS because it has a higher reducing power than PVP owing to the π–π bonds of the polythiophene moiety and also inspired by recent reports concerning the effect of polymers on shape control of gold nanoparticles (Han et al. 2010; Pardiñas-Blanco et al. 2008). In fact, the polymer may also partly act as a surface-active agent with its polythiophene backbone serving as a hydrophobic tail group and the sulfonate group acting as a hydrophilic head group (Laurenti et al. 2009). In addition, the thiophene ring can bind onto the surfaces of the Au nanoparticles (given the strong interaction of sulfur atoms with gold surfaces), stabilizing them in water more efficiently than PVP. Besides, using PTEBS for gold reduction and stabilization, we have also examined the effect of the simultaneous addition of heavy metal ions on the evolution of gold nanoparticle shapes.

7.2 AU^{III} REDUCTION BY PTEBS

The water-soluble conjugated polymer PTEBS was used to reduce HAuCl$_4$ and at the same time to stabilize gold nanostructures in water. We first studied the synthesis of Au nanocrystals using PTEBS. In a typical experiment, 1 mL of PTEBS (1 mM in monomer repeating units) is diluted with Millipore water to 9.8 mL and sonicated for 1 min. Subsequently, 186 µL of HAuCl$_4$ from a stock solution (26 mM) is added to the solution and stirred for 1 min at room temperature. After 1 day, the reaction is stopped by centrifugation at 4000 rpm and the solid precipitated is redispersed with 10 mL of Millipore water.

The TEM image of a grid prepared with an aliquot withdrawn from the HAuCl$_4$/ PTEBS solution after 10 min from the beginning of the reaction is shown in Figure 7.1a. The gold nanoparticles are aggregates and have sizes ranging between 10 and 300 nm. After 1 day, a new grid was prepared with an aliquot withdrawn from the same solution and the result is presented in Figure 7.1b. The nanoparticles are nearly monodisperse with diameters varying from 8 nm up to 12 nm.

Figure 7.1c shows the evolution of the absorbance when PTEBS was mixed with HAuCl$_4$ after 1 h (solid line) and after 1 day (dashed line). The absorbance of the pure PTEBS solution in water is at 425 nm (Lopez-Cabarcos and Carter 2005). After the addition of HAuCl$_4$ (0.5 mM), the peak at 425 nm instantaneously blue shifts at around 400 nm owing to the polymer oxidation (data not shown). The SPR of the

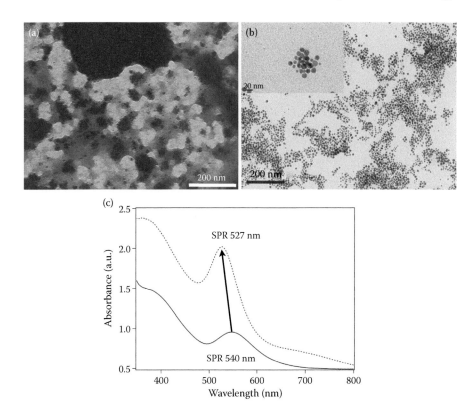

FIGURE 7.1 (a) TEM micrograph of the nanoparticles obtained from a PTEBS water solution (1 mM) and HAuCl$_4$ solution (0.5 mM) 10 min after the beginning of the reaction. (b) TEM image of the nanoparticles obtained after 1 day of reaction from the same solution. (c) UV-Vis experiment showing the evolution of the SPR of the same solution after 1 h (solid line) and after 1 day (dashed line).

Au nanocrystals changes after 1 day from the beginning of the reaction shifting from 540 to 527 nm (dashed line) as indicated by the black arrow. An explanation for this behavior could arise from the diminishing size of the aggregates of the gold nanoparticles as a function of time as is illustrated in the TEM micrographs presented in Figure 7.1. After 10 min from the beginning of the reaction, the sample is formed by aggregates of gold nanocrystals with different size embedded into the PTEBS polymer matrix. The former aggregates after 1 day evolved into separated gold nanocrystals (Figure 7.1b). Thus, the blue shift of the SPR could be assigned to the separation of the gold aggregates into individual gold nanoparticles.

7.3 SHAPE CONTROL OF GOLD NANOPARTICLES BY LEAD AND CADMIUM SALTS

The investigation of the influence of cadmium acetate and lead nitrate on the synthesis of gold nanocrystals using PTEBS and HAuCl$_4$ was carried out under the same reaction conditions used previously with PTEBS alone. Moreover, to understand the effects of the two metal ions, we initially studied them separately. The concentration of $Cd(COOCH_3)_2$ typically used was 0.78 mM, and the addition of the aliquot to the reaction medium was made before the addition of $HAuCl_4$ (0.5 mM). In a typical experiment, 1 mL of PTEBS (1 mM) is diluted with Millipore water to 9.5 mL. To this solution, 300 µL of $Cd(CH_3COO)_2$ from a stock solution (26 mM) is added and the resulting solution is sonicated for 1 min. One hundred eighty-six microliters µL of $HAuCl_4$ from a stock solution (26 mM) is added to the solution and stirred for 1 min. After 3 days, the reaction is stopped by centrifugation at 4000 rpm and the precipitated solid is redispersed with 10 mL of Millipore water. A parallel study was performed to investigate the effect of lead nitrate on the gold nanoparticles' synthesis.

The TEM micrograph presented in Figure 7.2a shows that the presence of Cd ions induces a dramatic change on the morphology of the Au nanostructures. In fact, compared to the synthesis performed with PTEBS alone, the colloidal gold loses its spherical shape and the majority of the structures do not have well-defined geometrical contours, and the nanostructures are largely anisotropic. Similar to cadmium, the effect of lead ions on the synthesis of Au nanocrystals has not been reported in the literature. Although lead nitrate has been used since 1930 in the gold cyanidation process, the knowledge of its effects on gold is still fragmentary. However, it seems that lead nitrate is able to induce the oxidative etching of gold nanocrystals in combination with the dissolved molecular oxygen (Deschenes et al. 2000). The result obtained is shown in Figure 7.2b. It can be seen that the lead nitrate effect on the synthesis is quite different from that of cadmium acetate or from PTEBS alone. The TEM picture shows irregular colloidal gold crystals with predominance of planar structure. The nanocrystal in the rectangle in Figure 7.2b illustrates the possible effect of gold re-dissolution and re-crystallization owing to oxidative etching and the evolution of the colloid structure from a two-dimensional (2D) platelet (gray zone with fringes) to a three-dimensional (3D) structure (dark zone).

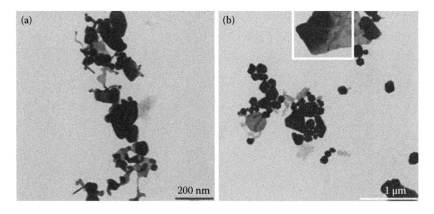

FIGURE 7.2 (a) The TEM picture shows Au nanocrystals obtained using 1 mM of PTEBS and 0.78 mM of cadmium acetate. (b) Au nanocrystals synthesized using 1 mM of PTEBS and 0.78 mM of lead nitrate. The nanocrystal in the rectangle illustrates the possible effect of the lead ions on the Au nanocrystals.

Since three very different nanostructures were obtained when using PTEBS alone or in combination with Cd^{2+} or Pb^{2+} ions, we decided to synthesize gold nanocrystals using PTEBS together with Cd^{2+} and Pb^{2+}. The reaction was carried out with the same recipe used for cadmium or lead with the difference that the lead nitrate and cadmium acetate aliquots were added together. Three concentrations of cadmium and lead were assayed (0.26, 0.53, and 0.78 mM) and a ratio of 1:1 between cadmium and lead was chosen for these experiments, after various previous investigations proved this is the optimal ratio to study morphology changes.

Figure 7.3a shows the Au nanostructures obtained mixing cadmium and lead at concentrations of 0.26 mM. Two colloid morphologies are observed: a 2D plate-let structure and a 3D colloidal structure. The hexagonal and triangular nanoplates observed are polydisperse with sizes between 150 nm and 1 μm and average size around 450 nm. The 3D gold particles are polydisperse with polyhedral shapes. Figure 7.3b shows the Au nanostructures obtained when the concentration of both salts is increased to 0.53 mM; as can be seen, the nanocrystals obtained under these conditions are different. Again, the structures can be separated into anisotropic platelets and 3D irregular colloids but they have much higher anisotropy with respect to the synthesis with 0.26 mM of Cd^{2+} and Pb^{2+}. Figure 7.4 shows gold nanocrys-tals produced when the concentrations of both salts are increased up to 0.78 mM. The obtained Au nanocrystals are highly anisotropic: they have plate-like hexago-nal, triangular, and truncated sections with tails attached to one edge of the planar structures.

The ion concentration effect on the synthesis of Au nanostructures was further examined looking at different ratios [Ions]/[Au^{III}]. A ratio close to 0.26:0.50 produces structures with platelet and 3D shapes, but when the ratio is increased to 0.53:0.50, shape anisotropy increases remarkably, indicating that the threshold ratio to observe anisotropy is around 1:1. An increase of the [Ions]/[Au^{III}] ratio to 0.78:0.50 induces dramatic changes and the structures obtained are highly anisotropic (Figures 7.4 and 7.5). It is important to point out the synergistic effect of both ions when they are used simultaneously in the synthesis of the gold nanoparticles. In fact, if we compare the

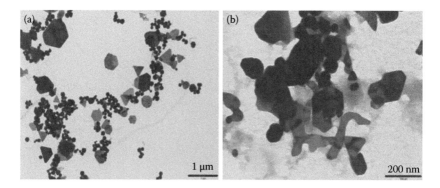

FIGURE 7.3 (a) TEM image of the synthesized Au nanostructures in the presence of 0.26 mM of both lead nitrate and cadmium acetate. (b) TEM image of the synthesized Au nanostructures in the presence of 0.53 mM of both lead nitrate and cadmium acetate.

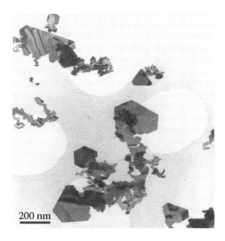

FIGURE 7.4 The TEM picture shows Au nanocrystals synthesized after 3 days of reaction using a concentration of 0.78 mM of Cd^{2+} and Pb^{2+}.

synthesized nanocrystals shown in Figure 7.2a with that shown in Figure 7.4, the amount of Cd^{2+} is the same in both cases, but the anisotropy of the nanostructures is much higher when Cd^{2+} is used in combination with Pb^{2+}. Furthermore, the presence of both ions modifies significantly the rate of the reaction, which now takes 3 days to give a high yield of anisotropic gold nanostructures, indicating a slowdown of the crystal growing rate.

A typical particle isolated from those presented in Figure 7.4 is shown in Figure 7.5. This image provides relevant information concerning the possible formation mechanism of these anisotropic nanostructures. The ellipse in Figure 7.4 highlights the presence of tiny holes and cracks, which seems to indicate that the triangular plate could be the product of the attachment of two Au plates or perhaps the result of the bending of a previous formed nanostructure. The black arrows illustrate the different

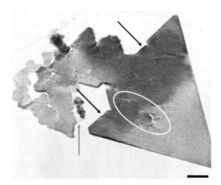

FIGURE 7.5 Single Au nanoplate obtained after 3 days with a tail. The oval indicates a part of the Au sheet where the structure is not totally formed. The gray arrow shows some nanoparticles close to the growing tail, while the black arrows show the different height of the triangle edge. The scale bar is 50 nm.

thickness on the same edge of the plate: in addition, the tail is present only on this edge of the nanoplate, supporting the hypothesis of preferential growing on this side of the nanoplate. The gray arrow indicates the presence of very small gold nanoparticles (diameter around 1.6 nm) close to the formed nanostructure, which would facilitate the formation process of these complicated nanostructures.

Similar Au structures as those shown in Figures 7.4 and 7.5 were reported in 2008 by Lim et al. using PVP as the gold reducing agent (Lim et al. 2008; Pardiñas-Blanco et al. 2008). In 2008, Muñoz et al. also synthesized core–shell silver@polypyrrole nanosnakes, the shapes of which they attributed to the presence of pyrrole (Muñoz-Rojas et al. 2008a). Their nanostructures are metastable at the reaction conditions and undergo bending and folding while they preserve the crystallographic coherence. Muñoz et al. proposed "nanomalleability" to explain how such extraordinary "coherent" shape evolution may be taking place (Muñoz-Rojas et al. 2008b). In our case, the shape cannot be attributed only to the presence of core–shell gold-PTEBS structures, since the synthesis with the conjugated polymer alone does not give the anisotropic shape. It seems that an important factor that induces these conformations is the combined effect of $Cd(COOCH_3)_2$, $Pb(NO_3)_2$, and the conjugated polymer PTEBS. In fact, we have investigated the synthesis of gold nanocrystals using the citrate method with the two salts at the same concentration used in the previous case and the reaction does not give any anisotropic nanostructures. The oriented attachment between adjacent snakes and small gold nanoparticles appears to be the basis of the formation of the triangular and hexagonal structures.

Extensive high-resolution TEM (HRTEM) investigation was performed on these later gold nanostructures. Figure 7.6 shows a typical HRTEM image taken by directing the electron beam perpendicular to the flat faces of a single anisotropic nanostructure. The inset in Figure 7.6b shows the related selected area electron diffraction (SAED) spot pattern obtained by focusing the electron beam on a nanoplate lying flat on the TEM grid. The hexagonal symmetry of the diffraction pattern is visible, demonstrating that the gold nanostructure is a single crystal with the preferential growth direction along the Au {111} plane (Sun and Xia 2004). Three sets of spots can be identified on the basis of d-spacing calculated from the SAED image. The inner set (circle) with a lattice spacing of 2.4 Å could correspond to the forbidden 1/3{422} reflection. The set (box) with a spacing of 1.4 Å could be assigned to the {220} reflection of the face-centered cubic (fcc) Au. It indicates that the prepared nanoplates are single crystalline with {111} lattice planes as the basal planes. The outer set (triangle) with a lattice spacing of 0.8 Å could be indexed to the {422} Bragg reflection. These two sets of reflection are both allowed by an fcc lattice (Jin et al. 2001; Washio et al. 2006). This phenomenon has been observed previously in Au or Ag plate-like crystals, and several explanations for the occurrence of such forbidden reflections have been suggested (see below).

HRTEM results confirmed that the nanoplates have the same crystallographic structure. According to the results of Germain et al., such 1/3{422} forbidden reflections observed on the plate-like structures of Au or Ag could be attributed to (111) stacking faults lying parallel to the (111) surface and extending across the entire nanosheet (Germain et al. 2003). The close-up in Figure 7.6b and c shows that the nanotail is attached to the truncated triangle forming a unique structure. In addition,

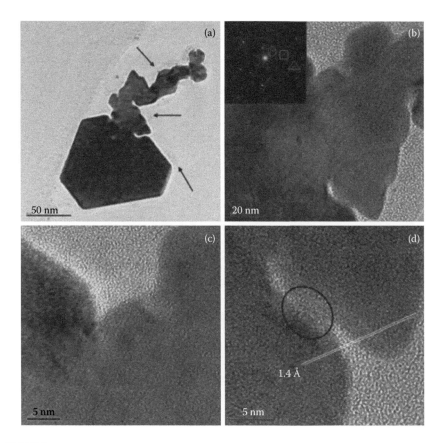

FIGURE 7.6 (a) HRTEM of a single Au anisotropic truncated triangle with a tail. The black arrows indicate the PTEBS covering the nanostructure and, at the end of the tail, small Au nanoparticles embedded into the polymer. (b) Close-up of panel (a) showing the part of the tail attached to the nanostructure. In the inset, the related SAED pattern obtained by focusing the electron beam on the nanoplate is shown. (c and d) The close-up of image panel (b) at the top and the bottom of the structure, respectively. The scale bar is 5 nm.

the fringes are separated by 1.4 Å (Figure 7.6d) that can be ascribed to the {220} reflection for an fcc lattice of gold and it is also visible on the partially formed Au layer between the tail and the truncated triangle (black ellipse). Moreover, the presence of the same distance between the fringes on the tail and the truncated triangle demonstrates crystallographic coherence, proven also by the SAED. X-ray fluorescence does not detect any trace of Pb^{2+} on the Au nanostructures, while it detects traces of Cd^{2+}, but its signal is so weak that quantitative estimation of Cd^{2+} cannot be performed. The effect of Pb^{2+} is therefore not related to the possibility of alloy formation with gold or of specific adsorption on gold crystal faces.

On the basis of the existing evidence—mostly from HRTEM—it seems that planar gold nanostructures are induced to a large extent when PTEBS is combined with Cd^{2+} salts. Nanosnakes may be partially formed in these solutions, but their formation is further promoted in the presence of Pb^{2+} ions, which could facilitate the oxidative etching of gold. The presence of these three compounds facilitates the folding and bending of the nanosnakes, probably with the contribution of small gold nanoparticles in solution, and helps create the impressive, highly anisotropic nanostructures of Figures 7.4 through 7.6. It is still unclear why a certain threshold concentration of the ions is requested in order to observe extensive gold nanostructure anisotropy ([Ions]/[Au^{III}] ratios larger than 1:1).

7.4   CONCLUSIONS

In conclusion, a one-pot method strategy for the fabrication of complex Au nano-architecture was developed using the water-soluble conjugated polymer PTEBS in conjunction with cadmium acetate or lead nitrate. When only the polymer is present, we obtained spherical Au nanoparticles with a diameter between 8 and 12 nm. The addition of 0.78 mM Cd(COOCH$_3$)$_2$ dramatically changes the morphology of the Au nanocrystals, inducing partial anisotropy, while the addition of 0.78 mM Pb(NO$_3$)$_2$ produces polydisperse Au colloids. However, when both ions are used together, their effect is to induce high anisotropy and the nanostructures obtained have planar triangular and hexagonal morphology with tails attached to one edge. The detailed TEM and HRTEM investigation of this system has revealed that the anisotropic particles are probably created by the folding and bending of the nanosnakes. On the existing evidence, we speculate that the planar gold structures are promoted mainly by the presence of Cd^{2+} ions, through their interaction with PTEBS. This investigation opens up an interesting way for the design of anisotropic gold (and probably also of other noble metal) nanoparticles. PTEBS combined with appropriate ions may lead to nanosnakes, and further combinations with additional ions that promote the bending of nanosnakes induce anisotropy. In this respect, it will be interesting to examine which ions coupled to PTEBS induce such behavior and also if similar effects can be observed for other polymers that can act as reducing agents for Au^{III} salts.

ACKNOWLEDGMENTS

The authors acknowledge financial support from the Spanish Science and Innovation Ministry (Grant MAT2010-15349) and from the COST Actions D43 and CM1101. Marco Laurenti acknowledges the Comunidad de Madrid for a fellowship to perform this work. The authors also thank Agustin Fernandez (UCM-Microscopy CAI) for his help with TEM measurements.

REFERENCES

Chen, H.M., Peng, H.C., Liu, R.S. et al. 2005a. Controlling the length and shape of gold nanorods. *J. Phys. Chem. B* 109:19553–5.
Chen, H.M., Liu, R.S., and D.P. Tsai. 2009. A versatile route to the controlled synthesis of gold nanostructures. *Cryst. Growth Des.* 9:2079–87.
Chen, J., Saeki, F., Wiley, B.J. et al. 2005b. Gold nanocages: bioconjugation and their potential use as optical imaging contrast agents. *Nano Lett.* 5:473–7.
Contreras-Caceres, R., Sanchez-Iglesias, A., Pastoriza-Santos, I. et al. 2008. Encapsulation and growth of gold nanoparticles in thermoresponsive microgels. *Adv. Mater.* 20:1666–70.

Contreras-Caceres, R., Abalde-Cela, S., Guardia-Giros, P. et al. 2011. Multifunctional microgel magnetic/optical traps for SERS ultradetection. *Langmuir* 27:4520–5.

Daniel, M.C. and D. Astruc. 2004. Gold nanoparticles: assembly, supramolecular chemistry, quantum size-related properties, and applications toward biology, catalysis, and nanotechnology. *Chem. Rev.* 104:293–346.

Deschenes, G., Lastra, R., Brown, J.R., Jin, S., May, O., and E. Ghali. 2000. Effect of lead nitrate on cyanidation of gold ores: progress on the study of the mechanisms. *Minerals Eng.* 13:1263–79.

Filankembo, A. and M.P. Pileni. 2000. Is the template of self-colloidal assemblies the only factor that controls nanocrystal shapes? *J. Phys. Chem. B* 104:5865–8.

Germain, V., Li, J., Ingert, D., Wang, Z.L., and M.P. Pileni. 2003. Stacking faults in formation of silver nanodisks. *J. Phys. Chem. B* 107:8717–20.

Grzelczak, M., Perez-Juste, J., Mulvaney, P., and L.M. Liz-Marzan. 2008. Shape control in gold nanoparticle synthesis. *Chem. Soc. Rev.* 37:1783–91.

Ha, T.H., Koo, H.J., and B.H. Chung. 2007. Shape-controlled syntheses of gold nanoprisms and nanorods influenced by specific adsorption of halide ions. *J. Phys. Chem. C* 111:1123–30.

Han, J., Li, L., and R. Guo. 2010. Novel approach to controllable synthesis of gold nanoparticles supported on polyaniline nanofibers. *Macromolecules* 43:10636–44.

Hu, J., Zhang, Y., Liu, B. et al. 2004. Synthesis and properties of tadpole-shaped gold nanoparticles. *J. Am. Chem. Soc.* 126:9470–1.

Huang, L., Wang, M., Zhang, Y., Guo, Z., Sun, J., and N. Gu. 2007. Synthesis of gold nanotadpoles by a temperature-reducing seed approach and the dielectrophoretic manipulation. *J. Phys. Chem. C* 111:16154–60.

Jana, N.R., Gearheart, L., and C.J. Murphy. 2001. Seed-mediated growth approach for shape-controlled synthesis of spheroidal and rod-like gold nanoparticles using a surfactant template. *Adv. Mater.* 13:1389–93.

Jin, R.C., Cao, Y.W., Mirkin, C.A., Kelly, K.L., Schatz, G.C., and J.G. Zheng. 2001. Photoinduced conversion of silver nanospheres to nanoprisms. *Science* 294:1901–3.

Laurenti, M., Benito-Retama, J., Garcia-Blanco, F., Frick, B., and E. Lopez-Cabarcos. 2009. Interpenetrated PNIPAM–polythiophene microgels for nitro aromatic compound detection. *Langmuir* 25:9579–84.

Leontidis, E., Kleitou, K., Kyprianidou-Leodidou, T., Bekiari, V., and P. Lianos. 2002. Gold colloids from cationic surfactant solutions. 1. Mechanisms that control particle morphology. *Langmuir* 18:3659–68.

Lim, B., Camargo, P.H.C., and Y. Xia. 2008. Mechanistic study of the synthesis of Au nanotadpoles, nanokites, and microplates by reducing aqueous $HAuCl_4$ with poly(vinyl pyrrolidone). *Langmuir* 24:10437–42.

Liz-Marzán, L.M. 2006. Tailoring surface plasmons through the morphology and assembly of metal nanoparticles. *Langmuir* 22:32–41.

Lopez-Cabarcos, E. and S.A. Carter. 2005. Effect of the molecular weight and the ionic strength on the photoluminescence quenching of water-soluble conjugated polymer sodium poly[2-(3-thienyl)ethyloxy-4-butylsulfonate]. *Macromolecules* 38:10537–41.

Magnussen, O.M. 2002. Ordered anion adlayers on metal electrode surfaces. *Chem. Rev.* 102:679–725.

Maier, S.A., Brongersma, M.L., Kik, P.G., Meltzer, S., Requicha, A.A.G., and H.A. At water. 2001. Plasmonics—a route to nanoscale optical devices. *Adv. Mater.* 13:1501–5.

Mayer, K.M. and J.H Hafner. 2011. Localized surface plasmon resonance sensors. *Chem. Rev.* 111:3828–57.

Millstone, J.E., Wei, W., Jones, M.R., Yoo, H., and C.A. Mirkin. 2008. Iodide ions control seed-mediated growth of anisotropic gold nanoparticles. *Nano Lett.* 8:2526–9.

Muñoz-Rojas, D., Oró-Solé, J., Ayyad, O., and P. Gómez-Romero. 2008a. Facile one-pot synthesis of self-assembled silver@polypyrrole core/shell nanosnakes. *Small* 4:1301–6.

Muñoz-Rojas, D., Oró-Solé, J., and P. Gómez-Romero. 2008b. From nanosnakes to nanosheets: a matrix-mediated shape evolution. *J. Phys. Chem. C* 112:20312–8.

Nikoobakht, B. and M.A. El-Sayed. 2003. Preparation and growth mechanism of gold nanorods (NRs) using seed-mediated growth method. *Chem. Mater.* 15:1957–62.

Pardiñas-Blanco, I., Hoppe, C.E., Piñeiro-Redondo, Y., López-Quintela, M.A., and J. Rivas. 2008. Formation of gold branched plates in diluted solutions of poly(vinylpyrrolidone) and their use for the fabrication of near-infrared-absorbing films and coatings. *Langmuir* 24:983–90.

Pastoriza-Santos, I. and L.M. Liz-Marzán. 2002. Synthesis of silver nanoprisms in DMF. *Nano Lett.* 2:903–5.

Sun, J., Guan, M., Shang, T., Gao, C., Xu, Z., and J. Zhu. 2008. Selective synthesis of gold cuboid and decahedral nanoparticles regulated and controlled by Cu^{2+} ions. *Cryst. Growth Des.* 8:906–10.

Sun, X.P., Dong, S.J., and E.K. Wang. 2004. Large-scale synthesis of micrometer-scale single-crystalline Au plates of nanometer thickness by a wet-chemical route. *Angew. Chem., Int. Ed.* 43:6360–3.

Sun, Y. and Y. Xia. 2004. Mechanistic study on the replacement reaction between silver nanostructures and chloroauric acid in aqueous medium. *J. Am. Chem. Soc.* 126:3892–3901.

Tang, X.L., Jiang, P., Ge, G.L., Tsuji, M., Xie, S.S., and Y.J. Guo. 2008. Poly(*N*-vinyl-2-pyrrolidone) (PVP)-capped dendritic gold nanoparticles by a one-step hydrothermal route and their high SERS effect. *Langmuir* 24:1763–8.

Taton, T.A., Mirkin, C.A., and R.L. Letsinger. 2000. Scanometric DNA array detection with nanoparticle probes. *Science* 289:1757–60.

Tkachenko, A.G., Xie, H., Coleman, D. et al. 2003. Multifunctional gold nanoparticle–peptide complexes for nuclear targeting. *J. Am. Chem. Soc.* 125:4700–1.

Turkevich, J., Stevenson, P.C., and J. Hillier. 1951. A study of the nucleation and growth process in the synthesis of colloidal gold. *Discuss. Faraday. Soc.* 11:55–75.

Washio, I., Xiong, Y.J., Yin, Y.D., and Y.N. Xia. 2006. Reduction by the end groups of poly(vinyl pyrrolidone): a new and versatile route to the kinetically controlled synthesis of Ag triangular nanoplate. *Adv. Mater.* 18:1745–9.

Zhai, L. and R.D. McCullough. 2004. Regioregular polythiophene/gold nanoparticle hybrid materials. *J. Mater. Chem.* 14:141–3.

Zhang, X., Young, M.A., Lyandres, O., and R.P. Van Duyne. 2005. Rapid detection of an anthrax biomarker by surface-enhanced Raman spectroscopy. *J. Am. Chem. Soc.* 127:4484–9.

8 Assembly of Non-Aqueous Colloidal Dispersions under External Electric Field

Halil Ibrahim Unal, Ozlem Erol, and Mustafa Ersoz

CONTENTS

8.1 INTRODUCTION

Electrorheological (ER) fluids, exhibiting electric field–induced rheological properties, are a kind of non-aqueous colloidal dispersions composed of polarizable particles dispersed in non-aqueous insulating media such as mineral or silicone oil (SO). These dispersed particles in non-aqueous media show a microstructural transition from a random state to a chain-like or column-like structure aligned to the direction of applied electric field rapidly and reversibly by controlling the applied electric field.

The significant changes in rheological properties under applied electric field strength are desirable characteristics for various applications such as clutches (Litvinov 2007), damping devices (Nguyen and Choi 2009), shock absorbers

(Wereley et al. 2004), haptic devices (Han and Choi 2008), microfluidic chips (Zhang et al. 2008), microfluidic pumps (Liu et al. 2006), ER polishing (Tsai et al. 2008), and a fuel injector with ER (Tao et al. 2008) for reducing the viscosity of petroleum fuels.

Thus, ER fluids have been increasingly studied since its discovery by Winslow (1949), owing to its extensive potential applications and scientific concern.

Different types of ER effect have been observed so far. If the rheological properties of ER fluid increase with E, this is called the positive ER effect. On the other hand, the reverse of the positive ER effect—a rather unusual behavior—is termed the negative ER effect, which can be described by saying that the rheological properties are reduced with applied electric field.

The principle behind the positive ER phenomenon originates from the electrostatic polarization mainly attributed to the field-induced polarization of the dispersed particles relative to the continuous phase, and this leads the dispersed particles to form fibrillated chains, which contribute to abrupt increases of the rheological parameters (Hao 2001). In the case of the negative ER effect, the dispersed particles attain a certain amount of surface charge caused by injection at the electrodes, producing a subsequent deposit of solids on the plate and leaving the interelectrode region formed by mostly pure liquid, which causes reduction in the rheological properties (Ramos-Tejada et al. 2010).

Among the ER materials, conducting polymer-based ER materials have been especially well used as dry-base ER materials to avoid device corrosion, water evaporation, narrow operational temperature range, and dispersion instability in applications (Cho et al. 1999). To overcome these shortcomings, materials intrinsically possessing polarizable species such as electrons and ions were introduced (Block et al. 1990). Besides, various conducting polymers have been combined with inorganic materials such as montmorillonite (Erol et al. 2010), mesoporous materials (Cho et al. 2004), TiO_2 (Fang et al. 2006), and SiO_2 (Liu et al. 2011) to improve their ER performance with the aid of synergistic characteristics from both polymers and inorganic compounds.

In this study, *in situ* polymerized PIn/O-MMT nanocomposites and PIn/colemanite were used as ER active materials. In this manner, after determining the antisedimentation ratios, the effects of applied electric field strength, shear rate, and frequency onto ER performance of the materials were investigated and creep-recovery characteristics were examined.

8.2 EXPERIMENTAL

8.2.1 MATERIALS

Colemanite ($2CaO.3B_2O_3.5H_2O$) was kindly supplied by ETI Mining Co. (Turkey). Natural bentonite was kindly supplied by Samas Co. of Istanbul, Turkey, enriched to increase the content of Na-MMT and then organically modified with cetyltrimethylammonium bromide $\left[CTAB, C_{16}H_{33}N(CH_3)_3^+ Br^- \right]$ to obtain organically modified O-MMT (Guzel et al. 2012a). Indene (Aldrich, Germany) was used after distillation. All the other chemicals were Aldrich products and used as received.

8.2.1.1 Synthesis of PIn/O-MMT Nanocomposite

PIn was *in situ* polymerized in the presence of O-MMT using $FeCl_3$ as an oxidizing agent in CH_3Cl at 15–20°C overnight, taking the ratio of monomer to oxidant as 1:1. Then, the crude product was washed with deionized water and diethyl ether, respectively, and vacuum dried before use. Details of the synthesis and full characterization of PIn and PIn/O-MMT nanocomposite may be found in the literature (Guzel et al. 2012a).

8.2.1.2 Synthesis of PIn/Colemanite Composite

PIn was *in situ* polymerized in the presence of colemanite using $FeCl_3$ as the oxidizing agent, taking the monomer-to-initiator ratio as 1:2 in $CHCl_3$, and the obtained PIn/colemanite was subjected to various characterization techniques, namely, Fourier transform infrared spectroscopy, particle size, magnetic susceptibility, density, conductivity, dielectric, thermogravimetry/differential scanning calorimetry, X-ray diffraction, scanning electron microscopy, and electrokinetic measurements (Cetin et al. 2012).

8.2.1.3 Preparation of Dispersions

Dispersions of PIn, PIn/O-MMT, and PIn/colemanite composite were prepared at $\varphi = 25$ V/V%, by dispersing definite amounts of solid particles in SO ($\rho = 0.965$ g cm^{-3}, $\eta = 1.0$ Pa s, $\varepsilon = 2.61$), which were both vacuum dried before mixing in an oven overnight to remove any moisture present. All the prepared dispersions were allowed to equilibrate overnight before ER measurements.

8.2.2 METHODS

8.2.2.1 Antisedimentation Ratio Measurements

Antisedimentation stabilities against gravitational forces of the dispersions were determined at $T = 25 \pm 0.1$°C. Glass tubes containing the above dispersions were immersed into a constant temperature water bath. During the neat eye observations, the height of phase separation between the particle-rich phase and the relatively clear oil-rich phase was recorded as a function of time by using a digital composing stick. The antisedimentation ratio was defined as the height of the particle-rich phase divided by the total height of dispersion.

8.2.2.2 ER Measurements

ER properties of the dispersions were determined with a Thermo-Haake RS600 parallel plate torque electrorheometer (Germany). The gap between the parallel plates was 1.0 mm and the diameters of the upper and lower plates were 35 mm. The potential used in these experiments was supplied by a 0–12.5 kV (with 0.5 kV increments) dc electric field generator (Fug Electronics, HCL 14, Germany), which enabled resistivity to be created during the experiments.

8.2.2.3 Creep and Recovery Measurements

During the creep-recovery experiment, a constant stress was applied instantaneously to the dispersions at constant conditions ($\varphi = 25$%, $T = 25$°C, $E = 0$ kV/mm, and

$E \neq 0$ kV/mm) and change in strain (γ) was measured over a period. Then, the stress was removed and the recoverable elastic portion of the deformation was determined.

8.3 RESULTS AND DISCUSSIONS

8.3.1 ANTISEDIMENTATION STABILITIES OF THE DISPERSIONS

Self-assembly of dispersed particles perpendicular to the direction of electric field to form chain-like structures between the upper and lower electrodes is valid if the dispersions are colloidally stable. Thus, gravitational stability is one of the most important desired parameter for colloidal dispersions. Various parameters affect the colloidal stability of dispersions, such as size and type of the dispersed particle (i.e., hollow, porous, lamellar), viscosity and density of the dispersant, and presence of any surfactant in the colloidal system. Antisedimentation ratio results obtained are depicted in Figure 8.1, and it was observed that PIn/O-MMT showed relatively highest antisedimentation ratio at the end of 25 days.

If the densities of the particles are compared with each other, it can be seen that the materials have very close density values (Table 8.1). On the other hand, the particle size of PIn is smaller than that of PIn/O-MMT and PIn/colemanite, whose particle sizes are similar. It can be concluded that the flake-like structure of MMT was inserted in the PIn chains in the nanocomposite and thus facilitated colloidal stability of PIn/O-MMT compared to PIn/colemanite.

8.3.2 ER STUDIES

Viscosity values of the dispersions at various electric field strengths are expressed in Figure 8.2. It was observed that an increase in externally applied electric field strength caused an increase in the viscosity of PIn/SO and PIn/O-MMT/SO colloidal systems, whereas it caused a decrease in the viscosity of the PIn/colemanite/SO

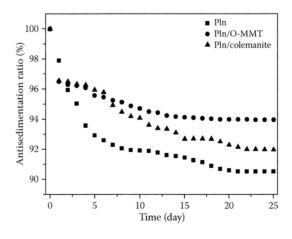

FIGURE 8.1   Antisedimentation ratio results of the dispersions against time ($\varphi = 25\%$, $T = 25°C$).

TABLE 8.1
Some Physical Characteristics of the Samples

Sample	Conductivity[a] $(\sigma$, S cm$^{-1})$	Dielectric Constant[a]	Density[a] $(\rho$, g cm$^{-3})$	Average Particle Size[a] $(d_{0.5}, \mu m)$
Pin	9.3×10^{-6}	4.3	1.04	1.1
PIn/O-MMT	5.1×10^{-6}	5.8	1.08	2.3
O-MMT	2.8×10^{-7}	6.5	1.58	5.5
PIn/colemanite	2.48×10^{-4}	83	1.02	2.2
Colemanite	3.14×10^{-4}	85	1.69	1.2

[a] See Guzel et al. (2012a,b) and Cetin et al. (2012).

colloidal system. When an electric field was applied, the particles became polarized and tended to attract each other, and consequently the fibril-like structures were formed aligned to the direction of the electric field, which provides additional resistance against flow. This situation is valid for PIn/SO and PIn/O-MMT/SO colloidal dispersions. But when the composition of composite was changed, the ER response was reversed to negative and a decrease in viscosity was observed.

The structural characteristics of MMT exhibit an octahedral aluminate sheet sandwiched between tetrahedral silicate layers, where the layer charge can easily be controlled by exchange of cations with different charges. It is generally accepted that particle polarization is responsible for the interaction force between the particles under the applied electric field. When the electric field is applied, cations or organic polymers adsorbed in the interlayer of MMT can move to induce interfacial polarization (Xiang and Zhao 2006). This leads to a positive ER response. Colemanite

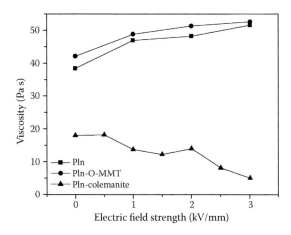

FIGURE 8.2 Viscosity versus applied electric field strength of the dispersions ($\varphi = 25\%$, $T = 25°C$).

has $[B_3O_4(OH)_3]_n^{2n-}$ polyanion groups and crystal water, which are held together by weak H-bonds involving both water molecules and hydroxyl groups in its structure. Upon the application of the electric field, charge injection at the electrodes may occur. Electrochemical reactions at the electrode–fluid interface cause charge accumulation on the particles and in the interparticle region (Felici 1997), which results with the dominated particle charging effect. The particle–particle interaction forces on PIn/O-MMT dispersion are large enough to hide the particle charging effect, and fibrillar chain-like structures between the dispersed particles produce a positive ER response.

Figure 8.3 shows shear stress (τ) and viscosity (η) versus shear rate $\dot{\gamma}$ data with and without the presence of applied electric field. Viscosities of PIn/O-MMT/SO and PIN/colemanite/SO colloidal dispersions show rapid decrease upon the application

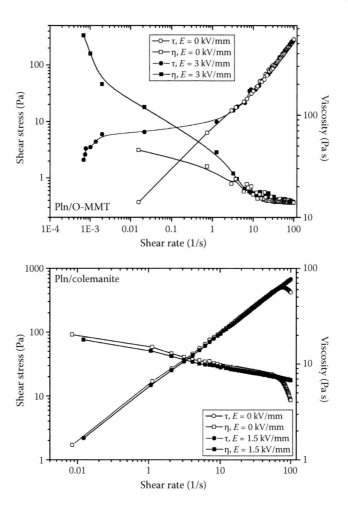

FIGURE 8.3 Flow curves of PIn/O-MMT and PIn/colemanite dispersions with and without the presence of electric field ($\varphi = 25\%$, $T = 25°C$).

of shear rate and result in shear-thinning behavior. On the other hand, the effect of the applied field is more pronounced at low shear rates, when the hydrodynamic interactions are weaker. Finally, at high shear rates, the curve corresponding to electric field strength tends to merge and reach the values in the absence of the electric field, showing that the electric field ceases to have an effect on the rheogram. But for PIn/colemanite, weaker rheological properties were observed under the electric field than without the electric field.

Creep and creep-recovery tests, which are described as the time-dependent increase in strain (γ) of a viscoelastic material under sustained stress (τ_0), are a critical method to obtain information on the deformation and recovery properties of the materials.

For viscoelastic materials, in the creep phase, strain (γ) increases with time under maintained stress (τ). In the recovery phase, when the applied stress is removed, the time-dependent deformation may be recoverable with time.

Figure 8.4 shows the change in strain with time for PIn/O-MMT/SO and PIn/colemanite/SO colloidal dispersions under $E = 0$ and $E \neq 0$ kV/mm conditions. PIn/colemanite/SO dispersion behaved like purely viscous material and no recovery took place after the stress was removed both under $E = 0$ kV/mm and $E = 3$ kV/mm conditions. Also, the strains that occurred in the PIn/colemanite dispersion with the applied stress ($\tau_0 = 5$ Pa) under $E = 3$ kV/mm were higher than those under $E = 0$

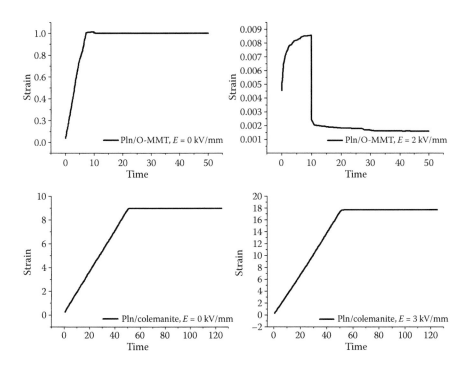

FIGURE 8.4 Creep-recovery responses of PIn/O-MMT/SO and PIn/colemanite/SO colloidal dispersions with and without the presence of electric field ($\varphi = 25\%$, $\tau_0 = 5$ Pa, $T = 25°C$).

kV/mm owing to the negative ER behavior shown. On the other hand, PIn/O-MMT dispersion behaved like a viscoelastic fluid under $E = 2$ kV/mm. For the $E = 0$ kV/mm condition, after the applied stress was removed ($\tau_o = 0$ Pa), there was no instantaneous elastic recovery, corresponding to a non-recoverable viscous deformation at the end of 10 s, which indicates that energy used for bond stretching is not stored and totally distributed in the PIn/O-MMT/SO colloidal system. Under $E = 2$ kV/mm, the following changes were recorded at the creep phase: (i) instantaneous elastic response, (ii) retarded elastic deformation, and (iii) viscous flow as a result of linear increase in strain. Besides, at the recovery phase under $E = 2$ kV/mm, the following changes were observed: (i) recovery of instantaneous strain and (ii) irreversible viscoelastic recovery. In conclusion, non-linear viscoelastic behaviors were recorded for the materials under the applied electric field. The recoverable strain is a measure of elasticity of materials, which is an indication of solid-like behavior of the suspensions under the applied electric field; the fibrillar-like aggregates of the dispersed particles caused solid-like structures resulting in viscoelastic response (Guzel et al. 2012b). These results showed that PIn/O-MMT nanocomposite became polarized under the applied electric field and stored the electric field–induced deformation, thus classified as a smart material.

8.4 CONCLUSIONS

ER properties of colloidal dispersions based on PIn/O-MMT nanocomposite and PIn/colemanite composite were investigated. It was found that the inorganic component in the nanocomposite or composite structure affected the ER response of the materials and the colloidal stability of the dispersions significantly. The PIn/colemanite/SO system showed negative ER behavior under all the applied electric field conditions, which was attributed to the migration of the dispersed particles to one of the electrodes, whereas the PIn/O-MMT/SO colloidal system showed positive ER response under applied electric field strength, which indicated self-assembly of the PIn/O-MMT particles perpendicular to the direction of the electric field between the electrodes.

ACKNOWLEDGMENTS

We are grateful to the European Science Foundation through COST Action D43, the Turkish Scientific and Technological Research Council (Grant No. 107T711), and the Gazi University Research Fund (Grant No. 05/2011-05, 05/2012-34) for the support of this work.

REFERENCES

Block, H., J.P. Kelly, A. Qin and T. Watson. 1990. Materials and mechanisms in electrorheology. *Langmuir* 6:6–14.
Cetin, B., H.I. Unal and O. Erol. 2012. Synthesis, characterization and electrokinetic properties of polyindene/colemanite conducting composite. *Clay. Clay Miner.* 60:300–14.

Cho, M.S., H.J. Choi and W.-S. Ahn. 2004. Enhanced electrorheology of conducting polyaniline confined in MCM-41 channels. *Langmuir* 20:202–7.

Cho, M.S., H.J. Choi, I.J. Chin and W.S. Ahn. 1999. Electrorheological characterization of zeolite suspensions. *Micropor. Mesopor. Mater.* 32:233–9.

Erol, O., H.I. Unal and B. Sari. 2010. Synthesis, electrorheology, and creep behaviors of in situ intercalated polyindole/organo-montmorillonite conducting nanocomposite. *Polym. Compos.* 31:471–81.

Fang, F.F., J.H. Sung and H.J. Choi. 2006. Shear stress and dielectric characteristics of polyaniline/TiO$_2$ composite-based electrorheological fluid. *J. Macromol. Sci. B* 45:923–32.

Felici, N.J. 1997. Interfacial effects and electrorheological forces: criticism of the conduction model. *J. Electrostat.* 40/41:561–72.

Guzel, S., H.I. Unal, O. Erol and B. Sari. 2012a. Polyindene/organo-montmorillonite conducting nanocomposites. I. synthesis, characterization, and electrokinetic properties. *J. Appl. Polym. Sci.* 123:2911–22.

Guzel, S., O. Erol and H.I. Unal. 2012b. Polyindene/organo-montmorillonite conducting nanocomposites. II. Electrorheological properties. *J. Appl. Polym. Sci.* 124:4935–44.

Han Y.-M. and S.-B. Choi. 2008. Control of an ER haptic master in a virtual slave environment for minimally invasive surgery applications. *Smart Mater. Struct.* 17:065012–21.

Hao, T. 2001. Electrorheological fluids. *Adv. Mater.* 13:1847–57.

Litvinov, W.G. 2007. Dynamics of electrorheological clutch and a problem for non-linear parabolic equation with non-local boundary conditions. *IMA J. Appl. Math.* 73:619–40.

Liu, L., X. Chen, X. Niu, W. Wen and P. Sheng. 2006. Electrorheological fluid actuated microfluidic pump. *Appl. Phys. Lett.* 89:083505–7.

Liu, Y.D., F.F. Fang, H.J. Choi and Y. Seo. 2011. Fabrication of semiconducting polyaniline/nano-silica nanocomposite particles and their enhanced electrorheological and dielectric characteristics. *Colloid. Surf. A* 381:17–22.

Nguyen, Q.-H. and S.-B. Choi. 2009. A new approach for dynamic modeling of an electrorheological damper using a lumped parameter method. *Smart Mater. Struct.* 18:115020–30.

Ramos-Tejada, M.M., F.J. Arroyo and A.V. Delgado. 2010. Negative electrorheological behavior in dispersions of inorganic particles. *Langmuir* 26:16833–40.

Tao, R., K. Huang, H. Tang and D. Bell. 2008. Electrorheology leads to efficient combustion. *Energy Fuels* 22:3785–88.

Tsai, Y.Y., C.H. Tseng and C.K. Chang. 2008. Development of a combined machining method using electrorheological fluids for EDM. *J. Mater. Process. Technol.* 201:565–9.

Wereley, N., J. Lindler, N. Rosenfeld and Y.-T. Choi. 2004. Biviscous camping behavior in electrorheological shock absorbers. *Smart Mater. Struct.* 13:743–52.

Winslow, W.M. 1949. Induced fibration of suspensions. *J. Appl. Phys.* 20:1137–40.

Xiang, L. and X. Zhao. 2006. Preparation of montmorillonite/titania nanocomposite and enhanced electrorheological activity. *J. Colloid Interface Sci.* 296:131–40.

Zhang, M., J. Wu, X. Niu, W. Wen and P. Sheng. 2008. Manipulations of microfluidic droplets using electrorheological carrier fluid. *Phys. Rev. E* 78:066305–9.

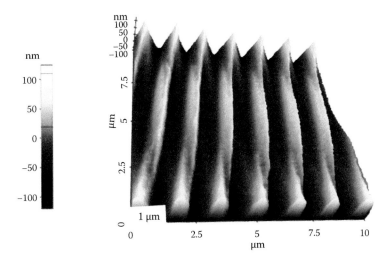

FIGURE 3.4 Atomic force microscopy image showing the channels of the commercial CD. (From *Nanotechnology*, IOP Publishing. With permission.)

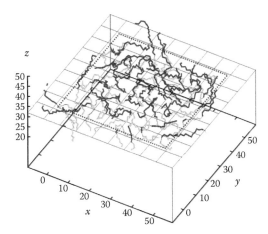

FIGURE 11.2 3D snapshot of the $C_{12}E_5$ monolayer structure at time $t = 800$ ps. The sections of these molecules underneath the interface is in green and those located above it are in red. The dotted line indicates the border of the SD in x and y directions. Hydrogen atoms and water molecules are not plotted.

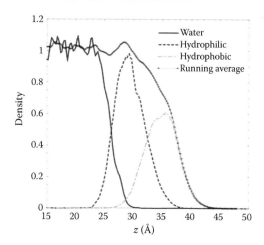

FIGURE 11.3 $C_{12}E_5$ monolayer structure. Water density profile (blue), hydrophilic region (red), hydrophobic region (green), and total mass density distributions (black). Averaging time is 1 μs.

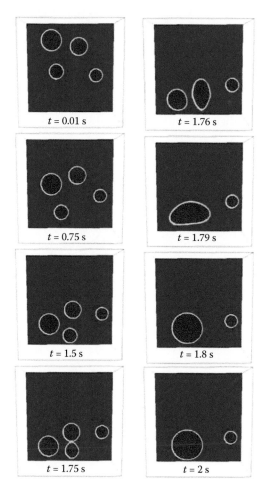

FIGURE 11.10 Snapshots of the time evolution of a system consisting initially of four distinct droplets. The time is indicated below each image. The parameters of the simulation are the same as in Figures 11.4 and 11.6.

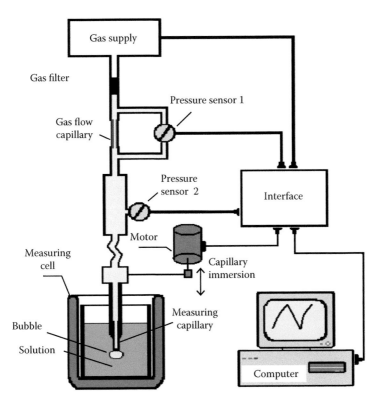

FIGURE 13.2 Principle of the maximum bubble pressure tensiometer BPA-1S (SINTERFACE Technologies, Berlin, Germany).

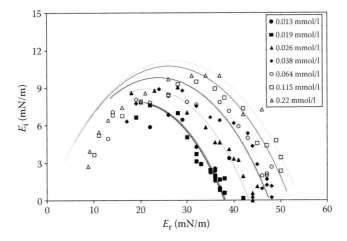

FIGURE 13.13 Cole–Cole diagram for the complex dilational viscoelasticity modulus for different $C_{12}DMPO$ concentrations obtained from oscillating bubble experiments, as presented in Kovalchuk et al. (2004). The lines are the theoretical predictions according to the Lucassen–van den Tempel model.

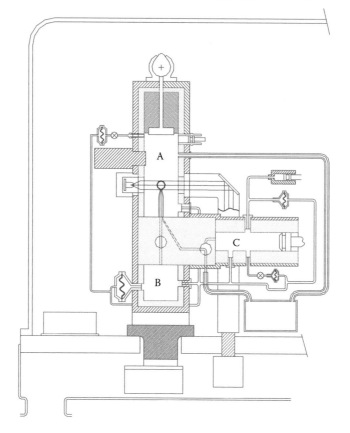

FIGURE 13.17 Scheme of the measurement cell for the generation and observation of thin liquid films (partial drawings). The main compartment (upper chamber, A) contains the matrix hydrocarbon and holds a coaxial double capillary. The lower chamber, B, contains the hydrocarbon for inflating the film. The aqueous surfactant solution is stored in the lateral chamber, C, featuring a piston device for delivering a definite amount of liquid to the capillary.

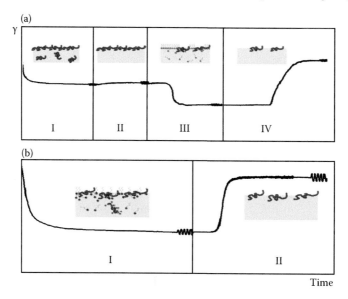

FIGURE 13.19 Experimental protocol for sequentially (a) and simultaneously (b) formed mixed layers performed with a coaxial double capillary to measure dynamic surface tensions.

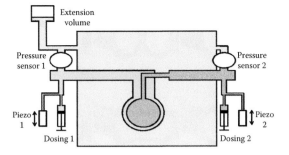

FIGURE 13.20 Schematic principle of concentric drops formation at the tips of the coaxial double capillary, linked with respective pressure sensors.

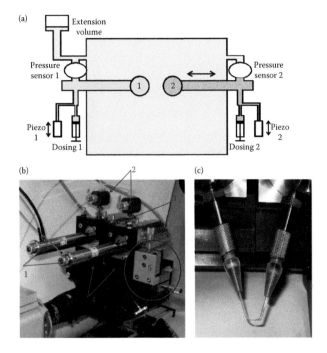

FIGURE 13.21 Schematic and photo of a DBMM with its main elements and capillaries. (a) Schematic of DBMM; (b) 1, piezo translator; 2, valve for injection of liquid from the syringe dosing system; 3, pressure sensor; 4, holder with capillary; 5, xyz micrometer stage. (c) Glass capillaries.

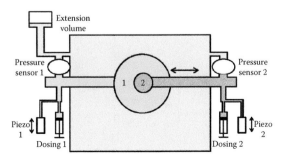

FIGURE 13.22 Two different ways of rearrangement of the DBMM: concentric drops using different sizes of capillaries.

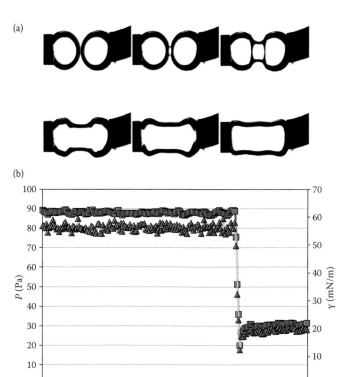

FIGURE 13.23 Steps of a coalescence process of two CTAB solution drops in hexane: (a) photos taken with a fast camera with time intervals of 500 μs between the images; (b) CP (ν) and interfacial tension (π) measured in the left drop before, during, and after the coalescence process.

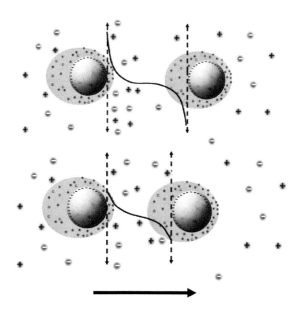

FIGURE 14.5 Schematic representation of the polarization clouds around negatively charged spherical particles for dilute (top) and concentrated (bottom) suspensions.

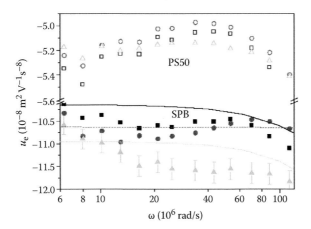

FIGURE 14.13 Dynamic mobility of polyelectrolyte brushes (SPBs) and polystyrene spheres 50 nm in radius (PS50) as a function of the frequency of the applied electric field for different particle concentrations in 0.5 mM NaCl. Symbols denote the experimental results. Lines are the result of model predictions with the following parameters: particle charge: -5×10^{-16} C, polymer charge plus condensed counterions charge: -8.5×10^{-14} C, dimensionless friction parameter: $\lambda R_c = 110$ at the particle surface. Squares and solid line: 5 wt%. Circles and dashed line: 6 wt%. Triangles and dotted line: 7 wt%. The dispersion in these measurements is around 2%–3%. (Details in Jimenez et al. 2011. http://pubs.rsc.org/en/content/articlelanding/2011/sm/c0sm01544j. Reproduced by permission of the Royal Society of Chemistry.)

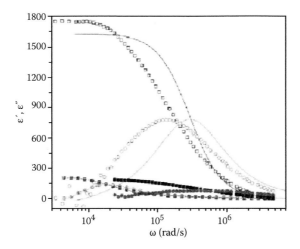

FIGURE 14.14 Real (squares) and imaginary (circles) parts of the relative permittivity of suspensions of SPBs (open symbols), polystyrene particles 50 nm in radius, PS50 (filled symbols), and 168 nm in radius, PS168 (crossed symbols), as a function of the frequency of the applied field. Particle concentration: 4 wt%. Ionic strength: 0.5 mM (NaCl). Lines are the model predictions with the same parameters as in Figure 14.13, and particle concentration is 4 wt%. (Details in Jimenez et al. 2011. http://pubs.rsc.org/en/content/articlelanding/2011/sm/c0sm01544j. Reproduced by permission of the Royal Society of Chemistry.)

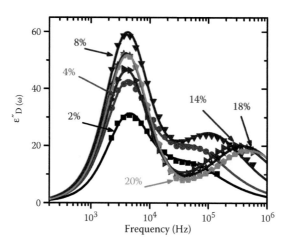

FIGURE 14.15 Spectra of the logarithmic derivative of the relative electric permittivity of suspensions of elongated hematite particles (semiaxes $a = 276 \pm 18$ nm, $b = 45 \pm 6$ nm) in a 0.5 mM KCl solution at pH 4, for the indicated values of the volume fraction of solids (ϕ). Symbols: experimental data. Lines: best fits to the logarithmic derivative of a combination of two Cole–Cole relaxation functions. (Details in Rica et al. 2011. http://pubs.rsc.org/en/content/articlelanding/2011/sm/c1sm05153a. Reproduced by permission of The Royal Society of Chemistry.)

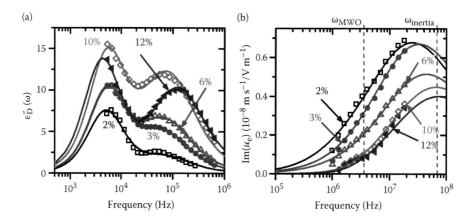

FIGURE 14.16 Spectra of (a) the logarithmic derivative of the relative electric permittivity and (b) the imaginary part of the dynamic electrophoretic mobility of suspensions of elongated hematite particles (semiaxes $a = 276 \pm 18$ nm, $b = 45 \pm 6$ nm) in a 0.5 mM KCl solution at pH 5.8, for the indicated values of the volume fraction of the solids (ϕ). Symbols: experimental data. Lines: best fits to the models. Dashed horizontal lines in panel (b) indicate the approximate position of the MWO relaxation and inertia decay in the electroacoustic spectrum of the suspension at 10% concentration. (Details in Rica et al. 2012. http://pubs.rsc.org/en/content/articlelanding/2012/sm/c2sm07365j. Reproduced by permission of the Royal Society of Chemistry.)

9 Oil-in-Water Microemulsions for the Synthesis of Nanocrystalline, Mesoporous, and Ultrafine CeO$_2$ Powders

Margarita Sanchez-Dominguez, Kelly Pemartin, Conxita Solans, and Magali Boutonnet

CONTENTS

9.1 INTRODUCTION

The use of surfactant-based systems for the synthesis of inorganic nanoparticles has been gaining interest owing to certain advantages such as the great control in particle size, dispersion, and porosity. Recently, we developed a novel method for the synthesis of inorganic nanoparticles based on oil-in-water (O/W) microemulsions (Sanchez-Dominguez et al. 2009), in contrast to the typically used water-in-oil (W/O) microemulsion reaction method (Boutonnet et al. 1982; Destrée and Nagy 2006; Eastoe et al. 2006). The aim of this chapter is to discuss our recent developments about the use of O/W microemulsions as confined reaction media for the synthesis of pure cerium oxide nanoparticles, as well as the synthesis of Eu-doped cerium oxide. In addition, the preparation of Au-impregnated cerium oxide and its use in the catalytic oxidation of CO is also discussed.

9.1.1 PROPERTIES AND APPLICATIONS OF CeO$_2$

There is an enormous interest in cerium oxide (ceria or CeO$_2$) due to the large number of applications in different areas: catalytic processes and fuel cells, photocatalysis, solar cells, gates for metal oxide semiconductor devices, phosphors, abrasive for planarization of silicon substrates, medicine, gas sensors, oxygen scavenger, and so on (Yuan et al. 2009). In addition, cerium oxide is used in selective solar control and low-emission windows in buildings in order to avoid incoming solar light and heat (Ohsaki et al. 1997). Most of these applications are due to its elevated oxygen storage capacity (OSC) combined with the ability to shift easily between the Ce^{4+} and Ce^{3+} oxidation states (Andersson et al. 2006; Tsunekawa et al. 2000).

It may be said that one of the most extended applications of cerium oxide is in the catalysis field, either as catalysts or as supports, owing to the unique redox properties on the surface of cerium oxide and its OSC, which is highly related to the oxygen vacancies in its crystalline structure (face-centered cubic, fluorite type). A high oxygen deficiency in the structure of cerium oxide is very likely, since these oxides are often exposed to different environments such as reducing atmospheres or high-temperature conditions. It has been used as an oxidative catalyst, for selective hydrogenation of unsaturated compounds, in photocatalytic oxidation of water and other catalytic reactions (Khaleer et al. 2010), for the oxidation of carbon monoxide (Sanchez-Dominguez et al. 2010), and in the water–gas shift reaction at low temperatures, which is an important reaction for the production of hydrogen in the reforming process (Djinovic et al. 2009; Kusar et al. 2006). In addition, a very important application of cerium oxide in the catalysis field is the three-way catalysts in the converters of automobiles, which convert toxic by-products in the exhaust of the internal combustion engine to less toxic substances, for example, oxidation of CO and hydrocarbons and reduction of NO$_x$. This ability is related to the excellent properties of cerium oxide as an oxygen buffer, which arises from its high tolerance to oxygenation/deoxygenation reversible cycles without changes in its cubic fluorite structure.

Other important properties of CeO$_2$ arise when it is synthesized as nanoparticles, such as the expansion of its crystalline lattice, increase in its electrical conductivity, and acceleration of CO oxidation. The achievement of high specific surface area (SSA) upon increasing the porosity and decreasing the particle size is of great interest for applications in heterogeneous catalysis. Several strategies have been employed in order to obtain a high SSA, such as the control of reaction time and the use of high temperatures and pressures in the hydrothermal method. Other methods include pyrolysis, thermal decomposition, combustion of solids, sol-gel, and gas condensation, among others. These methods have been used for the synthesis of pure as well as doped CeO$_2$. The characteristics of the obtained CeO$_2$ materials, in particular oxygen vacancies, the Ce^{3+}/Ce^{4+} ratio on the ceria structure, porosity, and SSA, depend strongly on the preparation method (Bumajdad et al. 2009).

9.1.2 MICROEMULSIONS AS REACTION MEDIA FOR THE SYNTHESIS OF CeO$_2$ NANOPARTICLES

It has been previously demonstrated that the use of surfactant-based self-assemblies as confined reaction media for nanoparticle synthesis is an important tool for the development of materials with controlled size, shape, surface area, and other properties. Microemulsions are one type of surfactant system used for that purpose. These are transparent and thermodynamically stable colloidal dispersions in which two liquids initially immiscible (typically water and oil) coexist in one phase owing to the presence of a monolayer of surfactant molecules (Danielsson and Lindman 1981). Depending on the ratio of oil and water and on the hydrophilic–lipophilic balance (HLB) of the surfactant(s), microemulsions can exist as oil-swollen micelles dispersed in water (O/W microemulsions) or water-swollen inverse micelles dispersed in oil (W/O microemulsions); at intermediate compositions and temperatures, microemulsions with both aqueous and oily continuous domains can exist as interconnected sponge-like channels (bicontinuous microemulsions). In contrast to emulsions, which require a considerable energy input for their formation, microemulsions form spontaneously upon gentle mixing of its components, once the thermodynamic conditions (composition and temperature) are appropriate.

As a result of their small droplet size (2–100 nm), discrete-type microemulsions (O/W and W/O) are subject to continuous Brownian motion; collisions between micelles are hence frequent, leading to the formation of transient dimers and continuous exchange of the interior of the droplets. These dynamic properties facilitate their use as confined reaction media. Since the early research of Boutonnet et al. (1982), where W/O microemulsions were used as nano-reactors for the synthesis of metallic nanoparticles, an increased interest in this approach has followed, given the advantages that this method offers such as control of the size, shape, and composition of the nanomaterials, as well as the use of simple equipment and, in general, soft reaction conditions. In addition, the presence of surfactant molecules can protect nanoparticles against agglomeration. The main strategy for the synthesis of nanoparticles in W/O microemulsions consists in mixing two microemulsions, one containing the metallic precursor and another one containing the precipitating agent. Upon mixing, both reactants will contact each other as a result of droplet collisions and coalescence and will react to form precipitates of nanometric size (Figure 9.1); the nanoparticles formed are stabilized by the surfactant molecules. The literature on this topic has grown steadily and there are excellent reviews about different aspects such as reactivity, mechanisms, and control of particle size and shape (Boutonnet et al. 2008; Destrée and Nagy 2006; Eastoe et al. 2006; Holmberg 2004; López-Quintela 2003; López-Quintela et al. 2004; Pileni 1997, 2003).

It has been reported that materials synthesized in W/O microemulsions exhibit unique surface properties; for example, nano-catalysts prepared by this method

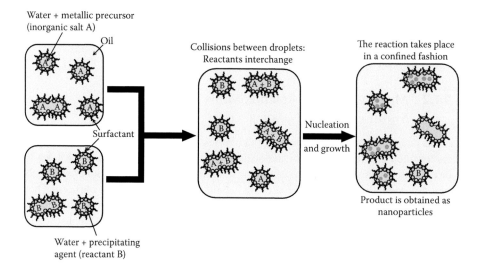

FIGURE 9.1 Synthesis of inorganic nanoparticles in W/O microemulsions.

show better performance (activity, selectivity) than those prepared by other methods (Boutonnet et al. 2008). But despite the superior properties and performance of nanoparticles obtained in W/O microemulsions, this method has not found good acceptance at the industrial level (Boutonnet et al. 2008), mainly because of the employment of large amounts of oils (solvents), which represent the continuous phase and, hence, the main component of these systems (Boutonnet et al. 2008). In addition, most studies employ relatively low concentration of the metal precursors, leading to small yields of nanoparticles per microemulsion volume. These drawbacks affect negatively from the economic and ecologic point of view.

With this in mind, a novel approach based on O/W microemulsions has been investigated in recent years for the synthesis of metal and metal oxide nanoparticles. The method consists of using organometallic precursors, dissolved in nanometer-scale oil droplets of O/W microemulsions (Figure 9.2). The precipitating agents, usually water

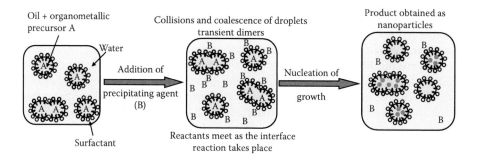

FIGURE 9.2 Synthesis of inorganic nanoparticles in O/W microemulsions (one-microemulsion approach).

soluble, can be added directly or as aqueous solutions (one-microemulsion approach), without compromising microemulsion stability and droplet size; alternatively, if oil-soluble precipitating agents are available, then a two-microemulsion approach may be used. Cerium oxide was the first type of metal oxide nanoparticles synthesized by this method as proof of the concept in 2009, along with metallic nanoparticles (Pt, Pd, Rh) (Sanchez-Dominguez et al. 2009). Since then, the synthesis of ceria nanoparticles in O/W microemulsions has been optimized; furthermore, hybrid ceria-based materials—Eu-doped and Au-impregnated ceria—have been produced as well.

9.2 PREPARATION OF CeO$_2$ NANOPARTICLES BY THE O/W MICROEMULSION REACTION METHOD

9.2.1 USE OF VARIOUS MICROEMULSION FORMULATIONS AND PRECIPITATING AGENTS

In the first stage of synthesis of CeO$_2$ nanoparticles by the O/W microemulsion reaction method, several microemulsion systems and compositions, cerium(III) precursors, and precipitation strategies were explored. The microemulsion systems studied were as follows: water/Tween 80/Span 20/1,2-hexanodiol/ethyl oleate (System A); water/Brij 96V/butyl-*S*-lactate (System B), and water/Synperonic 10/5/isooctane (System C). The following were the organometallic precursors used: cerium(III) acetylacetonate hydrate and cerium(III) 2-ethylhexanoate (Ce-AA and Ce-EH, respectively). The detailed microemulsion compositions, temperatures, and precipitation methods are shown in Table 9.1. The precursor Ce-AA could only be used with System A since this precursor was only soluble in the oil and cosurfactant comprised in this system (ethyl oleate and 1,2-hexanediol, respectively), although even in these solvents, its solubility was very limited. In contrast, the solubility of Ce-2EH was much higher in isooctane and butyl-*S*-lactate and low in ethyl oleate; hence, experiments with this precursor were carried out with Systems B and C. Synthesis was carried out by following different precipitation strategies. In one approach, concentrated H$_2$O$_2$ was added directly to the microemulsion, which resulted in the precipitation of yellow-orange particles. A variation of this approach was to add a second microemulsion in which some of the water had been replaced by H$_2$O$_2$ (and without precursor in the oil phase). The third strategy was to add concentrated ammonia, either directly to reach pH 11 or in a stoichiometric 1:1 ratio, in a second microemulsion; this approach in general resulted in yellow particles. Finally, another strategy was to combine the previous two, that is, to pH 11 with ammonia followed by a certain amount of H$_2$O$_2$.

Because of the low solubility of Ce-AA, investigation using this precursor was not continued further, since the maximum CeO$_2$ loading was only 0.1 g/kg of microemulsion. In contrast, the CeO$_2$ loading with the systems comprising Ce-2EH (Systems B and C) could reach up to 2.58 g of CeO$_2$ per kilogram of microemulsion, and such yield may be optimized further. Irrespective of the microemulsion system, the metal precursor, and the precipitating strategy used, the size observed by transmission electron microscopy (TEM) was below 5 nm, and generally approximately 2–3 nm, as observed in the sample TEM pictures from Figure 9.3; crystalline features could be detected by high-resolution TEM (HRTEM) (Figure 9.3a and b).

TABLE 9.1

Parameters of Microemulsion Composition and Precipitation Method Used for the Synthesis of CeO_2 Nanoparticles in the First Stage of the Investigation

Sample	Micro-emulsion System	Precursor Solution(s)	Oil Phase Content (wt%)	S^a Content (wt%)	CeO_2 Yield[b]	Precipitation Method
CeO_2-A1	A	Ce-AA in oil (0.004 wt% Ce) Ce-AA in CS2[c] (0.06 wt% Ce)	10.0	40.0[d]	0.1	H_2O_2 10× added directly
CeO_2-B1	B	Ce-EH in oil (0.98 wt% Ce)	5.0	10.0	0.50	H_2O_2 2× added in second microemulsion
CeO_2-B2	B	Ce-EH in oil (0.98 wt% Ce)	20.0	20.0	2.0	H_2O_2 2× added directly
CeO_2-B3	B	Ce-EH in oil (0.5 wt% Ce)	12.6	21.7	0.77	H_2O_2 2× added in second microemulsion
CeO_2-B4a	B	Ce-EH in oil (0.98 wt% Ce)	17	20.8	1.7	pH 11 with ammonia
CeO_2-B4b	B	Ce-EH in oil (0.98 wt% Ce)	17	20.8	1.7	pH 11 with ammonia and H_2O_2 2× added directly
CeO_2-B4c	B	Ce-EH in oil (0.98 wt% Ce)	17	20.8	1.7	pH 11 with ammonia and H_2O_2 4× added directly
CeO_2-B4d	B	Ce-EH in oil (0.98 wt% Ce)	17	20.8	1.7	pH 11 with ammonia and H_2O_2 10× added directly
CeO_2-B4e	B	Ce-EH in oil (0.98 wt% Ce)	17	20.8	1.7	pH 11.5 with ammonia and H_2O_2 10× added directly
CeO_2-B4f	B	Ce-EH in oil (0.98 wt% Ce)	17	20.8	1.7	H_2O_2 10× added directly
CeO_2-C1	C	Ce-EH in oil (0.1 wt% Ce)	14.0	21.5	0.26	H_2O_2 10× added in second microemulsion
CeO_2-C2	C	Ce-EH in oil (0.98 wt% Ce)	14.0	21.5	2.58	H_2O_2 2× added in second microemulsion
CeO_2-C3	C	Ce-EH in oil (0.98 wt% Ce)	14.0	21.5	2.58	Ammonia 1× in second microemulsion

Note: 10× and 2× indicate stoichiometry of precipitating agent with respect to Ce.

[a] S, surfactant.

[b] In grams of CeO_2 per kilogram of microemulsion.

[c] CS2, 1,2-hexanediol.

[d] S = S/CS1/CS2 (Tween 80/Span 20/1,2-hexanediol; weight ratio, 42:28:30).

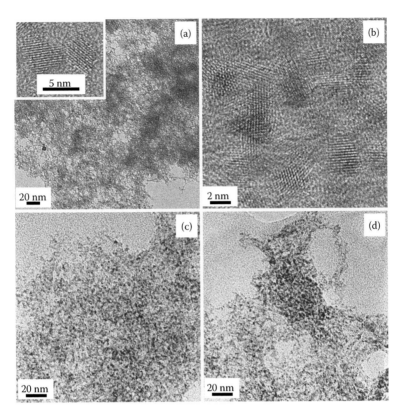

FIGURE 9.3   TEM images for (a) CeO$_2$-B1, (b) CeO$_2$-B2, (c) CeO$_2$-C1, and (d) CeO$_2$-C2. Inset to panel (a) is an HRTEM image of CeO$_2$-B1, showing lattice fringes of cubic fluorite–type ceria.

As reported previously (Sanchez-Dominguez et al. 2009), formation of nanocrystalline ceria for CeO$_2$-B3 and CeO$_2$-C3 was confirmed by x-ray diffraction (XRD) (Figure 9.4). From the reflection broadening (using the Debye–Scherrer equation), it was confirmed that small crystallite sizes were obtained (2.2 and 4.1 nm for CeO$_2$-B3 and CeO$_2$-C3, respectively). Hence, nanocrystalline ceria was obtained directly in the microemulsion at room temperature, without the need of calcination.

Figure 9.5 shows the XRD patterns of samples prepared using System B and the composition with 17 wt% oil phase (CeO$_2$-B4). The best-defined diffractogram was obtained for the sample precipitated at pH 11 with ammonia, without the use of H$_2$O$_2$, followed by the sample precipitated at pH 11 with ammonia and H$_2$O$_2$ in a 2:1 molar ratio with respect to Ce ions. The sample precipitated with only H$_2$O$_2$ was the most amorphous. However, the disadvantage of synthesis with System B and ammonia is that since the oil used is an ester (butyl-S-lactate), it is hydrolyzed in alkaline pH, forming by-products in the microemulsion media. Therefore, large amounts of ammonia are needed in order to reach pH 11. Hence, from the point of view of practicality, economy, ecology, and yield per microemulsion volume, the best

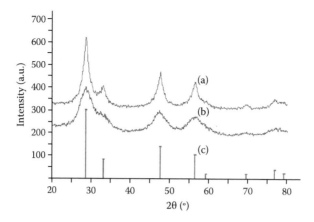

FIGURE 9.4 XRD patterns for (a) CeO_2-C3 and (b) CeO_2-B3. (c) Reference diffraction lines for cerianite with cubic fluorite–type structure. (With kind permission from Springer Science+Business Media: *J. Nanopart. Res.*, A novel approach to metal and metal oxide nanoparticle synthesis: the oil-in-water microemulsion reaction method, 11, 2009, 1823–1829, Sanchez-Dominguez, M. et al.)

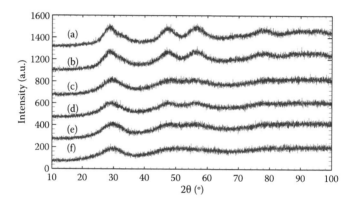

FIGURE 9.5 XRD patterns for (a) CeO_2-B4a, (b) CeO_2-B4b, (c) CeO_2-B4c, (d) CeO_2-B4d, (e) CeO_2-B4e, and (f) CeO_2-B4f.

approach developed in this preliminary study is to use the microemulsion CeO_2-C3, precipitated at pH 11 with ammonia (added directly to the microemulsion).

9.2.2 Optimized Synthesis Strategy and Systematic Microemulsion Composition

On the basis of the preliminary synthesis described above, a more systematic study was carried out. The surfactant Synperonic 10/5 was replaced with Synperonic 10/6 owing to unavailability (Synperonic 10/5 was no longer produced). In order to keep the temperature of microemulsion formation close to room temperature, hexane was used as oil instead of isooctane. The precursor used was Ce-2EH and the concentration of cerium in the oil phase was the same as in the preliminary study (1.48 wt% Ce). In this way, the composition with 14 wt% oil phase, 21.5 wt% surfactant, and 64.5 wt% aqueous phase formed a microemulsion at 35°C.

In order to select additional microemulsion compositions, the partial phase diagram (water corner) for water/Synperonic 10/6/hexane (System D) was studied at 35°C with and without precursor Ce-2EH, as shown in Figure 9.6a–c. According to this phase behavior study, Ce-2EH precursor may have a certain interfacial activity in O/W microemulsions, as it was observed that addition of the precursor slightly reduced the microemulsion region. This result suggests that the organometallic

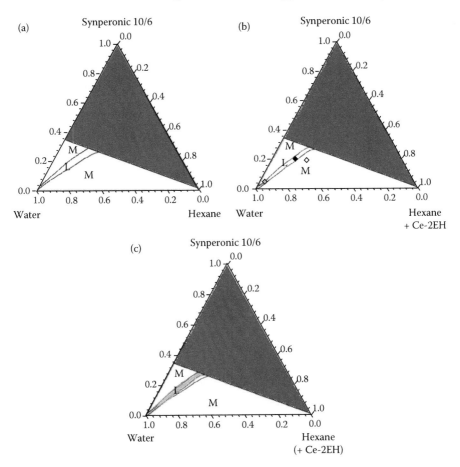

FIGURE 9.6 Equilibrium partial phase diagram of the system: water/Synperonic 10/6/hexane (35°C): (a) without precursor; (b) with Ce-2EH precursor; (c) diagrams from (a) and (b) overlapping; shaded area and shaded + non-shaded area represent the L (microemulsion) region for systems with and without Ce-2EH, respectively. M indicates the multiphasic region. Gray, black, and white diamonds in diagram (b) indicate compositions used for CeO$_2$ synthesis. (Adapted from *Curr. Opin. Colloid Interface Sci.*, 17, Sanchez-Dominguez, M. et al., Preparation of inorganic nanoparticles in oil-in-water microemulsions: A soft and versatile approach, 297–305. Copyright 2012, with permission from Elsevier).

precursor may be at least partially located at the interface, modifying the HLB and the curvature of the system. This is in agreement with the study by de Oliveira et al. (2011), in which spectroscopic evidence suggested that cobalt 2-ethylhexanoate doped in water/AOT/heptane W/O microemulsions resides at the droplet interface, allowing for hydration of the cobalt(III) ion by the water pool.

Additional interpretation is possible by inspecting the data shown in Figure 9.6a and b. One-phase microemulsions are formed when the surfactant/oil (S:O) ratio values are between 65:35 and 52:48 for the system with precursor, while without precursor, S:O ratio values range from 65:35 to 57:43. In addition, most of these microemulsions can be diluted to infinity with water, as the L region extends toward the water corner. This behavior is a fingerprint of O/W microemulsion droplets stabilized by non-ionic surfactants (Attwood et al. 1992; Garti et al. 2004); the microemulsion region is thin because its curvature is restricted possibly due to a rigid oil/water interface, resulting in well-defined droplet structures.

The compositions shown as gray, black, and white diamonds in Figure 9.6b were used for CeO_2 nanoparticle synthesis; these points correspond to water/Synperonic 10/6/hexane weight ratios 92.15/4.85/3, 64.5/21.5/14, and 60/20/20, named CeO_2-D1, CeO_2-D2, and CeO_2-D3, respectively. It must be highlighted that the working temperature for microemulsion CeO_2-D3 was 45°C; hence, even though this sample was multiphasic at the temperature used for the phase diagram of Figure 9.6b (35°C), at 45°C, the sample is a microemulsion and single phase. The precipitation method used was the direct addition of concentrated ammonia (30 wt%) to the microemulsion containing Ce-2EH, until pH 11 was reached, and the reaction mixture was left stirring at the appropriate temperature for 48 h. TEM analysis (Figure 9.7) has shown that particle size is similar, between 3 and 5 nm for CeO_2-D2 and CeO_2-D3, except that the nanoparticles of the sample synthesized with higher oil concentration (CeO_2-D3) appear better defined and more agglomerated, and this can be ascribed to the higher overall precursor concentration in the microemulsion. A similar trend was observed by XRD, as the diffractograms were better defined with the increase in oil (and hence precursor) concentration (Figure 9.8). The crystallite size

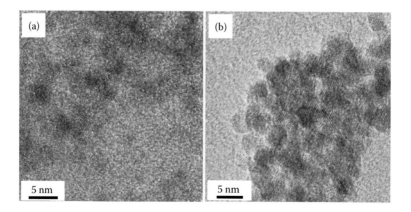

FIGURE 9.7   TEM micrographs for samples (a) CeO_2-D2 and (b) CeO_2-D3.

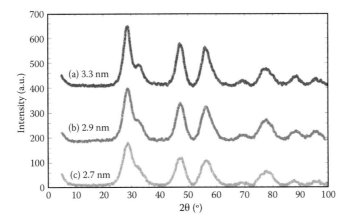

FIGURE 9.8   XRD patterns for (a) CeO₂-D3, (b) CeO₂-D2, and (c) CeO₂-D1.

(estimated using the Debye–Scherrer equation) increased slightly upon increase in oil concentration (from 2.7 to 3.3 nm).

On the other hand, according to the Brunauer–Emmett–Teller (BET) surface area analysis and the Barrett–Joyner–Halenda (BJH) pore size and pore volume analysis, the textural properties of the materials were highly dependent on the microemulsion composition used for synthesis (Table 9.2). The sample prepared with the highest concentration of surfactant (CeO₂-D2) had the largest SSA (252 m²/g). Interestingly, the sample with the highest concentration of oil (CeO₂-D3) presented similar textural properties as CeO₂-D1, the most diluted sample, except for the pore size distribution. As illustrated in Figure 9.9, CeO₂-D2 presented the narrowest pore size distribution, with a very sharp peak at 3.7 nm. CeO₂-D1 had a main peak around 4.5 nm and a less intense but wider peak around 500 nm. CeO₂-D3 had three peaks, the most intense around 3 nm, another around 6 nm, and a wide, less intense peak centered around 400 nm. The pore volume was similar for CeO₂-D1 and CeO₂-D3, while CeO₂-D2 presented the lowest value. By examining the TEM, XRD, and textural properties, it can be stated that even though the particle size and the crystallinity do not differ significantly when comparing the three samples synthesized using System D, the textural properties can indeed be tuned by varying the composition. Samples CeO₂-D1 and CeO₂-D2 have a similar S:O ratio value and, thus, a similar S:Ce ratio.

TABLE 9.2

Textural Properties of CeO₂ Samples Synthesized Using System D According to BET and BJH Analysis

Sample	Water/Synperonic 10/6/Hexane	SSA (BET) (m² g⁻¹)	Pore Volume (BJH) (cm³ g⁻¹)	Pore Size (BJH) (nm)
CeO₂-D1	92.15/4.84/3	198.7	0.24	4.6
CeO₂-D2	64.5/21.5/14	252.0	0.21	3.7
CeO₂-D3	60/20/20	190.2	0.24	5.4

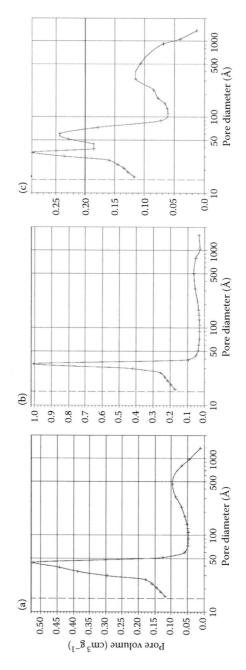

FIGURE 9.9 Pore size distribution from BJH desorption data for the following samples: (a) CeO_2-D1, (b) CeO_2-D2, and (c) CeO_2-D3.

However, CeO$_2$-D1 is more diluted (the concentration of water is higher), compared to CeO$_2$-D2; hence, it can be inferred that since the surfactant is less diluted in CeO$_2$-D2, this may help render the resulting material more porous. In other words, surfactant is more easily washed from sample CeO$_2$-D1; in addition, the highest concentration of oil droplets and surfactant in CeO$_2$-D2 allows for the formation of more particles, which may have more points of contact but at the same time are sufficiently protected against complete agglomeration by the surfactant, resulting in a more porous structure and a higher surface area. On the other hand, comparing CeO$_2$-D2 and CeO$_2$-D3, the surfactant/water ratio is the same; however, the oil concentration is higher in CeO$_2$-D3 and hence the surfactant/Ce ratio is lower. Hence, sample CeO$_2$-D3 results in more agglomerated particles, with less surfactant to protect them against such agglomeration. Overall, the material with the most superior characteristics is CeO$_2$-D2.

9.3 SYNTHESIS AND CHARACTERIZATION OF Eu-DOPED CERIA NANOPARTICLES

In a recent study by Tiseanu et al., a series of Eu-doped CeO$_2$ nanoparticles synthesized by the O/W microemulsion reaction method were investigated (Tiseanu et al. 2011). The CeO$_2$ nanoparticles were synthesized using the microemulsion composition and precipitation method of sample CeO$_2$-D2. Three samples were studied: pure CeO$_2$ (CE); Eu-doped CeO$_2$ (CEB), which was synthesized in the same way as CE but using 1 mol% of Eu(III) 2-ethylhexanoate (Eu-2EH) and 99 mol% of Ce-2EH; and Eu-impregnated CeO$_2$ (CEI), which was prepared by first synthesizing CeO$_2$ powder followed by wet impregnation with Eu-2EH (1 mol% Eu with respect to Ce). As mentioned in the Introduction, the cubic fluorite structure of ceria is able to incorporate an important amount of oxygen vacancies. Doping ceria with aliovalent cations such as trivalent lanthanide ions leads to the generation of more oxygen vacancies, which allows for high mobility of lattice oxygen and influences positively the thermal stability and the surface area compared to the pure oxide (Andersson et al. 2006). Among lanthanide ions, europium (Eu^{3+}) is a good option since it introduces luminescence properties, which makes it an effective probe for the local structure of CeO$_2$ (Bünzli and Choppin 1989).

It is well established that the nanocrystals with dopants incorporated within the lattice display different properties from those with dopants adsorbed on their surface (Erwin et al. 2005). Hence, the goals of the study were to investigate the following: the dopants' location as bulk sites versus surface/subsurface sites, the host's role in sensitization of the europium emission, and the relocation of the dopants on the various ceria sites with thermal treatment. Several techniques were used, including *in situ* and *ex situ* XRD and Raman spectroscopy as well as time-resolved photoluminescence spectroscopy. Dopant sites at the surface and in the interior of the particles were identified by time-resolved emission and excitation spectroscopy (Tiseanu and Lorenz-Fonfría 2010).

The characterization study shows that the ceria particles are nanocrystalline and small sized in the as-synthesized state exhibiting high surface area, and in fact,

TEM, XRD, and BET–BJH have shown basically the same results for CE and CEB but not for CEI. Characterization of the CEI sample showed that the only difference was the textural properties, with lower BET surface area, probably due to pore blockage arising from the organic part of the precursor. Upon calcination at 500°C and 1000°C, high surface area was maintained in all samples (Table 9.3), which demonstrates the good thermal stability of the materials synthesized by the O/W microemulsion method.

The high surface area obtained even after calcination is most likely related to the surfactant present in microemulsion (Bumajdad et al. 2004; Terribile et al. 1998; Trovarelli 1998), the small particle size, and the fact that the particles were nanocrystalline even before calcination. It should be highlighted that the thermal stability of the nanocerias reported in this study is higher than the stability of materials synthesized using W/O microemulsions stabilized by cationic surfactant didodecyldimethylammonium bromide (DDAB), reported by Bumajdad et al. (2004). Surface area in the order of 120 $m^2\,g^{-1}$ was maintained after calcination at 1000°C with nonionic O/W microemulsions, compared to 55 $m^2\,g^{-1}$ obtained with DDAB-based W/O microemulsions.

As illustrated in Figure 9.10, for the sample calcined at 500°C (CEB-500), the emission of europium in ceria displays an orange to red color emission tunable with the excitation wavelength. The tunable emission was explained by the heterogeneous distribution of the europium dopants within the ceria nanocrystals coupled with the progressive diffusion of the europium ions from the surface to the inner ceria sites and the selective participation of the ceria host in the emission sensitization. Hence, the coexistence of several europium species was established.

TABLE 9.3

Textural Characterization of the Investigated Pure CeO$_2$ (CE), Europium-Doped CeO$_2$ (CEB), and Impregnated CeO$_2$ (CEI) Nanocrystals

Sample	BET Surface Area ($m^2\,g^{-1}$)
CE	252
CE-500	246
CE-1000	121
CEI	152
CEI-500	141
CEI-1000	119
CEB	250
CEB-500	242
CEB-1000	110

Source: Reproduced from Tiseanu, C. et al., *Phys. Chem. Chem. Phys.*, 13, 17135–17145, 2011. With permission from the PCCP Owner Societies.

Note: The number in the sample code indicates the temperature at which the sample was calcined (lack of number indicates non-calcined sample).

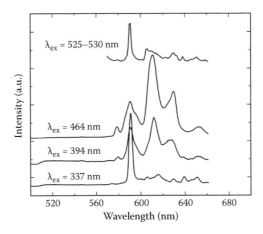

FIGURE 9.10 Steady-state emission spectra of CEB-500 nanocrystals after excitation at λ_{ex} = 290, 337, 394, 464, and 525 nm. (Reproduced from Tiseanu, C. et al., *Phys. Chem. Chem. Phys.*, 13, 17135–17145, 2011. With permission from the PCCP Owner Societies.)

The effects of the method for introducing Eu, that is, Eu doping versus impregnation, on the structural and optical properties of ceria nanocrystals were also studied. For the as-synthesized (non-calcined) nanocrystals, only a small europium fraction substitutes the Ce^{4+} sites from the bulk and this is observed only for the europium-doped ceria NCs. The europium ions are located mainly on the surface while the host sensitization of europium emission is not active. After calcination, defects and imperfections are removed from the lattice of the ceria nanocrystals whereas an enrichment of the surface by Ce^{4+} relative to Ce^{3+} species coupled with a migration of the surface europium ions inside the fluorite structure was evidenced.

In a more recent study (Tiseanu et al. 2012), focused only on CEI, a more detailed investigation of Eu^{3+} sites in CeO$_2$ nanocrystals was carried out by using spectrally and temporarily resolved photoluminescence spectroscopy at room temperature and 80 K. By carrying out the study on CEI only, it was ensured that prior to thermal treatment, Eu^{3+} resided mostly on the CeO$_2$ surface. After calcination (between 500°C and 1300°C), Eu^{3+} was distributed between the surface, the cubic lattice, and up to three additional crystalline sites, which was attributed to the oxygen vacancy charge-compensated defects. Overall, the results indicated that Eu^{3+}-induced oxygen vacancies are distributed around both Eu^{3+} and Ce^{4+}, and the Eu^{3+} oxygen vacancy interaction mode as nearest-neighbor or next-nearest-neighbor depends on calcination temperature.

9.4 IMPREGNATION OF CeO$_2$ WITH Au AND ITS USE IN CATALYSIS FOR CO OXIDATION: COMPARISON WITH OTHER OXIDES

It is well known that a key requirement for all catalysts in any application is a high active surface area or high metal dispersion. As it has been reported in the previous sections, the CeO$_2$ nanopowders synthesized in O/W microemulsions present a high surface area, mesoporosity, and good thermal stability; hence, these materials are good candidates for their application in catalysis. In the study by Sanchez-Dominguez et al. (2010), several oxides were synthesized in O/W microemulsions, and their potential as catalyst support was explored in the CO oxidation reaction by impregnation of the oxides with 2 wt% Au. In addition to CeO_2, the other oxides studied were $Ce_{0.5}Zr_{0.5}O_2$, ZrO_2, and TiO_2. The different materials were synthesized using the same microemulsion system (System D) and similar reaction conditions for the sake of comparison. CeO_2 was synthesized using the same formulation as CeO_2-D3, while $Ce_{0.5}Zr_{0.5}O_2$, ZrO_2, and TiO_2 were synthesized using the same formulation as CeO_2-D2, except that the appropriate metal 2-ethylhexanoate precursor (or combination of precursors) was used. Precipitation of oxide nanoparticles was carried out by adding concentrated ammonia to the microemulsion up to pH 11. By using these microemulsion compositions and reaction conditions, the SSA obtained after calcination at 400°C was in the same order for all the materials. Nanocrystalline cubic fluorite–type CeO_2 and $Ce_{0.5}Zr_{0.5}O_2$ were obtained under soft conditions, while ZrO_2 and TiO_2 presented wide XRD reflections. Crystallite size as estimated by the Debye–Scherrer equation and SSA (SSA_{BET}) are shown in Table 9.4. For the as-obtained materials, SSA_{BET} was on the order of 200–370 m²/g and the crystallite size was very small (~2–3 nm); this was in agreement with TEM results. The materials were calcined at 400°C after which a high SSA was maintained (100–150 m²/g) and the crystallinity was improved, yielding tetragonal phases for both TiO_2 (anatase) and ZrO_2; for these two materials, the crystallite size grew upon calcination (on the order of 16–18 nm), while CeO_2 and $Ce_{0.5}Zr_{0.5}O_2$ were thermally more stable and crystallite size only grew up to ~4 nm.

Representative HRTEM images and EDX analysis are shown in Figure 9.11 for Au/CeO_2, Au/ZrO_2, $Au/Ce_{0.5}Zr_{0.5}O_2$, and Au/TiO_2 catalysts obtained after Au

---

TABLE 9.4

Crystallite Size (Estimated from Debye–Scherrer Equation) and SSA (SSA_{BET}) of CeO_2, $Ce_{0.5}Zr_{0.5}O_2$, ZrO_2, and TiO_2 Nanoparticles Obtained in O/W Microemulsions, As-Obtained (Uncalcined) and Calcined at 400°C

Sample	Thermal Treatment	Crystallite Size (nm) (XRD)	SSA_{BET} (m² g⁻¹)
CeO_2 (CeO_2-D3)	As-obtained	3.3	190.2
CeO_2 (CeO_2-D3)-C4	Calcined 2 h at 400°C	4.5	159
$Ce_{0.5}Zr_{0.5}O_2$	As-obtained	1.9	207
$Ce_{0.5}Zr_{0.5}O_2$-C4	Calcined 2 h at 400°C	3.7	107
ZrO_2	As-obtained	1.1	377
ZrO_2-C4	Calcined 2 h at 400°C	18.5	116
TiO_2	As-obtained	1.0	297
TiO_2-C4	Calcined 2 h at 400°C	16.1	124

Source: Reproduced from *Catal. Today*, 158, Sanchez-Dominguez, M. et al., Synthesis of CeO_2, ZrO_2, $Ce_{0.5}Zr_{0.5}O_2$, and TiO_2 nanoparticles by a novel oil-in-water microemulsion reaction method and their use as catalyst support for CO oxidation, 35–43. Copyright 2010, with permission from Elsevier.

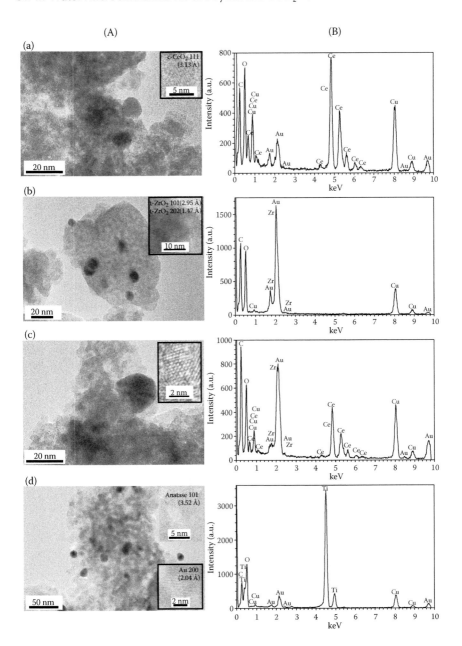

FIGURE 9.11 HRTEM pictures (column A) and EDX spectra (column B) of (a) Au/CeO$_2$, (b) Au/ZrO$_2$, (c) Au/Ce$_{0.5}$Zr$_{0.5}$O$_2$, and (d) Au/TiO$_2$. Insets in panels (a), (b), and (c) show high-resolution (HR) image of oxide; top inset in panel (d) shows HR image of oxide and bottom inset shows HR image of Au. (Reproduced from *Catal. Today*, 158, Sanchez-Dominguez, M. et al., Synthesis of CeO$_2$, ZrO$_2$, Ce$_{0.5}$Zr$_{0.5}$O$_2$, and TiO$_2$ nanoparticles by a novel oil-in-water microemulsion reaction method and their use as catalyst support for CO oxidation, 35–43. Copyright 2010, with permission from Elsevier.)

impregnation of calcined materials. The characteristic particle size of the support material for the different samples as observed by TEM is in agreement with the crystallite size estimated by XRD (small, around 3–5 nm support particles for Au/CeO_2 and $Au/Ce_{0.5}Zr_{0.5}O_2$, while larger, around 10–20 nm support particles for Au/TiO_2 and Au/ZrO_2). In terms of the crystalline versus amorphous phases, it was confirmed from HRTEM that the support materials in Au/CeO_2, $Au/Ce_{0.5}Zr_{0.5}O_2$, and Au/TiO_2 were, at least at a qualitative level, mainly crystalline, while for Au/ZrO_2, crystalline zones were less common than amorphous zones.

CO conversion curves as a function of temperature are displayed for Au-supported catalysts in Figure 9.12. At low temperature, Au/TiO_2 is the most active, with 50% conversion temperature at 44°C (T_{50}), followed by $Au/Ce_{0.5}Zr_{0.5}O_2$ and Au/CeO_2, giving T_{50} at 68°C and at 84°C, respectively. Au/ZrO_2 shows the lowest activity, with a T_{50} of 93°C. Moreover, Au/ZrO_2 does not reach complete CO conversion in the range of temperature explored; the maximum conversion of CO was 95% at 400°C.

The superior performance of $Au/Ce_{0.5}Zr_{0.5}O_2$ with respect to gold over the pure oxides is worth noticing. The XPS analysis revealed an increase of oxygen vacancies in the mixed oxides, since a higher percentage of Ce^{3+} is present with respect to pure ceria. For gold over ceria, a mechanism involving the support is proposed. Indeed, because of the OSC, the easy reducibility, and the high mobility of surface lattice oxide ions, ceria can act as an oxygen supplier with a Mars–van Krevelen mechanism (Gluhoi et al. 2005). Therefore, an increase of oxygen vacancies in the mixed oxide could be responsible for the enhanced catalytic performances.

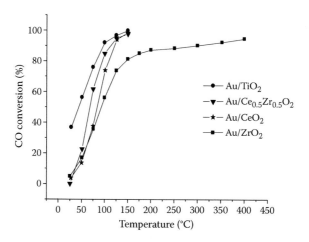

FIGURE 9.12 CO conversion (%) as a function of temperature for Au-supported catalysts. (Reproduced from *Catal. Today*, 158, Sanchez-Dominguez, M. et al., Synthesis of CeO_2, ZrO_2, $Ce_{0.5}Zr_{0.5}O_2$, and TiO_2 nanoparticles by a novel oil-in-water microemulsion reaction method and their use as catalyst support for CO oxidation, 35–43. Copyright 2010, with permission from Elsevier.)

9.5 CONCLUSIONS AND PERSPECTIVES

The synthesis of CeO$_2$ and related CeO$_2$-based materials in O/W microemulsions has been reviewed. Among the different O/W microemulsion systems used, the most appropriate for nanoparticle synthesis was water/Synperonic 10/6/hexane (System D) and this is due to both the highest level of Ce loading in the microemulsion, which allows for higher yields, and its convenience in the precipitating method applied, which is by adding concentrated ammonia up to pH 11. As a precursor, cerium(III) 2-ethylhexanoate is preferred over cerium(III) acetylacetonate because of the highest solubilization of the former in the oil phase. In terms of particle and crystallite size, most of the systems, compositions, and precipitating systems used resulted in particles around 2–3 nm and cubic fluorite–type structure. However, the textural properties can indeed be optimized by using System D with the following composition: 64.5 wt% water, 21.5 wt% surfactant, and 14 wt% oil phase. The CeO$_2$ powders present very good thermal stability; up to 1000°C, the materials retain an SSA on the order of 110 m^2/g. It has also been shown that the O/W microemulsion reaction method is suitable for the preparation of hybrid materials such as Eu-doped CeO$_2$, Ce-Zr oxide, and Au-impregnated CeO$_2$ and Ce-Zr oxide. The doping of CeO$_2$ with Eu was advantageous owing to the luminescence properties of the latter—it made it a very effective probe for the local structure of ceria; in addition, the use of luminescent CeO$_2$ could be investigated in biomedical imaging and optical applications. Finally, the investigation carried out by impregnating CeO$_2$, Ce$_{0.5}$Zr$_{0.5}$O$_2$, and other oxides with Au demonstrates the great potential of this approach for the preparation of various supports with high SSA for catalytic purposes.

ACKNOWLEDGMENTS

This research work has been performed in the framework of the D43/004/06 COST Programme Action. C. Tiseanu and V. Parvulescu are acknowledged for collaboration with Eu-doped ceria research (photoluminescence studies). A.M. Venezia, L.F. Liotta, and G. Di Carlo are acknowledged for collaboration with Au-impregnated oxides research (catalysis and XPS studies).

REFERENCES

Andersson, D.A., Simak, S.I., Skorodumova, N.V., Abrikosov, I.A., Johansson, B. 2006. Optimization of ionic conductivity in doped ceria. *Proc Natl Acad Sci USA* 103:3518–21.

Attwood, D., Mallon, C., Ktistis, G., Taylor, C.J. 1992. A study on factors influencing the droplet size in non-ionic oil-in-water microemulsions. *Int J Pharm* 88:417–22.

Boutonnet, M., Lögdberg, S., Svensson, E. 2008. Recent developments in the application of nanoparticles prepared from w/o microemulsions in heterogeneous catalysis. *Curr Opin Colloid Interface Sci* 13:270–86.

Boutonnet, M., Kizling, J., Stenius, P. 1982. The preparation of monodisperse colloidal metal particles from micro-emulsions. *Colloids Surf* 5:209–25.

Bumajdad, A., Eastoe, J., Mathew, A. 2009. Cerium oxide nanoparticles prepared in self-assembled Systems. *Adv Colloid Interface Sci* 56:147–8.

Bumajdad, A., Zaki, M.I., Eastoe, J., Pasupulety, L. 2004. Microemulsion-based synthesis of CeO$_2$ powders with high surface area and high-temperature stabilities. *Langmuir* 20:11223–33.

Bünzli, J.C., Choppin, G.R. 1989. Lanthanide probes in life. Chapter 7, *Chemical and Earth Sciences*. Elsevier, Amsterdam, pp. 219–93.

Danielsson, I., Lindman B. 1981. The definition of microemulsion. *Colloids Surf* 3:391–2.

Destrée, C., Nagy, J.B. 2006. Mechanism of formation of inorganic and organic nanoparticles from microemulsions. *Adv Colloid Interface Sci* 123–126:353–67.

Djinovic, P., Batista, J., Levec, J., Pintar, A. 2009. Comparison of water–gas shift reaction activity and long-term stability of nanostructured CuO–CeO$_2$ catalysts prepared by hard template and co-precipitation methods. *Appl Catal A Gen* 364:156–65.

Eastoe, J., Hollamby, M.J., Hudson, L. 2006. Recent advances in nanoparticle synthesis with reversed micelles. *Adv Colloid Interface Sci* 128–130:5–15.

Erwin, S.C., Zu, L., Haftel, M.I., Efros, A.L., Kennedy, T.A., Norris, D.J. 2005. Doping semi-conductor nanocrystals. *Nature* 43:91–5.

Garti, N., Yaghmura, A., Aserin, A., Spernath, A., Elfakess, R., Ezrahi, S. 2004. Solubilization of active molecules in microemulsions for improved environmental protection. *Colloids Surf A* 230:183–90.

Gluhoi, A.C., Vreeburg, H.S., Bakker, J.W., Nieuwenhuys, B.E. 2005. Activation of CO, O$_2$ and H$_2$ on gold-based catalysts. *Appl Catal A Gen* 291:145–50.

Holmberg, K. 2004. Surfactant-templated nanomaterials synthesis. *J Colloid Interface Sci* 274:355–64.

Khaleer, A., Shehadil, I., Al-Shamisi, M. 2010. Nanostructured chromium–iron mixed oxides: physicochemical properties and catalytic activity. *Colloids Surf A Physicochem Eng Aspects* 355:75–82.

Kusar, H., Hocevar, S., Levec, J. 2006. Kinetics of the water–gas shift reaction over nanostructured copper–ceria catalysts. *Appl Catal B Environ* 63:194–200.

López-Quintela, M.A. 2003. Synthesis of nanomaterials in microemulsions: formation mechanisms and growth control. *Curr Opin Colloid Interface Sci* 8:137–44.

López-Quintela, M.A., Tojo, C., Blanco, M.C., García Rio, L., Leis, J.R. 2004. Microemulsion dynamics and reactions in microemulsions. *Curr Opin Colloid Interface Sci* 9:264–78.

de Oliveira, R.J., Brown, P., Correia, G.B., Rogers, S.E., Heenan, R., Grillo, I. et al. 2011. Photoreactive surfactants: a facile and clean route to oxide and metal nanoparticles in reverse micelles. *Langmuir* 27:9277–84.

Ohsaki, H., Tachibana, Y., Kadowaki, K., Hayashi, Y., Suzuki, K. 1997. Bendable and temperable solar control glass. *J Non-Cryst Solids* 218:223–9.

Pileni, M.-P. 2003. The role of soft colloidal templates in controlling the size and shape of inorganic nanocrystals. *Nat Mater* 2:145–50.

Pileni, M.-P. 1997. Nanosized particles made in colloidal assemblies. *Langmuir* 13:3266–76.

Sanchez-Dominguez, M., Pemartin, K., Boutonnet, M. 2012. Preparation of inorganic nanoparticles in oil-in-water microemulsions: a soft and versatile approach. *Curr Opin Colloid Interface Sci* 17; DOI: 10.1016/j.cocis.2012.06.007.

Sanchez-Dominguez, M., Liotta, L.F., Di Carlo, G., Pantaleo, G., Venezia, A.M., Solans, C. et al. 2010. Synthesis of CeO$_2$, ZrO$_2$, Ce$_{0.5}$Zr$_{0.5}$O$_2$, and TiO$_2$ nanoparticles by a novel oil-in-water microemulsion reaction method and their use as catalyst support for CO oxidation. *Catal Today* 158:35–43.

Sanchez-Dominguez, M., Boutonnet, M., Solans, C. 2009. A novel approach to metal and metal oxide nanoparticle synthesis: the oil-in-water microemulsion reaction method. *J Nanopart Res* 11:1823–9.

Terribile, D., Trovarelli, A., Llorca, J., de Leitenburg, C., Dolcetti, G. 1998. The synthesis and characterization of mesoporous high-surface area ceria prepared using a hybrid organic/inorganic route. *J Catal* 178:299–308.

Tiseanu, C., Parvulescu, I.V., Sanchez-Dominguez, M., Boutonnet, M. 2012. Temperature induced conversion from surface to bulk sites in Eu^{3+} impregnated CeO$_2$. *J Appl Phys* 112:013521–29.

Tiseanu, C., Parvulescu, I.V., Boutonnet, M., Cojocaru, B., Primus, A.P., Teodorescu, M.C. et al. 2011. Surface versus volume effects in luminescent ceria nanocrystals synthesized by an oil-in-water microemulsion method. *Phys Chem Chem Phys* 13:17135–45.

Tiseanu, C., Lorenz-Fonfría, V.A. 2010. Time-resolved photoluminescence spectra, lifetime distributions and decay-associated spectra of lanthanide's exchanged microporous–mesoporous materials. *J Nanosci Nanotechnol* 10:2803–10.

Trovarelli, A. 1998. The preparation of high surface area CeO$_2$/ZrO$_2$ mixed oxides by a surfactant-assisted approach. *Catal Today* 43:79–88.

Tsunekawa, S., Ishikawa, K., Li, Z.Q., Kawazoe, Y., Kasuya, A. 2000. Origin of anomalous lattice expansion in oxide nanoparticles. *Phys Rev Lett* 85:3440–3.

Yuan, Q., Duan, H.H., Li, L.L., Sun, L.D., Zhang, Y.W., Yan C.H. 2009. Controlled synthesis and assembly of ceria-based nanomaterials. *J Colloid Interface Sci* 2:151–67.

10 Low-Density Solid Foams Prepared by Simple Methods Using Highly Concentrated Emulsions as Templates

Jordi Esquena

CONTENTS

10.1 PREFACE

The present chapter describes the use of highly concentrated emulsions as templates for the preparation of solid foams. These emulsions are characterized by having a volume fraction of the internal phase bigger than 0.74, which is the most compact arrangement of monodisperse rigid spheres. Because of this large volume fraction, highly concentrated emulsions consist of polyhedral droplets that are separated from each other by thin films of external phase. This morphology, resembling foams, is highly appropriate for the preparation of low-density and high-pore-volume materials, with well-interconnected pores. It is well known that solid foams with very low density, 0.05 g mL^{-1} or even lower, can be obtained by polymerizing or cross-linking in the external phase of highly concentrated emulsions. These organic polymer foams, commonly designated as polyHIPEs, can be prepared in a wide variety of different polymers and functionalities. Moreover, the preparation of inorganic low-density foams, in highly concentrated emulsions, has been extensively studied because of

199

its fundamental interest and technological applications, since these materials may possess simultaneously high pore volume and large surface area. In addition, the preparation of hybrid organic–inorganic monolithic materials has been described by many different methods on the basis of the use of highly concentrated emulsions as reaction media. Interesting hybrid monolithic foams have been prepared by incorporating nanoparticles to organic polymer foams, conferring new properties to these highly porous solid foams. For example, the use of particles with specific properties, such as superparamagnetism, allows obtaining porous materials with specific responses.

The different preparation methods, as well as the main characteristics and properties of the resulting materials, are reviewed in this chapter. First, a brief description of highly concentrated emulsions is provided (Section 10.2). Afterward, the preparation of macroporous organic polymer materials is described (Section 10.3). Macroporous and dual meso/macroporous inorganic oxides are the subject of the succeeding section (Section 10.4). Finally, hybrid organic–inorganic materials with high pore volume, prepared in highly concentrated (Pickering) emulsions, are reviewed (Section 10.5).

10.2 PROPERTIES AND MAIN ASPECTS OF HIGHLY CONCENTRATED EMULSIONS

Highly concentrated emulsions, also referred to in the literature as high-internal-phase ratio emulsions (HIPRE or HIPE), have been the subject of great attention for several decades, because of both theoretical interest and industrial applications. Highly concentrated emulsions have a volume fraction of the dispersed phase larger than 0.74, which is the maximum packing ratio for monodisperse spherical droplets [1–3]. For this reason, highly concentrated emulsions generally possess polyhedral and polydisperse droplets, surrounded by thin films of continuous phase, resembling the structure of gas–liquid conventional foams [4–6]. As illustration, an optical micrograph of a highly concentrated emulsion is shown in Figure 10.1.

FIGURE 10.1 Example of the typical structure of a highly concentrated emulsion (HIPE), as observed by optical microscopy.

Because of the structure of HIPEs, its viscosity is much higher than that of conventional emulsions. HIPEs are non-Newtonian fluids, showing a yield stress below which a solid-like behavior is observed [7–10]. The rheological behavior of highly concentrated emulsions, studied by dynamic oscillatory measurements, is viscoelastic [10–12], with properties that range from predominantly viscous to mainly elastic. Often, the values of the elastic modulus G are much higher than the viscous modulus G'', especially at large volume fractions of the dispersed phase [11,12]. The viscoelastic response can be fitted to the Maxwell model, where the elastic modulus depends on the interfacial area, whereas the viscous modulus is originated by the slippage of droplets against droplets [11].

Highly concentrated emulsions can be, as conventional emulsions, classified into two categories, water-in-oil (W/O) and oil-in-water (O/W). They also can be classified according to their nanostructure. Although it is generally assumed that emulsions consist of "simple" liquid phases, with the droplets being stabilized by a surfactant monolayer, surfactant self-aggregates can be present in the continuous phase, which can be a solution, a microemulsion [4–5,13], or a liquid crystalline phase [14–16]. As illustration, the ternary phase diagram of the $C_{12}E_8$/water/decane ternary system is shown in Figure 10.2. This phase diagram shows three biphasic regions, where highly concentrated O/W emulsions can be prepared, $W_m + O$, $I_1 + O$, and $H_1 + O$, where O is the excess oil, which may coexist with an aqueous micellar solution (W_m), a discontinuous cubic liquid crystal (I_1), and a hexagonal liquid crystal (H_1). Therefore, in this system, highly concentrated emulsions can be prepared with

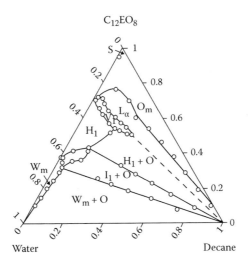

FIGURE 10.2 Ternary phase diagram of the water/C_{12}(EO)$_8$/decane system, at 25°C. W_m is an aqueous micellar solution phase. I_1, H_1, V_1, and L_α refer to a discontinuous micellar cubic liquid crystal, a normal hexagonal liquid crystal, a normal bicontinuous cubic liquid crystal, and a lamellar liquid crystal, respectively. O_m is a reverse micellar solution phase, O is an oil isotropic phase, and S is a surfactant solid phase. (Reproduced from Kunieda, H. et al., *J. Oleo Sci.*, 50, 633–639, 2001. With permission. Copyright [2001], Japan Oil Chemists's Society.)

three different nanostructures in the external phase, depending on the surfactant/water ratio.

Highly concentrated emulsions can also be classified according to the interaction forces between droplets, in two different categories: adhesive and non-adhesive [17]. The main feature of adhesive emulsions is that when placing emulsion droplets in contact with an excess of continuous phase, the droplets do not relax and maintain a polyhedral shape, with droplets adhered to each other owing to attractive forces between droplets. However, when non-adhesive emulsion droplets are brought into contact with an excess of its continuous phase, the droplets relax, reaching the state of spherical non-deformed droplets [17].

Because of the properties of highly concentrated emulsions, they are of great interest for both fundamental studies and technological applications. Currently, these emulsions are used in cosmetics, food products, pharmacy, and in many other fields. However, one of the most interesting applications is their use as templates for the preparation of low-density, highly porous solid foams, and this strategy has been used by many authors to obtain a great variety of different porous materials with controlled pore architecture [18–54]. The present chapter reviews these works on the preparation and properties of organic and inorganic solid foams, obtained by polymerization or cross-linking in the external phase of highly concentrated emulsions. Moreover, recent works dealing on the inclusion of nanoparticles in these materials will be described.

10.3   ORGANIC POLYMER FOAMS OBTAINED IN HIGHLY CONCENTRATED EMULSIONS

Solid low-density polymer foams, with interconnected sponge-like macropores, can be obtained by polymerization in the continuous phase of highly concentrated emulsions. This method, which allows obtaining densities smaller than 0.1 g mL^{-1}, was first disclosed in a patent from Unilever in 1982 [18], describing the preparation of polystyrene solid foams by polymerization in W/O emulsions stabilized by sorbitan fatty esters. This process, which in early works was studied mainly by Williams [19,20], Ruckenstein [21,23], and Cameron [24–26], has been applied to obtain a great variety of different macroporous organic polymer materials, which include styrene [24,55,56], styrene/divinylbenzene mixtures [24,33,55,57,58], divinylbenzene/ethylvinylbenzene [59], acrylamide and alkylmethacrylates [23,24,55], poly(furfuryl alcohol) [60], and chitosan [61], among many other organic polymers. Several review articles, describing this subject, can be found in the literature [51,57,62].

Macroporous polymer foams obtained by polymerizing in highly concentrated emulsions are usually known as "polyHIPEs." This term, which was coined as the abbreviation of "polymerized high internal phase emulsion," was first used in the 1982 patent [18], and it became a common denomination for such materials after its use by Hainey et al. in 1991 [22]. Since then, the term *polyHIPE* has been widely accepted, having been used in hundreds of papers published in academic journals. Therefore, in the present chapter, this term will also be used to denote macroporous organic polymer foams, obtained by polymerization in the external phase of highly concentrated emulsions.

The macropore morphology can replicate precisely the structure of highly concentrated emulsions used as templates, preserving the droplet size distribution, provided that the emulsions are stable during the formation of the polymer. As illustration, a typical example of a polystyrene polyHIPE, resulting from polymerization of styrene cross-linked with divinylbenzene, in the external phase of a W/O emulsion with 90 wt% of dispersed phase, is shown in Figure 10.3.

This figure shows an example of the internal topography of a polymer macroporous foam, with polyhedral macropores interconnected through narrower necks. Such porous texture results from the polymerization in the thin biliquid films that separate adjacent droplets in highly concentrated emulsions. The density of the foams can be as low as 0.02 g mL^{-1}, with pore volumes typically [57] larger than 15 mL g^{-1}. However, the specific surface area is usually low (<70 m^2 g^{-1}), since these polymer foams are mainly macroporous, with pore sizes typically in the range between 1 and 20 µm, with a negligible volume of smaller pores, such as micropores and mesopores. Nevertheless, the surface area can be increased to around 300 m^2 g^{-1}, by the addition of an inert oil into the monomer phase, which acts as a porogen [22,37]. Regarding the macroscopic size and shape of the monoliths, it can be easily controlled, since it depends on the container used as a mold. Figure 10.2b shows the example of a cylindrical monolith prepared by casting in a test tube.

In any case, the main requirement for the formation of polyHIPEs with interconnected macropores seems to be the formulation of highly concentrated emulsions properly stabilized, since emulsion stability is crucial to control the pore texture [63].

An interesting innovation, from an environmental point of view, was the use of supercritical CO$_2$-in-water highly concentrated emulsions, as described by Cooper et al. from 1991 [64–66]. Supercritical CO$_2$ has the advantage that solvent-free macroporous polymers can be obtained in the absence of organic solvents, either in the synthesis or in purification steps.

Polymerization of organic monomers in the external phase of highly concentrated emulsions can also be carried out in non-aqueous emulsions in mild conditions.

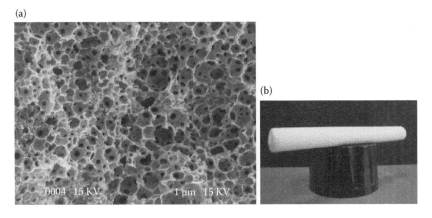

FIGURE 10.3 Examples of polystyrene polyHIPE solid foams. (a) Image acquired by scanning electron microscopy, showing the typical cellular structure of polyHIPEs. (b) An example of a macroscopic monolith, which was around 10 cm in length.

Recently, the use of oil-in-alcohol highly concentrated emulsions for the preparation of poly(furfuryl alcohol) polyHIPEs has been described [60]. Although these oil-in-alcohol emulsions are not very stable, polymerization takes place rapidly enough, and thus the texture of the porous foam replicates the morphology of the highly concentrated emulsion. Moreover, poly(furfuryl alcohol) is an organic polymer that can form carbon by pyrolization under inert atmosphere, with an acceptable yield on carbonization [67]. Therefore, macroporous carbonaceous materials can be obtained from highly concentrated emulsions [60], in a more simple preparation method than previously described [68–70].

Macroporous foams made of biopolymers have also been prepared. A recent example is the preparation of chitosan sponge-like foams [61], by cross-linking the polysaccharide derivative in the emulsion external phase. The resulting chitosan foams can possess an extremely low bulk density (<0.01 g mL^{-1}) and a very high pore volume (>100 mL g^{-1}) [61]. An example of such highly porous chitosan foams is shown in Figure 10.4.

The macropore morphology of polyHIPEs was studied in detail, in early works, by Williams et al. [19,20,63]. They found that increasing the volume fraction of the dispersed phase or the surfactant concentration produces an increase in the interconnectivity between macropores, because of thinning of the biliquid films that separate adjacent droplets. The formation of the interconnecting windows between macropores was also studied [19,20,63]. The monomers are adsorbed preferentially on surfactant monolayer regions located in the interstices between adjacent droplets, leading to open windows after polymerization [20]. The formation of such openings was observed in situ by Cameron et al. [71], using a cryo-transmission electron microscopy (TEM) technique. These observations proved that the formation of open windows between neighbor macropores is produced by the contraction of the thin polymer films during the conversion of monomer to polymer. This contraction seems to occur in a wide range of different organic polymers, and thus the open windows, which can be clearly observed in Figure 10.3a, are common in a wide variety of different polyHIPEs.

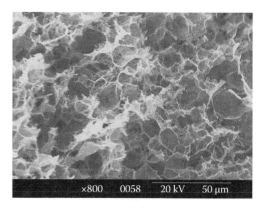

FIGURE 10.4   SEM image of a chitosan foam obtained by cross-linking in the external phase of a highly concentrated emulsion. The scale bar indicates 50 µm.

The high rugosity of the porous texture has interesting consequences on water wetting properties. Recent works by Molina et al. [72,73] have demonstrated that the internal surfaces of polystyrene polyHIPEs can be superhydrophobic, since the dynamic advancing contact angle against water is around 150°. This phenomenon, widely known as the "Lotus effect," means that water is a liquid that cannot penetrate into polyHIPE pores, under normal pressure, despite the huge pore volume of such materials. In fact, water has a very poor adhesion on the internal surface of polyHIPEs, and it can be observed that water drops roll off easily when the surface is tilted only a few degrees. However, the external surface of polyHIPE monoliths seems to be different. This surface has a rather smooth texture, since it was in contact with the mold surface during polymerization. As a result, the external surface of the monoliths is not superhydrophobic and it can be partly wetted with water. Therefore, polystyrene polyHIPEs possess dual wetting properties, different in the internal and in the external surfaces [72].

Nanoparticles can be incorporated into polyHIPE macroporous materials by polymerizing a monomer containing dispersions of hydrophobized nanoparticles. This subject has been extensively studied by Bismarck, Menner et al. [74–80]. In their earlier reports, they dispersed functionalized silica particles into styrene, which was polymerized in the external phase of W/O highly concentrated emulsions, obtaining polyHIPE macroporous polymer foams with imbibed nanoparticles [74,75]. The inclusion of these nanoparticles improved the mechanical strength of the foams, increasing toughness. A similar approach was used by Ghosh et al. [47,81], who incorporated Fe_3O_4 superparamagnetic nanoparticles into the polystyrene foams, and thus, low-density macroporous polyHIPEs were obtained with magnetic behavior.

In conclusion, a great variety of different organic polymer materials (polyHIPEs) can be obtained by using highly concentrated emulsions as templates. These emulsions are very versatile systems, which can provide a good control of macropore size, morphology, and functionalities. Moreover, inorganic materials with controlled porous textures can also be obtained using highly concentrated emulsions as reaction media. These inorganic materials, which are not included under the definition of polyHIPE, are reviewed in the next section.

10.4 INORGANIC FOAMS OBTAINED IN HIGHLY CONCENTRATED EMULSIONS

Inorganic solid foams, with low density and high pore volume, are very important in technological applications such as catalysis and molecular separation processes [82,83]. Moreover, great attention has been focused on materials with dual meso- and macroporous structures because bimodal pore size distributions combine the different advantages of simultaneously having mesopores and macropores. These dual materials not only possess high specific surface area, associated with mesopores, but also have the good diffusion properties associated with macropores. Because of these optimal properties, meso/macroporous dual materials can be used in a wide range of technological applications, which include separation processes [43,84], chromatographic adsorbents [85], and supports for photocatalytic processes [86]. Other recent studies include potential use in novel applications such as energy storage [87,88], fuel cells [89–91], or heterogeneous catalysis in biodiesel production [50].

The first inorganic material, which was obtained by carrying out sol-gel reaction in the external phase of concentrated and highly concentrated emulsions, was reported in 1997 by Imhof and Pine [27]. They described the preparation of silica, titania (shown in Figure 10.5), and zirconia, with narrow macropore size distribution, by condensing alkoxides in the external phase of isooctane-in-formamide emulsions [27].

A wide variety of different inorganic oxide materials can be prepared by applying sol-gel reactions in the external phase of concentrated and highly concentrated emulsions, obtaining macroporous materials that possess simultaneously both large surface areas and big pore volumes. In this subject, interesting contributions have been reported by Backov et al. [36,43,44,49,50,90–92], Stébé, Blin et al. [41,93,94], Tiddy et al. [35,39], Fournier and McGrath [46,53], and Esquena, Chmelka et al. [34,95,96], among other researchers. The preparation of hybrid organic–inorganic materials [43,44,49] can also be achieved by sol-gel processing in the external phase of highly concentrated emulsions.

The macropore texture of the inorganic foams is very similar to that previously described for organic polyHIPEs. Obviously, both types of materials possess a very open structure, with wide windows that interconnect the macropores. In highly concentrated emulsions, the biliquid films located between adjacent droplets are very thin, and consequently, most of the inorganic precursor molecules are in the plateau borders between adjacent droplets. Therefore, after sol-gel hydrolysis and condensation reactions, interconnecting windows appear between cellular macropores [39,43].

Dual meso/macroporous materials, in which mesopores are templated by surfactant self-aggregates while the macropores are templated by emulsion droplets,

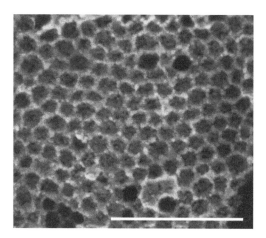

FIGURE 10.5 Porous TiO_2, obtained in emulsions with low polydispersity. The scale bar indicates 1 μm. (Reprinted by permission from Macmillan Publishers Ltd. *Nature* [27], copyright 1997.)

can be prepared by two-step methods, allowing an independent control of the size of macropores and mesopores [34,95]. In the first step, polyHIPE organic foams are previously prepared by polymerizing in highly concentrated emulsions [33,34,62]. These materials are macroporous, and mesoporosity is introduced in the second step, by impregnating the polyHIPEs with sol-gel solutions, which contain the inorganic precursor and the surfactant [34,95]. Self-assembled aggregates are formed, which direct the formation of mesopores. The final dual materials are obtained after calcination in the presence of air, in which surfactant and organic polymer, both acting as sacrificial templates, are removed.

These two-step processes are time-consuming and rather cost-ineffective. However, in industry, it is highly important to obtain dual meso/macroporous materials by methods that can be easily adapted to mass production at lower costs. Consequently, simple single-step methods should be preferred, in which the sol-gel reactions are carried out in the continuous phase of the emulsions [27,35,36,39,41,43,53,96]. For this purpose, many different surfactants can be used, including cationic [39,97] and nonionic surfactants such as ethoxylated alcohols [41,98] or block copolymers [27]. In any case, the selection of the surfactant system is very important, because hydrolysis of alkoxides (like TEOS, the most commonly used precursor for silica synthesis) release short-chain alcohols, which may reduce surfactant adsorption and therefore can decrease emulsion stability [99–101].

It is known that high specific surface areas can be obtained by carrying out sol-gel reactions directly in the external phase of highly concentrated emulsions [27,36,39,41,43]. Backov et al. have demonstrated that the presence of supramolecular aggregates, in the external phase of the highly concentrated emulsion, plays an important role in mesopore formation acting as templates [36]. As a result, dual meso/macroporous materials are obtained directly in highly concentrated emulsions by one-step processes. Indeed, the emulsion droplets template the formation of the macropores, and simultaneously the micelles and microemulsion droplets, present in the external phase, lead to the formation of mesopores. In this context, emulsions are certainly very versatile systems, because of this dual structure of micron-sized droplets in equilibrium to nanosized surfactant aggregates.

Recently, the formation of meso/macroporous silica in highly concentrated emulsions based on cubic liquid crystals has been recently proposed by Esquena et al. [96]. As also mentioned in Section 10.2, these emulsions can be in the form of oil droplets dispersed in an aqueous micellar cubic liquid crystal. Consequently, these systems can be regarded as hierarchically structured: micron-sized emulsion droplets coexisting with a dense packing of ordered nanometer-sized aggregates. The results have shown [96] that silica synthesis by TEOS hydrolysis and condensation in the external phase of O/W highly concentrated emulsions, with a cubic phase in that external phase, directly leads to the formation of dual meso/macroporous materials with high specific surface area, despite the fact that the mesopores are not ordered, probably due to the influence of ethanol released by TEOS hydrolysis. The surfactant used in this study is a polyoxyethylene-chain nonionic compound, which forms a discontinuous cubic phase, as described by Kunieda et al. [15]. The porous texture of a meso/macroporous silica material, prepared by this one-step method, is shown in Figure 10.6.

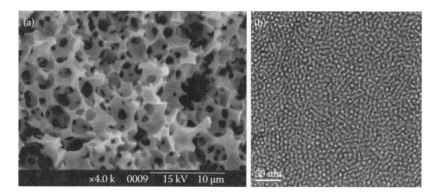

FIGURE 10.6 Images obtained by SEM (a) and TEM (b) of a silica sample prepared by TEOS hydrolysis and condensation in a highly concentrated emulsion with a cubic liquid crystal in the continuous phase. The composition was 7 wt% $C_{12}(EO)_8$, 0.4 wt% Pluronic F127, 11.1 wt% HCl(aq.) 0.5 mol dm^{-3}, 74.4 wt% decane, and 7.0 wt% TEOS. (Reprinted with permission from Esquena et al. 2012, 12334–12340. Copyright 2012, American Chemical Society.)

The scanning electron microscopy (SEM) image (Figure 10.6) shows the typical texture usually observed in polyHIPE materials. A closer look at higher magnification, by TEM (Figure 10.6b), also shows the presence of densely packed disordered mesopores, with an average size around 3.4 nm, as determined by nitrogen sorption porosimetry.

However, in a very recent work [102], monolithic dual meso/macroporous silica with ordered mesopores have been obtained by templating in highly concentrated emulsions based on a cubic liquid crystal. The mesopore ordering was preserved during reactions, by using a silica precursor, tetrahydroxyethyl orthosilicate, which releases ethylene glycol instead of an alcohol (ethanol in this case) during hydrolysis reactions [103–105]; diols, which are highly hydrophilic, do not greatly affect the liquid crystal structure. In this synthesis method, the emulsion droplets template the formation of macropores whereas the cubic liquid crystal, present in the continuous phase, templates the formation of the mesopores.

In the present section, methods for formation of inorganic oxide low-density materials by templating in highly concentrated emulsions have been reviewed. More complex materials, such as hybrid organic–inorganic materials, can also be prepared in highly concentrated emulsions. The next section summarizes the formation and properties of polyHIPE polymer foams in which nanoparticles are incorporated, enhancing mechanical properties and conferring new functionalities to the materials.

10.5 HYBRID ORGANIC–INORGANIC MATERIALS, PREPARED IN HIGHLY CONCENTRATED EMULSIONS STABILIZED WITH NANOPARTICLES

In all the examples described so far in this chapter, highly concentrated emulsions were prepared using surfactants as emulsifiers. However, it is also well known that emulsions can be prepared in the absence of surfactant molecules, stabilizing solely with particles adsorbed on the water–oil interface. These emulsions were first described by Ramsden [106] and are commonly known as Pickering emulsions [107]. In recent times, the formation and properties of such emulsions have been studied in detail by Binks [31,108–112] and other authors [113,114]. These emulsions are being used in a great variety of industrial applications such as mineral processing, food formulations, cosmetic products, and encapsulation of drugs and flavors. The stability mechanisms, because of particle adsorption, and the thermodynamic aspects of Pickering emulsions are now largely understood, thanks to the contributions by Kralchevsky and Danov [115–119], who described the interactions between particles adsorbed in liquid films.

Recently, the preparation of highly concentrated Pickering emulsions, stabilized solely with nanoparticles, has been achieved with volume percentages of the disperse phase above 90% [45]. Before then, there was a false perception that it was not possible because of catastrophic inversion, which often breaks Pickering emulsions when increasing the volume fraction of the disperse phase [109]. However, highly concentrated emulsions, stabilized only with silanized silica nanoparticles, have been obtained [45,78].

In Pickering emulsions, the particles used as stabilizers play an equivalent role as surfactant molecules in conventional emulsions. In order to obtain a stable Pickering emulsion, the particles have to be partially wetted by both water and oil phases. This partial wetting is generally quantified as the contact angle. A solid particle adsorbed at a water–oil interface forms a contact angle θ, defined as the angle between the water–oil interface and the tangent to the solid–water interface [109,110,120]. For a particle that is preferentially wetted by water, the contact angle is smaller than 90°, and the stable emulsion is O/W. The opposite occurs for particles that are preferentially wetted by oil. In this case, the contact angle in larger than 90° and W/O emulsions are obtained. This simple principle is illustrated in Figure 10.7, which shows

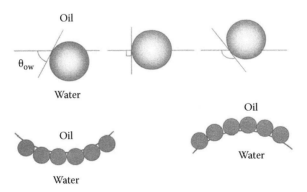

FIGURE 10.7 Schematic representation of particles adsorbed on water–oil interfaces, showing the contact angle in relation to the preferred of the interface. O/W emulsions are formed if particles are preferably wetted by water, while W/O emulsions are the result of particles that are mainly wetted by oil. (Reprinted from *Adv. Colloid Interface Sci.*, 100–102, Aveyard, R. et al., Emulsions stabilised solely by colloidal particles, 503–546. Copyright 2003, with permission from Elsevier.)

schematically the contact angle of spherical particles adsorbed on a planar water–oil interface, and the corresponding interface curvature.

In conclusion, the phase that preferably wets the particles becomes the external continuous phase of the emulsion. In this sense, the type of emulsion (O/W or W/O) is determined by a principle equivalent to the Bancroft rule: the continuous phase of the emulsion is the one in which the particles are preferentially dispersed [109]. In any case, strong particle adsorption is required to formulate stable Pickering emulsions [114].

The stabilizing mechanism in Pickering emulsions is complex, since the particle–particle interactions greatly contribute to the emulsion behavior. Flocculated dispersions are reported to be more efficient in stabilizing emulsions [113,120]. This increased stabilization arises from particle–particle weak attractions, which causes the formation of a three-dimensional network of particles in the continuous phase, surrounding the droplets. Consequently, two different mechanisms may lead to stable Pickering emulsions: (a) formation of a monolayer of adsorbed particles on the surface of emulsion droplets and (b) formation of a three-dimensional network of particles in the external phase around the emulsion droplets.

Highly concentrated emulsions, stabilized solely with nanoparticles, can be used to obtain low-density polymer foams (polyHIPEs). The preparation of such materials, using highly concentrated Pickering emulsions, which was first described by Binks [31], has been studied by Bismarck and Menner [48,76–80,121] and other authors [122]. Binks [31] demonstrated that a careful drying of these materials leads to the formation of monolithic macroporous silica foams with relatively high specific surface area, around 250 m²/g. Figure 10.8 shows the porous texture of silica monolithic materials obtained in Pickering emulsions.

One can observe (Figure 10.8) that the porous texture is different from that usually observed in polyHIPEs, with the absence of large open windows that connect adjacent macropores. Similar pore textures have also been observed by Bismarck

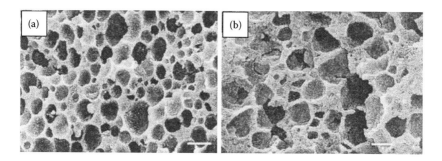

FIGURE 10.8 SEM images of porous silica obtained by evaporation of solvent from emulsions stabilized solely with nanoparticles (Pickering emulsions). The scale bar indicates 10 µm. (a) Prepared in 10 vol% water-in-hexane emulsions with 4 wt% silica functionalized particles. (b) 10 wt% water-in-hexane emulsions with 4 wt% silica. (Yi, G.R. et al.: Packings of uniform microspheres with ordered macropores fabricated by double templating. *J. Am. Chem. Soc.* 2002. 124. 13354–13355. Copyright Wiley-VCH Verlag GmbH & Co. KGaA. Reproduced with permission.)

et al. [78,79]. Most likely, this is the result of thicker biliquid films in Pickering emulsions.

The preparation of polyHIPE materials, in highly concentrated Pickering emulsions, has been achieved using many types of nanoparticles, which include silica [31], titania [76–78], polymer core–shell particles [123,124], and ferrite nanoparticles [48]. Another recent example of successful synthesis of a macroporous material, using highly concentrated emulsions stabilized solely with solid particles (Pickering HIPEs), has been described by Backov, Destribats et al. [125], in which limited coalescence was used to control the emulsion droplet size and the resulting macropore size of the materials. All these works indicate that the nanoparticles remain anchored at the surface, which has opened great possibilities for obtaining porous nanocomposite materials where nanoparticles are located mainly at the surface of the pores.

The crucial point for polyHIPE preparation in Pickering emulsions is the functionalization of nanoparticles, since surface properties should be intermediate between water and oil, required for optimal adsorption at the water–oil interface, and resulting in improved emulsion stability. Particles can be functionalized easily by using products such as alkylsilanes [31], oleic acid [112,126], or dodecylphosphonic acid [127]. This has allowed to control the contact angle of adsorbed particles and to obtain stable Pickering emulsions, allowing the preparation of new types of polyHIPE materials.

After particle functionalization, stable oil-in-monomer Pickering emulsions can be prepared. The polymerization in the external emulsion phase, induced by radical initiators, leads to the formation of polyHIPE materials in which nanoparticles remain on the polymer surface. As illustration, an example is shown in Figure 10.9.

In conclusion, highly concentrated Pickering emulsions are very useful and versatile systems, which can be used as a tool for the preparation of hybrid organic–inorganic, highly porous materials, in which nanoparticles are strongly attached to the pore

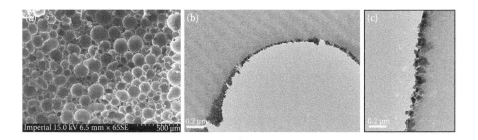

FIGURE 10.9 Electron microcopy images of a polyHIPE polystyrene material, prepared by polymerizing in the external phase of water-in-styrene highly concentrated emulsions, stabilized solely with Fe_3O_4 alkyl-functionalized nanoparticles. (a) SEM image, showing closed cell–type macropores, in the absence of connecting windows between macropores. (b) TEM image of the same sample, showing nanoparticles adsorbed at the macropore surface. (c) A detail of the macropore surface, by TEM, showing the nanoparticles at larger magnification. (Reprinted with permission from Vílchez, A. et al. *Langmuir*, 27, 13342–13352. Copyright 2011 American Chemical Society.)

surface. This type of materials could be very interesting in technological applications, since the particles are easily accessible for any product or reagent through the macroporous network.

10.6 CONCLUDING REMARKS

Highly concentrated emulsions are characterized by possessing a volume fraction of the dispersed phase bigger than 0.74, which is the maximum packing ratio for monodisperse spherical droplets. For this reason, the structure of highly concentrated emulsions generally consists of polyhedral and polydisperse droplets separated by a thin film of continuous phase. Because of such structure, highly concentrated emulsions can be used as templates for the preparation of low-density, highly porous solid foams, and many authors have considered this strategy for obtaining a wide variety of different porous materials with controlled pore architecture. Polymerization or cross-linking, performed in the external phase of the emulsions, followed by the removal of the internal phase, is a highly versatile approach, allowing preparation of organic, inorganic, or hybrid organic–inorganic monolithic materials with high pore volume. Organic polymer foams, with interconnected sponge-like macropores, can be obtained by polymerization of monomers in the continuous phase of highly concentrated emulsions. Moreover, inorganic oxide porous materials, with both large surface area and high pore volume, can be prepared by sol-gel reactions in the continuous phase. Nanoparticles can be incorporated to these materials, by polymerizing particle dispersions or using Pickering emulsions as reaction media.

REFERENCES

1. Lissant, K.J. 1966. The geometry of high-internal-phase-ratio emulsions. *J. Colloid Interface Sci.* 22:462–468.
2. Lissant, K.J. and K.G. Mayhan. 1973. A study of medium and high internal phase ratio water/polymer emulsions. *J. Colloid Interface Sci.* 42:201–208.
3. Princen, H.M. 1979. Highly concentrated emulsions. I. Cylindrical systems. *J. Colloid Interface Sci.* 71:55–66.
4. Kunieda, H., C. Solans, N. Shida, and J.L. Parra. 1987. The formation of gel-emulsions in a water/nonionic surfactant/oil system. *Colloids Surf.* 24:225–237.
5. Solans, C., R. Pons, and H. Kunieda. 1998. *Modern Aspects of Emulsion Science*, ed. B.P. Binks, Cambridge, UK: The Royal Society of Chemistry. 367–394.
6. Solans, C., J. Esquena, N. Azemar, C. Rodríguez, and H. Kunieda. 2004. Highly concentrated (gel) emulsion: Formation and properties. *Emulsions: Structure, stability and Interactions*, ed. D.N. Petsev, Amsterdam: Elsevier. 511–555.
7. Princen, H.M. 1983. Rheology of foams and highly concentrated emulsions. I. Elastic properties and yield stress of a cylindrical model system. *J. Colloid Interface Sci.* 91:160–175.
8. Princen, H.M. and A.D. Kiss. 1986. Rheology of foams and highly concentrated emulsions. III. Static shear modulus. *J. Colloid Interface Sci.* 112:427–437.
9. Princen, H.M. and A.D. Kiss. 1989. Rheology of foams and highly concentrated emulsions. IV. An experimental study of the shear viscosity and yield stress of concentrated emulsions. *J. Colloid Interface Sci.* 128:176–187.
10. Ravey, J.C., M.J. Stébé, and S. Sauvage. 1994. Water in fluorocarbon gel emulsions: Structures and rheology. *Colloids Surf.* A 91:237–257.

11. Pons, R., P. Erra, C. Solans, J.C. Ravey, and M.J. Stébé. 1993. Viscoelastic properties of gel-emulsions: Their relationship with structure and equilibrium properties. *J. Phys. Chem.* 97:12320–12324.

12. Babak, V.G., A. Langenfeld, N. Fa, and M.J. Stébé. 2001. Rheological properties of highly concentrated fluorinated water-in-oil emulsions. *Prog. Colloid Polym. Sci.* 118:216–220.

13. Pons, R., J.C. Ravey, S. Sauvage, M.J. Stébé, P. Erra, and C. Solans. 1993. Structural studies on gel emulsions. *Colloids Surf. A* 76:171–177.

14. Kunieda, H., M. Tanimoto, K. Shigeta, and C. Rodriguez. 2001. Highly concentrated cubic-phase emulsions: Basic study on D-phase emulsification using isotropic gels. *J. Oleo Sci.* 50:633–639.

15. Rodríguez, C., K. Shigeta, and H. Kunieda. 2000. Cubic-phase-based concentrated emulsions. *J. Colloid Interface Sci.* 223:197–204.

16. Uddin, M.D.H., H. Kunieda, and C. Solans. 2003. Highly concentrated cubic phase-based emulsions. *Structure–Performance Relationships in Surfactants*, eds. K. Esumi and M. Ueno, New York: Marcel Dekker. 70:599–626.

17. Babak, V.G. and M.J. Stébé. 2002. *J. Disper. Sci. Technol.* 1–3:

18. Barby, D. and Z. Haq. 1982. *European Patent 0060138*, Unilever, Editor.

19. Williams, J.M. and D.A. Wrobleski. 1988. Spatial distribution of the phases in water-in-oil emulsions. Open and closed microcellular foams from cross-linked polystyrene. *Langmuir* 4:656–662.

20. Williams, J.M. 1988. Toroidal microstructures from water-in-oil emulsions. *Langmuir* 4:44–49.

21. Ruckenstein, E. and J.S. Park. 1988. Hydrophilic–hydrophobic polymer composites. *J. Polym. Sci. C Polym. Lett.* 26:529–536.

22. Hainey, P., I.M. Huxham, B. Rowatt, D.C. Sherrington, and L. Tetley. 1991. Synthesis and ultrastructural studies of styrene-divinylbenzene polyhipe polymers. *Macromolecules* 24:117–121.

23. Ruckenstein, E. and J.S. Park. 1992. Stable concentrated emulsions as precursors for hydrophilic–hydrophobic polymer composites. *Polymer* 33:405–417.

24. Cameron, N.R. and D.C. Sherrington. 1996. High internal phase emulsions (HIPEs)—Structure, properties and use in polymer preparation. *Adv. Polym. Sci.* 126:162–214.

25. Cameron, N.R. and D.C. Sherrington. 1997. Preparation and glass transition temperatures of elastomeric PolyHIPE materials. *J. Mater. Chem.* 7:2209–2212.

26. Cameron, N.R. and D.C. Sherrington. 1997. Synthesis and characterization of poly(aryl ether sulfone) PolyHIPE materials. *Macromolecules* 30:5860–5869.

27. Imhof, A. and D.J. Pine. 1997. Ordered macroporous materials by emulsion templating. *Nature* 389:948–951.

28. Cameron, N.R. and A. Barbetta. 2000. The influence of porogen type on the porosity, surface area and morphology of poly(divinylbenzene) polyHIPE foams. *J. Mater. Chem.* 10:2466–2471.

29. Barbetta, A., N.R. Cameron, and S.J. Cooper. 2000. High internal phase emulsions (HIPEs) containing divinylbenzene and 4-vinylbenzyl chloride and the morphology of the resulting PolyHIPE materials. *Chem. Commun.* 3:221–222.

30. Busby, W., N.R. Cameron, and C.A.B. Jahoda. 2001. Emulsion-derived foams (PolyHIPEs) containing poly(Sε-caprolactone) as matrixes for tissue engineering. *Biomacromolecules* 2:154–164.

31. Binks, B.P. 2002. Macroporous silica from solid-stabilized emulsion templates. *Adv. Mater.* 14:1824–1827.

32. Yi, G.R., J.H. Moon, V.N. Manoharan, D.J. Pine, and S.M. Yang. 2002. Packings of uniform microspheres with ordered macropores fabricated by double templating. *J. Am. Chem. Soc.* 124:13354–13355.

33. Esquena, J., R. Sankar, and C. Solans. 2003. Highly concentrated W/O emulsions prepared by the PIT method as templates for solid foams. *Langmuir* 19:2983–2988.

34. Maekawa, H., J. Esquena, S. Bishop, C. Solans, and B.F. Chmelka. 2003. Meso/macroporous inorganic oxide monoliths from polymer foams. *Adv. Mater.* 15:591–596.

35. Sen, T., G.J.T. Tiddy, J.L. Casci, and M.W. Anderson. 2003. *Silica materials with meso- and macropores. WIPO Patent Application*, I.C.I. PLC, Editor.

36. Carn, F., A. Colin, M.F. Achard, H. Deleuze, E. Sellier, M. Birot, and R. Backov. 2004. Inorganic monoliths hierarchically textured via concentrated direct emulsion and micellar templates. *J. Mater. Chem.* 14:1370–1376.

37. Barbetta, A. and N.R. Cameron. 2004. Morphology and surface area of emulsion-derived (PolyHIPE) solid foams prepared with oil-phase soluble porogenic solvents: Three-component surfactant system. *Macromolecules* 37:3202–3213.

38. Brandhuber, D., N. Huesing, C.K. Raab, V. Torma, and H. Peterlik. 2005. Cellular mesoscopically organized silica monoliths with tailored surface chemistry by one-step drying/extraction/surface modification processes. *J. Mater. Chem.* 15:1801–1806.

39. Sen, T., G.J.T. Tiddy, J.L. Casci, and M.W. Anderson. 2005. Meso-cellular silica foams, macro-cellular silica foams and mesoporous solids: A study of emulsion-mediated synthesis. *Micropor. Mesopor. Mat.* 78:255–263.

40. Barbetta, A., R.J. Carnachan, K.H. Smith, C.T. Zhao, N.R. Cameron, R. Kataky, M. Hayman, S.A. Przyborski, and M. Swan. 2005. Porous polymers by emulsion templating. *Macromolec. Symp.* 226:203–211.

41. Blin, J.L., R. Bleta, J. Ghanbaja, and M.J. Stébé. 2006. Fluorinated emulsions: Templates for the direct preparation of macroporous–mesoporous silica with a highly ordered array of large mesopores. *Micropor. Mesopor. Mat.* 94:74–80.

42. Esquena, J. and C. Solans. 2006. Surfactant Science Series. *Emulsions and Emulsion Stability*, ed. J. Sjöblom, New York: Taylor & Francis.

43. Ungureanu, S., M. Birot, G. Laurent, H. Deleuze, O. Babot, B. Julián-López, M.F. Achard, M.I. Popa, C. Sanchez, and R. Backov. 2007. One-pot syntheses of the first series of emulsion based hierarchical hybrid organic–inorganic open-cell monoliths possessing tunable functionality (organo-Si(HIPE) series). *Chem. Mater.* 19:5786–5796.

44. Ungureanu, S., H. Deleuze, C. Sanchez, M.I. Popa, and R. Backov. 2008. First Pd@organo-Si(HIPE) open-cell hybrid monoliths generation offering cycling heck catalysis reactions. *Chem. Mater.* 20:6494–6500.

45. Ikem, V.O., A. Menner, and A. Bismarck. 2008. High internal phase emulsions stabilized solely by functionalized silica particles. *Angew. Chem. Int. Ed.* 47:8277–8279.

46. Fournier, A.C., H. Cumming, and K.M. McGrath. 2010. Assembly of two- and three-dimensionally patterned silicate materials using responsive soft templates. *Dalton Trans.* 39:6524–6531.

47. Ghosh, G., A. Vílchez, J. Esquena, C. Solans, and C. Rodríguez-Abreu. 2011. Preparation of porous magnetic nanocomposite materials using highly concentrated emulsions as templates. *Prog. Colloid Polym. Sci.* 138:161–164.

48. Vílchez, A., C. Rodríguez-Abreu, J. Esquena, A. Menner, and A. Bismarck. 2011. Macroporous polymers obtained in highly concentrated emulsions stabilized solely with magnetic nanoparticles. *Langmuir* 27:13342–13352.

49. Brun, N., S. Ungureanu, H. Deleuze, and R. Backov. 2011. Hybrid foams, colloids and beyond: From design to applications. *Chem. Soc. Rev.* 40:771–788.

50. Brun, N., A. Babeau-Garcia, M.F. Achard, C. Sanchez, F. Durand, G. Laurent, M. Birot, H. Deleuze, and R. Backov. 2011. Enzyme-based biohybrid foams designed for continuous flow heterogeneous catalysis and biodiesel production. *Energy Environ. Sci.* 4:2840–2844.

51. Kimmins, S.D. and N.R. Cameron. 2011. Functional porous polymers by emulsion templating: Recent advances. *Adv. Funct. Mater.* 21:211–225.

52. Lovelady, E., S.D. Kimmins, J. Wu, and N.R. Cameron. 2011. Preparation of emulsion-templated porous polymers using thiol-ene and thiol-yne chemistry. *Polym. Chem.* 2:559–562.

53. Fournier, A.C. and K.M. McGrath. 2012. Synthesis of porous oxide ceramics using a soft responsive scaffold. *J. Mater. Sci.* 47:1217–1222.

54. Kimmins, S.D., P. Wyman, and N.R. Cameron. 2012. Photopolymerised methacrylate-based emulsion-templated porous polymers. *React. Function. Polym.* 72:947–954.

55. Ruckenstein, E. 1997. Concentrated emulsion polymerization. *Adv. Polym. Sci.* 127:3–58.

56. Ruckenstein, E. and H. Li. 1997. The concentrated emulsion approach to toughened polymer composites: A review. *Polym. Compos.* 18:320–331.

57. Esquena, J. and C. Solans. 2006. Highly concentrated emulsions as templates for solid foams. *Emulsions and Emulsion Stability*, ed. J. Sjöblom, New York: Taylor & Francis. 132.

58. Bhumgara, Z. 1995. Polyhipe foam materials as filtration media. *Filtrat. Separ.* 32:245–251.

59. Mercier, A., H. Deleuze, and O. Mondain-Monval. 2000. High internal phase emulsions for the preparation of ultraporous functional polymers/emulsions inverses hautement concentrées pour la fabrication de polymères ultraporeux fonctionnalisés. *Actual. Chim.* 10–18.

60. Vílchez, S., L.A. Pérez-Carrillo, J. Miras, C. Solans, and J. Esquena. 2012. Oil-in-alcohol highly concentrated emulsions as templates for the preparation of macroporous materials. *Langmuir* 28:7614–7621.

61. Esquena, J., C. Solans, S. Vílchez, P. Erra, and J. Miras. 2009. *Materiales poliméricos macroporosos o meso/macroporosos obtenidos en emulsiones concentradas y altamente concentradas. P200930038 Spanish patent* CSIC, Editor. Spain.

62. Solans, C., J. Esquena, and N. Azemar. 2003. Highly concentrated (gel) emulsions, versatile reaction media. *Curr. Opin. Colloid. Interface Sci.* 8:156–163.

63. Williams, J.M., A. James Gray, and M.H. Wilkerson. 1990. Emulsion stability and rigid foams from styrene or divinylbenzene water-in-oil emulsions. *Langmuir* 6:437–444.

64. Butler, R., C.M. Davies, and A.I. Cooper. 2001. Emulsion templating using high internal phase supercritical fluid emulsions. *Adv. Mater.* 13:1459–1463.

65. Wood, C.D., R. Butler, A.K. Hebb, K. Senoo, H. Zhang, and A.I. Cooper. 2002. Synthesis and processing of porous polymers using supercritical carbon dioxide. *Prog. Rubber Plast. Recycl. Technol.* 18:247–258.

66. Butler, R., I. Hopkinson, and A.I. Cooper. 2003. Synthesis of porous emulsion-templated polymers using high internal phase CO_2-in-water emulsions. *J. Am. Chem. Soc.* 125:14473–14481.

67. Fitzer, E., W. Schaefer, and S. Yamada. 1969. The formation of glasslike carbon by pyrolysis of polyfurfuryl alcohol and phenolic resin. *Carbon* 7:643–646.

68. Álvarez, S., A.B. Fuertes, J. Esquena, and C. Solans. 2004. Meso/macroporous carbon monoliths from polymeric foams. *Adv. Eng. Mater.* 6:897–899.

69. Álvarez, S. and A.B. Fuertes. 2007. Synthesis of macro/mesoporous silica and carbon monoliths by using a commercial polyurethane foam as sacrificial template. *Mater. Lett.* 61:2378–2381.

70. Lépine, O., M. Birot, and H. Deleuze. 2009. Preparation of a poly(furfuryl alcohol)-coated highly porous polystyrene matrix. *Macromol. Mater. Eng.* 294:599–604.

71. Cameron, N.R., D.C. Sherrington, L. Albiston, and D.P. Gregory. 1996. Study of the formation of the open-cellular morphology of poly(styrene/divinylbenzene) polyHIPE materials by cryo-SEM. *Colloid Polym. Sci.* 274:592–595.

72. Molina, R., A. Vilchez, C. Canal, and J. Esquena. 2009. Wetting properties of polystyrene/divinylbenzene crosslinked porous polymers obtained using W/O highly concentrated emulsions as templates. *Surf. Interface Anal.* 41:371–377.

73. Canal, C., F. Gaboriau, A. Vílchez, P. Erra, M.J. Garcia-Celma, and J. Esquena. 2009. Topographical and wettability effects of post-discharge plasma treatments on macroporous polystyrene-divinylbenzene solid foams. *Plasma Proc. Polym.* 6:686–692.

74. Haibach, K., A. Menner, R. Powell, and A. Bismarck. 2006. Tailoring mechanical properties of highly porous polymer foams: Silica particle reinforced polymer foams via emulsion templating. *Polymer* 47:4513–4519.

75. Menner, A., K. Haibach, R. Powell, and A. Bismarck. 2006. Tough reinforced open porous polymer foams via concentrated emulsion templating. *Polymer* 47:7628–7635.

76. Menner, A., V. Ikem, M. Salgueiro, M.S.P. Shaffer, and A. Bismarck. 2007. High internal phase emulsion templates solely stabilised by functionalised titania nanoparticles. *Chem. Commun.* 4274–4276.

77. Menner, A., M. Salgueiro, M.S.P. Shaffer, and A. Bismarck. 2008. Nanocomposite foams obtained by polymerization of high internal phase emulsions. *J. Polym. Sci. Pol. Chem.* 46:5708–5714.

78. Ikem, V.O., A. Menner, and A. Bismarck. 2010. High-porosity macroporous polymers sythesized from titania-particle-stabilized medium and high internal phase emulsions. *Langmuir* 26:8836–8841.

79. Ikem, V.O., A. Menner, and A. Bismarck. 2011. Tailoring the mechanical performance of highly permeable macroporous polymers synthesized via Pickering emulsion templating. *Soft Matter* 7:6571–6577.

80. Wong, L.L.C., V.O. Ikem, A. Menner, and A. Bismarck. 2011. Macroporous polymers with hierarchical pore structure from emulsion templates stabilised by both particles and surfactants. *Macromol. Rapid Commun.* 32:1563–1568.

81. Ghosh, G., A. Vilchez, J. Esquena, C. Solans, and C. Rodríguez-Abreu. 2011. Preparation of ultra-light magnetic nanocomposites using highly concentrated emulsions. *Mater. Chem. Phys.* 130:786–793.

82. Scheffler, M. and P. Colombo. 2005. *Cellular Ceramics: Structure, Manufacturing, Properties and Applications.* Weinheim: Wiley-VCH.

83. Yang, X.Y., A. Léonard, A. Lemaire, G. Tian, and B.L. Su. 2011. Self-formation phenomenon to hierarchically structured porous materials: Design, synthesis, formation mechanism and applications. *Chem. Commun.* 47:2763–2786.

84. Chung, C.M., T.M. Chou, G. Cao, and J.G. Kim. 2006. Porous organic–inorganic hybrids for removal of amines via donor-acceptor interaction. *Mater. Chem. Phys.* 95:260–263.

85. Tanaka, N., H. Kobayashi, K. Nakanishi, H. Minakuchi, and N. Ishizuka. 2001. Monolithic LC columns. *Anal. Chem.* 73:420–429.

86. Chen, X., X. Wang, and X. Fu. 2009. Hierarchical macro/mesoporous TiO_2/SiO_2 and TiO_2/ZrO_2 nanocomposites for environmental photocatalysis. *Energ. Environ. Sci.* 2:872–877.

87. Frackowiak, E. and F. Béguin. 2001. Carbon materials for the electrochemical storage of energy in capacitors. *Carbon* 39:937–950.

88. Brun, N., R. Janot, C. Sanchez, H. Deleuze, C. Gervais, M. Morcrette, and R. Backov. 2010. Preparation of LiBH4@carbon micro-macrocellular foams: Tuning hydrogen release through varying microporosity. *Energ. Environ. Sci.* 3:824–830.

89. Mamak, M., N. Coombs, and G. Ozin. 2000. Mesoporous yttria-zirconia and metal-yttria-zirconia solid solutions for fuel cells. *Adv. Mater.* 12:198–202.

90. Flexer, V., N. Brun, R. Backov, and N. Mano. 2010. Designing highly efficient enzyme-based carbonaceous foams electrodes for biofuel cells. *Energ. Environ. Sci.* 3:1302–1306.

91. Flexer, V., N. Brun, O. Courjean, R. Backov, and N. Mano. 2011. Porous mediator-free enzyme carbonaceous electrodes obtained through integrative chemistry for biofuel cells. *Energ. Environ. Sci.* 4:2097–2106.

92. Brun, N., S.R.S. Prabaharan, C. Surcin, M. Morcrette, H. Deleuze, M. Birot, O. Babot, M.F. Achard, and R. Backov. 2012. Design of hierarchical porous carbonaceous foams

from a dual-template approach and their use as electrochemical capacitor and Li ion battery negative electrodes. *J. Phys. Chem. C* 116:1408–1421.

93. Du, N., J.L. Blin, and M.J. Stébé. 2010. Effect of hydrocarbon incorporation in the RH 12A(EO)9 system: Preparation of porous materials. *Micropor. Mesopor. Mater.* 135:149–160.

94. Du, N., M.J. Stébé, R. Bleta, and J.L. Blin. 2010. Preparation and characterization of porous silica templated by a nonionic fluorinated systems. *Colloids Surf. A* 357:116–127.

95. Chiu, J.J., D.J. Pine, S.T. Bishop, and B.F. Chmelka. 2004. Friedel–Crafts alkylation properties of aluminosilica SBA-15 meso/macroporous monoliths and mesoporous powders. *J. Catal.* 221:400–412.

96. Esquena, J., J. Nestor, A. Vílchez, K. Aramaki, and C. Solans. 2012. Preparation of mesoporous/macroporous materials in mighly concentrated emulsions based on cubic phases by a single-step method. *Langmuir* 28:12334–12340.

97. Hu, Y., M. Nareen, A. Humphries, and P. Christian. 2010. A novel preparation of macroscopic beads of porous silica. *J. Sol-gel. Sci. Technol.* 53:300–306.

98. Blin, J.L., J. Grignard, K. Zimny, and M.J. Stébé. 2007. Investigation of the C16(EO)10/decane/water system for the design of porous silica materials. *Colloids Surf. A* 308:71–78.

99. Yiv, S., R. Zana, W. Ulbricht, and H. Hoffmann. 1981. Effect of alcohol on the properties of micellar systems. II. Chemical relaxation studies of the dynamics of mixed alcohol + surfactant micelles. *J. Colloid Interface Sci.* 80:224–236.

100. Soni, S.S., G. Brotons, M. Bellour, T. Narayanan, and A. Gibaud. 2006. Quantitative SAXS analysis of the P123/water/ethanol ternary phase diagram. *J. Phys. Chem. B* 110:15157–15165.

101. Aramaki, K., U. Olsson, Y. Yamaguchi, and H. Kunieda. 1999. Effect of water-soluble alcohols on surfactant aggregation in the C12EO8 system. *Langmuir* 15:6226–6232.

102. Nestor, J., A. Vílchez, C. Solans, and J. Esquena. 2012. *Facile Synthesis of Meso/Macroporous Dual Materials with Ordered Mesopores Using Highly Concentrated Emulsions Based on a Cubic Liquid Crystalline Phase.* Unpublished work, IQAC-CSIC: Barcelona.

103. Sattler, K. and H. Hoffmann. 1999. A novel glycol silicate and its interaction with surfactant for the synthesis of mesoporous silicate. *Prog. Colloid Polym. Sci.* 112:40–44.

104. Hartmann, S., D. Brandhuber, and N. Hüssing. 2007. Glycol-modified silanes: Novel possibilities for the synthesis of hierarchically organized (hybrid) porous materials *Accounts Chem. Res.* 40:885.894.

105. Köhler, J., A. Feinle, M. Waitzinger, and N. Hüssing. 2009. Glycol-modified silanes as versatile precursors in the synthesis of thin periodically organized silica films *J. Sol-gel. Sci. Technol.* 51:256–263.

106. Ramsden, W. 1903. Separation of solids in the surface-layers of solutions and 'suspensions' (observations on surface-membranes, bubbles, emulsions, and mechanical coagulation). Preliminary account. *Proc. Royal Soc.* 72:156–164.

107. Pickering, S.U. 1907. CXCVI. Emulsions. *J. Chem. Soc.* 91:2001–2021.

108. Aveyard, R., B.P. Binks, J. Esquena, P.D.I. Fletcher, P. Bault, and P. Villa. 2002. Flocculation transitions of weakly charged oil-in-water emulsions stabilized by different surfactants. *Langmuir* 18:3487–3494.

109. Binks, B.P. 2002. Particles as surfactants—Similarities and differences. *Curr. Opin. Colloid. Interface Sci.* 7:21–41.

110. Aveyard, R., B.P. Binks, and J.H. Clint. 2003. Emulsions stabilised solely by colloidal particles. *Adv. Colloid Interface Sci.* 100–102:503–546.

111. Binks, B.P., J.A. Rodrigues, and W.J. Frith. 2007. Synergistic interaction in emulsions stabilized by a mixture of silica nanoparticles and cationic surfactant. *Langmuir* 23:3626–3636.

112. Binks, B.P., A. Desforges, and D.G. Duff. 2007. Synergistic stabilization of emulsions by a mixture of surface-active nanoparticles and surfactant. *Langmuir* 23:1098–1106.

113. Hunter, T.N., R.J. Pugh, G.V. Franks, and G.J. Jameson. 2008. The role of particles in stabilising foams and emulsions. *Adv. Colloid Interface Sci.* 137:57–81.

114. Frelichowska, J., M.A. Bolzinger, and Y. Chevalier. 2010. Effects of solid particle content on properties of o/w Pickering emulsions. *J. Colloid Interface Sci.* 351:348–356.

115. Kralchevsky, P.A. 1996. Conditions for stable attachment of fluid particles to solid surfaces. *Langmuir* 12:5951–5955.

116. Kralchevsky, P.A. 1997. Lateral forces acting between particles in liquid films or lipid membranes. *Adv. Biophys.* 34:25–39.

117. Kralchevsky, P.A., I.B. Ivanov, K.P. Ananthapadmanabhan, and A. Lips. 2005. On the thermodynamics of particle-stabilized emulsions: Curvature effects and catastrophic phase inversion. *Langmuir* 21:50–63.

118. Danov, K.D. and P.A. Kralchevsky. 2006. Electric forces induced by a charged colloid particle attached to the water–nonpolar fluid interface. *J. Colloid Interface Sci.* 298:213–231.

119. Danov, K.D., P.A. Kralchevsky, K.P. Ananthapadmanabhan, and A. Lips. 2006. Particle-interface interaction across a nonpolar medium in relation to the production of particle-stabilized emulsions. *Langmuir* 22:106–115.

120. Leal-Calderón, F. and V. Schmitt. 2008. Solid-stabilized emulsions. *Curr. Opin. Colloid. Interface Sci.* 13:217–227.

121. Ikem, V.O., A. Menner, A. Bismarck, and L.R. Norman. 2011. Liquid screen: Pickering emulsion templating as an effective route for forming permeable and mechanically stable void-free barriers for hydrocarbon production in subterranean formations. Proceedings of the SPE International Symposium on Oilfield Chemistry. 1:496–504

122. Studart, A.R., J. Studer, L. Xu, K. Yoon, H.C. Shum, and D.A. Weitz. 2010. Hierarchical porous materials made by drying complex suspensions. *Langmuir* 27:955–964.

123. Li, Z. and T. Ngai. 2010. Erratum: Macroporous polymer from core-shell particle-stabilized pickering emulsions (Langmuir (2010) 26 (5088)). *Langmuir* 26:16186.

124. Li, Z. and T. Ngai. 2010. Macroporous polymer from core-shell particle-stabilized pickering emulsions. *Langmuir* 26:5088–5092.

125. Destribats, M., B. Faure, M. Birot, O. Babot, V. Schmitt, and R. Backov. 2012. Tailored silica macrocellular foams: Combining limited coalescence-based pickering emulsion and sol–gel process. *Adv. Funct. Mater.* 22:2642–2654.

126. Lan, Q., C. Liu, F. Yang, S. Liu, J. Xu, and D. Sun. 2007. Synthesis of bilayer oleic acid-coated Fe_3O_4 nanoparticles and their application in pH-responsive Pickering emulsions. *J. Colloid Interface Sci.* 310:260–269.

127. Bachinger, A. and G. Kickelbick. 2010. Pickering emulsions stabilized by anatase nanoparticles. *Monatsh. Chem.* 141:685–690.

Section II

New Experimental Tools and Interpretations

11 Simulation of Interfacial Properties and Droplet Hydrodynamics

Adil Lekhlifi and Mickaël Antoni

CONTENTS

11.1 INTRODUCTION

The successful developments of computational techniques provide nowadays robust and stable algorithms for the simulation of systems involving interfaces. Molecular dynamics (MD) and computational fluid dynamics (CFD) are two techniques commonly used to numerically describe such systems. To illustrate the capabilities of both techniques, this chapter will present simulations of simplified systems consisting of surfactant monolayers at water/air interfaces and water droplets in a continuous paraffin oil phase. The goal of this chapter is hence not to present a detailed review of all the running challenges in the simulation of interfaces. It essentially aims to put emphasis on the opportunities offered by CFD and MD in the field of interfacial chemistry. Such systems have a major applicative importance since their understanding is fundamental in a large variety of research domains and in many industrial applications. The dispersion of two immiscible liquids, such as in emulsions, is, for example, commonly found in many technical processes with major importance in chemical, pharmaceutical, petroleum, and food industries (Binks 1998; Bourrel and Schechter 1988; Dickinson 2003; Léal-Calderon et al. 2007; Salager 2000). This is also the case for liquid droplets in a gaseous phase like in sprays (Charles 2012). Emulsions and sprays both involve liquid free interfaces and hence lead to the same difficulties when trying to simulate their behavior. Besides these practical issues, challenging applied mathematics questions also appear in the simulation of free-surface systems. In the simple case of evaporating sessile droplets, solid, liquid, and gaseous phases simultaneously come into play. The simulation of heat and mass transfers as well as contact line dynamics in such a basic system is still nowadays a difficult task (Girard and Antoni 2008; Girard et al. 2008). In all the aforementioned examples, interfaces are the regions where macroscopic and microscopic scales are intimately linked. They involve wetting properties, droplet coalescence, and evaporation, all of which rely on both hydrodynamic and molecular phenomena. The interplay between these interactions raises a number of open questions in the field of computer science that range from basic mass conservation constraints to the accurate description of moving free surfaces. Also still challenging is the numerical description of interfacial properties like interfacial tension or adhesion forces of liquids on solid substrates.

This chapter aims to give insights into the problems that appear when dealing with numerical interfaces. It specifically focuses on MD and CFD techniques. The objective of any numerical technique is to perform accurate simulations with minimal computational efforts. This goal is usually well achieved for CFD and MD approaches when considered separately. But when studying systems where both molecular and hydrodynamic scales have to be accounted for simultaneously, the first question is to find a way to associate these two techniques in a way to avoid prohibitive computer time. This problem is still one of the main limitations for a complete numerical description of interfaces, although multiscale approaches are available in the literature for turbulence (Weinan 2011; Wilcox 2006) and for a large family of systems (Horstemeyer et al. 2010). The example of evaporating sessile droplets given above is a typical multiscale system where mass and heat transfers can be described by CFD using the Navier–Stokes and the heat equations together with continuity and specific boundary condition equations. But when considering contact line dynamics, adhesion forces have to be accounted for. As they act at the molecular level between the droplet and the substrate, they require the use of molecular-based techniques. As a result, a simple system such as the unique sessile droplet involves almost all possible spatial and temporal scales, that is, from a macroscopic down to a microscopic scale: macroscopic, for hydrodynamics (convection, heat transfer), and microscopic, for molecular contributions (liquid–solid adhesion forces, surfactant layer dynamics). This has led to the development of numerical methods for intensive calculations to simulate either hydrodynamic phenomena, with CFD (Scardovelli and Zaleski 1999), or molecular scales with MD or quantum-based techniques (Schlick 2002).

Systems such as droplets combine different complexity levels: free surfaces, unsteadiness, chemistry, and at least two (or more) separated phases. All these properties involve specific time or space scales that can differ from one each other by orders of magnitude. This problem seems at first sight simple, but from a numerical point of view, it is highly non-trivial. A moving surface indeed "disappears" upstream and is "created" downstream. This simple propagation problem remains unresolved and affects numerical stability. At the macroscopic scale, models are continuous and it is clear that numerical algorithms have to keep track of the interfaces to correctly catch the overall evolution of each phase. CFD is then the most adapted technique although interfaces are regions where sharp gradients are present in the physical and chemical properties. But in CFD, interfaces are considered as two-dimensional (2D) regions, while at the microscopic scale, interfaces present a finite width particularly in the presence of surfactant layers. Classical molecular techniques (like MD) or quantum chemistry approaches (such as density functional theory, molecular mechanics) then have to be used. Consequently, depending on the scale that is described, specific numerical methods are likely to yield more accurate results with less computational efforts. There is still no systematic possibility nowadays to describe all the phenomenology of systems from a macroscopic down to a microscopic scale even with massively parallel supercomputer facilities. It is for this reason that the physics and chemistry involved eventually dictate which method will be used. As a result, numerical descriptions require system-specific techniques, each having its advantages and drawbacks.

Despite intense algorithmic research to speed up MD simulations, computing power is still too limited to precisely simulate molecular systems with an explicit molecular description for more than a few tens of thousands of molecules and for more than a few hundreds of microseconds. This is still too limited to be efficiently coupled with CFD simulations. To overcome this difficulty, the usual approach consists in the identification of time (space) domains of interest and to separate the fast (small scale) evolving properties from the slow (large scale) ones. This is somehow anecdotal evidence but it constitutes, in the simulation of interfacial phenomena, the key argument to justify the use of either MD or CFD techniques. Unless such scale separation is possible, it is difficult to systematically obtain results that are, even qualitatively, coherent with experimental observations. This is actually why equilibrium or quasi-steady hypotheses are generally invoked. They present the important advantage of allowing the dissociation of hydrodynamic scales from the molecular ones. Such quasi-steady approximations are often well justified in MD simulations as molecular phenomena are very fast when compared to hydrodynamic ones. The latter hence do not need to be described explicitly and implicit or averaged models can then be sufficient (Roux and Simonson 1999). In continuum models, on the other hand, all molecular properties are smeared out and contribute only through equations of state that are included as coupling contributions into the Navier–Stokes or heat equations for CFD simulations. But such scale separation techniques clearly reach their limits with interest in the colloid community for micro- or nanoparticles (Ravera et al. 2008; Santini et al. 2011). The properties of the latter at liquid/liquid or liquid/gas interfaces introduce new constraints owing to the appearance of contact line regions between the different coexisting phases where wetting properties might come into play. Although reasonably justified in most of the studies, scale separation might here become problematic since both microscopic and macroscopic phenomena have to be considered simultaneously. Nanoparticles are indeed large objects for MD and small objects for CFD. As they might generate granular-like properties of the bulk liquid, such particles can still be described with CFD techniques provided adapted viscosity models are implemented (Murshed et al. 2008). But such nanoparticles can also be surface active and hence potentially modify interfacial properties such as in Ramsden–Pickering emulsions (Binks 2002; Binks Rodrigues 2007; Ferrari et al. 2012; Pickering 1907; Schmitt-Rozières et al. 2009). When adsorbed on interfaces, they then generate a contact line where wetting and adhesion forces have to be accounted for. These last forces act at the molecular level and should hence be investigated with MD or any other molecular-based simulation technique.

The explicit simulation of surfactant layers from first principles is a straightforward application field for MD. Adsorbed surfactant molecules indeed often do not present any chemical reactivity. Quantum effects are then absent and it is therefore possible to assume that atoms are classical point-like massive and charged particles. As molecules are not transformed into other compounds, the overall chemical composition remains unchanged even though surfactant transfer mechanisms between bulk phases and interfaces are present. In this context, it is actually the structure of the interfaces and their underlying kinetics that are usually under focus with the objective to understand mass transfer mechanisms as well as properties such as surface tension or elasticity (Miller and Liggieri 2011). But slow diffusive mechanisms are not accessible with explicit MD approaches and need the introduction of implicit models to speed up simulations (Panczyk et al. 2012; Roux and Simonson 1999). The use of such techniques for the description of surfactant migration from the bulk phase into the interfaces (and vice versa) is hence not easily feasible because of demanding computer power. This is why the simulation of surfactant layers with explicit MD simulations is often restricted to the description of the organization of surfactant molecules inside these layers and on their short time evolution. Still, these simulations have to be long enough to ensure steady configurations to be reached. This is a fundamental and necessary condition to allow relevant numerical predictions.

Surfactant layers have long been the object of many experimental studies. A very rich literature still focuses on surface tension measurements and on the organization of surfactants at liquid/liquid and liquid/gas interfaces (Ferrari et al. 2012). Neutron reflection experiments were also carried out and showed that molecules are not vertically anchored on interfaces but appear to be tilted with angles that depend on their concentration and structure (Lu et al. 1993a,b, 1994, 1998, 2000). The influence of temperature and headgroup size has also been investigated experimentally with film balance and Brewster angle microscopy (Islam and Kato 2005, 2006), and the hydration state of non-ionic monolayers was characterized with vibrational spectroscopy (Tyrode et al. 2005). Besides all these experimental studies, important literature has also been devoted to the simulation of surfactant layers. MD numerical investigations with monolayers consisting of polyethylene glycol molecules at water/air and air/water/oil interfaces, for example, have confirmed the tilted geometry of these molecules (Kuhn and Rehage 1998, 1999, 2000a,b). This behavior indicates that the traditional picture of surfactant monolayers consisting of linearly shaped and unfolded molecules vertically anchored in interfaces has to be revisited. These results were recently completed with explicit MD simulations accounting for all the degrees of freedom coming into play in the motion of the atoms. It has been demonstrated that molecules were not only displaying tilted geometries but also exhibiting highly folded structures for both their hydrophilic and hydrophobic groups (Cuny et al. 2004, 2008). This has been further used to explain the narrower intrinsic width of the surfactant monolayers observed experimentally and moreover suggested that monolayers have an important structural roughness owing to molecular entanglement. It is important to note here that explicit MD outputs are not always in good agreement with experiments although they often provide interesting qualitative views of the actual molecular arrangements. One reason for this is the treatment of long-range forces that are often artificially damped to zero with a direct cutoff treatment (Huang et al. 2010). Other techniques like Ewald summation procedures can be employed for a better treatment of long-range interactions. But they require larger computational efforts for almost similar outputs. Overall, although explicit MD techniques constitute a very useful tool for qualitative understanding and interpretation, the predictions they provide are not systematically in good qualitative agreement with experimental measurements. Besides cutoff problems, the limited size of the systems, the choice of the initial condition, and the choice of the force field that governs the time evolution of the molecules can also explain this discrepancy. These last points will be discussed in more detail in Section 11.2.

The limited size of the systems that can be investigated with explicit MD techniques prevents a straightforward access to other fundamental aspects of interface properties such as Marangoni circulation flows and more generally to hydrodynamic phenomena in both interfaces and bulk. Simulations of multiphase flows, including interfacial phenomena, hence need to change perspective and to go for continuum descriptions and CFD. Such approaches rely on discretized partial differential equations (PDEs), and the first task in this context is to properly monitor the spatial derivatives they introduce. This implies creating mesh grids to partition the simulation domain (SD) in an ensemble of cells and to carefully treat boundary conditions. The presence of free-moving interfaces introduces an additional level of complexity since it requires adapted front localization techniques. From the numerical point of view, interfaces correspond to discontinuities where derivatives have to be handled and computed carefully. Their accurate description is actually one of the main concerns in the simulation of multiphase systems. It is indeed of first importance to properly capture interfaces, to preserve their sharpness, and, while running simulations, to ensure mass conservation. Two approaches are possible for this: Lagrangian or "front tracking" methods using markers with unstructured mobile meshes and Eulerian or "front capturing" methods based rather on fixed structured meshes. These two approaches were confronted to experimental observations and it was established that Lagrangian approaches are generally more accurate than Eulerian methods and can therefore be used as references for Eulerian methods (Rider and Kothe 1995). Eulerian techniques have the advantage of avoiding the reconstruction of a new mesh grid at each time step and, hence, limiting computational power. They moreover rely on integration schemes that are robust in time and do rapidly converge (Hirt et al. 1970; Hirt and Nichols 1981). But they have the drawback of being computationally demanding for the description of interfacial phenomena because of artifacts.

The simulation of multiphase flows with free interfaces hence remains a difficult task as will be discussed in Sections 11.3 and 11.4 in the case of water droplets in paraffin oil systems. This system will be considered in detail here since it can be seen in the context of free-surface CFD as a paradigm system where hydrodynamics, unsteadiness, and interface properties have to be simultaneously described. As just mentioned, one critical problem here is the stability of the numerical schemes that can potentially generate artifacts. The appearance of spurious currents owing to the implementation of capillary forces is one example of artifacts that will be discussed in Section 11.3. Overall, the performances of CFD approaches to simulate multiphase systems with free interfaces rely on their ability to maintain the thickness of interfaces as small as possible, to preserve conservation laws at acceptable levels, and finally to damp as much as possible all spurious phenomena generated by the integration algorithms. All these conditions need the development of solvers able to obtain well-converged hydrodynamic solutions. Important efforts are still devoted to the optimization of solvers but, among the actual challenges in computer science, the elaboration of optimal methods for the description of sharp free-moving surfaces is still one of the main concerns.

This chapter is organized in five sections. After this introduction, Section 11.2 will put the emphasis on the ability of MD simulations to describe the structure of pentaethylene glycol monododecyl ether ($C_{12}E_5$) monolayers adsorbed at water/air interfaces. Section 11.3 proposes a fast overview of CFD techniques for the simulation of free interface systems in the context of continuum mechanics with a special focus on volume of fluid (VOF) techniques. Section 11.4 will illustrate the ability of CFD to describe the dynamics of water droplets in paraffin oil. Finally, Section 11.5 will be devoted to conclusions.

11.2   MD APPLIED TO SURFACTANT MONOLAYERS

From the very first principles' point of view, interfaces consist of interacting atoms that generate macroscopic properties resulting from the time evolution of all the atomic degrees of freedom. Ideally, when simulating a system, the numerical technique should be based on these microscopic interactions. They indeed contain all the information to describe its complete phenomenology at all time and spatial scales. This could be particularly interesting for the description of turbulence in fluids for which effective, although well-validated, models are invoked (Wilcox 2006). But such an explicit atomic-based approach is still not yet feasible even if the increase of computer power, boosted by important improvements in models and algorithms and by the emergence of massively parallel supercomputers and clusters, provides nowadays huge computer capacities that allow the simulation of very complex systems. Nowadays, these large instruments make possible MD simulations of a large number of atoms on reasonable time scales and therefore allow accurate investigations of liquid/liquid or liquid/air interfacial properties. But their capacities still remain very limited for an explicit atomic-based approach of realistic systems. This section intends to give insights and simple illustrations into the ability of MD to describe interfacial properties and to show to what extent it can be used for surfactant-stabilized systems. It hence does not aim to give a full overview of all the capabilities of molecular modeling. It instead proposes to illustrate the interest of this specific technique with an emphasis on the different problems that are usually encountered. Moreover, still with the idea of keeping the presentation as illustrative as possible, this section will concentrate on monolayers consisting of $C_{12}E_5$ molecules (see Figure 11.1a). The reason for considering this surfactant is that it is widely used and, moreover, has the advantage of being non-ionic. The description of counterion contributions in the simulations is hence not necessary.

In explicit MD simulations, all interactions are accounted for as long as they do not involve quantum effects. Depending on the level of accuracy of the simulations, bonded (i.e., intramolecular) and non-bonded (i.e., intermolecular) interactions explicitly contribute to the motion equation of each atoms. In the particular case of water molecules, bond–bond O–H distance and dihedral angle HOH are two bonded interactions whereas non-bonded ones are described by electrostatic and van der

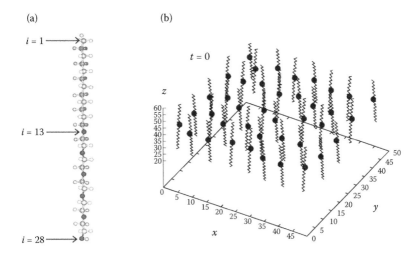

FIGURE 11.1 (a) Minimal energy conformation of a $C_{12}E_5$ molecule. Oxygen atoms and carbon are indexed from 1 to 28. Hydrogen atoms are also displayed but not indexed. (b) Initial configuration of the $C_{12}E_5$ monolayer considered in this section. The black circles represent the first oxygen atom of the glycol chain [i.e., atom $i = 13$ in (a)].

Waals couplings. As all degrees of freedom are coupled together, electrostatic and van der Waals interactions will modify the time evolution of bonded degrees of freedom and vice versa. For molecules like $C_{12}E_5$, torsion and out-of-plane contributions also have to be considered. Therefore, large molecules need an increasing number of degrees of freedom to be accurately described. Another important parameter here is the level of distortion of the molecules. When they are close to their minimal energy conformation, the use of harmonic potentials is sufficient to describe their potential energy. But far-from-equilibrium conformations can also show up. It is then important to complete these harmonic contributions with non-harmonic ones as well as direct couplings between the different bonded degrees of freedom. A consequence of this is an increasing complexity of the interaction potentials that turn out to be highly complex mathematical objects (Wendy et al. 1995). In the frame of explicit MD, such potentials are called force fields and constitute the essential ingredients of this technique. They indeed give the expression of the potential energy of the considered system and, as the motion equation of each atom derives from this force field, also govern the overall time evolution. Besides force field, two other elements have to be carefully addressed. First the thermodynamic ensemble within which the simulations are performed and the necessity to keep constant specific associated physical quantities (like for example temperature for simulations in the canonical ensemble). Then, as the geometry of the molecules can become highly distorted, it might become important to not only to include non linearities but also to compute the coupling constants of bounded interactions with quantum based techniques. The latter is important when the deformations of the molecules become so large that partial atomic charges need to be recomputed. Concerning conserved quantities, for microcanonical simulations (mass, volume, and energy constants), motion equations have to be integrated with specific symplectic (i.e., phase space area preserving) integrators (Yoshida 1990). But in most cases, molecular systems are open and, in the common case of canonical simulations (constant mass, volume, and temperature), motion equations of the atoms are coupled to an extra equation that acts like a viscous contribution describing the effect of a thermostat (Andersen 1980; Evans and Holian 1985). When modifications in the partial atomic charges have to be included, a coupling with quantum algorithms or the use of databases becomes necessary to obtain their actual values and the ones of the modified coupling constants. Dedicated additional techniques are then to be introduced.

The choice and the treatment of boundary conditions is also an important issue in MD. In the example discussed below, the size of the SD is given in the x and y directions by $L_x = L_y = 50.6$ Å and $L_z = 100$ Å where z is the coordinate normal to the interface plane (see Figure 11.1b). Boundary conditions are periodic in x and y while the atoms near the bottom of the SD (i.e., $z = 0$) have a fixed position. Such an SD is sufficiently large to avoid artifacts that could show up owing to uncontrolled interactions induced by the boundary conditions (Kuhn and Rehage 1998, 1999, 2000a,b; Tarek et al. 1995).

One important problem when performing MD simulations is the solvent modeling. This problem becomes particularly acute for water because of polarizability. Since it constitutes for many systems the main chemical compound, it is often described as a way to minimize computation time and to account for basic dielectric and thermodynamic properties (Bandyopadhyay et al. 2000; Kuhn and Rehage 1998, 1999, 2000a,b; Schweighofer et al. 1997; Shelley et al. 2000; Tarek et al. 1995; Van Buuren et al. 1993; Van Buuren 1995). Several models—rigid, flexible, or polarizable—have been developed to describe water as accurately as possible (Guillot 2002). The most widely used are SPC, SPC/E, or TIP3P models (Berendsen et al. 1981, 1987; Jorgensen et al. 1983). They are all based on Lennard-Jones interaction potentials that adequately describe the thermodynamic properties of water under ambient temperature and pressure conditions (Vorholz et al. 2000). Furthermore, these models have the advantage of giving good estimations of thermodynamic properties, particularly when liquid and gaseous phases coexist. To simulate a solute in water, the SPC model seems the most appropriate. This is also the case for water/decane interfaces where it gives good predictions of the solubility of water in decane (Van Buuren et al. 1993; Van der Spoel et al. 1998). Investigations dedicated to polyoxyethylene surfactants in lamellar phases have successfully used either SPC/E or TIP3P (Bandyopadhyay et al. 2000). It follows from this work that the interlamellar spacing of surfactant bilayers and the values of the area per surfactant molecule do not significantly depend on the considered water model. Similar observations were made for water/decane interfaces with SPC/E and SPC models (Van Buuren et al. 1993). But this is not always the case. For complex systems like the ones involving, for example, proteins, unsuitable water models can potentially modify the overall protein structure (Tarek and Tobias 2000).

Besides the choice of the force field and the water (or solvent) model, another fundamental problem is the choice of the initial condition. The necessity to carefully manage the construction of the initial conditions is essential when considering liquid systems. In MD simulations, temperature is finite (i.e., non-zero) and the initial velocities of the atoms are naturally set in a way to obey a Maxwell–Boltzmann distribution. But the choice of their initial positions is very tricky and needs to be carefully handled for the simulations to properly converge. One potential risk here is indeed to consider initial positions that are so non-typical that it will not be possible to relax them to steady configurations without prohibitive central processing unit (CPU) time. This is why the initial position of water molecules is so critical when investigating surfactant properties with explicit MD techniques. As they constitute the solvent, they can be at the origin of completely artificial and non-typical molecular structures not only for the surfactants but also for the water sample itself. To avoid this problem, initial conditions are first submitted to energy minimizations or randomization procedures associated subsequently to MD runs. Introducing the surfactant molecules is the last step in the construction of initial conditions. The most straightforward procedure is to place them randomly on the interface and perpendicularly to it (see Figure 11.1b). This requires the removal of the water molecules that prevent this placement owing to limited space. A short randomization, energy minimization, or MD run is then again necessary to get a steady configuration for the complete surfactant monolayer + water system. Such a randomization has been used for $C_{12}E_2$ molecule layers but with a procedure where water molecules were constrained to remain fixed (Bandyopadhyay et al. 2000). Other authors originally proposed to immerse the surfactant molecules completely and randomly in a water sample and to let them migrate to the interface until a stationary configuration is reached (Tomassone et al. 2001a,b). Initial conditions obtained from preliminary runs with randomization algorithms together with MD simulations have also been used for sodium dodecyl sulfate monolayers for a water/CC_{14} mixture (Schweighofer et al. 1997). This procedure presents the important advantage of starting with an initial condition that can be considered as being very close to a steady-state configuration, hence significantly speeding up the simulations although it can become CPU time consuming when explicit MD is used. Conversely to the previous discussion, where relaxed configurations are searched, there are cases where initial conditions are based on the importance of the arrangement of the molecules. It has been proposed, for example, that monolayers with an initial hexagonal distribution be considered in order to be as close as possible to known thermodynamic phases (Karaborni 1993).

To illustrate the capabilities of explicit MD simulations, this chapter will now focus on a water/air interface with adsorbed $C_{12}E_5$ surfactant molecules. All molecules are assumed to be non-rigid with periodic boundary conditions. The time evolution of water and $C_{12}E_5$ molecules is handled on the same basis and motion equations are derived from a second-generation force field noted CFF91 (Maple et al. 1994). CFF91 includes all the relevant degrees of freedom: bond and bond angle coupling, the first three Fourier components for dihedral angles, torsion, and all the cross contributions between bonded degrees of freedom. The partial atomic charges are all fixed since it is assumed that the deformations of the molecules remain in the range where non-harmonic potentials are sufficient for their description. Moreover, 1–4 non-bonded interactions are scaled by 0.5 with a 9-4 Lennard-Jones potential. The parameters of this last potential (radii, coupling constants, etc.), when the effects of two atoms overlap, are deduced from combination rules (Waldman and Hagler 1993). The validity of the choice of such a force field is still questionable, particularly regarding the chosen values of coupling constants, to conservation laws and even to the different contributions it includes. Therefore, comparison with experiments is essential. CFF91 is used here on the strength of the fact that it is widely applied in the structural optimization of hydrocarbon chains and their derivatives. It is a second-generation force field derived from ab initio models and was parameterized against a wide range of experimental observables for organic compounds. Moreover, it gives reasonable agreement between the computed $C_{12}E_5$ monolayer structure and neutron reflection measurements (Lu et al. 1998).

All the simulations presented in this section were performed with the Cerius² Package (Cerius2 2000) in the canonical ensemble with a time step of 0.1 fs and a maximal integration time of 1.2 µs. The $C_{12}E_5$ molecules are initially introduced vertically in a water sample obtained from preliminary simulations and containing approximately 2400 molecules. Water molecules that prevent this introduction are removed and the hydrocarbon chains are initially located completely above the interface. The hydrophilic ones are hence completely immersed in water. A new energy minimization and a subsequent MD round (for a few tenths of microseconds) is then carried out before starting the monolayer MD simulation. Figure 11.1a illustrates the conformation of a $C_{12}E_5$ molecule in the minimal energy configuration and Figure 11.1b is a plot of the initial monolayer structure. It consists of 40 surfactant molecules. The black circles represent the first oxygen atom of the glycol chain (i.e., atom $i = 13$ in Figure 11.1a). It is located for all molecules at height $z \approx 44$ Å. For clarity, hydrogen atoms and water molecules are not represented in this figure.

A snapshot of the monolayer steady configuration at time $t = 800$ ps is shown in Figure 11.2 in a three-dimensional (3D) representation (Cuny et al. 2004). It appears that the conformation of each surfactant molecule is far from its initial configuration in Figure 11.1b and that the structure of the monolayer exhibits an important molecular entanglement. This is a consequence of the use of CFF91 that includes all the bound–bound couplings and therefore eases energy transfers between intramolecular degrees of freedom. This enhances the mobility of all the atoms, allowing a larger

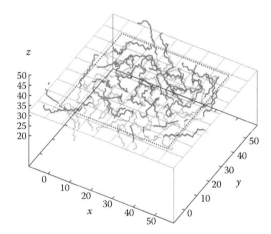

FIGURE 11.2 (See color insert.) 3D snapshot of the $C_{12}E_5$ monolayer structure at time $t = 800$ ps. The sections of these molecules underneath the interface is in green and those located above it are in red. The dotted line indicates the border of the SD in x and y directions. Hydrogen atoms and water molecules are not plotted.

variety of possible conformations for the surfactant molecules. As time evolves, they support distorted conformations (distorted but in a way to keep coherent with the fixed atomic partial charges approximation) with, for example, more than one anchor point into water or even, from time to time, complete immersion into it. This variety of conformations gives a picture of surfactant monolayers consisting of highly entangled molecules. This is consistent with neutron reflection measurements indicating that the intrinsic width of such surfactant monolayers are smaller than the one expected from fully stretched molecular conformations as the one illustrated in Figure 11.1b (Lu et al. 1993a,b, 1994, 1998, 2000).

One way to characterize the monolayer in Figure 11.2 is to study time-averaged mass densities and to focus particularly on one of those generated by the hydrophobic and hydrophilic chains of the surfactant molecules. Figure 11.3 is a plot of these mass densities obtained from time averages that run over approximately 1 μs. Such an averaging procedure is necessary to reduce fluctuations induced by finite size effects and to reasonably smooth out the curves. Three different domains can be identified in this figure when crossing the interface from the bottom of the SD to its top along the z direction: First, a pure water phase, then a domain where hydrophilic behavior prevails, and finally a domain where it is the hydrophobic character that becomes dominant. This is a quite traditional expectation when considering non-ionic surfactants at water/air interfaces. But here, because of the entanglement of the molecules, these domains are now overlapping and do not constitute well-separated regions as one would expect with the molecular arrangement in Figure 11.1b. The mass density profiles in Figure 11.3 do not evolve anymore with time since surfactant molecules keep on moving in a way to preserve the overall monolayer average properties. This results from the fact that 40 surfactant molecules, the size of the solvent (approximately 2400 water molecules), and the averaging time are sufficient

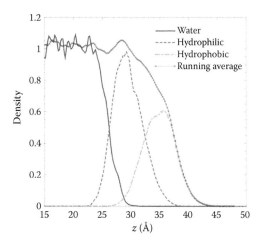

FIGURE 11.3 **(See color insert.)** $C_{12}E_5$ monolayer structure. Water density profile (blue), hydrophilic region (red), hydrophobic region (green), and total mass density distributions (black). Averaging time is 1 μs.

to circumscribe finite size effect and to ensure that the "memory" of the initial condition is lost. The overlap between the different density distributions is clearly visible in this figure, which shows that there is no abrupt change in the monolayer properties. They instead successively and smoothly change from pure water to hydrophobic behaviors and finally to hydrophilic (air can be seen as vacuum with regard to the size of the SD). This figure also indicates that because of the folding of the $C_{12}E_5$ molecules, hydration does not involve all their hydrophilic groups. Only five to six water molecules actually contribute to their adsorption (Cuny et al. 2008).

The study proposed in this section was aimed to illustrate how explicit MD simulations can help the understanding of the structure of molecular monolayers and to give an idea of all the conditions that have to be monitored to simulate such systems. The choice of the force field and the initial condition problem here are essential. The model for water description is also fundamental since, as time elapses, it can generate artifacts. There is actually no systematic rule that indicates which model is best suited for MD simulations of interfaces with adsorbed surfactant molecules. It is actually only the confrontation of the numerical outputs with experiments that can help decide whether one model is better than another. It also appears from this section that the size of the system that can be reasonably described with MD and the simulated periods are limited. Explicit MD simulations can therefore not address phenomena involving interfacial hydrodynamics. Other approaches based on continuous media and CFD techniques are then better adapted.

11.3 NUMERICAL APPROACH OF HETEROGENEOUS CONTINUOUS MEDIA

Multiphase flows and continuous media with free surfaces introduce a set of constraints that raise important issues in numerical science particularly when investigating the physical and chemical properties of interfaces. For such systems, the numerical technique must be able to simulate sharp interfaces and simultaneously ensure total mass conservation. The performance of a numerical method in this context is therefore mainly associated to its ability to maintain a thickness of the interfaces as small as possible and to respect elementary conservation laws. Keeping sharp interfaces is not obvious here since basic models in continuous media rely on PDEs and hence on derivatives that have to remain bounded for numerical stability. When simulating free-surface systems, one usual problem is the artificial "smearing" of the interfaces owing to numerical diffusion phenomena. It is thus fundamental to find appropriate discretization procedures of the PDEs under consideration and to circumvent the problem of divergences with specific free interface-preserving techniques, among them, boundary integral approaches. They allow a high order of accuracy for the description of interfaces but are unfortunately not applicable to viscous flows such as the ones governed by the Navier–Stokes equation (Hou et al. 2001). On the other hand, Lagrangian approaches allow the description of free-surface systems with mesh grids that map precisely the interfaces (Hirt et al. 1970). However, they impose a frequent remeshing of the SD and interpolations of all the physical properties to unstructured meshes. This introduces noise in the simulations that has to be carefully controlled in order to prevent numerical artifacts like unphysical mass variations. Another drawback of Lagrangian descriptions is the non-trivial handling of meshes when complex geometries have to be treated such as in evaporating or coalescing droplets. Cusps or topological changes cannot be addressed here without important computational effort. Lagrangian methods are hence well adapted for interface tracking as long as they keep smooth geometries. For complex ones, VOF approaches are potentially better adapted (Kleefsman et al. 2005). They are mass conservative and allow the description of systems with non-trivial geometries without complex grid nodes restructuring (Hirt and Nichols 1981). They belong to the family of Eulerian techniques and treat the SD as a whole, which makes them easier to formulate. But the price to pay for this simplicity is the loss of interface sharpness and the necessity to introduce specific methods for its preservation.

VOF approaches applied to multiphase flows raise a series of challenges like the description of the physical properties of the coexisting phases and the even more critical surface tension modeling. In Lagrangian methods, both tasks can be managed quite naturally since remeshing procedures allow keeping track of the interfaces and the integration of interfacial forces can be achieved with basic techniques like finite differences algorithms. When using VOF techniques, there are two ways of treating multiphase systems: Eulerian–Lagrangian approaches, where the coexisting fluids are considered as separate systems with moving markers to account for the time evolution of the boundary conditions, and Eulerian–Eulerian approaches, where there is no necessity to introduce any kind of particles or markers. One natural way to simulate binary systems like emulsions or sprays is to use two-fluid interphase slip algorithms that are based on the description of the two fluids with two distinct Navier–Stokes equations: one for the continuous phase and the other for the dispersed one (Latsa et al. 1999; Miller and Miller 2001). But, although it is more intuitive and seems to better fit two-fluid systems, this technique yields a more important level of diffusion of the interfaces. Single-fluid approaches on the other hand consider the complete system as being a unique fluid but with properties (density, viscosity, etc.) that depend on space. It is then necessary to introduce a space- and time-dependent function acting like a fluid tracer that takes a specific value in each phase to make the fluids distinguishable. This function, noted C in the following, is often called color function and employed to compute the interface curvature. It is propagated along the velocity field obtained from the Navier–Stokes equation by the advection equation (Bonometti and Magnaudet 2007; Nichols and Hirt 1973; Yang et al. 2009):

$$\frac{\partial C}{\partial t} + V.\nabla C = 0, \tag{11.1}$$

where V is the velocity vector. Ideally, for two-fluid systems, the color function should show a sharp decay from 1 (in the dispersed phase) to 0 (in the continuous phase). The density and dynamic viscosity of the complete system are then obtained from the following equations:

$$\rho = \rho_c + (\rho_d - \rho_c) \times H(C) \tag{11.2}$$

$$\mu = \mu_c + (\mu_d - \mu_c) \times H(C), \tag{11.3}$$

where ρ_d and μ_d (ρ_c and μ_c) are the density and dynamic viscosity of the dispersed (continuous) phase and H is a smoothed Heaviside. As discussed above, because of the spatial discretization and the use of finite size meshing, interfaces slowly broaden as simulations run even if fluids are assumed to be immiscible. This broadening is illustrated in Figure 11.4. It is a typical finite size effect and actually constitutes one of the critical problems here. This numerical diffusion has therefore to be controlled with adapted algorithms (Bonometti and Magnaudet 2007; Chang et al. 1996; Sussman et al. 1998). This diffusion first appears at the apex of the droplet where singularities in the velocity field show up owing to the presence of a stagnation point that cannot be resolved with enough accuracy (when using a 200 × 200 mesh grid). This is a consequence of the use of structured mesh grids that do not exactly follow the interfaces.

The reason for introducing Equation 11.1 in VOF is the need to capture interfaces in Eulerian–Eulerian approaches. This can be seen as a weak solution when compared to direct Lagrangian front tracking techniques. But the simplicity of VOF methods, their robustness, and the fact that all phases can be treated on the same basis (that is to say, with a single Navier–Stokes equation) are all important advantages. It is also important here to be, again, precise because VOF techniques preserve the conservative character of the overall dynamics even for very basic Cartesian mesh grids. This makes an important difference with other techniques relying on iterative remeshing such as front tracking techniques (Hirt et al. 1970; Tome and Mckee 1994; Welch et al. 1966), which can use grid nodes involving separate boundary fitted mesh grids for each phase (Lötstedt 1982; Ryskin and Leal 1984; Unverdi and Tryggvason 1992) or moving Lagrangian grids together with fixed Eulerian ones. All these methods are also more complex to implement numerically and to upgrade to 3D problems. This is actually not the case for VOF techniques that are mainly limited by the available CPU power and memory capacities. As discussed above, difficulties persist when trying to maintain sharp interfaces. The resolution of Equation 11.1 is indeed not completely free from numerical diffusion. Higher-order weighted

(a)   (b)

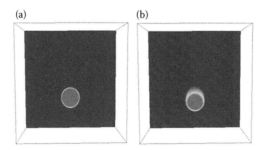

FIGURE 11.4   One-millimeter droplet radius in a 1 cm × 1 cm 2D SD with a 200 × 200 mesh grid. (a) Controlled interface sharpness obtained when using higher-order schemes for the integration of Equation 11.1. (b) Diffusing interface owing to the use of first-order schemes. The droplet was initially in the upper part of the SD.

essentially non-oscillatory (WENO) (Zahran 2009) and total variation diminishing (TVD) schemes can help tackle this problem (Titarev and Toro 2005). Several VOF approaches are proposed in the literature (Hirt and Nichols 1981). Most of them allow the simulation of heterogeneous systems without demanding CPU efforts (up to some extent) for the simulation of free interface problems.

In Eulerian–Eulerian single-fluid methods, the criterion of interface sharpness can be accurately tackled with higher-order WENO schemes (Harvie et al. 2006; Zahran 2009). They are well adapted to hyperbolic problems like the one of Equation 11.1 and only require limited additional computational effort. But when trying to account for interfacial tension, unwanted numerical artifacts consisting of spurious velocities (the so-called parasitic currents) show up in the neighborhood of the interfaces. An illustration of these artifacts for a single droplet is given in Figure 11.5. They are non-negligible for interface locations that are slowly evolving, for example, in the solid/liquid/gas contact line region of evaporating sessile droplets. In pinned evaporative regimes, the contact line is not moving while the droplet volume decreases. The hydrodynamics in the vicinity of the contact line is then strongly affected by artifacts. This is in particular the case when using VOF techniques together with the continuous surface force (CSF) model (Brackbill et al. 1992). The magnitude of the parasitic currents can be decreased by both the inertial and viscous terms of the Navier–Stokes equation but not with increased mesh refinement or decreased computational time step (Harvie et al. 2006). Efforts are still devoted to this problem with improved interface reconstruction algorithms (PLIC-VOF), second-order gradient techniques or hybrid methods (Meier et al. 2002; Ménard et al. 2007), and height function formulation (Afkhami and Bussmann 2008; Popinet 2009). Approaches coupling VOF techniques together with an immersed set of connected marker points for explicit interface tracking have also been developed (Tryggvason et al. 2001).

At the origin of artifacts, the implementation of capillary forces in VOF techniques is essential when investigating interfacial phenomena. Capillary phenomena like surface tension indeed play a fundamental role in the geometry of droplets and bubbles (Bonometti and Magnaudet 2007; Bordère et al. 2007) and in the possibility

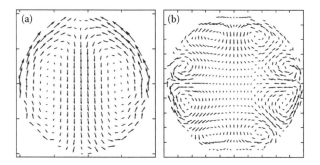

FIGURE 11.5 Circulation flow without parasitic currents (a) and with parasitic currents (b). Simulation conditions are the same as in Figure 11.4 and images correspond to hydrodynamics shortly after the simulation started. Resolution is larger in panel (b) to better visualize the parasitic currents.

for them to flocculate and coalesce such as in emulsions or boiling flows. Surface tension contributions in VOF simulations are directly included in the Navier–Stokes equation employing the CSF approach already introduced above (Brackbill et al. 1992). The main idea here is to map surface forces into volume ones. In liquid interfaces, capillary forces that govern the pressure jump across an interface are described by the Laplace law. It is proportional to the interface curvature and therefore needs the computation of second-order spatial derivatives that are highly sensitive to spatial discretization. The latter often impose the use of high-resolution approaches implying the development of higher-order algorithms to manage numerical divergences and keep numerical outputs coherent. Flux limiters are then often necessary to ensure that the numerical solution is TVD (Yee 1987). In practice, capillary forces act like source terms in the Navier–Stokes equation similarly to hydrostatic pressure gradients or viscous forces. With surface tension contributions, it writes:

$$\rho\frac{\partial V}{\partial t}+\rho V.\nabla V = -\nabla P+\rho g+\mu\Delta V+\sigma\kappa n.\nabla\bar{C}, \tag{11.4}$$

where σ is the interfacial tension, $\bar{C}$ is an averaged color function, $n = \nabla\bar{C}/\|\nabla\bar{C}\|$ is the normal to the interface and $\kappa = -\nabla.n$ is the curvature. The use of the averaged color function $\bar{C}$ for the computation of κ is necessary here to ensure derivability as CSF involves gradients. P and g are respectively the pressure and the gravity constant. Several averaging techniques can be used to compute $\bar{C}$. In Section 11.4, it is computed using the nearest neighbors of each mesh point and, despite this smoothing, the curvature remains a quite rough function. It takes its greatest magnitude in the area where $\bar{C}$ has intermediate values and falls to zero everywhere else. The possibility to obtain smoother curvatures employing, for example, color function contouring procedures have been tried with homemade programs (Lekhlifi 2011). But, although more sophisticated, none of them showed the numerical stability of CSF or of height function approaches (Popinet 2009; Afkhami and Bussmann 2008).

The capillary force in Equation 11.4 is normal to the interface and hence perpendicular to the local velocity field. It hence does not contribute directly to it and consequently only comes into play for the computation of the pressure field. Regarding the latter, it is important to emphasize that pressure is not fixed here by inlet or outlet boundary conditions. As it appears through a gradient in Equation 11.4, it is defined up to a constant that has undetermined value. To overcome this problem, a gauge point is used with, at this point, a pressure fixed to a known value. This method of managing pressures avoids unwanted fluctuations and gives the correct hydrostatic pressure differences between the top and the bottom of the system as well as between inside and outside the droplets in Figure 11.4a.

The simulation of interfacial chemistry like surfactant dynamics is clearly one important issue in the context of interfacial phenomena. Besides the still ongoing algorithmic improvements for surface tension modeling, another growing research domain is the development of CFD descriptions of surfactants (Alke and Bothe 2009; Ashley and Lowengrub 2004; Javadi et al. 2010; Muradoglu and Tryggvason 2008; Tasoglu et al. 2008). From Section 11.2, it is clear that explicit MD simulations are not adapted for the description of such systems at large scales and a continuous media formulation remains here the easiest way to investigate their properties. But while a rich literature has focused on the numerical investigation of clean interfaces, only limited efforts have been devoted to the description of interfaces with surfactants.

Like for the previously discussed interface tracking procedures or parasitic currents, computation of interfacial flows with surfactants also raises a number of challenging questions at both the algorithmic and modeling levels. The first condition here is clearly the preservation of interface sharpness. Unless this condition is fulfilled, finite size effects will come into play and numerical diffusion will take over and finally prevent any accurate description of surfactant effects. Computational studies involving surfactants have first focused on Stokes flows using hybrid methods with unstructured grids (Yon and Pozrikidis 1998) or boundary element techniques (Cheng and Cheng 2005) and it is only one decade ago that interest has really grown for the use of VOF methods. The surfactant concentration at liquid interfaces can be deduced, in the simplest case, from the Langmuir isotherm. But in practice, for moving interfaces, such surfactant layers are submitted to a tangential shear stress that can make them unstable in particular if this stress brings them close to their maximal packing concentration. Beyond this threshold, surfactant layers become stiff and, in the specific case of water–air surfactant monolayers, are known to collapse through a fracturing process (Smith and Berg 1980). In the presence of surfactants, interfaces can thus behave as inelastic systems including rupture thresholds. This clearly means that it is not always sufficient to assume convective and diffusive contributions when simulating surfactant layers. Such behaviors as well as interfacial viscosity effects will be neglected hereafter although they both constitute very important properties of interfaces. To our knowledge, no stable VOF algorithms are presently proposed to address such problems. The problem of surfactant incorporation in CFD models is therefore reduced to a convective–diffusive one that is coupled with the Navier–Stokes equation through the velocity field.

In liquid/liquid or liquid/gas systems, insoluble surfactant molecules at low concentrations are mostly collected on the interfaces where they form molecular layers. Changes in the value of the interfacial surfactant concentration (noted Γ in the following) will modify interfacial tension that will act back onto the interface through Marangoni shear stress. The consequence of this is similar to what can be observed in thermocapillary convection (Tsuji et al. 2008). The interfacial shearing will produce a tangential force onto the adjacent liquid and will generate flows at the origin of macroscopic effects like droplet motion. The previous discussion suggested that the evolution of Γ can be described, in first approximation, by the 2D convection–diffusion equation that writes:

$$\frac{\partial \Gamma}{\partial t} + \nabla_s (V_s . \Gamma) = D_s \Delta_s \Gamma + j_s \qquad (11.5)$$

where $\nabla_s = \nabla - n(n.\nabla)$ is the gradient along the interface, $\Delta_s = \nabla_S^2$, D_s is the diffusion coefficient of the adsorbed surfactant, and V_s is the tangential velocity field at the interface obtained from Equation 11.2. j_s is the mass flux in the vicinity of the interface. It is vanishing for insoluble surfactants, whereas for soluble ones, it can be modeled by an adsorption/desorption kinetic model (Tasoglu et al. 2008). In the case of soluble surfactants, Equation 11.5 has to be completed with another equation describing the dissolved surfactant properties. For the surfactant dissolved in one of the considered continuous phases, this equation writes:

$$\frac{\partial [C]}{\partial t} + V.\nabla[C] = D_b \Delta[C], \tag{11.6}$$

where $[C]$ is the surfactant concentration in the continuous phase and D_b is its diffusion coefficient. In the vicinity of the interface, $[C]$ and j_s are linked by Fick's first law. Both Equations 11.5 and 11.6 are hence coupled together by this boundary equation that actually appears as a mixed Neumann–Dirichlet boundary condition. Finally, a closed set of equations is obtained when introducing a constitutive law for the surface tension. One possibility is to use the expression of the Langmuir equation of state for the dependence of σ in Γ and in temperature (Tasoglu et al. 2008).

The actual diversity of free-surface numerical techniques indicates that the present modeling of interfacial phenomena is still a maturing domain in computer science. Many new improvements in numerical algorithms are still to come and will most probably provide in the near future new techniques to better overcome the crucial problem of parasitic currents in VOF descriptions. Presently, VOF methods have the notable advantage of robustness and fast convergence in particular when using implicit solvers (Sweby 1984). This is particularly true when associating implicit integration and higher-order schemes to update the velocity field. They are also well adapted to catch topological transitions in the geometry of interfaces such as in coalescing droplets. The use of implicit solvers yields several new constraints, in particular for the value of time steps, which have to obey the CFL condition. The necessity to fulfill this condition for numerical convergence can lead to prohibitive CPU time if not carefully treated. But satisfactory compromises between time and space discretization can usually be found to overcome this difficulty. Conversely to MD, CFD simulations of interfaces give access to the hydrodynamics. In this context, the molecular processes occurring at interfaces are not modeled at the molecular scale but rely on equations of state. Although interesting because of their robustness, VOF techniques still have limitations in particular when introducing capillarity. Spurious currents appear and lead to important artifacts in circulation flows. But their relative intensity reduces when the position of free interfaces is evolving sufficiently fast such as in the case of falling viscous droplets, which will be described in the next section.

11.4   SIMULATION OF DROPLET HYDRODYNAMICS

To illustrate the capabilities of VOF techniques, it is proposed in this section to focus on a simplified system consisting of water droplets in paraffin oil. Water and paraffin oil are two Newtonian and incompressible fluids that are described with an Eulerian–Eulerian single-fluid model based on Equations 11.1 through 11.4. The role

TABLE 11.1

Properties of Water and Paraffin Oil

ρ_d (kg/m³)	ρ_c (kg/m³)	μ_d (Pa s)	μ_c (Pa s)	σ (N/m)
1000	860	0.00106	0.099	0.02

Note: σ is the water–paraffin oil surface tension. All the data are given at $T = 298$ K.

of surfactants will not be addressed hereafter; this is why Equations 11.5 and 11.6 will not be further considered. Both fluids are submitted to gravity, confined in the 1 cm × 1 cm SD already presented in Figure 11.4. The model under consideration is 2D and the SD is therefore a portion of a plane within which water droplets evolve and settle owing to density differences and gravity. The spatial resolution in x and y directions is 5×10^{-5} m and boundary conditions are fixed in time. They are set to non-slip conditions at the left side wall ($x = 0$ cm), at the right side wall ($x = 1$ cm), and at the bottom ($y = 0$ cm). At all these three boundaries, the velocity is hence fixed to zero. At the top of the SD ($y = 1$ cm), a slip boundary condition is applied with zero normal and free tangential velocity coordinates. Finally, at the water/paraffin oil interface, boundary properties are computed, employing Equations 11.2 and 11.3. Working temperature is fixed to 298 K and the corresponding physical properties are given in Table 11.1.

Figure 11.6 is a snapshot of the system at time $t = 0.2$ s showing a 1 mm radius droplet. Initially (i.e., at time $t = 0$), the center of the droplet was located at $x_{cent}(0) = 0.2$ cm and $y_{cent}(0) = 0.8$ cm. This is why no defined symmetry can be observed in this figure. Symmetric configurations are obtained only when the droplet center is initially set to $x_{cent}(0) = 0.5$ cm as represented in Figure 11.4. Simulations of nonsymmetric configurations are actually one of the reasons for considering a 2D model. Indeed, realizing a full 3D description of such a system with the same numerical accuracy would need prohibitive CPU time. In 2D, however, a two-processor quad core workstation is sufficient to get the overall droplet drainage dynamics with limited CPU time (see below). The resolution of Equation 11.1 uses the Phoenics package (CHAM Ltd). Time step has a constant value and is set to $\Delta t = 10^{-3}$ s. This value

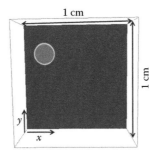

FIGURE 11.6 Example of color function $C(x,y,t)$ for a 200 × 200 mesh grid at time $t = 0.2$ s. Orientation of gravity field is top–down and droplet radius is 1 mm. The water/paraffin oil interface shows a good spatial resolution. Initially, the center of the droplet is located at $x_{cent}(t = 0) = 0.2$ cm and $y_{cent}(t = 0) = 0.8$ cm and velocities are all set to zero.

is sufficiently small to ensure the convergence of the calculations. The following simulations hence do not make use of the CFL condition mainly for simplicity reasons. With this time step, typical CPU time required for a run is approximately 10 h and corresponds to the simulation of 2 s of the droplet's evolution. Regarding initial conditions, it is also important to premise here that for the simulations presented in this section, the velocity field is everywhere set to zero initially. As temperature is fixed to 298 K in all the SD and for all time, local heating due to fluid friction and viscous dissipation effects are neglected. This is fully justified by the typical flow velocities that show up in this system. As a result, heat equation can be neglected.

In Section 11.3, it was mentioned that single-fluid techniques rely on the advection of the color function introduced in Equation 11.1. As the dispersed phase is water in this section, the color function is set to $C(x,y,t) = 1$ [$C(x,y,t) = 0$] in water [in paraffin oil]. In the water/paraffin oil interface, C takes intermediate values between 0 and 1 as illustrated by the circle surrounding the water droplet in Figures 11.4 and 11.6. Although surface tension is fully accounted for (see Equation 11.4), it is important to remind here that the model does not yet include all interfacial properties. Surface viscosity is, for example, not considered.

Even though higher-order schemes are employed for the resolution of Equation 11.1, a residual numerical diffusion still shows up in C in the neighborhood of the paraffin oil/water interface. This residual diffusion is controlled by a correction that consists in imposing abruptly $C(x,y,t) = 1$ [$C(x,y,t) = 0$] in the interfacial region when water [paraffin oil] is the majority phase. This rough correction procedure perturbs slightly the velocity field in the immediate neighborhood of the water/paraffin oil interface and generates small fluctuations in the mass of the droplet (approximately 2% for a 200 × 200 mesh grid). The amplitude of these perturbations can be reduced by increasing spatial resolution. For the 200 × 200 mesh grid used here, the color function is corrected every 50 time steps (i.e., 0.05 s). After such a correction is operated, the velocity field is slightly modified and a few time steps are necessary to recover the non-perturbed one. Larger values have been used and 200 time step periods between two successive corrections seem to be an upper limit. Velocity field recovery and fluctuation amplitudes then become critical. Although quite rough, it has been verified, by lowering both spatial resolution and time step, that this correction procedure does not significantly change the overall results of the simulations.

The droplet average settling velocity $\bar{v}(t)$ is computed using all the velocities in the domain defined by $C(x,y,t) > 0.9$ that can be considered as the SD region within the droplet. Initially, the droplet is at rest [$\bar{v}(0) = 0$ m/s] but because of gravity and the smaller paraffin oil density, it gets off its initial position and accelerates toward the bottom of the SD making $\bar{v}(t)$ time dependent. Examples of the time evolution of $\bar{v}(t)$ are displayed in Figure 11.7 for three different values of $x_{cent}(0)$ [with $y_{cent}(0) = 0.8$ cm in all cases]. When $x_{cent}(0) = 0.5$ cm, Figure 11.7a indicates that it is possible to identify three different times (t_{R1}, t_{R2}, and t_{R3}) corresponding to three different regimes (R_1, R_2, and R_3) (Lekhlifi et al. 2010). In R_1 ($0 < t < t_{R1}$), the droplet is essentially submitted to gravity acceleration since viscous drag forces have not yet come into play. In R_2 ($t_{R1} < t < t_{R2}$), the droplet has constant velocity and gravity forces are compensated by the increasing drag forces that now fully act on it because of its larger average velocity. Finally, in R_3 ($t_{R2} < t < t_{R3}$), the droplet slows down owing to

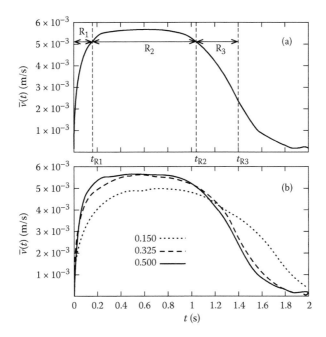

FIGURE 11.7 Time evolution of the average velocity $\bar{v}(t)$ for a droplet with radius $r = 1$ mm with $y_{cent}(0) = 0.8$ cm. Panel (a) displays the evolution of $\bar{v}(t)$ for $x_{cent}(0) = 0.5$ cm, whereas in panel (b), several values of $x_{cent}(0)$ are considered. Regimes R_1, R_2, and R_3 are represented in panel (a) as well as corresponding times t_{R1}, t_{R2}, and t_{R3}.

paraffin oil recirculation in the neighborhood of the droplet as the proximity with the bottom of the SD increases. This recirculation generates an overall force that finally overcomes gravity. The droplet evolution in R_2 is comparable to the one known in 3D axisymmetric case in an infinite continuous phase where the average velocity is predicted by the Hadamard–Rybczynski law (Levich 1962). The expectation from the Hadamard–Rybczynski law for a 1 mm radius water droplet in paraffin oil is 5.3×10^{-3} m/s (see Table 11.1). Surprisingly, this value turns out to be within 20% of the velocity deduced from Figure 11.7a. For the initial condition considered in this figure $[x_{cent}(0) = 0.5$ cm], the Hadamard–Rybczynski law hence makes possible a reasonable estimation of the drainage velocity although the droplet is a 2D object and the SD is limited in size. For $t > t_{R3}$, the droplet arrives closer to the bottom of the SD and finally comes in contact with it. Wetting phenomena of water are then coming into play and require not only the modeling of solid/liquid wetting properties but also careful control of the spurious currents.

Although VOF algorithms allow a complete description of droplet deformations, no noticeable deformations caused by either shear forces (when droplets are close to the domain walls) or flattening (when droplets are in the neighborhood of the bottom of the domain) have been observed. This is due to the small value of $\bar{v}(t)$ making the capillary number small enough for interfacial tension to prevail. This is also the reason why it is sufficient to follow the evolution of the droplet center to obtain its full trajectory as seen in Figure 11.7. Figure 11.7b shows that the time evolution of $\bar{v}(t)$ depends on the initial location of the droplet. For non-symmetric initial conditions [i.e., $x_{cent}(0) \neq 0.5$ cm], the three regimes R_1, R_2, and R_3 can still be identified (Lekhlifi et al. 2010). Conversely to 3D geometries, 2D models allow investigations of such non-symmetric configurations with quite limited computational efforts. As expected, walls have an important impact on the motion of droplets and what Figure 11.7 clearly evidences is that a quantitative analysis is now possible. Such wall effects are most probably also present in 3D geometries and of particular relevance when studying the hydrodynamics of emulsions. Experimentally, emulsions are indeed usually confined in small containers that obviously modify their overall hydrodynamics.

For a non-symmetric initial condition as the one in Figure 11.6, the two side walls of the SD act differently on the droplet. One way to probe the consequences of this is to follow the time evolution of $x_{cent}(t)$ for different initial values of $x_{cent}(0)$. Figure 11.8 represents such time evolutions. When $x_{cent}(0) = 0.5$ cm, the droplet settles to the SD bottom with $x_{cent}(t) = 0.5$ cm for all times because of the symmetry of this configuration. But lateral drifting motions toward the center of the SD show up for all other values. These last motions result from water and paraffin oil incompressibility and from the increased shear stress they apply on the droplet interface when in the vicinity of the wall. The paraffin oil tends indeed to circulate in between the droplet and its nearest side wall, without the possibility of flowing away faster from this region owing to incompressibility. The resulting effect is a force acting on the droplet perpendicularly to the side wall that strengthens with increasing values of $\bar{v}(t)$. The drift of the trajectories in the x direction toward $x_{cent}(0) = 0.5$ cm is hence a consequence of droplet drainage and incompressibility. Because of the non-slip conditions at the walls, a reduction of this effect is expected when droplets become too close to them. A careful study of the curves in Figure 11.8 indicates moreover that for intermediate values of $x_{cent}(0)$ [typically for $0.2 < x_{cent}(0) < 0.4$], a minimum value of $x_{cent}(t)$ different from $x_{cent}(0)$ also shows up. The droplet first tends to evolve toward its closest wall and then moves apart from it. This is a consequence of the droplet acceleration

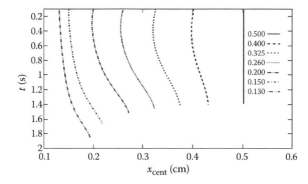

FIGURE 11.8 Time evolution of $x_{cent}(t)$ for different initial values of $x_{cent}(0)$ given in the legend. $y_{cent}(0) = 0.8$ cm for all trajectories. Time evolves from top to bottom. The representation of the trajectories is interrupted at time t_{R3}.

in R_1 that first balances the lateral forces owing to the non-symmetric hydrodynamics in both paraffin oil and water. The circulation of paraffin oil in between the droplet and the nearest wall is indeed slower than the one at the opposite side. This is why the balance of the viscous forces and gravity is first in favor to bring them closer.

The interaction between droplet drainage and circulation flows in paraffin oil generates important modifications in the droplets' hydrodynamics as illustrated in Figure 11.9. The images of this figure represent the velocity field inside the droplet

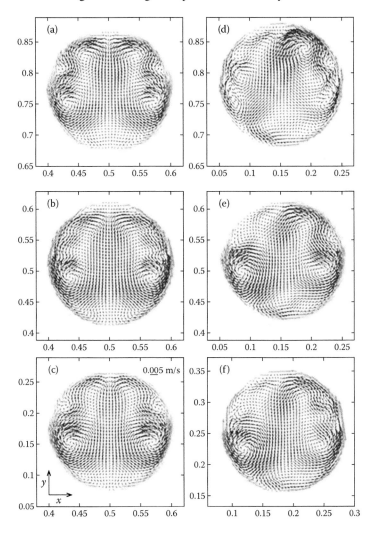

FIGURE 11.9 Velocity field in the frame moving with the droplet at time $t = 0.1$ s (top line), $t = 0.6$ (middle lines), and $t = 1.3$ s (bottom line) for two different values of $x_{cent}(0)$. Droplet radius is set to 1 mm and mesh grid is 200×200. For panels (a), (b), and (c), $x_{cent}(0) = 0.5$ cm. For panels (d), (e), and (f), $x_{cent}(0) = 0.15$ cm. The lengths of the arrows give the velocity intensity and can be estimated from the reference in the upper right corner of panel (c). Interfacial velocity field is represented with blue arrows.

in the frame moving with it, at several times and for two different values of $x_{cent}(0)$ [still with $y_{cent}(0) = 0.8$ cm]. In Figure 11.9a, b, and c, $x_{cent}(0) = 0.5$ cm and symmetric velocity fields show up. After a short transcient regime ($t < 0.05$ s), convective cells appear in the circulation of water. They are generated by the shearing forces at the water/paraffin oil interface owing to the non-zero value of $\bar{v}(t)$. As the latter is not constant, this interfacial shear stress evolves, making the velocities unsteady. In R_1, four convective cells show up as illustrated in Figure 11.9a. These cells are robust in the sense that better resolutions in both time and space do not modify them significantly. They are hence either actual properties of the flows inside the droplet or a residual effect of parasitic currents. In R_2, only two symmetric convective cells are observable (see Figure 11.9b). These cells persist in R_3 (Figure 11.9c) but with new small ones in the upper half section of the droplet. The number of cells observed here is a consequence of the 2D geometry of the model. For 3D axisymmetric droplets, the two convective cells in Figure 11.9b would actually belong to a unique convective torus (Levich 1962).

For non-symmetric initial configurations, the velocity fields have no symmetry and display circulation flows that become more and more complex as $x_{cent}(0)$ is decreased. Figure 11.9d, e, and f illustrate such circulation flows when $x_{cent}(0) = 0.15$ cm, that is to say, for a droplet next to the left side wall of the SD (see Figure 11.6). When compared to the previous observations, velocities are now strongly modified. The two convective cells visible in the lower half section of Figure 11.9b are still present in Figure 11.9e, but in the upper section, the situation has completely changed. The velocities are smaller in the left section of the droplet owing to the wall proximity, and this effect is clearly visible in the upper left section of the droplet in Figure 11.9e and f. These complex circulation flows are a consequence of the interfacial shear stress differences between the left- and the right-hand side of the droplet and clearly demonstrate the fundamental importance of the description of wall contributions.

Figure 11.8, together with Figure 11.9, actually summarizes the interest to focus on 2D systems. They both indicate how complex the hydrodynamics can get when accounting for the effective role of capillarity and boundary conditions. All this phenomenology probably also exists in 3D geometries with higher complexity. But performing 3D simulations with the spatial resolution used here, though feasible, is unfortunately not yet easy to access with standard workstations, because of excessive computational effort.

The extension of the above single-droplet model to 2D emulsions is easily feasible when employing Eulerian–Eulerian methods and VOF techniques. When several droplets coexist, coalescing and flocculating phenomena occur in addition to drainage. Both mechanisms are fundamental for emulsion aging. If all the droplets have the same composition, it is possible to use the same approach as compared with single-droplet systems and to describe them with a unique color function obeying again Equation 11.1. In this case, the simulation of two coalescing droplets can be seen, from a numerical point of view, as nothing but a fast propagation of the color function, propagation that will end up with a new droplet having a volume identical to the one of the two coalesced droplets but with a smaller interfacial curvature. Careful computation of curvature and high time resolution are therefore sufficient for a first approach of coalescence. But this is clearly not sufficient to precisely account for all the complexity of coalescence mechanisms. They imply not only macroscopic mechanisms but also microscopic ones (such as Newton black film formation) that are not accessible with CFD. This problem also shows up when considering flocculation mechanisms where it is necessary to implement specific interfacial forces that make the simulation of aggregation kinetics more arduous. Coalescence has indeed to be inhibited with repulsive and steric phenomena that have to be included in the interface models. One straightforward possibility to address this problem consists in using as many color functions as there are droplets contained in the emulsion. This has been tried recently but considerably slows down the simulations as important memory capacities are needed (Lekhlifi 2011). Unless using several color functions, no VOF method is available in the literature to our knowledge to describe flocculation in emulsions. Front tracking with mesh refinement techniques is here the easiest way to catch the details of the interactions between neighboring droplets. Lagrangian approaches with mesh refinement at the interfaces can, for example, help solve this problem on safe ground. Such an approach has already been used but in a different context, that is, for the investigation of nucleated vapor bubble dynamics (Bonometti and Magnaudet 2007; Bunner and Tryggvason 2002). But, as already discussed, the main drawback here is the important CPU time that is needed. Algorithms have to track moving free interfaces and to maintain global numerical convergence; in addition, they are also asked to make the simulation of realistic liquid/liquid or liquid/gas interfacial properties possible. Other methods have been employed to numerically investigate emulsion aging. Hard-sphere models (Marnette et al. 2009) or lattice Boltzmann approaches (Dupin et al. 2006; Onishi et al. 2008; Van der Zwan et al. 2009), for example, have given fundamental results although these methods do not always provide complete descriptions. Emulsions indeed consist of deformable droplets, each having its own hydrodynamics that is generated by complex moving boundary conditions that often involve specific liquid/liquid interfacial properties. All these phenomena should be included in the models. However, to be simple, the idea of considering a single Navier–Stokes equation together with adapted interfacial boundary conditions with VOF techniques, as it has been done for the simulations in Figure 11.10, is very tempting.

As stated above, single-fluid Eulerian–Eulerian models with VOF can be interesting when studying emulsions if all the details of coalescence are not taken into account and if flocculation is assumed to be negligible. Figure 11.10 shows an example of the evolution of such a simplified model consisting initially of four water droplets in a continuous paraffin oil phase. As drainage takes place, droplets come together and coalescence shows up. When describing the dispersed phase with a single color function like in this figure, coalescence occurs when the interface of droplets starts to overlap. In such an overlapping region, the forces acting on the interface are outward (i.e., in the direction of the paraffin oil) and intense since curvature is large. Such a contact situation is illustrated at time $t = 1.75$ s in Figure 11.10. Right after this contact is established, droplets merge and give rise to a new elongated one ($t = 1.76$ s). As the latter relaxes to its circular equilibrium shape, it can come into contact with other neighboring droplets and generate new coalescences ($t = 1.79$ s). Clearly, the model used here to describe coalescence is not consistent with what is known

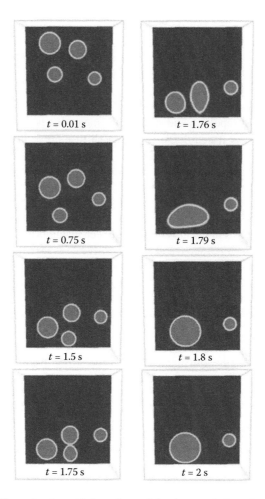

FIGURE 11.10 **(See color insert.)** Snapshots of the time evolution of a system consisting initially of four distinct droplets. The time is indicated below each image. The parameters of the simulation are the same as in Figures 11.4 and 11.6.

to occur in experimental conditions. The system is assumed to be a 2D continuous medium and chemistry at interfaces is not properly described. But as discussed in Section 11.1, there is no way to keep track of all the physics and chemistry from macroscopic to microscopic scales when dealing with capillarity problems with common numerical means. What the images in Figure 11.10 actually indicate is that CFD can offer interesting numerical insights into the dynamics of droplet systems with a moderated computational effort (approximately 12 h CPU time for this figure). It is also important to note here that such simulations give access to all the hydrodynamics in the SD and hence provide the possibility to study the circulation flows inside and in the neighborhood of each droplet. This can be particularly interesting when studying the effect of soluble surfactants. Their concentration is indeed modified because of surfactant release when coalescence occurs and because of the overall hydrodynamics in the SD. Consequently, when considering, for example, paraffin oil–soluble surfactants, droplets will evolve in a non-homogeneous surfactant concentration field. They will experience surfactant gradients that will create gradients in their interfacial tension and hence generate Marangoni shear stress. The latter will come in addition to droplet drag and, hence, has the tendency to reduce drainage. This interplay between drainage and Marangoni effects is still an open problem in emulsion science. For such a fundamental question, the use of CFD simulations with simplified models can constitute a very useful approach to evidence new mechanisms in emulsion aging.

11.5 CONCLUSIONS

The aims of this chapter were to give insights into the ability of MD and CFD to tackle multiphase systems and to show that simulations of such systems still raise many challenging questions that can be considered at first sight as very basic: the propagation of free surfaces in immiscible liquids, the choice of initial condition, and the effect of boundary conditions are, among others, typical problems that are almost systematically encountered when dealing with numerical approaches. Another aspect that was discussed in this chapter was the description level of interfacial phenomena. From a chemical point of view, interfaces consist of molecules and hence need explicit molecular-based algorithms. But when investigating larger-scale phenomena, such as hydrodynamics, molecular descriptions cannot be adapted and continuous media techniques have to be invoked.

To illustrate how microscopic and macroscopic scales can be simulated, specific techniques and examples are presented. Explicit MD approaches are first discussed and illustrated with a surfactant monolayer at a water/air interface. The problem of the choice of the force field is addressed together with the necessity for the simulations to converge to steady configurations. This is mandatory since there is no possibility to describe typical interfacial properties unless such steady configurations are reached. A careful construction of the initial condition is necessary as well as approximations for the treatment of long-range interactions. When a steady configuration is reached, it is possible to investigate properties of monolayers. It was shown here that surfactant molecules display highly entangled structures that confer to the latter a smaller intrinsic width than the one expected from fully stretched molecular conformations.

A simple 2D hydrodynamic model is also presented to illustrate to what extent continuous media can catch the time evolution of deformable free-surface systems. To this end, a 1 mm radius water droplet evolving in a continuous paraffin oil phase is considered. It is simulated in a closed domain with non-slip boundary conditions and an Eulerian–Eulerian single-fluid model. The velocity field is computed from the Navier–Stokes equation and the treatment of free interfaces on the basis of the evolution of a color function. The latter propagates the water/paraffin oil interface along the velocity field through an advection equation and hence presents the advantage to avoid interface reconstruction. The continuum surface force model is used to account for surface tension phenomena. Because of buoyancy forces, the droplets show an unsteady hydrodynamics and settle down to the bottom of the SD experiencing three different regimes. Depending on their initial location, droplets moreover display modified trajectories that result from the balance between buoyancy and drag forces that are also strongly dependent on boundary conditions. Finally, simulations show that the hydrodynamics inside the droplets display modified circulation flows when initially close to the boundaries.

This chapter also suggests one possible extension of single-droplet models to emulsions. New effects like droplet coalescence and flocculation come then in addition to buoyancy and drainage forces. Although not complete, since microscopic effects are not accounted for, there is a straightforward possibility to obtain interesting numerical insights for coalescing droplets with Eulerian–Eulerian approaches. Such simulations allow the description of the hydrodynamics all over the SD and in particular the one showing up inside and in the neighborhood of each droplet. They can moreover be very useful when investigating the effect of soluble surfactants. Surfactants are indeed expected to show a non-homogeneous concentration in the SD owing to the release generated by droplet coalescence. As a result, droplets will experience surfactant gradients that can generate Marangoni-driven motions. This raises the fundamental question of the interplay between drainage, drag forces, and Marangoni shear stress in confined emulsions. In this context, free-surface simulations are important tools since they can give access to new mechanisms and help evidence original and unexpected phenomena in emulsion aging.

ACKNOWLEDGMENTS

The authors acknowledge financial support from the European Space Agency (FASES MAP AO-99-052), Centre National d'Etudes Spatiales (CNES), and COST actions D43, P21, MP1106 and CM1101.

REFERENCES

Afkhami, S., Bussmann, M. 2008. Height functions for applying contact angles to 3D VOF simulations. *Int. J. Numer. Methods Fluids* 61:827–847.

Alke, A., Bothe, D. 2009. 3D numerical modelling of soluble surfactant at fluidic interfaces based on the volume-of-fluid method. *Fluid Dynam. Mater. Process.* 5:345–372.

Andersen, H.C. 1980. Molecular dynamics simulations at constant pressure and/or temperature. *J. Chem. Phys.* 72:2384–2393.

Ashley, J.J., Lowengrub, J. 2004. A surfactant-conserving volume-of-fluid method for interfacial flows with insoluble surfactant *J. Comp. Phys.* 201:685–722.

Bandyopadhyay, S., Tarek, M., Lynch, M.L., Klein, M.L. 2000. Molecular dynamics study of the poly(oxyethylene) surfactant C12E2 and water. *Langmuir* 16:942–946.

Berendsen, H.J.C., Grigera, J.R., Straatsma, T.P. 1987. The missing term in effective pair potentials. *J. Phys. Chem.* 91:6269–6271.

Berendsen, H.J.C., Postma, J.P.M., Van Gunsteren, W.F., Hermans, J. 1981. Interaction models for water in relation to protein hydration. *Intermolecular Forces.* B. Pullman (ed.). D. Reidel Publishing Company, 331–342.

Binks, B.P. 1998. *Modern Aspects of Emulsion Science.* The Royal Society of Chemistry, Cambridge.

Binks, B.P. 2002. Particles as surfactants—similarities and differences. *Curr. Opin. Colloid Interface Sci.* 7:21–41.

Binks, B.P., Rodrigues, J.A. 2007. Enhanced atabilization of emulsions due to surfactant-induced nanoparticle flocculation. *Langmuir* 23:7436–7439.

Bonometti, T., Magnaudet, J. 2007. An interface-capturing method for incompressible two-phase flows. Validation and application to bubble dynamics. *Int. J. Multiphase Flow* 33:109–133.

Bordère, V., Vincent, S., Caltagirone, J.P. 2007. Stochastic energetic approach devoted to the modeling of static two-phase problems dominated by surface tension. *Comput. Fluids* 39:392–402.

Bourrel, M., Schechter, R.S. 1988. *Microemulsions and Related Systems. Surfactant Science Series*. Marcel Dekker, New York.

Brackbill, J.U., Kothe, D.B., Zemach, C. 1992. A continuum method for modeling surface tension. *J. Comp. Phys.* 100:335–354.

Bunner, B., Tryggvason, G. 2002. Dynamics of homogeneous bubbly flows. Part 1. Rise velocity and microstructure of the bubbles. *J. Fluid. Mech.* 466:17–52.

Cerius2. 2000. Molecular Simulations Inc. Cerius² version 4.2, San Diego, CA.

Chang, Y.C., Hou, T.Y., Merriman, B., Osher, S. 1996. A level set formulation of eulerian interface capturing methods for incompressible fluid flows. *J. Comp. Phys.* 124:449–464.

Charles, C.W. 2012. *Practical Spray Technology: Fundamentals and Practice*, Lake Innovation LLC, 1st edition, Lake Jackson, Texas.

Cheng, A.H.D., Cheng, T.D. 2005. Heritage and early history of the boundary element method. *Eng. Anal. Boundary Elem.* 29:268–302.

Cuny, V., Antoni, M., Arbelot, M., Liggieri, L. 2004. Numerical analysis of non ionic surfactant monolayers at water/air interfaces. *J. Phys. Chem. B.* 108:13353–13363.

Cuny, V., Antoni, M., Arbelot, M., Liggieri, L. 2008. Structural properties and dynamics of $C_{12}E_5$ molecules adsorbed at water/air interfaces: A molecular dynamic study. *Colloids Surf. A* 323:180–191.

Dickinson, E. 2003. Hydrocolloids at interfaces and the influence on the properties of dispersed systems. *Food Hydrocolloids* 17:25–39.

Dupin, M.M., Halliday, I., Care, C.M. 2006. A multi-component lattice Boltzmann scheme: Towards the mesoscale simulation of blood flow. *Med. Eng. Phys.* 28:13–18.

Evans, D.J., Holian, B.L. 1985. The Nosé-Hoover thermostat. *J. Chem. Phys.* 83:4069–4074.

Ferrari, M., Liggieri, L., Miller, R., eds. 2012. Drops and bubbles in contact with solid surfaces. *Progress in Colloid and Interface Science*. CRC Press.

Girard, F., Antoni, M. 2008. Influence of substrate heating on the evaporation dynamics of pinned water droplets. *Langmuir* 24:11342–11345.

Girard, F., Antoni, M., Faure, S., Steinchen, A. 2008. Influence of heating temperature and relative humidity in the evaporation of pinned droplets. *Colloids Surf. A* 323:36–49.

Guillot, B. 2002. A reappraisal of what we have learnt during three decades of computer simulations of water. *J. Mol. Liquids* 101:219–260.

Harvie, D.J.E., Davidson, M.R., Rudman, M. 2006. An analysis of parasitic current generation in volume of fluid simulations. *Appl. Math. Model.* 30:1056–1066.

Hirt, C.W., Cook, J.L., Butler, T.D. 1970. A Lagrangian method for calculating the dynamics of an incompressible fluid with free surface. *J. Comp. Phys.* 5:103–124.

Hirt, C.W., Nichols, B.D. 1981. Volume of fluid (VOF) method for the dynamics of free boundaries. *J. Comp. Phys.* 39:201–225.

Horstemeyer, M.F., Leszczynski, J., Shukla, M.K. 2010. Multiscale modeling: A review. *Practical Aspects of Computational Chemistry*. Springer, 87–135.

Hou, T.Y., Lowengrub, J.S., Shelley, M.J. 2001. Boundary integral methods for multicomponent fluids and multiphase materials. *J. Comp. Phys.* 169:302–362.

Huang, C., Li, C., Choi, P.Y.K., Nadakumar, K., Kostiuk, L.W. 2010. Effect of cut-off distance used in molecular dynamics simulations on fluid properties. *Molec. Simul.* 36:856–864.

Islam, N., Kato, T. 2005. Influence of temperature and headgroup size on condensed phase patterns in Langmuir monolayers of some oxyethylenated nonionic surfactants. *Langmuir* 21:2419–2424.

Islam, N., Kato, T. 2006. Influence of temperature and alkyl chain length on phase behaviour in Langmuir monolayers of some oxyethylenated nonionic surfactants. *J. Colloid Interface Sci.* 294:288–294.

Javadi, A., Ferri, J.K., Karapantsios, Th., Miller, R. 2010. Interface and bulk exchange: Single drops experiments and CFD simulations. *Colloids Surf. A* 365:145–153.

Jorgensen, W.L., Chandrasekhar, J., Madura, J.D., Impey, R.W., Klein, M.L. 1983. Comparison of simple potential functions for simulating liquid water. *J. Chem. Phys.* 79:926–935.

Karaborni, S. 1993. Molecular dynamics simulations of long-chain amphophilic molecules in Langmuir monolayers. *Langmuir* 9:1334–1343.

Kleefsman, K.M.T., Fekken, G., Veldman, A.E.P., Iwanowski, B., Buchner, B. 2005. A volume-of-fluid based simulation method for wave impact problems *J. Comp. Phys.* 206:363–393.

Kuhn, H., Rehage, H. 1998. Orientation and structure of monododecyl pentaethylene glycol adsorbed at the air/water interface studied by molecular dynamics computer simulation. *Tenside Surf. Det.* 35:448–453.

Kuhn, H., Rehage, H. 1999. Molecular dynamics computer simulations of surfactant monolayers: monododecyl pentaethylene glycol at the surface between air and water. *J. Phys. Chem. B.* 103:8493–8501.

Kuhn, H., Rehage, H. 2000a. Molecular dynamics computer simulation of a $C_{12}E_5$ surfactant monolayer at the water/air and water/octane interface. *Transp. Mech. Fluid Interfaces.* 136:69–80.

Kuhn, H., Rehage, H. 2000b. Molecular orientation of monododecyl pentaethylene glycol at water/air and water/oil interfaces. A molecular dynamics computer simulation study. *Colloid Polym. Sci.* 278:114–118.

Latsa, M., Assimacopoulos, V., Stamou, A., Markatos, V. 1999. Two-phase modeling of batch sedimentation. *Appl. Math. Model.* 23:881–897.

Leal-Calderon, F., Schmitt, V., Bibette, J. 2007. *Emulsion Science: Basic Principles.* 2nd ed. Springer, New York.

Lekhlifi, A. 2011. Étude numérique de l'hydrodynamique de drainage de gouttes d'eau dans l'huile de paraffine. PhD Thesis, Université Aix Marseille.

Lekhlifi, A., Antoni, M., Ouazzani, J. 2010. Numerical simulation of the hydrodynamics of a falling water droplet in paraffin oil. *Colloids Surf. A.* 365:70–78.

Levich, V.G. 1962. *Physicochemical Hydrodynamics.* Prentice-Hall, Englewood Cliffs, NJ.

Lötstedt, P. 1982. A front tracking method applied to Burger's equation and two-phase porous flow. *J. Comp. Phys.* 47:211–228.

Lu, J.R., Li, Z.X., Thomas, R.K., Staples, E.J. 1993a. Neutron reflection from triethylene glycol monododecyl ether adsorbed at the air/liquid interface: The variation of the hydrocarbon chain distribution with surface concentration. *Langmuir* 9:2417–2425.

Lu, J.R., Li, Z.X., Thomas, R.K., Staples, E.J. 1993b. Neutron reflection from a layer of monododecyl hexaethylene glycol adsorbed at the air/liquid interface: The configuration of the ethylene glycol. *J. Phys. Chem.* 97:8012–8020.

Lu, J.R., Li, Z.X., Thomas, R.K., Staples, E.J. 1994. Neutron reflection from a layer of monododecyl octaethylene glycol adsorbed at the air/liquid interface: The structure of the layer and the effects of temperature. *J. Phys. Chem.* 98:6559–6567.

Lu, J.R, Li, Z.X., Thomas, R.K., Binks, B.P., Crichton, D. 1998. The structure of monododecyl pentaethylene glycol monolayers with and without added dodecane at the air/solution interface: A neutron reflection study. *J. Phys. Chem.* 102:5785–5793.

Lu, J.R., Thomas, R.K., Penfold, V. 2000. Surfactant layers at the air/water interface: structure and composition. *Colloid Interface Sci.* 84:143–304.

Maple, J. R. et al. 1994. Derivation of class II force fields. I. Methodology and quantum force field for the alkyl functional group and alkane molecules. *J. Comp. Chem.* 15: 162–182.

Marnette, O., Perez, E., Pincet, F., Bryant, G. 2009. Two-dimensional crystallization of hard sphere particles at a liquid–liquid interface. *Colloids Surf. A* 346:208–212.

Meier, M., Yadigaroglu, G., Smith, B.L. 2002. A novel technique for including surface tension in PLIC-VOF methods. *Eur. J. Mech.* 21:61–73.

Ménard, T., Tanguy, S., Berlemont, A. 2007. Coupling level set/VOF/ghost fluid methods: Validation and application to 3D simulation of the primary break-up of a liquid jet. *Int. J. Multiphase Flow* 33:510–524.

Miller, R., Liggieri, L., Eds. 2011. Bubble and drop interfaces. *Progress in Colloid and Interface Science*. CRC Press.

Miller, T.F., Miller, D.J. 2001. A Fourier analysis of the IPSA/PEA algorithms applied to multiphase flows with mass transfer. *Comput. Fluids* 32:197–221.

Muradoglu, M., Tryggvason, G. 2008. A front-tracking method for computation of interfacial flows with soluble surfactants. *J. Comp. Phys.* 227:2238–2262.

Murshed, S.M.S., Leong, K.C., Yang, C. 2008. Investigations of thermal conductivity and viscosity of nanofluids. *Int. J. Thermal Sci.* 47:560–568.

Nichols, B.D., Hirt, C.W. 1973. Calculating three-dimensional free surface flows in the vicinity of submerged and exposed structures. *J. Comp. Phys.* 12:234–246.

Onishi, J., Kawasaki, Y., Chen, H. 2008. Lattice Boltzmann simulation of capillary interactions among colloidal particles. *Comput. Math. Appl.* 55:1541–1553.

Panczyk, T., Szabelski P., Drach M. 2012. Implicit solvent model for effective molecular dynamics simulations of systems composed of colloid nanoparticles and carbon nanotubes. *J Colloid Interface Sci.* 383:55–62.

Pickering, S.U. 1907. Emulsions. *J. Chem. Soc.* 91:2001–2021.

Popinet, S. 2009. An accurate adaptive solver for surface-tension-driven interfacial flows. *J. Comp. Phys.* 228:5838–5866.

Ravera, F., Ferrari, M., Liggieri, L., Loglio, G., Santini, E., Zanobini, A. 2008. Liquid–liquid interfacial properties of mixed nanoparticle–surfactant systems. *Colloids Surf. A* 323:99–108.

Rider, W.J., Kothe, D.B. 1995. Stretching and tearing interface tracking methods. In *AIAA Computational Fluid Dynamics Conference, 12th, and Open Forum*. San Diego, CA, pp. 806–816.

Roux, B., Simonson, T. 1999. Implicit solvent models. *Biophys. Chem.* 78:1–20.

Ryskin, G., Leal, L.G. 1984. Numerical solution of free-boundary problems in fluid mechanics. Part 2. Buoyancy driven motion of a gas bubble through a quiescent liquid. *J. Fluid Mech.* 148:19–35.

Salager, J.L. 2000. *Pharmaceutical Emulsions and Suspensions*. F. Nielloud, G. Marti-Mestres (eds.). Marcel Dekker, New York.

Santini, E., Kragel, J., Ravera, F., Liggieri, L., Miller, R. 2011. Study of the monolayer structure and wettability properties of silica nanoparticles and CTAB using the Langmuir trough technique. *Colloids Surf. A* 382:186–191.

Scardovelli, R., Zaleski, S. 1999. Direct numerical simulation of free-surface and interfacial flow. *Annu. Rev. Fluid Mech.* 31:567–603.

Schlick, T. 2002. *Molecular Modeling and Simulation: An Interdisciplinary Guide*, Springer-Verlag New York, Inc. Secaucus, NJ, USA.

Schmitt-Rozières, M. et al. 2009. From spherical to polymorphous dispersed phase transition in water/oil emulsion. *Langmuir* 25:825–831.

Schweighofer, K.J., Essmann, U., Berkowitz, M. 1997. Simulation of sodium dodecyl sulfate at the water–vapor and water–carbon tetrachloride interfaces at low surface coverage. *J. Phys. Chem. B.* 101:3793–3799.

Shelley, M.Y., Sprik, M., Shelley, J.C. 2000. Pattern formation in a self-assembled soap mono-layer on the surface of water: A computer simulation study. *Langmuir* 16:626–630.

Smith, R. D., Berg, J. C. 1980. The collapse of surfactant monolayers at the air–water inter-face. *J. Colloids Interface Sci.* 74:273–286.

Sussman, M., Fatemi, E., Smereka, P., Osher, S. 1998. An improved level set method for incompressible two-phase flows. *Comput. Fluids* 27:663–680.

Sweby, P.K. 1984. High resolution schemes using flux-limiters for hyperbolic conservation laws. *SIAM J. Numer. Anal.* 21:995–1011.

Tarek, M., Tobias, D.J. 2000. The dynamics of protein hydration water: A quantitative compar-ison of molecular dynamics simulations and neutron-scattering experiments. *Biophys.* 79:3244–3257.

Tarek, M., Tobias, D.J., Klein, M.L. 1995. Molecular dynamics simulation of tetradecyl-trimethylammonium bromide monolayers at the air/water interface. *J. Phys. Chem.* 99:1393–1402.

Tasoglu, S., Demirci, U., Muradoglu, M. 2008. The effect of soluble surfactant on the transient motion of a buoyancy-driven bubble. *Phys. Fluids* 20:040805–040815.

Titarev, V.A., Toro, E.F. 2005. WENO schemes based on upwind and centred TVD fluxes. *Comput. Fluids* 34:705–720.

Tomassone, M.S., Couzis, A., Maldarelli, C.M., Banavar, J.R., Koplik. J. 2001a. Phase transi-tions of soluble surfactants at a liquid–vapor interface. *Langmuir* 17:6037–6040.

Tomassone, M.S., Couzis, A., Maldarelli, C.M., Banavar, J.R., Koplik, J. 2001b. Molecular dynamics simulation of gaseous–liquid phase transitions of soluble and insoluble sur-factants at a fluid interface. *J. Chem. Phys.* 115:8634–8642.

Tome, M.F., Mckee, S. 1994. GENSMAC: A computational marker and cell method for free surface flows in general domains. *J. Comp. Phys.* 110:171–186.

Tryggvason, G. et al. 2001. A front-tracking method for the computations of multiphase flow. *J. Comp. Phys.* 169:708–759.

Tsuji, M., Nakahara, H., Moroi, Y., Shibata, O. 2008. Water evaporation rates across hydro-phobic acid monolayers at equilibrium spreading pressure. *J. Colloid Interface Sci.* 318:322–330.

Tyrode, E., Johnson, C.M., Kumpulainen, A., Rutland, M.W., Claesson, P.M. 2005. Hydration state of nonionic surfactant monolayers at the liquid/vapor interface: Structure determi-nation by vibrational sum frequency spectroscopy. *J. Am. Chem. Soc.* 127:16848–16859.

Unverdi, S.O., Tryggvason, G. 1992. A front-tracking method for viscous, incompressible, multi-fluid flow. *J. Comp. Phys.* 100:25–37.

Van Buuren, A.R. 1995. Characterization of oil/water interfaces. A molecular dynamics study. PhD Thesis, University Library Groningen.

Van Buuren, A.R., Marrink, S.J, Berendsen, H.J.C. 1993. A molecular dynamics study of the decane/water interface *J. Phys. Chem.* 97:9206–9212.

Van der Spoel, D., Van Maaren, P.J., Berendsen, H.J.C. 1998. A systematic study of water models for molecular simulation: Derivation of water models optimized for use with a reaction field. *J. Chem. Phys.* 108:10220–10230.

Van der Zwan, E., van der Sman, R., Schroen, K., Boom, R. 2009. Lattice Boltzmann simula-tions of droplet formation during microchannel emulsification. *J. Colloid Interface Sci.* 335:112–122.

Vorholz, J., Harismiadis, V.I., Rumpf, B., Panagiotopoulos, A.Z., Maurer, G. 2000. Vapor plus liquid equilibrium of water, carbon dioxide, and the binary system, water plus carbon dioxide, from molecular simulation. *Fluid Phase Equilib.* 170:203–234.

Waldman, M., Hagler, A.T. 1993. New combining rules for rare gas van der Waals parameters. *J. Comp. Chem.* 14:1077–1084.

Weinan, E. 2011. *Principles of Multiscale Modeling*. Cambridge University Press.

Welch, J.E., Harlow, F.H., Shannon, J.P., Daly, B.J. 1966. The MAC method: A computing technique for solving viscous, incompressible, transient fluid flow problems involving free surfaces. Los Alamos Scientific Laboratory, University of California.

Wendy, D.C. et al. 1995. A second generation force field for the simulation of proteins, nucleic acids, and organic molecules. *J. Am. Chem. Soc.* 117:5179–5197.

Wilcox, D.C. 2006. *Turbulence Modeling for CFD*. 3rd ed. D. C. W. Industries, California.

Yang, W., Liu, S.H., Wu, Y.L. 2009. An unsplit Lagrangian advection scheme for volume of fluid method. *J. Hydrodynam.* 22:73–80.

Yee, H.C. 1987. Construction of explicit and implicit symmetric TVD schemes and their applications. *J. Comp. Phys.* 68:151–179.

Yon, S., Pozrikidis, C. 1998. A finite-volume/boundary-element method for flow past interfaces in the presence of surfactants, with application to shear flow past a viscous drop. *Comput. Fluids* 27:879–902.

Yoshida, H. 1990. Construction of higher order symplectic integrators. *Phys. Lett. A* 150:262–268.

Zahran, Y.H. 2009. An efficient WENO scheme for solving hyperbolic conservation laws. *Appl. Math. Comput.* 212:37–50.

12 The Contact Angle as an Analytical Tool

Victoria Dutschk and Abraham Marmur

CONTENTS

12.1 INTRODUCTION

12.1.1 WETTING AND WETTABILITY

Wetting is the process of making contact between a liquid and a solid. The term *wetting* describes a displacement of a solid–gas (air) interface with a solid–liquid interface, that is, a process in which Gibbs energy decreases in a system consisting of three contacting phases. The term *wettability* describes the ability of a surface to maintain contact with a liquid. The degree of wetting and wettability is determined by a force balance between adhesive and cohesive molecular forces. Both wetting and de-wetting of liquids on different surfaces play an important role in many natural and technological processes. Typical examples for wetting-dependent processes are printing, cleaning, painting, detergency, and lubrication. Wetting is also known to be a necessary condition for good adhesiveness. Since Gibbs elaborated the fundamentals of the thermodynamic theory of capillarity [1], diligent work has been performed to describe the wetting behavior of heterogeneous systems,

255

thereby determining surface energies of liquid and solid surfaces and, in this manner, predicting their adhesion behavior. Over a period, plenty of literature data have been accumulated, proposing various measurement techniques and different evaluation possibilities including criticism of one or the other computational algorithm or fundamental idea [2–22].

12.1.2 WHAT IS A CONTACT ANGLE?

The central property that characterizes wetting systems is the equilibrium contact angle Θ. It is defined as the angle between the tangents to the liquid–vapor (in general, liquid–fluid) interface and the solid surface at the contact line between the three involved phases. By convention, it is measured on the liquid side of the liquid–fluid interface, as shown in Figure 12.1. When the system involves two immiscible liquids, it is usually measured on the denser liquid side. For an ideal solid surface, that is, smooth, rigid, chemically homogeneous, insoluble, and non-reactive, and a pure liquid, the final state of a wetting process can be characterized by the Young equation [23]:

$$\gamma_{SV} = \gamma_{SL} + \gamma_{LV} \cdot \cos \Theta, \qquad (12.1)$$

where Θ is the equilibrium contact angle, and γ_{LV}, γ_{SV}, and γ_{SL} are the liquid–vapor, solid–vapor, and solid–liquid interfacial tensions, respectively. The Young contact angle Θ depends only on the physicochemical nature of the three phases involved and is independent of the system geometry and gravity. In most cases, the Young contact angle cannot be measured because an ideal surface rarely exists. Analysis of Young's contact angles as well as of problems of experimental and theoretical verification of equilibrium contact angle is provided in the work of Chibowski [24].

12.1.3 STATIC AND DYNAMIC CONTACT ANGLES

If the three-phase contact (TPC) line spontaneously moves relative to an adjacent solid surface (spreading), a dynamic contact angle will be observed. The dynamic contact angle can significantly differ from the static contact angle. Moreover, the thermodynamic consideration of wetting processes is necessary, but it is not sufficient to describe a lot of technological processes as kinetic aspects are not considered.

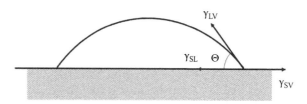

FIGURE 12.1   Liquid drop on a solid surface: Θ is the equilibrium contact angle; γ_{LV}, γ_{SV}, and γ_{SL} are liquid–vapor, solid–vapor, and solid–liquid interfacial tensions, respectively.

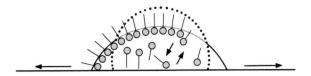

FIGURE 12.2   Water drop containing surfactant molecules simultaneously spreading over a surface.

In such applications, dynamic wetting and de-wetting processes are of crucial importance. Usually, dynamic contact angles are measured with liquids containing surface-active agents [25–27], as shown in Figure 12.2.

Characterization of materials on the basis of wetting measurements is the most frequently used method in technological applications. Basically, there are two different types of materials characterization: (i) characterization of a solid surface using a model liquid in order to detect changes after solid surface modification [28] and (ii) characterization of surfactant solutions using model solid surfaces in order to obtain the efficiency of surfactants [29].

12.1.4   POTENTIAL OF CONTACT ANGLE AS A MACROSCOPIC TOOL SENSITIVE TO "NANO-DEFECTS"

Real surfaces are usually heterogeneous, chemically and topographically. On rough surfaces, the local inclination of the solid surface with respect to some reference plane may change from one point to another. Similarly, the chemical composition may be different at various locations. Therefore, it is of fundamental interest to understand the relationship between the Young contact angle, Θ, and the measured contact angle, θ.

A theoretical feasibility study of detecting chemical nano-heterogeneities on solid surfaces by measurement of contact angles was described in the study of Bittoun and Marmur [30]. Two simplified models of cylindrical (two-dimensional) and axisymmetric (three-dimensional) drops on chemically heterogeneous, smooth solid surfaces were considered. This feasibility depends on the ratio between the external energy input to the drop and the energies needed to deform its liquid–vapor interface and move the TPC line across energy barriers. The variations of the liquid–gas interfacial energy are discussed in terms of orders of magnitude. By comparing these energies, it was concluded that under regular building vibrations, contact angle measurements can detect chemical heterogeneities at a few-nanometer scale.

12.2   A GUIDE TO CONTACT ANGLE MEASUREMENT

When a drop of a liquid is put in contact with a flat surface, two distinct equilibrium regimes may be found [31]: *partial wetting* with a finite contact angle or *complete wetting* with a zero contact angle. A low contact angle means that the solid surface is well wetted by the liquid, that is a *hygrophilic* solid surface, while a high contact angle indicates a preference for solid–liquid contact, that is a *hygrophobic* solid

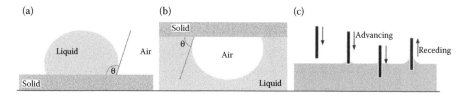

FIGURE 12.3   (a) Sessile drop, (b) captive bubble, and (c) Wilhelmy plate methods to measure contact angles. θ, h, and d are the contact angle, drop base, and drop height, respectively.

surface. Both terms were introduced by Marmur [32,33]. When the process of interest is static, equilibrium contact angles are discussed; when it involves very slow motion, quasi-equilibrium may exist, which closely approximates the equilibrium state; when the process requires high speeds or is forced by spontaneous spreading, dynamic contact angles are considered (cf. Section 12.1.3). There are many possible configurations of a solid–liquid–fluid system: a liquid drop on a solid surface (sessile drop), a fluid bubble underneath a solid surface (captive bubble), a solid plate partially dipped into a liquid (Wilhelmy plate [34]), a liquid inside a porous medium (Washburn method [35]), and a particle floating on a liquid–fluid interface.

Three methods have proven to be useful for characterization of surfaces—sessile drop, captive bubble, and Wilhelmy plate methods (Figure 12.3). With the sessile drop method, a drop of a test liquid is placed on the solid. After reaching equilibrium, the contact angle is read on an enlarged picture of the resting drop. In the captive bubble measurement, a bubble (of vapor or a fluid lighter than the liquid) is held captive against a solid surface The fluid phases and capture surface are all contained within a cell that permits imaging under conditions of equilibration over a range of temperatures and pressures. With the Wilhelmy plate method, a sample with well-determined geometry is dipped in the test liquid and then the contact angle is determined from the changes in force occurring in the process. This method is especially suitable to characterize wetting kinetics on fibers by polymer melt. Velocity and degree of fiber wetting by a polymer melt can be directly followed, allowing consideration of the rheological aspect when characterizing strength properties of actual composite materials [36].

For the first two methods mentioned above—sessile drop and captive bubble—drop shape analysis is a convenient way to measure contact angles. The essential assumptions are (i) the drop is symmetric about a central vertical axis and (ii) the drop is not in motion in the sense that viscosity or inertia are playing a role in determining its shape. Contact angles are calculated by fitting a mathematical expression to the shape of the drop and then calculating the slope of the tangent to the drop at the TPC line.

12.2.1   IDEAL SURFACES

For ideal surfaces, that is, mathematically smooth, chemically homogeneous, insoluble, and (non-)reactive, the *ideal* contact angle θ_i is considered as the equilibrium contact angle that a liquid makes with an ideal solid surface (cf. Figure 12.4a). This

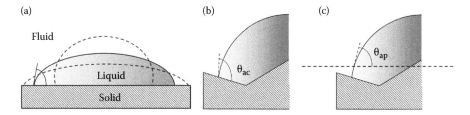

FIGURE 12.4 Contact angles: (a) various geometric contact angles for the same drop volume, (b) the actual contact angle, and (c) the apparent contact angle. (Reprinted from Marmur, A., *Contact Angle, Wettability and Adhesion*, Vol. 6, p. 5. With permission from Koninklijke Brill NV, Leiden. Copyright 2009.)

contact angle is sometimes referred to as the intrinsic or inherent contact angle [9]. For most macroscopic systems, the Young contact angle Θ (Equation 12.1) represents the ideal contact angle. Equilibrium of a wetting system at constant pressure and temperature is achieved when the Gibbs energy is minimal [37]. The Gibbs energy G of a wetting system with an ideal solid surface is given by

$$G = \gamma_{LV} A_{LV} + \gamma_{SL} A_{SL} + \gamma_{SV} A_{SV} = \gamma_{LV} A_{LV} + (\gamma_{SL} - \gamma_{SV}) A_{SL} + \gamma_{SV} A_{total}, \quad (12.2)$$

where A_{LV}, A_{SV}, and A_{SL} are the liquid–vapor, solid–vapor, and solid–liquid areas, respectively, and A_{total} is the total area of the solid surface, which is constant. For calculating all possible Gibbs energy states of the system, the *geometric* contact angle θ is used as an independent variable calculated using geometric consideration only, regardless of whether the system is at equilibrium. In the case of an ideal solid surface, the Gibbs energy of a wetting system, as a function of the geometric contact angle, has a single minimum, as shown in Figure 12.5a. The geometric contact angle

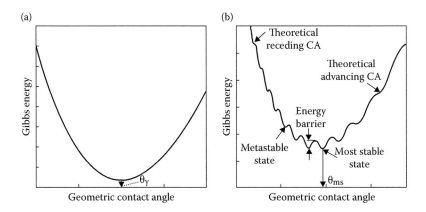

FIGURE 12.5 Gibbs energy for a liquid drop (a) on an ideal solid surface and (b) on a rough or chemically heterogeneous solid surface. (Reprinted from Marmur, A., *Contact Angle, Wettability and Adhesion*, Vol. 6, p. 7. With permission from Koninklijke Brill NV, Leiden. Copyright 2009.)

leading to this minimum is defined as the ideal contact angle. If the interfacial tensions are constant over the whole corresponding surfaces, the ideal contact angle is equal to the Young contact angle.

12.2.2   SMOOTH, CHEMICALLY HETEROGENEOUS SURFACES

In reality, surfaces are rough or chemically heterogeneous, or both. The influence of surface roughness and surface chemical heterogeneity is shown in Figure 12.6a, where the water droplets on a smooth hydrophobized glass surface do not have a spherical shape. On chemically heterogeneous surfaces, the chemical composition is different at various locations as it can be seen in Figure 12.6b. In general, for non-ideal solid surfaces, the actual contact angle (Figure 12.4b) varies along the TPC line, according to the local chemical composition and inclination of the solid surface as shown in Figure 12.7. Consequently, the contact line is "wavy" and the shape of the liquid–air interface has to adjust itself accordingly. However, the actual contact angle is usually not a measurable quantity, since it may be very difficult or even impossible to make local contact angle measurements.

It is well known that the apparent contact angle, as schematically shown in Figure 12.4c, is the measurable quantity that can be observed macroscopically. The relationship between the apparent contact angle θ_{ap} and Young contact angle Θ depends on the nature of the solid surface. For an ideal surface, the apparent contact angle is, obviously, identical to the actual contact angle as well as to the ideal contact angle. For a smooth heterogeneous surface, the apparent contact angle may vary from one point on the solid surface to the other. For perfectly smooth chemically heterogeneous surfaces, it may be possible to measure the various local values for the apparent contact angle.

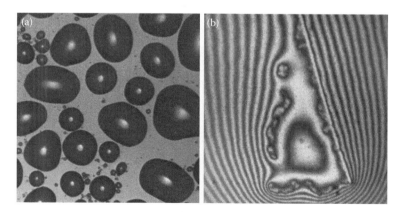

FIGURE 12.6   (a) Water droplets on smooth hydrophobized chemically heterogeneous glass surface. (b) Water droplet spreading over a hydrophilic smooth and chemically heterogeneous glass surface.

(a)   (b)

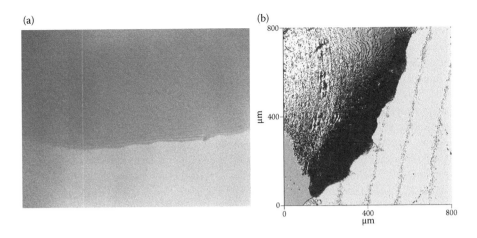

FIGURE 12.7 Water droplet on cleaned (a) glass and (b) steel surfaces. (From Radoev, B. Unpublished results.)

12.2.3 ROUGH, CHEMICALLY HOMOGENEOUS SURFACES

Surface roughness can be viewed as surface energy fluctuations [31], which act as barriers for the TPC line propagation. From the viewpoint of fluid mechanics, there is the *no-slip* boundary condition [39]. In wetting experiments, suitable precautions have to be used to separate dynamic effects from static hysteresis. For this purpose, a thorough topographic characterization of a surface to be evaluated with respect to its wetting properties is absolutely essential.

On rough surfaces, the local contact angle of inclination of the solid surface with respect to some reference plane varies from point to point. It is essential to understand how to meaningfully measure contact angles on rough surfaces and what kind of information can be derived from each apparent contact angle. This understanding can be derived from the Gibbs energy curve for rough surfaces (see Figure 12.5b). The Gibbs energy for such systems is calculated for all possible states of the system, which are defined by the geometric contact angle. The most prominent feature of the Gibbs energy curve for a wetting system with a real solid surface is the existence of multiple minima [40], that is, in contrast to the case of an ideal solid surface, which is characterized by a single minimum. Each such minimum defines a (metastable) equilibrium state. The minimum that is associated with the lowest Gibbs energy values defines the most stable contact angles. The other minima in the Gibbs energy curve represent metastable equilibrium states. In between every pair of minima points there is, obviously, a maximum point. The difference in Gibbs energy between a minimum and the following maximum defines the energy barrier to be overcome when moving from one metastable equilibrium state to the next, as shown in Figure 12.5b.

In general, surfaces can be either *randomly* rough or *periodically* rough. A liquid drop placed on a rough surface can sit on peaks or wet grooves depending on the geometry of surface roughness as well as on surface tension of a liquid. Existing theories for pure liquids can be taken into account in case of different wetting regimes [4]: (i) homogeneous wetting according to Wenzel [2] and (ii) heterogeneous wetting according to Cassie and Baxter [3]. A textured solid can be considered as a two-dimensional porous material in which liquid can be absorbed by surface wicking [41]. When the contact angle is smaller than a critical value θ_{cr}, a film propagates from a deposited droplet, a small amount of liquid is sucked into the texture, and the remaining drop sits on a patchwork of a solid and a liquid. It is very common that fibrous materials encounter roughness on surfaces and walls of pores. The driving force for such surface wicking depends on the geometry of the grooves, the surface tension of the liquid, and the Gibbs energies of the solid–gas and solid–liquid interfaces. An excellent overview of wetting regimes on fibrous surfaces is given in the work of Rengasamy [42].

New topographic concepts for a mechanistic understanding of wetting phenomena were presented by one of the authors in Refs. [43–46]. Moreover, in the study of Calvimontes et al. [47], it was shown that there are significant differences between the soiling behavior and cleanability of polyester textile materials with different topographic structures despite the similarity of their chemical nature.

12.2.4 CONTACT ANGLE HYSTERESIS

As Figure 12.5b shows, the minima in the Gibbs energy exist only over a finite range of apparent contact angles. This is the so-called contact angle hysteresis range. The apparent contact angle θ_{ap} as shown in Figure 12.4c is an equilibrium contact angle at a point on a solid surface that may be rough or chemically heterogeneous. The apparent contact angle is defined as the angle between the tangent to the liquid–vapor interface and the apparent solid surface, as macroscopically observed. θ_{ap} is actually an equilibrium value of the geometric contact angle.

The advancing contact angle θ_a is the highest possible contact apparent angle that can be achieved for a given wetting system. The advancing contact angle is an equilibrium contact angle, since θ_a is by definition an equilibrium contact angle. The receding contact angle is the lowest possible apparent contact angle that can be achieved for a given wetting system. Similarly to the advancing contact angle, the receding contact angle is an equilibrium contact angle. The difference between the advancing (θ_a) and receding (θ_r) contact angles is called contact angle hysteresis

$$\Delta\theta = \theta_a - \theta_r. \tag{12.3}$$

It is essential to measure both contact angles and report the contact angle hysteresis to fully characterize a surface. The hysteresis can be classified in thermodynamic and kinetic terms. Roughness and heterogeneity of the surface are sources for hysteresis in a thermodynamic sense. Kinetic hysteresis is characterized by time-dependent changes in contact angle, which depends on deformation, reorientation, and mobility of the surface as well as liquid penetration. Advancing and receding contact angle measurements are possibly force driven if the drop volume will be increased or decreased. Unfortunately, a comprehensive theory of contact angle hysteresis has not yet been developed, although some initial steps have been taken [17,48–50]. Therefore, the hysteresis range can currently be used in two ways only: (i) as an indication of the heterogeneity extent of the solid surface and (ii) as a means to estimate the most stable contact angle. Two suggestions for the latter option have been given in the literature. One is [51]

$$\cos\theta_{ms} = \frac{(\cos\theta_a + \cos\theta_r)}{2},$$ (12.4)

where θ_{ms} is the most stable contact angle.

Another correlation is [52]

$$\theta_{ms} = \frac{(\theta_a + \theta_r)}{2}.$$ (12.5)

However, neither of these originates from any fundamental theory. Moreover, in practice, the differences between two types of averages may not be very meaningful.

12.3 ANALYTICAL CAPABILITIES BASED ON CONTACT ANGLE MEASUREMENT

12.3.1 CONTACT ANGLE MEASUREMENTS

After the placement of a drop on an engineered solid surface, it acquires an apparent contact angle somewhere within the hysteresis range. This "as is" apparent contact angle depends on the balance between the energy barriers and available drop energy. The latter is determined by the dynamic process that the drop undergoes during touching the solid surface and the oscillations that follow [53,54].

Thus, the state of the drop when it lands on a surface is rather randomly determined. Consequently, it is extremely difficult, if not impossible, to gain useful information from an apparent contact angle measured "as is" (sometimes referred to as the "static contact angle"). The only uniquely defined apparent contact angles are the advancing, receding, and most stable ones. The former two are easier to measure than the latter. As is well known, they can be experimentally determined by increasing or decreasing the drop volume until the maximum (advancing) or minimum (receding) contact angle is reached, respectively. However, there is always a difference between the theoretical and practical values of these contact angles, and there is yet no substantiated theory available for their interpretation. In contrast, the most stable contact angle is the one that can be theoretically interpreted in terms of Young contact angles, using the Wenzel or Cassie equation, but its experimental determination (e.g., using vibrations) is still under development [55].

Regardless of which of these apparent contact angles is measured, there are some common requirements that have to be carefully fulfilled in order to perform meaningful measurement and interpretation. The most important requirement that is extracted from the equilibrium contact angle theory is that the ratio of the drop size to the scale of roughness or chemical heterogeneity must be sufficiently large. This requirement is essential for the theoretical interpretation of the most stable contact angle (in order to be able to use the Wenzel or Cassie equation); it is also crucial for ensuring axisymmetry of the drop, without which the apparent contact angle measurement may be meaningless in many situations. In addition, it is advantageous also for advancing contact angle measurements, since this angle is less sensitive to the drop volume if the latter is sufficiently large. The next requirement is closely related to the former: the drop must be axisymmetric in order for the contact angle measurement and interpretation to be meaningful. This requirement necessitates an experimental measurement of axisymmetry. The easiest way to do it is by taking pictures of the drop from above [55]. The contact angles can still be measured from pictures taken from the side, using a second camera, or calculated from the maximum drop diameter captured from above, using the Young–Laplace equation with input values for the drop volume, liquid surface tension, and its density [55]. As explained above, the only substantiated theoretical correlation presently available links the Young contact angle with the most stable contact angle, using the Wenzel or Cassie equation. Thus, one needs the most stable contact angle in order to calculate the Young contact angle. This can be achieved by two approaches, each having its drawbacks. The first approach suggests calculating the most stable contact angles from the advancing and receding ones, using various averages as shown in Equations 12.4 and 12.5.

This approach is experimentally convenient; however, its main drawback is the lack of theoretical substantiation. The second approach advocates a direct measurement of the most stable contact angles by applying, for example, vibrations. The main disadvantage of this approach is the uncertainty in the experimental definition of the most stable contact angle. However, once the most stable contact is calculated for a rough surface, the Wenzel equation can be used to calculate Θ. Of course, the roughness ratio must also be known. For a chemically heterogeneous surface, the most stable contact angle of a sufficiently large drop (the Cassie contact angle) serves as an average Θ for the surface. Though a theory for contact angle hysteresis is still in its infancy, measurement of the contact angle hysteresis range may be informative as an empirical measure for surface roughness or chemical heterogeneity.

12.3.2 SURFACE TENSION OF A SOLID

The most important application of contact angle measurement of pure liquids on engineered surfaces is the assessment of the solid surface tension γ_{SV}. At present, the only way to assess γ_{SV} is through the Young equation (Equation 12.1), which, in principle, enables this calculation if Θ and γ_{SL} are known. Since γ_{SL} cannot be experimentally measured, it has to be calculated from γ_{SV} and γ_{LV} (i.e., independently known), using semi-empirical correlations presented in Refs. 56–62. The main experimental challenge here is measuring the apparent contact angle that can yield the Young contact angle. From a theoretical point of view, the best option is to measure the most stable contact angle of a large drop because there are theories to calculate the Young contact angle. It is also important to understand that to calculate the Young contact angle from the most stable contact angle, additional information on the heterogeneity of the surface is required. For a rough but chemically homogeneous surface, a measurement of the roughness ratio according to Wenzel [2] is needed. This may depend on the resolution of the measurement; thus, understanding roughness and its measurement is essential.

12.4   DYNAMIC CONTACT ANGLES OF SURFACTANT SOLUTIONS

As mentioned in Section 12.1.3, besides characterization of an engineered solid surface with a pure liquid, it is also important to characterize surfactant solutions using model (well characterized) solid surfaces, in order to obtain the efficiency of surfactants.

For many applications, surfactants are introduced into the aqueous phase to increase the rate and uniformity of wetting. Despite their enormous technical importance, there is limited information in the literature about the spreading dynamics of aqueous surfactant solutions. The knowledge of how surfactant adsorption at the surfaces involved affects the spreading mechanism and dynamics is also limited. Characterization of materials on the basis of wetting measurements/interfacial thermodynamics is the most frequently used method in technological applications.

To explain and predict the behavior of aqueous surfactant solutions on the solid–liquid interface, information of their dynamic behavior at the liquid–vapor interface is absolutely necessary. The dynamic surface tension of aqueous solutions can be measured according to different time windows by suitable methods: (i) bubble pressure tensiometry, (ii) drop volume tensiometry, and (iii) drop/bubble profile analysis, described in more detail elsewhere [63–65].

Dynamic contact angle measurements are possible either as force driven, if the drop volume will be increased/decreased, or as time-dependent contact angle measurements with a constant volume. In the former case, advancing or receding angles are formed to analyze the *contact angle hysteresis*, that is, analysis of chemical and mechanical heterogeneities (see Section 12.2.4). In the latter case, the temporal contact angle change because of spontaneous spreading of the liquid is measured. The contact angle dependence on the contact time of the solid surface with the measuring liquid is called *dynamic contact angle* as shown in Figure 12.8.

The dynamic behavior of a pure liquid on an ideal solid surface can be successfully mathematically described by the equilibrium contact angle Θ (cf. Sections 12.1.2 and 12.2.1), the dynamic (time-dependent) contact angle $\theta(t)$, and the spreading velocity dr/dt, where r is the base radius of a spreading drop. The *spreading velocity* or *spreading rate* as a time-dependent drop radius variation is often an important criterion on which basis the efficiency of surface-active substances (surfactants) can be estimated.

In the hydrodynamic consideration, the spreading force is

$$\gamma_{LV} \left(\cos \theta_0 - \cos \theta(t)\right), \tag{12.6}$$

where θ_0 is initial contact angle. The most popular hydrodynamic models gave a successful interpretation of experimental data on complete wetting and propagation at capillary numbers [39,66–68]. Such a theory that considers slippage of the liquid

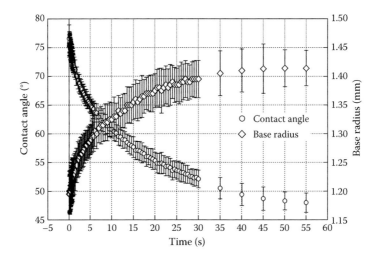

FIGURE 12.8 Contact angle and drop base radius of a spreading water drop containing surfactant molecules on a smooth and homogeneous hydrophobic surface.

with respect to the solid in the TPC line vicinity is applicable to the description of a slow spreading near equilibrium. The molecular-kinetic theory [19,69] assumes, however, particular displacements on a molecular level at the TPC line as a possible reason for the spreading force; it is suitable for describing high spreading velocities far from equilibrium. Although this theory, in contrast to the hydrodynamic theory, includes surface effects, its application to predict the spreading velocity is rather problematic since the molecular parameters such as the density of the adsorption centers and the distance between them on real surfaces are unknown and generally inaccessible to experiments. By completely neglecting the viscous drag, a theoretical dependence of the dynamic contact angle $\theta(t)$ on the TPC line velocity is determined from the balance of the driving force and the friction force in the TPC zone. This approach is based on Eyring et al.'s theory for transport phenomena [70].

The reduction of water surface tension by adsorption of surfactant molecules on a water–vapor interface and adsorption of surfactant molecules on solid–liquid and solid–vapor interfaces alters a non-wetting behavior of aqueous solutions on hydrophobic substrates into a partial or even complete wetting behavior. Surfactants have been used for a long time and their influence on surface wettability is well known and widely used. However, employing surfactants to enhance spreading complicates the wetting process through time-dependent diffusion and adsorption processes at the involved interfaces. The same processes are important in the case of water penetration into hydrophobic porous media. Aqueous surfactant solutions can spontaneously penetrate into hydrophobic porous substrates and the penetration rate depends on both the surfactant type and its concentration. Both the liquid–vapor interfacial tension, γ_{LV}, and the contact angle of moving meniscus, θ_a (advancing contact angle), become concentration dependent. Despite the enormous technical importance of spreading of aqueous surfactant solutions over solid surfaces, information on possible spreading mechanisms is still limited in the literature. It was shown in the work by Starov et al. [22] that the Young equation for the equilibrium contact angle is an empirical one and should be replaced by the Derjaguin–Frumkin equation [71,72]. The latter equation expresses the equilibrium contact angle via measurable physical properties, which are surface forces acting in the TPC line vicinity. An overview on some dynamic wetting and spreading phenomena in the presence of surfactants was recently described [26] for both smooth and rough hydrophobic substrates. A methodology to analyze and compare aqueous solutions of different surfactants with respect to their wetting and spreading on smooth and rough, chemically homogeneous and heterogeneous polymer surfaces was provided [29]. In order to better distinguish between very similar surfactants, highly hydrophobic polymer samples with very smooth and homogeneous surfaces are recommended to use as model surfaces.

ACKNOWLEDGMENT

The authors thank the COST D43 program for support of this collaboration.

REFERENCES

1. Gibbs, J.W. 1876. On the equilibrium of heterogeneous substances. *Trans. Connecticut Acad.* 3:108–248.
2. Wenzel, R.N. 1936. Resistance of solid surfaces to wetting by water. *Ind. Eng. Chem.* 28:988–994.
3. Cassie, A.B.D., Baxter, S. 1944. Wettability of porous surfaces. *Trans. Faraday Soc.* 40:546–551.
4. Johnson, R.E., Dettre, R.H. 1969. Wettabillity and contact angle. In *Surface and Colloid Science* Vol. 2, ed. E. Matijevic, 85–153. New York: Wiley-Interscience.
5. Huh, C., Mason, S.G. 1977. Effects of surface roughness on wetting (theoretical). *J. Colloid Interface Sci.* 60:11:38.
6. Good, R.J. 1979. Contact angles and the surface free energy of solids. In *Surface and Colloid Science* Vol. 11, eds. R.J. Good, R.R. Stromberg, 1–29. New York: Plenum Press.
7. Neumann, A.W., Good R.J. 1979. Techniques of measuring contact angles. In *Surface and Colloid Science* Vol. 11, eds. R.J. Good, R.R. Stromberg, 31–91. New York: Plenum Press.
8. Petrov, J., Radoev, B. 1981. Steady motion of the 3 phase contact line in model Langmuir–Blodgett systems. *Colloid Polym. Sci.* 259:753–760.
9. Marmur, A. 1983. Equilibrium and spreading of liquids on solid surfaces. *Adv. Colloid Interface Sci.* 19:75:102.
10. Joanny, J.F., de Gennes, P.G. 1984. A model for contact angle hysteresis. *J. Chem. Phys.* 81:552–562.
11. Schwartz, L.W., Garoff, S. 1985. Contact-angle hysteresis on heterogeneous surfaces. *Langmuir* 1:219–230.
12. Shanahan, M.E.R. 1991. A simple analysis of local wetting hysteresis on a Wilhelmy plate. *Surf. Interface Anal.* 17:489–495.
13. Di Meglio, J.M. 1992. Contact angle hysteresis and interacting surface defects. *Europhys. Lett.* 17:607–612.
14. Churaev, N.V, Zorin, Z.M. 1992. Wetting films. *Adv. Colloid Interface Sci.* 40:109–146.
15. Petrov, P.G., Petrov, J.G. 1992. A combined molecular-hydrodynamic approach to wetting kinetics. *Langmuir* 8:1762–1767.

16. Petrov, J.G., Petrov, P.G. 1992. Forced advancement and retraction of polar liquids on a low-energy surface. *Colloid Surf.* 64:143–149.

17. Marmur, A. 1994. Thermodynamic aspects of contact angle hysteresis. *Adv. Colloid Interface Sci.* 50:121–141.

18. Churaev, N.V., Zorin, Z.M. 1995. Penetration of aqueous surfactant solutions into thin hydrophobized capillaries. *Colloids Surf. A* 100:131–138.

19. Ruckenstein, E. Effect of short-range interactions on spreading. 1996. *J Colloid Interface Sci.* 179:136–142.

20. Churaev, N.V., Esipova, N.E., Hill, R.M., Sobolev, V.D., Starov, V.M., Zorin, Z.M. The superspreading effect of trisiloxane surfactant solutions. 2002. *Langmuir* 17:1338–1348.

21. Marmur, A. 2006. Soft contact: measurement and interpretation of contact angles. *Soft Matter* 2:12–17.

22. Starov, V., Velarde, M., Radke, C. 2007. Dynamics of wetting and spreading. In *Surfactant Sciences Series*, Vol. 138. Boca Raton, FL: Taylor & Francis.

23. Young, T. 1805. An essay on the cohesion of fluids. *Phil. Trans. R. Soc. Lond.* 95:65–87.

24. Chibowski, E. 2007. On some relations between advancing, receding and Young's contact angles. *Adv. Colloid Interface Sci.* 133:51–59.

25. Marmur, A., Lelah, M.D. 1981. The spreading of aqueous surfactant solutions on glass. *Chem. Eng. Commun.* 13:133–143.

26. Dutschk, V., Sabbatovskiy, K.G., Stolz, M., Grundke, K., Rudoy, V.M., 2003. Unusual wetting dynamics of aqueous surfactant solutions on polymer surfaces. *J. Colloid Interface Sci.* 267:456–462.

27. Lee, K.S., Ivanova, N., Starov, V.M., Hilal, N., Dutschk, V. 2008. Kinetics of wetting and spreading by aqueous surfactant solutions, *Adv. Colloid Interface Sci.* 144:54–65.

28. Marmur, A. 2009. Solid-surface characterization by wetting. *Annu. Rev. Matter Res.* 39:473–489.

29. Dutschk, V. 2011. Wetting dynamics of aqueous solutions on solid surfaces. In *Drop and Bubble Interfaces, Progress in Colloid and Interface Science*, eds. R. Miller, L. Liggieri, 2:223–242. Leiden-Boston: Brill.

30. Bittoun, E., Marmur, A. 2009. Chemical nano-heterogeneities detection by contact angle hysteresis: Theoretical feasibility. *Langmuir* 25:1277–1281.

31. De Gennes, P.G. 1985. Wetting and static dynamics. *Rev. Modern Phys.* 57:827–863.

32. Marmur, A. 2009. A guide to the equilibrium contact angle maze. In *Contact Angle, Wettability and Adhesion*, 6:3–18. Leiden: Brill.

33. Marmur, A. 2012. Hydro- hygro-oleo-omni-phobic? Terminology of wettability classification. *Soft Matter* 8:2867–2870.

34. Wilhelmy, L. 1983. Über die Abhängigkeit der Capillaritäts-Konstanten des Alkohols von Substanz und Gestalt des benetzten festen Körper. *Ann. Phys.* 119:177–217.

35. Washburn, E.W. 1921. The dynamics of capillary flow. *Phys. Rev.* 17:273–283.

36. Brantseva, T., Gorbatkina, Yu., Dutschk, V., Vogel, R., Grundke, K., Kerber, M.L. 2003. Modification of epoxy resin by polysulfone to improve the interfacial and mechanical properties in glass fibre composites. I. Study of processes during matrix/glass fibre interface formation *J. Adhesion Sci. Technol.* 17:2047–2063.

37. Gibbs, J.W. 1961. *The Scientific Papers of J. Willard Gibbs* Vol. 1, 288. New York: Dover Publications.

38. Radoev, B. unpublished results.

39. Huh, C., Scriven, L.E. 1971. Hydrodynamic model of steady movement of a solid/liquid/fluid contact line. *J. Colloid Interface Sci.* 35:85–101.

40. Marmur, A. 1992. Contact angle and thin film equilibrium. *J. Colloid Interface Sci.* 148:541–550.

41. Kissa, E. 1996. Wetting and wicking. *Textile Res. J.* 66:660–668.

42. Rengasamy, R.S. 2006. Wetting phenomena in fibrous materials. In *Thermal and Moisture Transport in Fibrous Materials*. Cambridge, England: Woodhead Publishing.

43. Hasan Badrul, M.M., Calvimontes, A., Dutschk, V. 2009. Correlation between wettability and cleanability of polyester fabrics modified by a soil release polymer and their topographic structure. *J. Surfact. Deterg.* 12:285–294.

44. Calvimontes. A., Dutschk, V., Stamm, M. 2010. Advances in topographic characterization of textile materials. *Textile Res. J.* 80:1004–1015.

45. Calvimontes, A., Hasan Badrul, M.M., Dutschk, V. 2010. Effects of topographic structure on wettability of differently woven fabrics. In *Woven Fabric Engineering*. Rijeka: SCIYO.

46. Calvimontes, A., Saha, R., Dutschk, V. 2011. Topographical effect of O_2- and NH_3-plasma teratment on woven plain polyester fabric in adjusting hydrophilicity. *AUTEX Res. J.* 11:24–30.

47. Calvimontes, A., Dutschk, V., Koch, H., Voit, B. 2005. New detergency aspects through visualization of soil release polymer films on textile surfaces. *Tenside Surf. Det.* 42:210–216.

48. He, B., Lee, J., Patankar, N.A. 2004. Contact angle hysteresis on rough hydrophobic surfaces. *Colloid Surf. A* 248:101–104.

49. Li, W., Amirfazli, A. 2005. A thermodynamic approach for determining the contact angle hysteresis for superhydrophobic surfaces. *J. Colloid Interface Sci.* 292:195–201.

50. Vedantam, S., Panchagnula, M.V. 2008. Constitutive modeling of contact angle hysteresis. *J. Colloid Interface Sci.* 321:393–400.

51. Decker, E.L., Garoff, S. 1996. Using vibrational noise to probe energy barriers producing contact angle hysteresis. *Langmuir* 12:2100–2110.

52. Andrieu, C., Sykes, C, Brochard, F. 1994. Average spreading parameter on heterogeneous surfaces. *Langmuir* 10:2077–2080.

53. Roisman, I.V., Prunet-Foch, B., Tropea, C., Vignes-Adler, M. 2002. Multiple drop impact onto a dry solid substrate. *J. Colloid Interface Sci.* 256:396–410.

54. Rozhkov, A., Prunet-Foch, B., Vignes-Adler, M. 2002. Impact of water drops on small targets, *Phys. Fluids* 14:3485–3501.

55. Meiron, T.S., Marmur, A., Saguy, I.S. 2004. Contact angle measurement on rough surfaces. *J. Colloid Interface Sci.* 274:637–644.

56. Girifalco, L.A., Good, R.J. 1957. A theory for the estimation of surface and interfacial energies. I. Derivation and application to interfacial tension. *J. Phys. Chem.* 61:904–909.

57. Fowkes, F.M. 1963. Additivity of intermolecular forces at interfaces. I. Determination of the contribution to surface and interfacial tensions of dispersion forces in various liquids. *J. Phys. Chem.* 67:2538–2541.

58. Owens, D.K., Wendt, R.C. Estimation of the surface free energy of polymers, *J. Appl. Polym. Sci.* 13:1741–1747.

59. van Oss, C.J., Good, R.J., Busscher, R.J. 1990. Estimation of the polar surface tension parameters of glycerol and formamide, for use in contact angle measurements on polar solids. *J. Dispers. Sci. Technol.* 11:75–81.

60. Della Volpe, C., Siboni, S.J. 1997. Some reflections on acid–base solid surface free energy theories, *J. Colloid Interface Sci.* 195:121–136.

61. Shalel-Levanon, S., Marmur, A. 2003. Validity and accuracy in evaluating surface tension of solids by additive approaches. *J. Colloid Interface Sci.* 262:489–499, 268:272.

62. Marmur, A., Valal, D. 2010. Correlating interfacial tensions with surface tensions: A Gibbsian approach. *Langmuir* 26:5568–5575.

63. Fainerman, V.B., Miller, R. 2011. Maximum bubble pressure tensiometry: Theory, analysis of experimental constrains and applications. In *Bubble and Drop Interfaces*, eds. R. Miller, L. Liggieri, 75–118. Leiden-Boston: Brill.

64. Miller, R., Dutschk, V., Fainerman, V.B. 2004. Influence of molecular processes at liquid interfaces on dynamic surface tension and wetting dynamics. *J. Adhesion Sci. Technol.* 80:549–561.

65. Kotsmar, C., Grigoriev, D.O., Makievski, A.V., Ferri, J.K., Krägel, J., Miller, R. 2008. Drop profile analysis tensiometry with drop bulk exchange to study the sequential and simultaneous adsorption of a mixed β-casein/C_{12}DMPO system. *Colloid Polym. Sci.* 286:1071–1077.

66. Cox, R.G. 1986. The dynamics of spreading of liquids on a solid-surface. *J. Fluid Mech.* 168:169–194, 195–220.

67. Voinov, O.V. 1976. Hydrodynamics of wetting. *Fluid Dyn.* 11:714–721.

68. Shikhmurzaev, Y.D. 2008. *Capillary Flows with Forming Interfaces.* Boca Raton: Chapman & Hall/CRC.

69. Blake, T.D., Haynes, J.M. Kinetics of liquid/liquid displacement. 1969. *J. Colloid Interface Sci.* 30:421–423.

70. Glasstone, S., Laidler, K.J., Eyring, H.J. 1941. *The Theory of Rate Processes.* New York: McGraw-Hill.

71. Churaev, N.V. 1995. The relation between colloid stability and wetting. *J. Colloid Interface Sci.* 172:479–484.

72. Churaev, N.V. 1995. Contact angles and surface forces. *Adv. Colloid Interface Sci.* 58:87–118.

13 Capillary Pressure Experiments with Single Drops and Bubbles

Aliyar Javadi, Jürgen Krägel, Mohsen Karbaschi,
Jooyoung Won, Adhijit Dan, Geogi Gochev,
Alexander V. Makievski, Guiseppe Loglio,
Libero Liggieri, Francesca Ravera,
Nina M. Kovalchuk, Marzieh Lotfi,
Vamseekrishna Ulaganathan,
Volodymyr I. Kovalchuk, and Reinhard Miller

CONTENTS

13.1   INTRODUCTION

Products based on foams and emulsions are omnipresent in our modern world and we see them from morning till night. We enjoy foams and emulsions during our meals not only in the form of ice cream or mousse au chocolate. Moreover, we are impressed with foamed metals, which have an enormously lower weight than solid materials and often have even much better mechanical properties.

Foams and emulsions can easily be formed; however, the control of their properties is not trivial. So far, most of the successful technologies are based on empirical knowledge. However, experience and intuition are more and more replaced by fundamental knowledge based on measured quantities of real foams or emulsions and the search for quantitative relationships to properties of the corresponding liquid films and adsorption layers.

The top-down approach allows us, for example, to see foams as an arrangement of bubbles that are in an ensemble and contact each other. The stability of foam therefore depends directly on the stability of the many foam lamellae between the bubbles. If these foam films are not stabilized by the two adsorption layers, there is no chance of having stable foam. In turn, what are the required properties of the surfactant adsorption layers that allow having stable foam films? The same questions can be formulated for the formation and stabilization of emulsions. In summary, the quantitative understanding of foams or emulsions is a multi-scale subject that requires fundamental knowledge on all three levels from single adsorption layers via liquid films up to the behavior of a real liquid disperse system.

The successful formation of liquid disperse systems is directly linked to the fast adsorption of the stabilizing surfactants at the freshly formed bubbles or drops. Hence, the capillary pressure (CP) techniques are the respective powerful tools that allow measuring the required quantities in terms of dynamic surface and interfacial tensions.

For the stabilization of foams/emulsions, additional interfacial quantities are needed. Most frequently discussed here are the dilational viscoelastic properties. For their measurement, single drop and bubble methods are again the methods of choice. While in the past wave damping techniques were mainly applied for the characterization of the dilational elasticity and viscosity, in recent years, the oscillating drop and bubble tensiometry was further developed and now represents the state of the art.

This chapter not only is dedicated to the description of CP methods but also includes drop and bubble profile techniques. The short history of the drop/bubble profile tensiometry and CP measurements in bubbles and drops intends to give a brief insight into the various stages of their development. The bubble pressure tensiometry is obviously the method for which the first commercial instruments were on the market. The profile methods became routine instruments only later, with the availability of affordable video cameras, and are now obviously the most frequently applied methods in each interfacial science laboratory.

Yet quite new are the CP setups, recently available as commercial instruments, which typically allow for measurements of dynamic surface and interfacial tensions at shortest adsorption times. In addition, these instruments can be used for investigations of the dilational viscoelasticity of interfacial layers. While drop and bubble profile tensiometers give access to the dilational elasticity and viscosity at low oscillation frequencies (1 mHz to about 0.2 Hz), the CP tensiometers allow to investigate the interfacial tension response to harmonic perturbations in the frequency range from 0.1 to about 100 Hz, for bubbles, even up to a few hundred hertz.

After a short historical survey of CP techniques, in the subsequent sections, more details about the classical bubble pressure tensiometry and about growing and oscillating drops and bubbles will be given. Additional sections are dedicated to the tensiometry of spherical films and to rising bubbles in surfactant solutions. The final sections deal with two very special applications of the CP technique. One is the so-called coaxial double capillary, which provides the opportunity to exchange the volume of a drop during the experiments. Moreover, the combination of two CP units provides the option for studies on the direct interaction between two drops or bubbles, mimicking the elementary processes happening in emulsions or foams. As a kind of outlook, the possibility of model investigations for double emulsions is proposed, that is, the simultaneous study of the CP in a drop that is formed in another, slightly larger drop.

13.2   HISTORY OF CP EXPERIMENTS

The CP tensiometry exploits the direct application of the Young–Laplace equation for a spherical interface

$$\Delta P_{\mathrm{cap}} = \frac{2\gamma}{R}. \tag{13.1}$$

From this equation, the surface/interfacial tension γ can be derived by the simultaneous measurement of the CP ΔP_{cap} and radius of curvature, R, of the interface. Under suitable mechanical quasi-equilibrium conditions, such equation also holds for dynamic systems, which makes the method also attractive for measurements of dynamic surface/interfacial tensions.

The possibility to use the CP measurement to infer the surface or interfacial tension according to the above equation attracted the interest of scientists already from the beginning of the 20th century. However, because of lack of accurate and automated sensors for the measurement of low pressures in liquids—down to 1 Pa—these efforts lead to the development of the maximum bubble pressure method, exploiting in fact only the maximal pressure achieved in a small bubble formed continuously at the tip of a capillary for different gas flow rates. The theoretical basis for the method, together with its practical application, was established first for static surface tension studies in the early years of the last century (Jäger 1917; Schrödinger 1915; Sugden 1922) and much later for dynamic surface tensions by Fainerman (1979).

The full development of the CP tensiometry was possible only when low-range pressure transducers for liquids became available, that is, about 30 years ago. In a typical CP tensiometer (Liggieri and Ravera 1998), a droplet is formed inside a liquid, at the tip of a capillary. The pressure difference across the interface is monitored by a pressure transducer while the drop radius is measured by direct imaging or it is calculated from the injected liquid volume; then, interfacial tension is derived according to Equation 13.1.

Relying on the sphericity of the interface, the technique requires negligible interface deformation, which can be obtained by using very small drops or for small differences of the density between the two fluids. The method is therefore particularly suitable for liquid–liquid interfaces. The maximum bubble pressure method, in addition, is used routinely for measuring adsorption data at the shortest adsorption time of less than 1 ms using bubbles small enough that any deviation from sphericity is negligible (Fainerman and Miller 2011).

CP tensiometry is a powerful tool for studies of equilibrium and dynamic surface and interfacial tensions. In principle, the same experimental apparatus can be used to measure with different experimental methodologies various physicochemical aspects characterizing the interface and the features of adsorption layers, including surface dilational viscoelasticity.

A first example of using such type of tensiometer was reported in Passerone et al. (1991). There, a method utilizing the values of the pressure measured inside a droplet slowly growing at a constant flow rate at the tip of a small capillary was proposed to measure the interfacial tension between pure immiscible liquids. The method was called the pressure derivative method since it was actually based on the linear best fit of Equation 13.1 to a set of $(\Delta P, 2/R)$ data measured during the droplet growth. The interfacial tension was then obtained as the slope via a best-fit procedure. It is remarkable to underline that this method does not require any image acquisition and analysis. The drop volume at each time is in fact calculated after knowing the volume dosing rate and the time at which the droplet passes the hemispherical shape, clearly identified by a maximum in the measured pressure (cf. Figure 13.1). If the adsorption kinetics is much faster than the drop growth, the above method can also be utilized to measure the equilibrium interfacial tension of surfactant solutions.

In practice, this is the case for concentrated surfactant solutions. Alternatively, the equilibrium interfacial tension of surfactant solutions can be obtained in a CPT according to the pressure–radius step method discussed in Liggieri et al. (2002). To this aim, the area of the droplet is increased in small steps. At each step, the drop radius and the pressure after achieving the adsorption equilibration are measured. Finally, similarly to the above procedure, a linear relationship is fitted to the obtained set of data to obtain the interfacial tension.

The concept of growing drops was then proposed (MacLeod and Radke 1993; Nagarajan and Wasan 1993; Soos et al. 1994; Zhang et al. 1994) for investigations of the adsorption dynamics. A CP tensiometer was utilized to measure the dynamic interfacial tension while increasing continuously the droplet area. In this case, the adsorption, Γ, varies under the effect of both the surface dilation $d(\ln A)/dt$ and the adsorption flux Φ_{net}, according to

$$\frac{d\Gamma}{dt} = \Phi_{net} - \Gamma \frac{d\ln A}{dt}. \qquad (13.2)$$

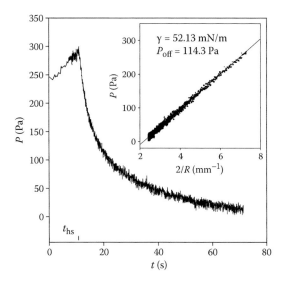

FIGURE 13.1 Pressure signal acquired during the continuous growth of a droplet of hexane in water. The maximum in the signal corresponds to the transition through the hemisphere. The inset shows the same pressure data plotted versus the droplet total curvature, according to the pressure derivative method (Passerone et al. 1991). The interfacial tension is calculated as the slope of the linear relationship.

The choice of the dilation rate allows different adsorption characteristic times to be investigated. Under the hypothesis of diffusive surfactant transport, the variation of the adsorption for a droplet with changing (growing) radius $R(t)$ is given by (MacLeod and Radke 1993)

$$\frac{d\Gamma}{dt} = -R^2 \sqrt{\frac{D}{\pi}} \int_0^t \left(\frac{dC_0}{dt}\right)_\tau \frac{d\tau}{\sqrt{\int_\tau^t [R(\xi)]^4 \, d\xi}} - \Gamma \frac{d\ln A}{dt}. \tag{13.3}$$

This equation can be utilized as the basis for the interpretation of results from growing drop experiments.

The CP tensiometry has also been applied to classical investigations of the adsorption kinetics, by measuring the dynamic interfacial tension during the aging of a "fresh" interface, that is, at $\Gamma(t = 0) = 0$. In a first approach, the fresh interface is obtained with good approximation by a fast and large expansion of the interfacial area: the expanded drop (ED) method (Liggieri et al. 1995; Ravera et al. 1991). After the expansion, the relaxation of the dynamic tension of the interface at rest is the result of the adsorption kinetics. An effective way to obtain such an expansion exploits a particular fluid-dynamic instability (Liggieri et al. 1990) that develops when the phase forming the droplet is sufficiently compressible. This can be achieved by trapping a small volume of gas in the liquid forming the droplet. By the ED method, it is possible to investigate adsorption processes with characteristic times from a few seconds to several minutes. In another approach, the fast formed drop method (Horozov and Arnaudov 1999), the liquid forming the drop is injected at a high flow rate though a capillary into the external fluid phase. After the abrupt termination of the flow, the liquid jet breaks, leaving a fresh and nearly hemispherical interface at the capillary tip. This method is more efficient in achieving data at short adsorption times. In addition, the smaller radius of the droplet, in comparison with the ED method, enhances the CP values. The method has, however, some drawbacks owing to the large flow rate and the invasiveness of the liquid jet, which can cause turbulence in the fluids, affecting the adsorption process. The utilization of CP tensiometers for the investigation of properties of droplet interfaces created at high flow rate will be discussed in more detail further below.

Finally, the possibilities offered by modern electronics to elaborate images in real time and control accurately the injection of small liquid volumes, for example, by piezoelectric actuators, made the CP tensiometer an ideal tool for investigations on interfacial dilational rheology, which is the response of interfacial tension to perturbations of the interfacial area. The most important application is based on the oscillating drop/bubble analyzer (ODBA) method for measurements of the surface dilational viscoelasticity as a function of the perturbation frequency.

The method relies on measurements of the interfacial tension response to sinusoidal low-amplitude perturbations of the interfacial area and can be in principle be implemented in any type of dynamic tensiometer. In practice, however, the fluid-dynamic constraints limit its applicability to low-frequency perturbations (below 0.1 Hz) in all tensiometers except the CP. The implementation of the method using CP measurements was conceived 40 years ago (Lunkenheimer and Kretzschmar 1975), but effectively applied only after the development of efficient CP tensiometers (Fruhner and Wantke 1996; Liggieri et al. 2002). Since then, the ODBA method has been upgraded by many authors (Alexandrov et al. 2009; Kovalchuk et al. 2000; Ravera et al. 2010), developing improved experimental setups and different approaches for the extraction of the viscoelasticity from the raw pressure data obtained by the experiments. In fact, the major difficulty in the high-frequency measurements arises from the presence of spurious pressure effects superimposing the CP. Suitable experiment models and specific calibration procedures are therefore needed to obtain reliable values of the rheological parameters. So far, the ODBA method implemented in CP provides accurate measurements of the dilational viscoelasticity up to frequencies on the order of 100 Hz for liquid–air interfaces and about 20 Hz for liquid–liquid interfaces. The limitations are given mainly by presently unavailable theoretical models suitable to interpret the pressure signals in the non-linear regimes of oscillations. This fills partially the measurement gap previously existing in the frequency range between 0.1 Hz (attainable with drop profile or Langmuir trough instruments) and 1000 Hz (about the bottom limit for thermo wave techniques).

In the last few years, different setups have been developed integrating CP tensiometry with drop profile tensiometers, mostly in commercial instruments (Georgieva et al. 2009; Javadi et al. 2012; Russev et al. 2008). Such solutions can be very convenient to exploit the capabilities of both techniques, for example, in order to measure the surface dilational viscoelasticity in different frequency ranges. Some attempts have also been made to apply the CP tensiometry to investigations of the tension and dilational properties of spherical liquid films (Bianco and Marmur 1993; Georgieva et al. 2009; Kim et al. 1997).

Finally, the CP tensiometry is probably the sole technique that does not rely on weight effects. For this reason, it is suitable to measure interfacial tensions at interfaces between isodense liquids or under weightlessness conditions, such as the conditions available on spacecrafts. In fact, on the basis of CP tensiometry, experiments on surfactant adsorption at liquid–liquid and liquid–air interfaces have been performed under microgravity conditions, during some NASA Space Shuttle missions. The experiments were performed with the fully automated instrument FAST (Facility for Adsorption and Surface Tension) (Krägel et al. 2005; Liggieri et al. 2005a). Exploiting the purely diffusive conditions in the absence of any convections and simplified fluid-dynamic conditions, these experiments have confirmed specific hypothesis and checked models on dilational rheology and adsorption kinetics (Kovalchuk et al. 2010; Liggieri et al. 2005b). Such investigations are going to be continued with new experiments onboard the International Space Station.

13.3 BUBBLE PRESSURE TENSIOMETRY

As mentioned above, the maximum bubble pressure tensiometry (MBPT) already belongs to the classical methods, although modern commercial instruments have been introduced on the market only about 20 years ago. Since then, the quality of data and also the scientific basis have been significantly improved. The best instruments of this measuring principle now provide adsorption data for less than 1 ms surface age.

As an example for the design of a bubble pressure tensiometer, the scheme of the BPA-1S is shown in Figure 13.2 (Fainerman et al. 2004; Fainerman and Miller 2004). This instrument is equipped with two sensors, one to measure the gas flow and the other to determine the CP and consequently the surface tension. The air is pumped through a filter and pneumatic system and finally forms the bubbles at the capillary tip immersed into the solution. All elements of the instrument are controlled by the interface and software running on a computer.

The typical measurement procedure consists in measuring the pressure as a function of time during a continuous formation of bubbles at a given gas flow rate. A mathematical algorithm allows extracting the maximum pressure values P from this signal, which then provides the surface tension γ at the given bubble formation time:

$$P = \frac{2\gamma}{r_{cap}} + \Delta\rho gh + P_{dyn}, \qquad (13.4)$$

where r_{cap} is the radius of the capillary, $\Delta\rho$ is the density difference between the solution and air, g is the acceleration due to gravity, and h is the immersion depth. The term P_{dyn} on the right-hand side of Equation 13.4 is a dynamic pressure contribution.

The corresponding time at the moment of pressure maximum is determined differently in the various instruments. The BPA-1S analyzes the pressure $P(t)$ and the gas flow rate $L(t)$ in the intervals between two successive maxima or minima, from which the total bubble time $t_b = t_l + t_d$, the lifetime t_l, and deadtime t_d of the bubbles

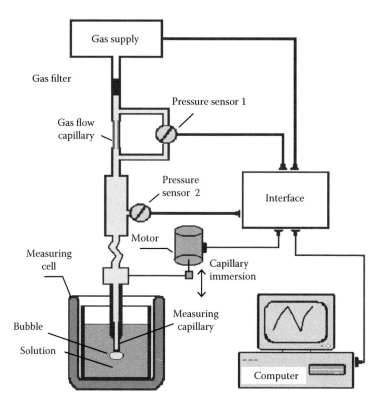

FIGURE 13.2 (See color insert.) Principle of the maximum bubble pressure tensiometer
BPA-1S (SINTERFACE Technologies, Berlin, Germany).

at the given experimental conditions are determined. In Figure 13.3, this procedure
is shown for the case of gas flow rate $L(t)$.

In the BPA-1S discussed here, a special procedure that gives access to measurements at very short adsorption times is implemented (Fainerman et al. 2004;
Fainerman and Miller 2004). In this procedure, the pressure P is measured as a
function of the gas flow rate L, which leads to results of the type given in Figure
13.4. As one can see, there are two regimes in this dependency. At high flow rates,
we have the so-called jet regime where single bubbles are only formed as the result
of a disintegration of the liquid jet in a certain distance from the capillary tip. With a
decrease in the gas flow rate, we reach a critical point from which individual bubbles
are formed. This bubble regime is suitable to determine the surface tension from the
measured maximum pressure value, using Equation 13.4. At the kink point, where
the two regimes meet, we can say that the bubble time is identical to the deadtime,
$t_b = t_d$, which provides us with a tool to determine the value of the deadtime. This
idea was first proposed by Kloubek (1972). The dependencies of $P(L)$ for capillaries
of different lengths are shown in Figure 13.4. The parameters on which the deadtime
directly depends have been discussed for example by Fainerman and Miller (2011).

All curves show a transition point, named critical point at P_c and L_c. The linear sections of the curves at $L > L_c$ can be described by the Poiseuille equation

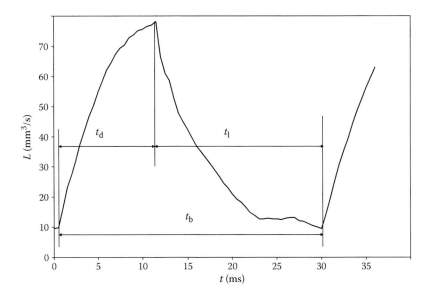

FIGURE 13.3 Dependence of $L(t)$ for a given experimental condition with the definition of the respective bubble times.

and correspond to the jet regime of gas expansion from the capillary. For $L < L_c$, the injection of the gas into the liquid results in the formation and separation of individual bubbles with $t_1 > 0$. In the transition point ($L = L_c$), the lifetime vanishes and the time interval between two successive bubbles becomes equal to the deadtime.

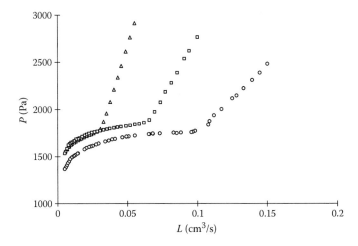

FIGURE 13.4 Dependence of pressure in the measuring system on the air flow rate for a 0.2% Triton X-100 solution, $r_0 = 0.0084$ cm, $l = 6$ cm ($\triangle$), 3 cm ($\square$), and 1.5 cm ($\bigcirc$). (From Fainerman, V.B. and Miller, R., *Bubble and Drop Interfaces, Vol. 2, Progress in Colloid and Interface Science*, pp. 75–118, Brill Publishing, Leiden, 2011.)

As mentioned above, the measured pressure P is influenced by aerodynamic and hydrodynamic effects summarized in P_{dyn}, which depends on the experimental conditions. If not taken into account, we obtain apparent dynamic surface tensions that are typically higher than the correct values. For pure water at 20°C, different various volumes of the measuring system are shown in Figure 13.5 (Fainerman and Miller 2004). To eliminate errors caused by an incorrect capillary radius and bubble nonsphericity, a calibration with respect to water can be performed, using the known reference value of the surface tension of pure water, 72.75 mN/m, at 20°C.

For lifetimes longer than 300 ms, the measured surface tensions of pure water are correct within 0.1 mN/m. A much more detailed analysis of all possible effects is given by Fainerman and Miller (2011).

Figure 13.6 shows the dynamic surface tensions of micellar solutions of $C_{14}EO_8$ as functions of the effective lifetime (Fainerman et al. 2006a). The CMC of this surfactant is 9 μmol/l at 25°C (Ueno et al. 1981). These data demonstrate impressively the effect of micelle dissolution on the adsorption dynamics of surfactants from micellar solutions. A quantitative analysis of these effects was given recently in Danov et al. (2006a,b). Most probable are these theories involving at least two stages: a fast process governed by the separation of monomers from the micelle and a slow process corresponding to the complete dissolution of micelles. Hence, maximum bubble pressure, applied to the range of very short adsorption times, allows for an analysis of the micelle kinetics via measurements of dynamic surface tensions.

Another spectacular example of results obtained by bubble pressure tensiometry is those for short-time adsorption of highly concentrated protein solutions. Although discussed from time to time in literature, the phenomenon of negative surface pressure at short adsorption times was only recently investigated

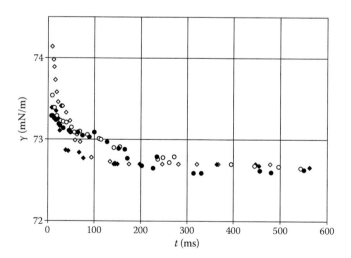

FIGURE 13.5 Dependence of apparent dynamic surface tension of water at 20°C and measuring system volume $V_s = 1.5$ ml ($\diamond$), 3.7 ml ($\blacklozenge$), 4.5 ml ($\circ$) and 20.5 ml ($\bullet$). (From Fainerman, V.B. and Miller, R., *Bubble and Drop Interfaces, Vol. 2, Progress in Colloid and Interface Science*, pp. 75–118, Brill Publishing, Leiden, 2011.)

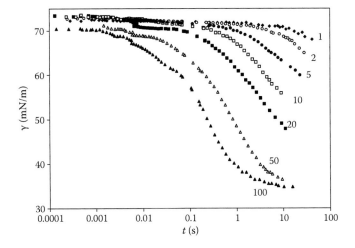

FIGURE 13.6 Dynamic surface tension for micellar solutions of C14EO8 as a function of the effective surface lifetime; the labels at the curves correspond to the surfactant concentrations in multiples of the CMC, which is 9×10^{-6} mol/l. (From Fainerman et al. 2006a.)

more systematically by using the BPA-1S. Selected data of the measurements for β-casein solutions are shown in Figure 13.7 (Ulaganathan et al. 2012). As one can unambiguously see, at short adsorption times, the surface first increases, passes through a maximum, and then decreases as it is expected for systems containing surface-active molecules.

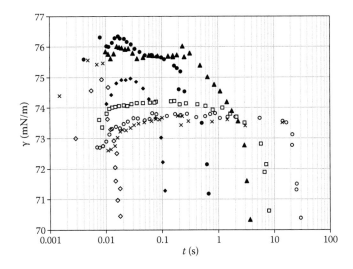

FIGURE 13.7 Dynamic surface tension of β-casein solutions at pH 7 for the following concentrations: (×) 10^{-6} mol/l, (○) 2×10^{-6} mol/l, (□) 5×10^{-6} mol/l, (▲) 10^{-5} mol/l, (●) 2×10^{-5} mol/l, (◆) 5×10^{-5} mol/l, (◇) 10^{-4} mol/l. (From Ulaganathan, V. et al., *Colloids Surf. A*, submitted for publication.)

13.4 GROWING DROPS AND BUBBLES

Growing drops and bubbles is one of the most important and interesting methods for interfacial property measurements according to the physics of the technique, which corresponds to a similar phenomenon in real processes. Two-phase flow processes such as dispersions, extraction, foam and emulsification, spraying, and printing are involved in drop and bubble formation phenomena. Therefore, the experimental results based on this technique, in addition to standard output after data analysis, also provide direct understanding of the effects of interfacial properties on drop and bubble shape formation, size distribution, and the detachment process.

13.4.1 THE DROP VOLUME METHOD

The drop volume or weight method can be considered as the simplest traditional method on the basis of this approach (Javadi et al. 2011). This method was derived first as the stalagmometer counted only the number of drops and estimated the liquid's composition according to the effective surface tension. It was possibly used for the first time by the pharmacist Tate in 1864, who used the number of drops as a measure of a certain liquid volume in order to dose liquid medicine. Tate postulated that the weight W of a drop detaching from a capillary of the size r_{cap} is proportional to the product of capillary radius and surface tension γ, known now as the law of Tate:

$$W = 2\pi r_{cap}\gamma. \tag{13.5}$$

This law has been used for a long time to determine the surface tension of liquids despite the fact that it is only a rough estimation. Lohnstein (1906a,b, 1907, 1908, 1913) particularly criticized it and made a series of calculations to establish a basis for an accurate theory. Lord Rayleigh (1899) has already suggested that the basic equation (Equation 13.5) has to be corrected by a factor F in order to obtain a more accurate relationship:

$$W = 2\pi r_{cap}\gamma F. \tag{13.6}$$

This approximation was developed to the following formulation for the modern drop volume tensiometry:

$$\gamma = \frac{\Delta\rho g V}{2\pi r_{cap} F}. \tag{13.7}$$

The measuring procedure is realized by a precise dosing system that allows an accurate measurement of the drop volume V during the continuous formation of drops at a capillary, which is needed to determine the surface tension γ.

Because of its simplicity and applicability for both liquid–gas and liquid–liquid interfaces, the drop volume method is still a frequently used experimental methodology. However, one of the strongest limitations is the short time range in which it provides reliable experimental data. Therefore, there is a wide variety of drop volume tensiometers described in literature.

13.4.2 GROWING DROP CP MEASUREMENTS

Nowadays, advanced experimental growing drop instruments that directly measure the CP as shown by Javadi et al. (2010) recently make them an alternative tool for measuring at fast dynamic conditions. CP measurements during the growth or oscillation of drops are most suitable for fast dynamic measurements, as it was already summarized above in Section 13.2. Studies under highly dynamic conditions require a lot of effort in the development of the experimental hardware/software and of the theoretical basis for extracting accurate CP data from the measured total pressure, in particular for high liquid flow rates (e.g., $Q > 2$ mm^3/s, for capillary tip diameters of 0.5 mm). Here, the hydrodynamic pressure loss due to viscosity, inertia and drag forces, and the deformation of the droplet affects the measured total pressure significantly (Javadi et al. 2010, 2012). Mechanical and electric noise can also produce large scattering of experimental data and are additional obstacles to overcome. Therefore, for obtaining high-quality data, an optimized instrument is required, which includes optimum capillary size, shape, and material, and efficient experimental protocols, as discussed recently (Javadi et al. 2010, 2012), and three different experimental protocols were proposed: (1) continuously growing drop (CGD), (2) pre-aged growing drop (PGD), and (3) stopped growing drop (SGD). The CGD is a procedure analogous to the well-known MBPT for liquid–gas interfaces; however, the processes of drop detachment and residual drop formation require a rather complex data analysis. For the PGD protocol based on a growing drop with an initially pre-established equilibrium adsorption layer, the complexities are less, but for very high surfactant concentrations, it is not optimal. The SGD protocol provides a drop with an almost fresh surface and the dynamic interfacial tension can be monitored in the absence of hydrodynamic effects. All three protocols complement each other and provide a good set of data for dynamic interfacial tensions at short adsorption times.

The interfacial tension values are determined from the measured total pressure P_{total} according to the Laplace equation,

$$\gamma = \frac{\Delta P_{total} - \Delta P_{static} - \Delta P_{hyd}}{2/r}. \tag{13.8}$$

The static pressure ΔP_{static} refers to the liquid level from the drop pole to the averaged position of the pressure sensor. The hydrodynamic pressure losses ΔP_{hyd} include the viscosity effects inside the capillary tip and the connected tubing and also the drag force of the second phase on the surface of the growing drop. While the drop volume method provides just one interfacial tension value for a drop at the final situation before detachment, the CP technique delivers a full set of data during the growth process at different drop age and size. Figure 13.8 shows the measured CP P_c and the drop radius

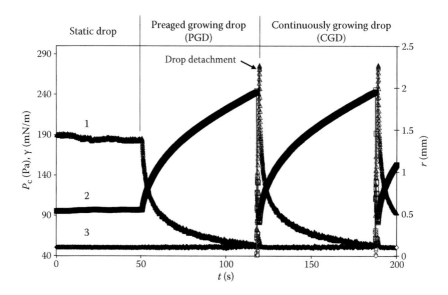

FIGURE 13.8 Water drop growing in hexane: CP P_c (1), drop radius r (2), and interfacial tension γ (3).

r for a water drop growing in hexane, and the resulting constant interfacial tension γ of 51 mN/m is observed. Figure 13.9 shows the results of a similar experiment when a water drop is continuously formed in hexane + 10^{-4} m/l span80, demonstrating the significant differences in the evolution of CP and interfacial tension values The results for fast dynamic measurements at high liquid flow rates are presented in Figure 13.10.

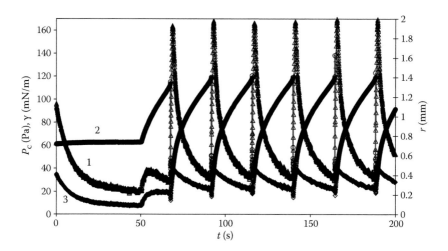

FIGURE 13.9 Water drop growing in hexane + 10^{-4} m/l span80: CP P_c (1), drop radius r (2), and interfacial tension γ (3). For the slowly growing drops presented in Figure 13.8 and in this figure, the hydrodynamic contributions are negligible; however, the drop cannot reach the state of a clean surface (negligible surface coverage).

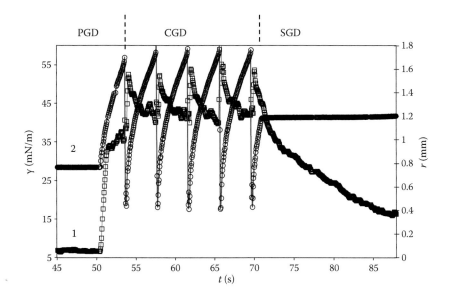

FIGURE 13.10 Water drop growing in hexane + 10^{-4} m/l span80: fast dynamic measurements at high growth rate, drop radius r (2), and interfacial tension γ (1).

13.4.3 THEORETICAL MODELING OF ADSORPTION MEASUREMENTS

A theoretical modeling for the CP method is required for a quantitative evaluation of the hydrodynamic contribution to the measured total pressure. However, additional efforts are required for the consideration of the adsorption process at and possible mass transfer across the interface. In some cases, this is also influenced by Marangoni convections. For SGD data, in the absence of hydrodynamic effects, these data can be described by the set of equations based on Fick's diffusion law in a spherical geometry (droplet) as discussed recently (Javadi et al. 2010). For growing drops, the convection term plays an important role in the mass transfer mechanism. The available theories are generally based on a radial flow for a non-aggregating, non-ionic surfactant inside and outside of a spherical drop (MacLeod and Radke 1993, 1994) as below:

$$\frac{\partial c}{\partial t} - v_r \frac{\partial c}{\partial r} = D \left(\frac{\partial^2 c}{\partial r^2} + \frac{2}{r} \frac{\partial c}{\partial r} \right), \tag{13.9}$$

where r is the drop radius, c is the concentration in the bulk, v_r is the fluid radial velocity, and D is the diffusion coefficient. This equation should be solved for both phases considering the related parameters (except of a negligible solubility in one of the phases). The effects of adsorption and surface coverage are involved in the modeling considering relevant initial and boundary conditions. Assuming a diffusion-controlled adsorption at the interface, and the possibility of transfer of surfactant between the two phases, the variation of adsorption Γ with time caused by the incoming/outgoing diffusive fluxes is

$$\frac{d\Gamma}{dt} = -D_1 \frac{\partial c}{\partial r}\bigg|_{r=R_1^-} + D_2 \frac{\partial c}{\partial r}\bigg|_{r=R_1^+}. \tag{13.10}$$

This equation is the boundary condition at the interface located at $r = R_1$. The boundary conditions at $r = 0$ and $r = R_2$ follow from the symmetry of the system and the infinity boundary condition:

$$\frac{\partial c}{\partial r}\bigg|_{r=0} = 0, \quad \frac{\partial c}{\partial r}\bigg|_{r=R_2} = 0. \tag{13.11}$$

There are some numerical solutions to this problem; however, an approximate comparison of the experimental results with the theory can be obtained considering the proposed model by van Uffelen and Joos (1994), which can be reasonable for low growth rates but not for fast dynamic conditions. This approximate solution for the adsorption process at the surface of a growing drop is based on the diffusion relaxation time defined by

$$\tau_D = \frac{1}{D}\left(\frac{d\Gamma}{dc}\right)^2, \tag{13.12}$$

where $d\Gamma/dc$ is obtained from the adsorption isotherm of the respective surfactant. The approximate solution was derived for linearized conditions and is therefore applicable for small deviation from equilibrium. With a radial flow $V_r = r\theta$, θ is the surface dilatation rate. For growing spherical drops with a constant flow rate Q, the expansion rate α can be estimated by $\alpha = Q/V_0$ (V_0 is the initial drop volume). The final equation for the variation of interfacial tension during the drop growth reads

$$\Delta\gamma(t) = \varepsilon_0 \sqrt{\tau_D} \sqrt{\frac{7\pi\alpha}{12}} \frac{(1+\alpha t)^{2/3} - 1}{\sqrt{(1+\alpha t)^{7/3} - 1}}, \tag{13.13}$$

with $\varepsilon_0 = -(d\gamma/d \ln \Gamma)$ being the Gibbs elasticity.

A better description of the data can be expected via a direct numerical solution as proposed by Macleod and Radke (1993, 1994) or by a rigorous analysis solving the Navier–Stokes equations together with the transport of surfactant in the bulk and at the interface, as discussed in general by Alke and Bothe (2009).

13.5   OSCILLATING DROPS AND BUBBLES

It is seen from the previous consideration that the dynamics of a drop or bubble interface is determined by a coupling of the transport of surfactant with the transport of momentum. The description of this complex problem can be simplified essentially when small-amplitude oscillations of the interface are considered. In this case, the governing equations can be linearized with respect to small deviations of the system parameters from their equilibrium values and then a linear analysis can be applied. An advantage of this approach is that it allows to obtain such important characteristic of the interfacial layer as its dynamic viscoelasticity modulus $E = d\gamma/d\ln A$, where $d\gamma$ is the surface/interfacial tension variation induced by the relative surface area variation $d\ln A$. The modulus E represents the dynamic properties of the interface and depends on the relaxation processes within the adsorption layer. It is intensively studied in the recent years for various types of interfaces formed between different liquid media in the presence of different surface-active substances (Miller and Liggieri 2009; Miller et al. 2010). It is important that the modulus E obtained from oscillating drop or bubble experiments can be compared with the same modulus measured by other methods, such as elastic ring (Loglio et al. 1979, 1986; Miller et al. 1991), oscillating barrier (Lucassen and van den Tempel 1972a,b), and longitudinal or capillary waves (Lucassen-Reynders and Lucassen 1969; Noskov 1995; Noskov et al. 1999; Stenvot and Langevin 1988). Combining the results of several methods allows extending the frequency limits in which the dynamic viscoelasticity can be determined.

In the oscillating drop or bubble method, the liquid meniscus is forced to small-amplitude volume oscillations that result in corresponding CP oscillations. The CP, $P_c = \dfrac{2\gamma}{r}$, varies because of both the variation in the radius of curvature and the variation in interfacial tension:

$$\delta P_c(t) = -\frac{2\gamma_0}{a_0^2}\delta r(t) + \frac{2}{a_0}\delta\gamma(t), \qquad (13.14)$$

where a_0 and γ_0 are the equilibrium curvature radius and interfacial tension, whereas $\delta r(t) = r(t) - a_0$ and $\delta\gamma(t) = \gamma(t) - \gamma_0$ are the respective variations.

Applying the Fourier transform to Equation 13.14, we can write for the frequency domain:

$$\delta P_c(i\omega) = \left(-\frac{2\gamma_0}{a_0^2}\frac{dr}{d\ln A} + \frac{2E(i\omega)}{a_0}\right)\delta\ln A(i\omega), \qquad (13.15)$$

where the viscoelasticity modulus $E(i\omega)$ substitutes $\delta\gamma/\delta\ln A$ and $dr/d\ln A$ substitutes $\delta r/\delta\ln A$. Equation 13.15 shows that the dilational viscoelasticity modulus, $E(i\omega)$, can be obtained from the measured CP response, $\delta P_c(i\omega)$, and the relative surface area variation, $\delta\ln A(i\omega)$, as

$$E(i\omega) = \frac{a_0}{2}\frac{\delta P_c(i\omega)}{\delta\ln A(i\omega)} + \frac{\gamma_0}{a_0}\frac{dr}{d\ln A}, \qquad (13.16)$$

provided that the geometrical characteristics of the meniscus (radius and area) and the equilibrium interfacial tension are known.

Historically, the first oscillating bubble instrument was proposed in 1970 by Kretzschmar and Lunkenheimer (1970). Initially the meniscus oscillations were induced by pressure variations in a closed gas chamber connected to a capillary that was submerged into the studied solution. Photometric detection of meniscus oscillations was employed (Figure 13.11). Several modifications of the oscillating bubble instrument were proposed in Chang and Franses (1994a,b), Johnson and Stebe (1994, 1996), and Karapantsios and Kostoglou (1999). With the availability of precise electric pressure sensors and piezoelectric translators, the measurements became much more efficient (Javadi et al. 2012; Kovalchuk et al. 2004; Liggieri et al. 2002; Wantke et al. 1998). In the later versions of the instrument, the meniscus oscillations are excited by pressure variations in a closed liquid chamber instead of via the gas chamber (Figure 13.12). Herewith, the meniscus volume variation is not directly measured but is determined in an indirect way from the piezoelectric rod movement (Figure 13.12).

For sufficiently small oscillation frequencies (usually up to several tens of hertz), the meniscus volume variation is approximately equal to the volume variation produced by the piezo-piston. However, with increasing frequency, they become different because of the volume compressibility of the liquid and deformability of the cell walls (Kovalchuk et al. 2002). In this case, the meniscus volume oscillates with amplitude and phase that are different from the piston oscillations. Quantitatively, the liquid compressibility and deformability of walls can be accounted for by using either the effective cell elasticity (Kovalchuk et al. 2002, 2004) or the compressibility of an effective amount of gas (Liggieri et al. 2002; Ravera et al. 2005), which can be measured in a separate calibration experiment. When the meniscus volume variation is obtained, the relative surface area variation, $\delta \ln A$, used in Equations 13.15 and 13.16, can also be found from the known geometry of the meniscus.

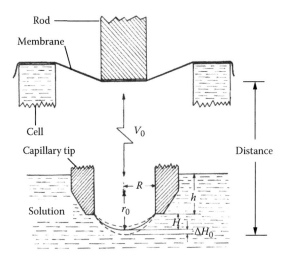

FIGURE 13.11 Oscillating bubble instrument with excitation in a closed gas chamber. (From Lunkenheimer, K. et al., *Colloids Surf.*, 8, 271, 1984.)

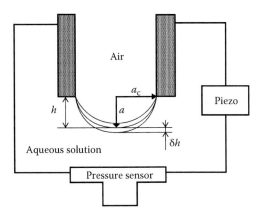

FIGURE 13.12 Oscillating bubble instrument with excitation in a closed liquid chamber. (From Kovalchuk, V.I. et al., *J. Colloid Interface Sci.*, 252, 433, 2002.)

The CP variation in the drop or bubble, δP_c, is determined also in an indirect way because the pressure measured by the pressure sensor additionally includes the pressure drop on the capillary tip and the dynamic pressure of the liquid adjacent to the meniscus. These additional pressure contributions can be either estimated theoretically, when the capillary tip geometry is precisely known, or obtained from another set of calibration experiments with pure liquids (Kovalchuk et al. 2004; Wantke et al. 1998).

The considered procedure requires that the meniscus shape is close to spherical, not deformed because of gravity. Therefore, the drops and bubbles used in these experiments should be small, about 500 μm in diameter or smaller. Precise manipulation with such small volumes is rather difficult and requires special equipment and software (Javadi et al. 2012).

The meniscus shape should remain close to a sphere also during the oscillations; that is, only zero-mode (radial) oscillations are allowed and any higher-mode oscillations should be avoided. According to Rayleigh's equation (Shen et al. 2010), the respective frequency limit depends on the drop radius, surface tension, and liquid density. For an aqueous solution with a surface tension of 50 mN/m, the free oscillation for the mode with $l = 2$ has a frequency that increases from 285 to 805 Hz with the decrease of the drop radius from 0.5 to 0.25 mm. The frequency of forced drop oscillations in surface rheological experiments should be about one order of magnitude smaller.

It is also important that the three-phase contact line remains fixed at the capillary tip during the meniscus oscillations. This can be achieved by making a sharp edge at the tip and high wetting contrast between the external and internal surface of the capillary.

As an example, Figure 13.13 shows the results of the measured dilational viscoelasticity modulus of dodecyl dimethyl phosphine oxide (C_{12}DMPO) solutions obtained with the oscillating bubble method under microgravity conditions (STS-107 NASA mission, see also Section 13.7.2) (Kovalchuk et al. 2004). The lines in this

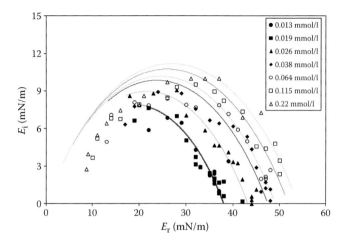

FIGURE 13.13 **(See color insert.)** Cole–Cole diagram for the complex dilational visco-elasticity modulus for different $C_{12}DMPO$ concentrations obtained from oscillating bubble experiments, as presented in Kovalchuk et al. (2004). The lines are the theoretical predictions according to the Lucassen–van den Tempel model.

figure show the theoretical predictions according to the Lucassen–van den Tempel model that assumes diffusion-limited adsorption of the surfactant. A rather good correspondence between the experimental results and theoretical prediction shows that this assumption is valid for $C_{12}DMPO$ solutions with concentrations sufficiently far from the CMC.

13.6 RISING BUBBLES

13.6.1 FORMULATION OF THE PROBLEM

Studies on bubble motion in aqueous media could be traced back to Leonardo Da Vinci who has probably been the first to report on the helical motion of rising bubbles (Prosperetti 2004). The rising of bubbles in liquid media is an important step in the flotation process as it involves the adsorption–desorption kinetics onto the bubble surface that differs from that of a static bubble. The convective-diffusion kinetics that involves adsorption and desorption exchange with the subsurface is accompanied by the hydrodynamics of the subsurface around the bubble. Frumkin and Levich (1947) have shown that the adsorption layer of a rising bubble has a surface concentration gradient. The physicochemical nature of this phenomenon has been further elaborated by Levich (1962). The movement of the bubble induces a non-uniform distribution of adsorbed material over the bubble surface: the rear pole is enriched (highest surface coverage) as compared to the leading pole. The surface concentration gradient generates a Marangoni stress that results in a retardation of the bubble motion. A stagnant cap is formed at the bottom pole once the bubble reaches a steady-state motion (Cuenot et al. 1997; Malysa et al. 2005). Such layer is termed dynamic adsorption layer (DAL) as it differs from that of a stationary bubble surface. The understanding of the dynamics of such systems has been extended by Derjaguin and Dukhin in the 1960s, who have described the heterogeneity of DAL at lower surface coverage owing to a weakly retarded surface (Dukhin et al. 1998). The theory of the stagnant cap formation and its angular dependence has been further developed by Sadhal and Johnson (1983) and latest developments in the DAL theory have been addressed by He et al. (1991). The most recent detailed review by Malysa et al. (2011) highlights the milestones of the knowledge on rising bubbles gained so far and represents an excellent guide for the reader.

13.6.2 Experimental Setup

The setup consists of a square glass column with a cross-section area of 40×40 mm and a height of 50 cm, as shown schematically in Figure 13.14. A capillary with an inner diameter of 0.075 mm is mounted to the bottom and connected to a syringe pump for air inflow. A stroboscope that produces flashes of a certain frequency is used as a light source and a digital camera records the video information. The extracted frames are analyzed by an image analysis software. The distance between the bottom poles of subsequent images of a bubble in single frames L_b is the distance traveled by the bubble in the time interval of subsequent flashes; for example, for

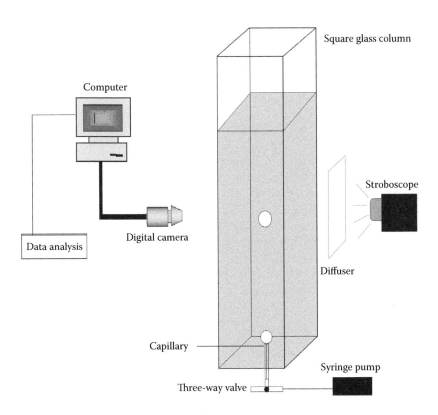

FIGURE 13.14 Schematic of a rising bubble column.

a frequency of 100 Hz, it gives 0.01 s, and $U_L = L_b/0.01$ s gives the local velocity. Image analysis allows the additional estimation of the vertical d_v and horizontal d_h diameters of a non-spherical bubble. Therefore, the deformation of the bubble at each height can be analyzed. The time of bubble formation and its diameter can be controlled by the gas flow rate and the size of the capillary, respectively.

The detachment of the bubble occurs once the buoyancy force overcomes the adhesion force between the bubble and the orifice. Tate (1864) (Adamson 1990) has derived a law "…the weight of the drop of liquid is in proportion to the diameter of the tube in which it is formed," which relates the detaching force of a pending drop to the adhering capillary force at the orifice. After modification, taking into account the buoyancy force and size of the capillary orifice, it can be used for estimation of the diameter of detaching spherical bubble d_b as the rising bubble. d_b depends on the diameter of the capillary d_c and the surface tension γ under the assumption that the buoyancy force is in equilibrium to the surface tension at the moment of detachment:

$$d_b^3 = 6d_c \gamma / g \Delta \rho, \tag{13.17}$$

where $\Delta \rho$ is the density difference between the gas and the liquid phases.

During the bubble growth in a surfactant solution, adsorption/desorption processes occur on the surface; that is, a DAL is formed (Dukhin et al. 1998). The kinetics of these processes depends on the concentration and the diffusion coefficient of the surfactant. Warszynski et al. (1998) and Jachimska et al. (1998) elaborated an adsorption kinetics model for growing bubbles, by means of which the degree of adsorption θ at the detaching bubble surface can be calculated. The surface tension of the growing bubble is expressed by

$$\gamma = \gamma_0 + RT\Gamma_\infty [\ln(1 - \theta) + \theta^2 H/RT], \tag{13.18}$$

where γ_0 is the surface tension of water, Γ_∞ is the maximum surface concentration as $\theta = \Gamma/\Gamma_\infty$, and H is the Frumkin interaction parameter. A more detailed description of this model can be found in Dukhin et al. (1998), Jachimska et al. (1998), and Warszynski et al. (1998).

From a high-speed camera recording, the diameter of the detaching bubble (assumed to be spherical) can be accurately measured, and therefore, γ and θ can be calculated using Equations 13.17 and 13.18. The formation of a DAL and the degree of surface coverage for a detaching bubble depend on the bubble growth rate and the surfactant adsorption kinetics.

13.6.3 LOCAL VELOCITY PROFILES

13.6.3.1 Rising Bubbles in Water

As explained above in the experimental section, the local velocity U_L of the rising bubble at distance L (or time t) from the capillary tip is measured, thus plotting U_L as a function of L (or t) gives the local velocity profile of the rising bubble. Once the bubble is detached in clean water, it accelerates rapidly and its spherical shape is deformed. At a certain distance, the bubble reaches a terminal (constant) velocity U_T (corresponding to a plateau in the velocity profile, see Figure 13.15) that is achieved when the drag force acting against the buoyancy force attains equilibrium. The period of acceleration has been quantitatively studied by Krzan and Malysa (2002), Krzan et al. (2007), and Malysa et al. (2011), and values in the range between 600 and 900 cm/s^2 have been reported. The bottom graph in Figure 13.15 shows profiles of the bubble deformation (defined by the ratio d_h/d_v). A good correlation between both dependences is observed, illustrating that bubble deformation in water increases during the acceleration and remains constant at terminal velocity. As it will be discussed below, this finding is valid for surfactant solutions as well.

For calculating U_T, contributions of different origin such as gravity, drag, lift, surface tension, viscous effects, virtual mass, and history force should be considered in the force balance equation. Concerning this subject, the main efforts in literature aim at finding a proper and general drag coefficient equation that can be inserted into the force balance equation. Rising bubble modeling has started with very simple assumptions for getting an analytical solution of the Navier–Stokes equation. Hadamard (1911) and Rybczynski (1911) independently applied an approach based on Stokes' law considering a small bubble with a rigid shape in a laminar liquid flow (Re ≪ 1) and estimated the terminal velocity as

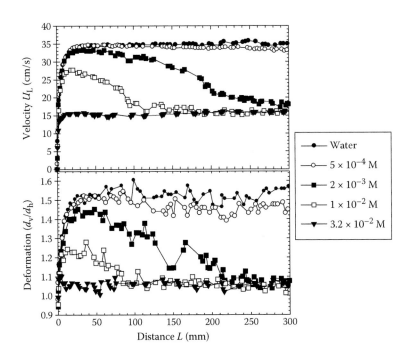

FIGURE 13.15 Local velocity and deformation profiles for rising bubbles in water and *n*-butanol solutions. (Redrawn from Krzan, M. and Malysa, K., *Colloids Surf. A*, 207, 279, 2002.)

$$U_T = \frac{g d_e^2 (\rho_l - \rho_g)}{6\mu_l} \frac{\mu_l + \mu_g}{2\mu_l + 3\mu_g},$$

(13.19)

where g is the gravity acceleration, d_e is the equivalent diameter of the spherical bubble, and μ_l, μ_g, ρ_l, and ρ_g are the densities and viscosities of the liquid and the gas phase, respectively.

In the laminar regime (Re $\ll$ 1), the Hadamard–Rybczinski equation (Equation 13.19) predicts well the terminal velocity (Hadamard 1911; Rybczynski 1911). Bubbles of diameters <0.1 mm satisfy this condition and Parkinson et al. (2008) have reported a very good fit between Equation 13.19 and experimental data measured in "ultraclean" water.

However, there is no exact solution of the Navier–Stokes equations for more intensive flows (i.e., for Re > 1), which is the case for larger bubbles. Nevertheless, there are attempts in literature that deal with this problem (Clift et al. 1978; Davies and Taylor 1950; Duineveld 1995; Levich 1962; Magnaudet and Eames 2000; Moore 1963, 1965; Rodrigue 2001). For instance, Figure 13.16 illustrates some of these models in terms of the drag coefficient $C_D \sim 1/U_T^2$ as a function of the Reynolds number (Malysa et al. 2011).

Figure 13.15 shows the local velocity profiles $U_L(L)$ for rising bubbles with $d_b = 1.5$ mm and a bubble formation time of 1.6 s in water and n-butanol solutions of different concentrations (Krzan and Malysa 2002). At certain times after detachment, terminal velocities of $U_T = 34.8 \pm 0.2$ cm/s for water and $U_T = 15$ cm/s for n-butanol concentrations beyond 32 mM are attained. A further increase in the concentration has a negligible effect on U_T caused by the rigidity of the surface, making the bubble behave like a solid sphere. The profile for the lowest measured concentration

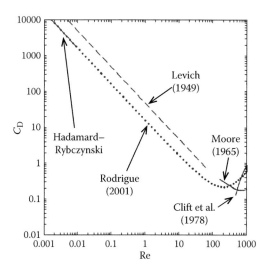

FIGURE 13.16 Drag coefficient versus Reynolds number plot for rising bubble in pure liquids. (From Malysa, K. et al., *Progress Colloid Interface Science*, Vol. 2, p. 243, Brill, 2011.)

(Krzan and Malysa 2002) follows an identical course as that for pure water, showing that no detectable adsorption occurs in this case. For intermediate concentrations, the velocity profiles exhibit a maximum followed by a monotonous decrease of U_L until a certain U_T is attained. An increase of the surfactant concentration leads to a decrease of the height of the maximum and its position is shifted toward shorter distances (times).

13.6.3.2 Rising Bubbles in Surfactant Solutions: Experimental Example and Discussion

The bubble motion in liquids can be divided into two limiting cases: "clean" bubble (in water) and bubble with steady-state DAL immobilizing the bubble surface. The first case, as discussed above, can be analytically described by the Hadamard–Rybczinski equation (Equation 13.19) for small bubbles and the second case has been described by Frumkin and Levich (1947), Levich (1962), and Dukhin et al. (1998). Intermediate cases correspond to a DAL formation. An experimental evidence for that is the existence of a maximum followed by a deceleration stage in the velocity profiles. During the DAL formation, a stationary non-uniform distribution of the surface coverage at the gas/solution interface is not established, while beyond a certain concentration, this does happen and the bubble reaches a terminal velocity right after the acceleration stage.

The adsorption at the leading pole and the desorption from the rear pole of the bubble occur during the bubble motion and are caused by interfacial convection, and consequently, the adsorbed surfactant molecules are pushed toward the bottom pole, generating a surface tension gradient. If Γ_{top}, Γ_{bottom}, and Γ_{eq} are the surface concentrations at the top and bottom poles, and at equilibrium, respectively, then $\Gamma_{top} = \Gamma_{bottom} = \Gamma_{eq}$ is valid for the stationary bubble condition, whereas for rising bubbles, we have $\Gamma_{top} < \Gamma_{eq} < \Gamma_{bottom}$ (Dukhin et al. 1998; Frumkin and Levich 1947). Therefore, the bubble deceleration can be explained by a Marangoni stress that supports the drag force exerted on the bubble (Dukhin et al. 1998; He et al. 1991; Levich 1962; Stebe and Maldarelli 1994).

The shape deformation of bubbles rising in pure liquids ($\gamma = \text{const}$) depends on its size. According to the Laplace law, the smaller the radius, the higher is the CP inside the bubble that resists a deformation. At high surfactant concentrations, the deformation is smaller as the bubble surface is immobile and acts like a rigid sphere. Velocity profiles and shape oscillations have been extensively studied by Krzan and Malysa (2002) and Krzan et al. (2004, 2007) for bubbles of various sizes and times of bubble formation in different surfactant solutions.

13.7 FILM TENSIOMETRY

13.7.1 State of the Art

Thin liquid films are the main elements of foams and emulsions. For sufficiently thick films, where the disjoining pressure is negligible (usually for thicknesses $h \geq 30$ nm), the film tension is equal to the sum of interfacial tension acting at the two single interfaces (Georgieva et al. 2009; Kim et al. 1997; Soos et al. 1994). Expansion of a film results in expansion of its interfaces, which is accompanied by an increase of the interfacial tension and subsequent relaxation to a new equilibrium state. It is important, however, that the film elasticity cannot be reduced simply to the sum of two surface elasticities as, in contrast to single interfaces being in contact with an infinite volume of solution, the number of surfactant molecules (or other surface-active materials) in a thin film is limited (Georgieva et al. 2009; Kovalchuk et al. 2009; Lucassen 1981). Moreover, the relaxation processes in thin liquid films are coupled with hydrodynamics. Thus, though showing some common features with single interfaces, thin liquid films are more complicated systems and require specific methods for their investigations.

It is widely recognized that the viscoelastic properties of liquid films are very important for the stability of foams and emulsions. Unfortunately, despite their importance, studies on the viscoelasticity of liquid films are rather scarce. The idea that thin liquid films should be characterized by certain elasticity goes back to the works of Gibbs (1961). In the simplest form, neglecting the solution non-ideality and exact positions of the Gibbs dividing surfaces, the film elasticity can be written as (Kruglyakov and Exerowa 1997; Lucassen 1981):

$$E_f = \frac{2E_0}{1 + \dfrac{h}{2}\dfrac{dc}{d\Gamma}}, \tag{13.20}$$

where E_0 is the limiting surface elasticity, h is the film thickness, and $\Gamma(c)$ is adsorption, which is a function of the bulk concentration c.

Initial experiments were performed with flat liquid films (Krotov et al. 1972a,b; Mysels et al. 1961; Prins et al. 1967; van den Tempel et al. 1965). They confirmed in general the validity of Equation 13.20, though the experiments met some difficulties related to the surfactants' purity—the influence of the Plateau borders supporting the films and other problems. It was shown recently that measuring the CP of spherical foam or emulsion films formed at a capillary tip is a very promising method to study their elasticity (Bianco and Marmur 1993; Gabrieli et al. 2012; Kim et al. 1997; Makievski et al. 2005; Soos et al. 1994). Such methodology should allow more precise experiments under properly selected conditions.

It is seen from Equation 13.20 that similar to the viscoelasticity of single interfaces, the film elasticity depends on $E_0 = d\gamma/d\ln\Gamma$ and $d\Gamma/dc$, which are determined by the particular surface tension isotherm $\gamma(c)$. Thus, to predict the film elasticity, we should precisely know this isotherm. In particular, as shown by surface rheological studies, the intrinsic compressibility of the adsorption layer should be taken into account (Kovalchuk et al. 2009). It is also very important to know the surfactant concentration in the film, which depends on the way the film is formed and can be different from the initial concentration of the original solution owing to the depletion effect (Kovalchuk et al. 2009). These depletion effects are more significant for low surfactant concentrations, which probably explains the smaller stability of films and foams produced from solutions of surfactants with longer hydrophobic chains.

A very big problem for studies of film rheology is the simultaneous expansion of the film and the surfaces of the Plateau borders supporting the film (i.e., pulling the new film from the solution). The relative expansion in the film and in the Plateau borders is determined by the ratio of the film elasticity and the surface elasticity of the solution (Lucassen 1981). As the surface elasticity is usually smaller than the film elasticity (because of larger solution depth), the expansion in the Plateau borders can represent a significant part of the total expansion. A non-uniformity of real films can create additional difficulties (Kovalchuk et al. 2009). Therefore, it is very important to have the possibility to measure independently the local thickness of the film. Such a possibility is foreseen in future microgravity experiments discussed in the next subsection.

13.7.2 APPLICATIONS IN MICROGRAVITY EXPERIMENTS

Transient and oscillatory CP experiments were conducted aboard orbiting vehicles, during the STS-95 and the STS-107 NASA missions, by using the flight module FAST, within the framework of the project FASES (Fundamental and Applied Studies of Emulsion Stability) promoted by the European Space Agency (Kovalchuk et al. 2010; Loglio et al. 2005). Such microgravity experiments pursue the objective of establishing a link between emulsion stability and the physicochemical characteristics of droplet interfaces and of emulsion films, taking advantage of the favorable weightlessness conditions in space.

Future microgravity experiments aboard the International Space Station are foreseen to be accomplished by means of an improved CP flight module (FASTER). An additional module (LIFT, acronym for LIquid Film Tensiometer) is also under development. The LIFT and FASTER combined modules are designed for performing measurements of interfacial rheology on thin spherical aqueous emulsion films (Makievski et al. 2005), generated inside a hydrocarbon matrix, at different surfactant concentrations and at various temperatures, with the aim of further pursuing fundamental achievements in the field of thin emulsion films.

The value of the film thickness is determined by two complementary methods, specifically by the well-established interferometric technique and by a new application of the optical evanescent wave effect (Gabrieli et al. 2012). Such simultaneous methodologies provide a plurality of consistent data on the non-homogeneous film thickness and on the film thinning trend, at the nanoscale level.

The scheme in Figure 13.17 shows the special designed cell dedicated to the formation of a thin spherical emulsion film. The film is formed from an aqueous phase inside a hydrocarbon matrix. The cell consists essentially of three chambers. The main chamber (upper chamber, A) contains the matrix hydrocarbon and holds a coaxial double capillary in a central position. The other two chambers, lower chamber B and lateral chamber C, connected to the coaxial capillary by tubing and valves, contain the aqueous surfactant solution and the reservoir hydrocarbon, respectively. Two optical windows, at the opposite sides of the cell, allow visualization of the film aspect. An additional optical path, orthogonal in respect to the visualization optical pathway, conveys the coherent laser light beam of a Mach–Zehnder interferometer.

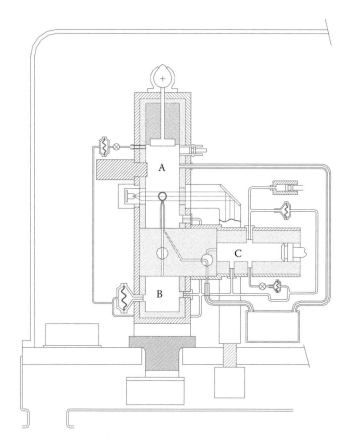

FIGURE 13.17 **(See color insert.)** Scheme of the measurement cell for the generation and observation of thin liquid films (partial drawings). The main compartment (upper chamber, A) contains the matrix hydrocarbon and holds a coaxial double capillary. The lower chamber, B, contains the hydrocarbon for inflating the film. The aqueous surfactant solution is stored in the lateral chamber, C, featuring a piston device for delivering a definite amount of liquid to the capillary.

The coaxial capillary has the purpose of transporting proper amounts of the two liquids (e.g., aqueous SDS solution and *n*-decane) for the generation of a liquid–liquid emulsion film, in the form of a spherical aqueous phase inside the hydrocarbon matrix. The coaxial capillary has a crucial partial surface coating, obtained with a hydrophobic perfluoropolymer. The external capillary has a hydrophobic external surface, up to the capillary tip, and an internal hydrophilic surface. In opposite mode, the internal capillary has a hydrophobic internal surface and a hydrophilic external surface, up to the capillary tip. A dedicated software generates the film and controls the film radius.

The synchronous measurement of the CP of the film, by means of a pressure transducer, and of the drop radius, either through image analysis technique or by extrapolation from the moved liquid volume, allows the dynamic interfacial tension to be easily calculated. This measurement technique offers a great flexibility and allows different aspects of interfacial physical chemistry and of the adsorption processes to be investigated by applying different controlled perturbations to the film. In particular, the measurement of the interfacial tension response, excited by harmonic variations of the interfacial area, gives access to a relevant parameter of emulsion stability, which is the steady-state (or transient) linear (or non-linear) dilational viscoelasticity.

13.8 COAXIAL DOUBLE CAPILLARY

This paragraph deals with coaxial double capillaries as a tool to form droplets, foam lamellas, or thin oil films at a capillary tip. The common uses of concentric capillaries in microfluidic devices to produce monodisperse multiple emulsions will not be touched here. The general idea is the drop formation at capillary tip by two different fluid delivering systems. This includes that one of the fluid can be a gas, for example, air. Therefore, the technique is very flexible and can be used to form drops of one liquid in another liquid, multilayered drops, thin liquid lamellas, or oil films, and allows for the internal liquid exchange inside a single drop. The advantage of this technique is that all processes can be controlled quantitatively by the drop shape analysis. Interfacial tension, interfacial area, and drop volume are the output parameters of such an analysis. This makes the technique suitable for use as a kind of mini-Langmuir trough, a tensiometer, and with sinusoidal drop/bubble/film oscillations as a dilatational interfacial rheometer. The first time this technique was described was by Cabrerizo-Vilchez et al. (1999), when they used it for subphase exchange experiments (Wege et al. 1999). Figure 13.18 shows the general concept for the formation of thin liquid films in air or in another liquid as an example proposed by Makievski et al. (2005).

For the applications of coaxial double capillaries in interfacial studies, only small amounts of liquids are needed. Therefore, the dosing of liquids is normally done by micro syringe pumps. In a previous step, first a drop of a certain size will be formed at the capillary tip pumping liquid 1 through the outer capillary. After this, an internal drop (step 1 in Figure 13.18) can be formed by pumping liquid 2 through the inner capillary up to a certain size. Simultaneously, liquid 1 can be sucked off by the outer capillary to control a constant total drop size (step 2 in Figure 13.18). Because of the mutual use of syringe pumps, one has the opportunity for an internal liquid exchange

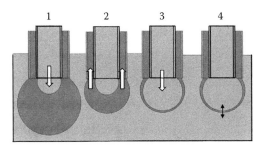

FIGURE 13.18 Stages of formation of an emulsion film at the tip of a coaxial double capillary system.

in the drop or to form thin layers of controlled thickness of liquid 1 in liquid 2 (step 3 in Figure 13.18). This can be done when the surrounding fluid is air or a third liquid. The internal fluid pumped through the inner capillary can be also air or any other gas. Any changes in drop size and shape can be controlled via the drop profile analysis. The general principle of this technique is very flexible and offers different experimental applications to study complex systems. In combination with the drop or bubble profile analysis, it can be used for the controlled formation of thin films, as shown in Figure 13.18. Once the thin film is formed, sinusoidal oscillations can be generated via the micro syringe pumps at low frequencies or via a piezo translator at higher frequencies (step 4 in Figure 13.18). Such experiments allow a direct characterization of foam or emulsion films under various conditions (Kovalchuk et al. 2009; Makievski et al. 2005). Another opportunity is the formation of a drop inside a drop, all this in a third matrix liquid, to mimic the situation in multiple emulsions. When the drops are formed via a special adapted CP cell, the coalescence behavior between two drops or bubbles can be studied. Such information is important to understand the complex processes in multi-interfacial systems of multiple emulsions.

The general idea of the internal subphase exchange in a single pendent drop is applied to numerous experimental protocols, such as film balance studies (Wege et al. 1999), sequential and simultaneous formation of mixed adsorption layers (Dan et al. 2012; Ganzevles et al. 2006; Kotsmar et al. 2008), desorption studies (Loglio et al. 2001), penetration experiments (Ferri et al. 2010), wash-off studies (Fainerman and Miller 2011), or even for multilayer formation (Ferri et al. 2005). For example, Wege et al. (1999) used the technique as a film balance for penetration studies of soluble surfactants into insoluble monolayers at the liquid–liquid interface. An insoluble monolayer is spread on the pure water surface of a previously formed drop by the outer capillary. Then, the monolayer-coated drop is immersed into an oil phase. Thereafter, the monolayer is brought to a certain compression state by reducing the interfacial area via the outer capillary. In this state, the subphase is exchanged by injection through the coaxial capillary system. Adding, for example, surfactants to the exchanged liquid allows studies of different interactions with the monolayer.

Another application for the coaxial double capillary pendent drop technique is the formation of multilayers at the drop surface. A strategy for the preparation of freestanding ultrathin nanocomposite films and elucidation of time scales required for layer-by-layer adsorption of polyelectrolytes onto a charged insoluble monolayer template has been proposed by Ferri et al. (2005). The advantage of this method is that the geometry of the drop interface permits the highly precise measurement of the dynamic surface tension and, therefore, the evolution of the surface free energy during multilayer assembling, providing insight into the dynamics of polyelectrolyte assembly. First, a lipid monolayer is deposited onto a pendent drop of aqueous saline solution by spreading and then compressed to acquire a defined surface charge density to the subphase of the drop. The subphase is then exchanged by injecting, in alternating sequence, solutions of polycations, saline, and polyanions, while maintaining constant drop surface area. This results in freestanding polymeric nanocomposite membrane of a thickness defined by the number of adsorption layers (Ferri et al. 2005).

Such internal liquid exchange experiments can also be applied to interfacial active systems that are soluble in water and form interfacial layers by adsorption (e.g., proteins). The general experimental procedure and protocol allowing drop volume exchanges during the experiments have been described in detail recently (Kotsmar et al. 2008). Briefly, a drop profile analysis tensiometer (PAT 1D, SINTERFACE Technologies, Berlin, Germany) is used for this kind of experiment. First, a drop of a solution drop-let is formed using the outer, primary syringe and allowed to equilibrate. The droplet subphase is then exchanged by injecting a second solution via the inner capillary by the secondary syringe. The PAT 1D software allows keeping the drop volume or interfacial area constant via feedback control using the drop profile and withdrawal of the liquid from the droplet interior at the same volumetric flow rate via the primary syringe. As the exchange proceeds, the concentration of the second liquid in the drop increases, while decreasing the concentration of the first liquid. A complete exchange can take sev-eral seconds, minutes, or even hours depending on the liquid exchange rate. The great advantage of the double capillary technique for the investigation of interfacial properties is based on the assumption that the material exchange between the bulk and interface is governed by an adsorption/desorption mechanism. Clearly, the diffusion and con-vection transport mechanisms support the adsorption process via exchanging the bulk. However, during the exchange process, the sublayer should not be disturbed by forced convection or turbulences; otherwise, the adsorbed layer cannot be formed correctly.

The method allows surfactant desorption kinetics and adsorption reversibility studies by desorption experiments (Kovalchuk et al. 2009). This can be accom-plished by forming a drop of surfactant solution via the outer capillary, allow-ing it to reach adsorption equilibrium and subsequently exchange the subphase by pure water via the inner capillary, which induces desorption of surfactant from the interface to the bulk, leading to an increasing surface tension. Although the subphase exchange rate is finite, the process of convection, which attenuates the diffusion barrier in the bulk, accelerates the rate of desorption kinetics and hence the surface tension increases.

For sequential and simultaneous formation of mixed adsorption layers, the coaxial double capillary offers unique protocols for mixed protein/surfactant systems. The timelines of the respective experiments are shown in Figure 13.19. In a sequential adsorption (Figure 13.19a), a droplet is formed with the outer capillary from a pure protein solution and allowed to reach the equilibrium state (stage I). The interface is completely covered by adsorbed protein molecules. The first bulk exchange with the pure buffer solution washes the protein molecules from the drop bulk, while keeping the drop volume constant (stage II). Because of strong adsorption of protein molecules at the interface (Fainerman et al. 2006b), desorption into the bulk of the solution is negligible and no significant increase in surface tension is observed. The result of this exchange is a drop with a protein-covered surface but containing no protein molecules in the drop bulk. In the second bulk exchange, the surfactant solution is injected into the drop through the inner capillary, replacing the buffer solution (stage III). The sur-factant molecules penetrate into the pre-adsorbed protein layers, modifying the surface layer structure. Hence, the protein/surfactant complexes are formed only at the inter-face. This leads to a decrease in surface tension, the absolute value of which depends on the type and concentration of the surfactant injected. A third bulk exchange is

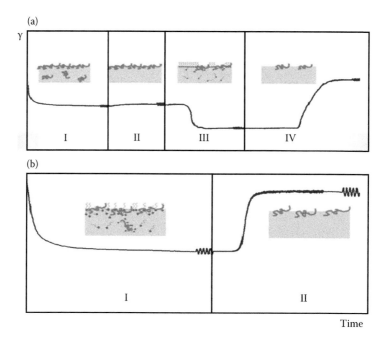

FIGURE 13.19 (See color insert.) Experimental protocol for sequentially (a) and simultaneously (b) formed mixed layers performed with a coaxial double capillary to measure dynamic surface tensions.

performed again with a pure buffer solution (stage IV), which replaces any molecules from the solution bulk. All protein molecules not displaced by the surfactant molecules still remain in the adsorbed state after this washing-off experiment. In contrast to proteins, the adsorption of surfactants is reversible, and hence, they desorb completely from the drop surface after the washing-off experiment, as it was described by Dan et al. (2012). After each stage of experiment, low-frequency surface layer oscillation can be performed, which provides additional information of the surface layer composition.

In case of simultaneous adsorption (Figure 13.19b), the protein and surfactant solutions are mixed together and a droplet is formed with the outer capillary from their mixed solution. Thus, protein/surfactant complexes are already formed in the bulk and they adsorb with the free surfactant molecules in a competitive manner (stage I). After this competitive adsorption, when the adsorption kinetics reached equilibrium, a washing-off experiment is performed against a pure buffer solution (stage II) so as to understand if the location of interaction has an impact on the nature and structure of the adsorption layers. Here, also, harmonic oscillations are made after each stage of the experiment in order to estimate the surface composition in terms of surface dilatational viscoelasticity.

A further application of coaxial double capillaries is to use them for the controlled formation of drops in another to mimic the situation in multiple emulsions. Figure 13.20 shows the schematic principle. The main idea of this experimental technique is to estimate the coalescence stability by measuring the CP in both droplets, as further discussed below.

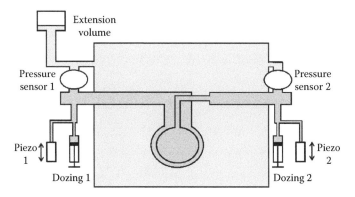

FIGURE 13.20 **(See color insert.)** Schematic principle of concentric drops formation at the tips of the coaxial double capillary, linked with respective pressure sensors.

13.9 DIRECT INTERACTION OF SINGLE DROPS AND BUBBLES

Emulsions and foams are dispersions of two immiscible fluids. Much work is dedicated to studies of the stabilization of emulsions and foams (Loglio et al. 2011). Emulsion stability is controlled by different processes, for example, Ostwald ripening, creaming, flocculation, and coalescence. Therefore, droplet coalescence is one topic studied intensively by different techniques. The basic concept of coalescence is to understand the stability of the thin liquid film formed between the two droplets or bubbles. The mechanisms for foam stability are similar to emulsions. For example, in a foamed liquid, thin foam lamellas are formed between gas bubbles. Because of the approach of the bubbles, the liquid is squeezed out from the lamella. The drainage of liquid in such thin films is partly controlled by the corresponding interfacial properties. Obviously, the lifetime of bubbles or droplets is directly related to the stability of the thin films formed between them. The evolution of the many contacts between the bubbles or droplets is a key factor for the behavior of the respective foams or emulsions.

The Drop Bubble Micro Manipulator (DBMM) is a new experimental tool for the quantitative analysis of interaction between two droplets or two bubbles or even between a single droplet and a bubble in a liquid medium. It was designed as an additional module for the standard commercial drop profile tensiometer PAT-1 (del Gaudio et al. 2008; Javadi et al. 2010). In brief, it consists of two so-called oscillating drop and bubble pressure analyzers, also available as a single additional module for PAT-1 (see the scheme of Figure 13.21a). Each of the two cells is equipped with a specially designed capillary, pressure sensor, piezoelectric translator, and a syringe dosing system, as was already described above in detail (Javadi et al. 2012). The two cells are mounted in such a way that one cell has a fixed position to the video camera of the PAT while the second cell can be moved in all directions by an xyz stage in order to bring the two capillaries face to face (Figure 13.21b).

The two capillaries are mounted in the way shown in Figure 13.21c and can be prepared in different ways depending on the purpose or materials used in the experiments. With controllable volume using the dosing system and piezo, droplets or

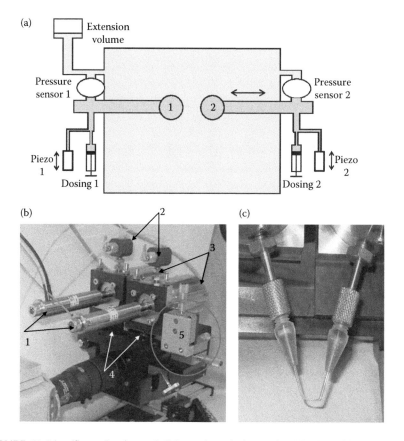

FIGURE 13.21 (See color insert.) Schematic and photo of a DBMM with its main elements and capillaries. (a) Schematic of DBMM; (b) 1, piezo translator; 2, valve for injection of liquid from the syringe dosing system; 3, pressure sensor; 4, holder with capillary; 5, xyz micrometer stage. (c) Glass capillaries.

bubbles can be formed at both capillary tips. The exact positioning of the spherical objects in the focus of the PAT-1 camera is done first by the instrument's stages for the fixed capillary and then for the movable capillary by the additional xyz stage.

The two independent sets of dosing systems consist each of a rough syringe pump (ILS, Stützerbach, Germany) with an accuracy between 50 and 5000 μl and an additional fine dosing system realized by a piezo translator (P-843.40, Physik Instrumente, Germany). The accuracy of this piezo translator was improved by replacing the moving rod of 5 mm diameter by a thinner one of 2 mm diameter. In this way, the volume for a full stroke of 60 μm was significantly reduced down to about 0.2 mm³ = 200 nl.

The measuring range available from the pressure sensor (PDCR-4000, GE-Sensing, Groby, UK) is up to 7000 Pa. The sensor can be overloaded by a 10 times higher pressure. To avoid any damage, the maximum total pressure is limited to a value of 10,000 Pa by a software setting. The sensor's accuracy for a single reading is ±3 Pa and can be improved by averaging several readings on the expense of the data acquisition rate (Won et al. 2012).

There are several ways to use the DBMM. The simplest experiment is the approach of two droplets or bubbles against each other. Once the two droplets of a desired size are formed, the xyz stage allows the second droplet to move toward the first one until both are positioned, for example, just opposite to each other or also out of a common axis. Instead of approaching two droplets or two bubbles, an asymmetric system of a droplet approaching a bubble can also be arranged. Asymmetry can be arranged also by using different fluids for the drop/bubble immersed into the same matrix liquid.

The provided software routines give access to the radii of the two menisci and the respective CPs in real time. These values can be recorded in a file with a selected data acquisition rate. Also, the video sequence can be recorded simultaneously. The standard protocol is the following. Once the two drops, for example, are positioned and brought into contact, the lifetime of the contact is determined. Depending on the system, this can be seconds, minutes, or hours. After the recorded experimental data, this lifetime and the corresponding interfacial tensions on either side can be obtained.

Another experiment is the stepwise (manually driven) forced approach of the two drops against each other with simultaneous recording of the drop radii and CP in order to probe a critical deformation to induce coalescence. This can be of special interest when a small droplet is moved toward a bigger one.

For very stable foam or emulsion systems, a superimposed perturbation can be generated in order to cause the rupture of the liquid film and hence coalescence of the two bubbles/drops. This protocol mimics an external perturbation that is generally assumed to be the reason for the instability of liquid films. Before generating such a perturbation, the two drops/bubbles are brought into the pole-to-pole position and moved into contact. Then, a harmonic oscillation is generated in one drop with a given amplitude and frequency. Keeping the amplitude constant, the critical frequency for film rupture can be found. For a constant frequency, one can also search for the critical amplitude at which the film ruptures.

There are more experimental options for the DBMM. One was shown above in Figure 13.20 designed to mimic a multiple emulsion. The two capillaries are connected to the inner and outer capillary of a so-called coaxial double capillary (Wege et al. 1999). Another arrangement that can be proposed is the following procedure. While a drop is first formed within the matrix liquid through the outer capillary, with the inner capillary, we are able to form a secondary drop inside the primary drop. The liquid of this inner drop can be identical to or different from the external matrix liquid, as it is the typical case in multiple emulsions. Using the respective dosing systems, the volume ratios of the primary and secondary drops can be tuned. Recording the CP in the two drops, the process of adsorption of the added emulsifiers can be monitored. The arrangement of Figure 13.22 is applicable for the formation of a spherical foam lamella as well as by choosing gas as the inside and outside fluid. One of the two sets of dosing systems allows controlling the thickness of the lamella. It is technically feasible to measure in vivo the thickness of the film, as it was proposed by Gabrieli et al. (2012).

The process of the approach of two CTAB solution droplets against each other and immersed in hexane is shown as an example in Figure 13.23. An important point to notice here is that the CP decreases identically upon coalescence at both pressure

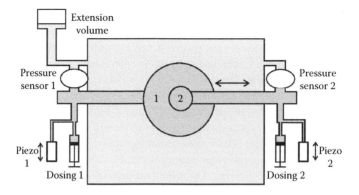

FIGURE 13.22 (See color insert.) Two different ways of rearrangement of the DBMM: concentric drops using different sizes of capillaries.

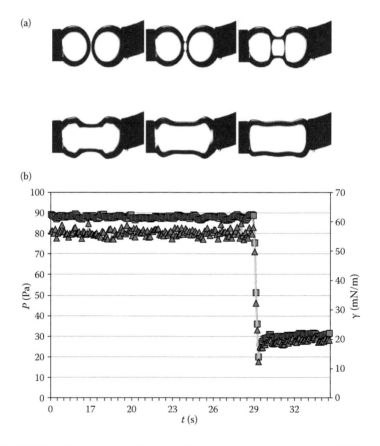

FIGURE 13.23 (See color insert.) Steps of a coalescence process of two CTAB solution drops in hexane: (a) photos taken with a fast camera with time intervals of 500 μs between the images; (b) CP (ν) and interfacial tension (π) measured in the left drop before, during, and after the coalescence process.

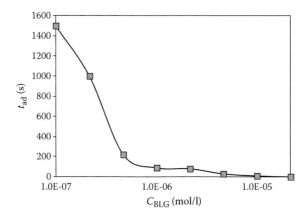

FIGURE 13.24 Coalescence time of two bubbles in BLG solutions as a function of concentration, measured upon direct contact between the two bubbles.

sensors. However, the calculated surface/interfacial tension has no physical meaning after coalescence, because the corresponding radius of curvature cannot be gained by the software. The ongoing process can be recorded also in a movie at standard frame rates of 25 fps. Combined with a high-speed camera, which is provided and embedded into the DBMM software (HSC from SINTERFACE Technologies), it can provide more information about specific interfacial dynamic processes.

Another group of experiments is performed with air bubbles in aqueous solutions containing food proteins such as β-lactoglobulin (BLG), β-casein, and different types of added surfactants. In Figure 13.24, it is shown which pre-adsorption time t_{ad} is required to avoid a spontaneous coalescence of the two approaching air bubbles formed in BLG solutions of different concentration and then brought in contact.

13.10 CONCLUSIONS

A new level of investigating dynamic interfacial properties has been established in recent years. The MBPT is well established now, after it became a routine method about 25 years ago. The standard performance of most commercial instruments is now such that a range of about 10 ms up to about 10 s of adsorption time is covered. Special instrument designs made this methodology the fastest technique for adsorption layer studies of surfactants at the water/air interface. The principle of the equivalent CP methods, developed for investigations of the dynamics of liquid–liquid interfaces, is similar to that of the MBPT; however, because of the viscosity of liquids, the shortest adsorption time accessible is on the order of 10 ms.

The CP methodology is suitable for giving access to the dynamics of the adsorption process. Generating harmonic perturbations, it also allows studying the relaxation behavior of interfacial layers. The most advanced setups work for up to a few hundred hertz for gas bubbles in a liquid, and up to about 100 Hz for drops in a second fluid. The target is still to master the involved hydrodynamics in order to refine the methods for even higher frequencies. With a special capillary, it is even feasible to form spherical films and apply harmonic expansions and compressions in order to study the viscoelasticity of these liquid films. Because of gravity effects, such films are usually inhomogeneous in thickness and therefore experiments on weightlessness conditions are underway.

The experiment of rising bubbles was so far more a qualitative than a quantitative experimental tool. With the opportunity of quantitative simulations using computational fluid dynamics, it will be possible to use this experiment as a quantitative method for determining the presence of surfactants and their interfacial properties under various conditions relevant for practical situations.

The two very specific methods, only recently developed to become routine techniques, are the double capillary for drop bulk exchange and the drop/bubble manipulator that allows mimicking the elementary processes in foams and emulsions. Here, we can say that such methods are already well established; however, the question as to which experimental protocols are most favorable for optimum data analysis remains unanswered.

Besides the large number of new concepts, instruments, and experimental protocols that broaden the hardware basis for investigations of liquid interfacial dynamics, we can also claim that we have reached a new level of theoretical simulations. In particular, the high quality of CFD simulations for growing/oscillating drops and bubbles opens a new era of interfacial dynamics of liquids. Specifically, the limits of oscillating drops, bubbles, and spherical films can be better analyzed via such simulations. The same is true for the growing drops and bubble experiments. In all these cases, the impact of the hydrodynamics on the measured quantities has to be quantitatively understood. In addition, all these experimental protocols are based on the assumption of homogeneous adsorption layers at the interface, as the measured pressure is not a local quantity along the liquid interface.

ACKNOWLEDGMENTS

The research was financially supported by the European Space Agency (FASES, PASTA), the Deutsche Luft- und Raumfahrt (DLR 50WM1129), the Italian Space Agency (LIFT), and the Deutsche Forschungsgemeinschaft SPP 1506 (Mi418/18-1). Financial assistance from the Bundesministerium für Bildung, Wissenschaft, Forschung und Technologie (BMBF) and the Ukrainian Ministry of Education and Science (common project UKR 10/039) is also gratefully acknowledged. The work was supported by COST actions D43, CM1101, and MP1106.

REFERENCES

Adamson, A.W., 1990, *Physical Chemistry of Surfaces*, John Wiley & Sons Inc., New York p. 21.

Alexandrov, N., Marinova, K.G., Danov, K.B. and I.B. Ivanov, 2009, *J. Colloid Interface Sci.* 339, 545.

Alke, A. and D. Bothe, 2009, *Fluid Dynam Mater. Process.*, 5, 345.

Bianco H. and A. Marmur, 1993, *J. Colloid Interface Sci.*, 158, 295.

Cabrerizo-Vilchez, M.A., Wege, H.A., Holgado-Terriza, J.A. and A.W. Neumann, 1999, *Rev. Sci. Instrum.*, 70, 2438–2444.

Chang C.-H. and E.I. Franses, 1994a, *J. Colloid Interface Sci.*, 164, 107.

Chang C.-H. and E.I. Franses, 1994b, *Chem. Eng. Sci.*, 49, 313.

Clift, R., Grace, J.R. and M.E. Weber, 1978, *Bubbles, Drops and Particles*, Academic Press, New York.

Cuenot, B., Magnaudet, J. and B. Spennato, 1997, *J. Fluid Mech.*, 339, 25.

Dan, A., Kotsmar, Cs., Ferri, J.K., Javadi, A., Karbaschi, M., Krägel, J., Wüstneck, R. and R. Miller, 2012, *Soft Matter*, 8, 6057.

Danov, K.D., Kralchevsky, P.A., Denkov, N.D., Ananthapadmanabhan, K.P. and A. Lips, 2006a, *Adv. Colloid Interface Sci.*, 119, 1–16.

Danov, K.D., Kralchevsky, P.A., Denkov, N.D., Ananthapadmanabhan, K.P. and A. Lips, 2006b, *Adv. Colloid Interface Sci.*, 119, 17–33.

Davies, R.M. and G. Taylor, 1950, *Proc. R. Soc. Lond. A*, 200, 375.

del Gaudio, L., Pandolfini, P., Ravera, F., Krägel, J., Santini, E., Makievski, A.V., Noskov, B.A., Liggieri, L., Miller, R. and G. Loglio, 2008, *Colloids Surf. A*, 323, 3.

Duineveld, P.C., 1995, *J. Fluid Mech.*, 292, 325.

Dukhin, S.S., Miller, R. and G. Loglio, 1998, Physico-chemical hydrodynamics of rising bubble. In *Studies in Interface Science*, Elsevier, Amsterdam, Vol. 6.

Fainerman, V.B., 1979, *Koll. Zh.*, 41, 111.

Fainerman, V.B. and R. Miller, 2004, *Adv. Colloid Interface Sci.*, 108–109, 287–301.

Fainerman, V.B. and R. Miller, 2011, Maximum bubble pressure tensiometry: theory, analysis of experimental constrains and applications. In *Bubble and Drop Interfaces, Vol. 2, Progress in Colloid and Interface Science*, R. Miller and L. Liggieri (eds.), Brill Publ., Leiden, pp. 75–118.

Fainerman, V.B., Makievski, A.V. and R. Miller, 2004, *Rev. Sci. Instrum.*, 75, 213–221.

Fainerman, V.B., Mys, V.D., Makievski, A.V., Petkov, J.T. and R. Miller, 2006a, *J. Colloid Interface Sci.*, 302, 40–46.

Fainerman, V.B., Miller, R., Ferri, J.K., Watzke, H., Leser, M.E. and M. Michel, 2006b, *Adv. Colloid Interface Sci.*, 163, 123.

Ferri, J.K., Dong, W.F. and R. Miller, 2005, *J. Phys. Chem. B*, 109, 14764.

Ferri, J.K., Kotsmar, Cs. and R. Miller, 2010, *Adv. Colloid Interface Sci.*, 161, 29.

Fruhner, H. and K.-D. Wantke, 1996, *Colloids Surf. A*, 114, 53.

Frumkin, R.B. and V.G. Levich, 1947, *Zh. Phys. Chim.*, 21, 1183.

Gabrieli, R., Loglio, G., Pandolfini, P., Fabbri, A., Simoncini, M., Kovalchuk, V.I., Noskov, B.A., Miller, R., Ravera, F. and L. Liggieri, 2012, *Colloids Surf. A*, 413, 101.

Ganzevles, R.A., Zinoviadou, K., van Vliet, T., Cohen Stuart, M.A. and H.H.J. de Jongh, 2006, *Langmuir*, 22, 10089.

Georgieva, D., Cagna, A. and D. Langevin, 2009, *Soft Matter*, 5, 2063.

Gibbs, J.W., 1961, *The Scientific Papers*, Vol. 1, Dover Publ., New York.

Hadamard, J.S., 1911, *Acad. Sci., Paris, C. R.*, 152, 1735.

He, Z., Maldarelli, C. and Z. Dagan, 1991, *J. Colloid Interface Sci.*, 146, 442.

Horozov, T. and L. Arnaudov, 1999, *J. Colloid Interface Sci.*, 219, 99.

Jachimska, B., Warszynski, P. and K. Malysa, 1998, *Colloids Surf. A*, 143, 429.

Jäger, F.M., 1917, *Anorg. Allgem. Chem.*, 100, 1.

Javadi, A., Krägel, J., Pandolfini, P., Loglio, G., Kovalchuk, V., Aksenenko, E., Ravera, F., Liggieri, L. and R. Miller, 2010, *Colloids Surf. A*, 365, 62–69.

Javadi, A., Miller, R. and V.B. Fainerman, 2011, Drop volume tensiometry. In *Bubble and Drop Interfaces, Vol. 2, Progress in Colloid and Interface Science*, R. Miller and L. Liggieri (eds.), Brill Publ., Leiden, p. 119.

Javadi, A., Krägel, J., Makievski, A.V., Kovalchuk, N.M, Kovalchuk, V.I., Mucic N., Loglio, G., Pandolfini, P., Karbaschi, M. and R. Miller, 2012, *Colloids Surf. A*, 407, 159–168.

Johnson, D.O. and K.J. Stebe, 1994, *J. Colloid Interface Sci.*, 168, 21.

Johnson, D.O. and K.J. Stebe, 1996, *J. Colloid Interface Sci.*, 182, 526.

Karapantsios, T.D. and M. Kostoglou, 1999, *Colloids Surf. A*, 156, 49.

Kim, Y.H., Koczo, K. and D. Wasan, 1997, *J. Colloid Interface Sci.*, 187, 29.

Kloubek, J., 1972, *J. Colloid Interface Sci.*, 41, 7.

Kotsmar, Cs., Grigoriev, D.O., Makievski, A.V., Ferri, J.K., Krägel, J., Miller, R. and H. Möhwald, 2008, *Colloid Polym. Sci.*, 286, 1071.

Kovalchuk, V.I., Krägel, J., Miller, R., Fainerman, V.B., Kovalchuk, N.M., Zholkovskij, E.K., Wüstneck, R. and S.S. Dukhin, 2000, *J. Colloid Interface Sci.*, 235, 232.

Kovalchuk, V.I., Krägel, J., Makievski, A.V., Loglio, G., Ravera, F., Liggieri, L. and R. Miller, 2002, *J. Colloid Interface Sci.*, 252, 433.

Kovalchuk, V.I., Krägel, J., Makievski, A.V., Liggieri, F., Ravera, L., Loglio, G., Fainerman, V.B. and R. Miller, 2004, *J. Colloid Interface Sci.*, 280, 498.

Kovalchuk, V.I., Krägel, J., Pandolfini, P., Loglio, G., Liggieri, L., Ravera, F., Makievski, A.V. and R. Miller, 2009, In *Progress in Colloid and Interface Science*, Vol. 1, Brill, Leiden, pp. 476–518.

Kovalchuk, V.I., Ravera, F., Liggieri, L., Loglio, G., Pandolfini, P., Makievski, A.V., Vincent-Bonnieu, S., Krägel, J., Javadi, A. and R. Miller, 2010, *Adv. Colloid Interface Sci.*, 161, 102.

Krägel, J., Kovalchuk, V.I., Makievski, A.V., Simoncini, M., Ravera, F., Liggieri, L., Loglio, G. and R. Miller, 2005, *Microgravity Sci. Technol. J.*, 16, 186.

Kretzschmar, G. and K. Lunkenheimer, 1970, *Ber. Bunsenges. Phys. Chem.*, 74, 1064.

Krotov, V.V., Rusanov, A.I. and N.A. Ovrutskaya, 1972a, *Kolloid. Zh.*, 34, 528.

Krotov, V.V., Rusanov, A.I. and N.D. Rjasanova, 1972b, *Kolloid. Zh.*, 34, 534.

Kruglyakov, P.M. and D.R. Exerowa, 1997, Foams and foam films. In *Studies of Interface Science*, Vol. 5, D. Möbius and R. Miller (eds.), Elsevier, Amsterdam.

Krzan, M. and K. Malysa, 2002, *Colloids Surf. A*, 207, 279.

Krzan, M., Lunkenheimer, K. and K. Malysa, 2004, *Colloids Surf. A*, 250, 431.

Krzan, M., Zawala, J. and K. Malysa, 2007, *Colloids Surf. A*, 298, 42.

Levich, V. G., 1949, *Zh. Eksp. Teor. Fiz.* 19, 18.

Levich, V.G., 1962, *Physicochemical Hydrodynamics*, Prentice-Hall, Englewood Cliffs, NJ.

Liggieri, L., and F. Ravera, 1998, Drops and bubbles in interfacial research. In *Studies in Interface Science*, Vol. 6, D. Möbius and R. Miller (eds.), Elsevier, Amsterdam, p. 239.

Liggieri, L., Ravera, F. and A. Passerone, 1990, *J. Colloid Interface Sci.*, 140, 436.

Liggieri, L., Ravera, F. and A. Passerone, 1995, *J. Colloid Interface Sci.*, 169, 226.

Liggieri, L., Attolini, V., Ferrari, M. and F. Ravera, 2002, *J. Colloid Interface Sci.*, 252, 225.

Liggieri, L., Ravera, F., Ferrari, M., Passerone, A., Loglio, G., Miller, R., Krägel, J. and A.V. Makievski, 2005a, *Microgravity Sci. Technol.*, 16, 196.

Liggieri, L., Ravera, F., Ferrari, M. and A. Passerone, 2005b, *Microgravity Sci. Technol. J.*, 16, 201.

Loglio, G., Tesei, U. and R. Cini, 1979, *J. Colloid Interface Sci.*, 71, 316.

Loglio, G., Tesei, U. and R. Cini, 1986, *Colloid Polym. Sci.*, 264, 712.

Loglio, G., Pandolfini, P., Miller, R., Makievski, A.V., Ravera, F., Ferrari, M. and L. Liggieri, 2001, In *Studies in Interface Science*, D. Möbius and R. Miller (eds.), Vol. 11, Elsevier, Amsterdam, p. 439.

Loglio, G., Pandolfini, P., Miller, R., Makievski, A., Krägel, J., Ravera, F. and L. Liggieri, 2005, *Microgravity Sci. Technol.*, 16, 205–209.

Loglio, G., Pandolfini, P., Ravera, F., Pugh, R., Makievski, A.V., Javadi, A. and R. Miller, 2011, Experimental observation of drop-drop coalescence in liquid-liquid systems: Instrument design and features. In *Bubble and Drop Interfaces, Vol. 2, Progress in Colloid and Interface Science*, R. Miller and L. Liggieri (eds.), Brill Publ., Leiden, pp. 384–400.

Lohnstein, T., 1906a, *Ann. Phys.*, 20, 237.

Lohnstein, T., 1906b, *Ann. Phys.*, 20, 606.

Lohnstein, T., 1907, *Ann. Phys.*, 21, 1030.

Lohnstein, T., 1908, *Z. Phys. Chem.*, 64, 686.

Lohnstein, T., 1913, *Z. Phys. Chem.*, 84, 410.

Lord Rayleigh, 1899, *Phil. Mag.*, 48, 321.

Lucassen, J., 1981, Dynamic properties of free liquid films and foams. In *Anionic Surfactants. Physical Chemistry of Surfactant Action. Surfactant Science Ser.*, Vol. 11, E.H. Lucassen-Reynders (ed.), Marcel Dekker Inc., New York, pp. 217–265.

Lucassen, J. and M. van den Tempel, 1972a, *Chem. Eng. Sci.*, 27, 1283.

Lucassen, J. and M. van den Tempel, 1972b, *J. Colloid Interface Sci.*, 41, 491.

Lucassen-Reynders, E.H. and J. Lucassen, 1969, *Adv. Colloid Interface Sci.*, 2, 347.

Lunkenheimer, K. and G. Kretzschmar, 1975, *Z. Phys. Chem. (Leipzig)*, 256, 593.

Lunkenheimer, K., Hartenstein, C., Miller, R. and K.-D. Wantke, 1984, *Colloids Surf.*, 8, 271.

Macleod, C. and C. Radke, 1993, *J. Colloid Interface Sci.*, 160, 435.

Macleod, C. and C. Radke, 1994, *J. Colloid Interface Sci.*, 166, 73.

Magnaudet, J. and I. Eames, 2000, *Ann. Rev. Fluid Mech.*, 32, 659.

Makievski, A.V., Kovalchuk, V.I., Krägel, J., Simoncini, M., Liggieri, L., Ferrari, M., Pandolfini, P., Loglio, G. and R. Miller, 2005, *Microgravity Sci. Technol.*, 16, 215–218.

Malysa, K., Krasowska, M. and M. Krzan, 2005, *Adv. Colloid Interface Sci.*, 114–115, 205.

Malysa, K., Zawala, J., Krzan, M. and M. Krasowska, 2011, Bubbles rising in solutions; local and terminal velocities, shape variations and collisions with free surface. In *Progress Colloid Interface Science*, Vol. 2, Brill, Leiden, p. 243.

Miller, R. and L. Liggieri (eds.), 2009, Interfacial rheology. In *Progress in Colloid and Interface Science*, Vol. 1, Brill, Leiden-Boston.

Miller, R., Loglio, G., Tesei, U. and K.-H. Schano, 1991, *Adv. Colloid Interface Sci.*, 37, 73.

Miller, R., Fainerman, V.B., Kovalchuk, V.I., Liggieri, L., Loglio, G., Noskov, B.A., Ravera, F. and E.V. Aksenenko, 2010, Surface dilational rheology, *Encyclopedia of Surface and Colloid Science*, P. Somasundaran and A. Hubbard (eds.), Taylor & Francis, Boca Raton, FL, pp. 1–18.

Moore, D.W., 1963, *J. Fluid Mech.*, 16, 161.

Moore, D.W., 1965, *J. Fluid Mech.*, 23, 749.

Mysels, K.J., Cox, M.C. and J.D. Skewis, 1961, *J. Phys. Chem.*, 65, 1107.

Nagarajan, R. and D.T. Wasan, 1993, *J. Colloid Interface Sci.*, 159, 164.

Noskov, B.A., 1995, *Colloid. Polym. Sci.*, 273, 263.

Noskov, B.A., Alexandrov, D.A. and R. Miller, 1999, *J. Colloid Interface Sci.*, 219, 250.

Parkinson L., Sedev R., Fornasiero, D. and J. Ralston, 2008, *J. Colloid Interface Sci.*, 322, 168.

Passerone, A., Liggieri, L., Rando, N., Ravera, F. and E. Ricci, 1991, *J. Colloid Interface Sci.*, 146, 152.

Prins, A., Arcuri, C. and M. van den Tempel, 1967, *J. Colloid Interface Sci.*, 24, 84.

Prosperetti, A., 2004, *Phys. Fluids*, 16, 1852.

Ravera, F., Liggieri, L., Passerone, A. and A. Steinchen, 1991, In *Proceedings of the First European Symposium on Fluids in Space*, ESA SP-353, p. 213.

Ravera, F., Ferrari, M., Santini, E. and L. Liggieri, 2005, *Adv. Colloid Interface Sci.*, 117, 75.

Ravera, F., Loglio, G., Pandolfini, P., Santini, E. and L. Liggieri, 2010, *Colloids Surf. A*, 365, 2.

Rodrigue, D., 2001, *Can. J. Chem. Eng.*, 79, 119.

Russev, S.C., Alexandrov, N., Marinova, K.G., Danov, K.D., Denkov, N.D., Lyutov, L., Vulchev, V. and C. Bilke-Krause, 2008, *Rev. Sci. Instrum.*, 79, 104102.

Rybczynski, W., 1911, *Bull. Acad. Sci. Cracow*, A40, 40.

Sadhal, S.S. and R.E. Johnson, 1983, *J. Fluid Mech.*, 126, 237.

Schrödinger, E., 1915, *Ann. Phys.*, 46, 413.

Shen, C.L., Xie, W.J. and B. Wei, 2010, *Phys. Rev. E*, 81, 046305.

Soos, J.M., Koczo, K., Erdos, E. and D.T. Wasan, 1994, *Rev. Sci. Instrum.*, 65, 3555.

Stebe, K.J. and C. Maldarelli, 1994, *J. Colloid Interface Sci.*, 163, 177.

Stenvot, C. and D. Langevin, 1988, *Langmuir*, 4, 1179.

Sugden, S., 1922, *J. Chem. Soc.*, 121, 858.

Tate, T., 1864, *Phil. Mag.*, 27, 176.

Ueno, M., Takasawa, Y., Miyashige, H., Tabata, Y. and K. Meguro, 1981, *Colloid Polym. Sci.*, 259, 761.

Ulaganathan, V., Fainerman, V.B., Gochev, G., Aksenenko, E.V., Gehin-Delval, C. and R. Miller, 2013, *Colloids Surf. A*, submitted for publication.

van den Tempel, M., Lucassen, J. and E.H. Lucassen-Reynders, 1965, *J. Phys. Chem.*, 69, 1798.

van Uffelen, M. and P. Joos, 1994, *Colloids Surf. A*, 85, 107.

Wantke, K.-D., Fruhner, H., Fang, J. and K. Lunkenheimer, 1998, *J. Colloid Interface Sci.*, 208, 34.

Warszynski, P., Wantke, K.D. and H. Fruhner, 1998, *Colloids Surf. A*, 139, 137.

Wege, H.A., Holgado-Terriza, J.A., Neumann, A.W. and M.A. Cabrerizo-Vilchez, 1999, *Colloids Surf. A*, 156, 509.

Won, J.Y., Krägel, J., Makievski, A.V., Javadi, A., Gochev, G., Loglio, G., Pandolfini, P., Leser, M.E., Gehin-Delval, C., Gunes, D.Z. and R. Miller, 2013, *Colloids Surf. A*, submitted for publication.

Zhang, X., Harris, T. and O.A. Basaran, 1994, *J. Colloid Interface Sci.*, 168, 47.

14 AC Electrokinetics in Concentrated Suspensions

A. V. Delgado, Raúl A. Rica, Francisco J. Arroyo,
Silvia Ahualli, and Maria L. Jiménez

CONTENTS

14.1 INTRODUCTION

That colloidal dispersions of nanoparticles are often used as concentrated slurries is a well-known fact, particularly when industrial applications are in mind. Suffice it to say that volume fractions as high as 50% can be found in ceramic suspensions, and that when drug particles are administered as dispersions, therapeutic levels require values as high as 30% (Alejo and Barrientos 2009; Beirowski et al. 2012; Johnson et al. 2000; Vauthier et al. 2008). Although a number of techniques are available for evaluation of the characteristics of such systems, notably rheological determinations (Alejo and Barrientos 2009; Derkach 2009; Hobbie 2010; Melito and Daubert 2011; Siebenburger et al. 2012), in this chapter, we will limit our study to electrokinetic techniques as potential methods. Recall that these are methods based on the determination of the response of the solid/solution interface (or, from a more macroscopic point of view, of the whole system) to the application of an external field, and this may be gravitational, electric, concentration gradient, pressure gradient, and their combinations thereof (Delgado 2002; Hunter 1987; Lyklema 1995; Ohshima 2006; Stoilov and Stoimenova 2007). The response mentioned is relatively easy to access experimentally in the case of single particles (dilute suspensions in fact) or isolated interfaces: for instance, (micro)electrophoresis and streaming potential are well-established methods since the 19th century (Wall 2010). Concentrated suspensions are less prone to evaluation by means of those techniques: their optical turbidity makes it impossible to use methods based on light scattering, and only indirect techniques based on the measurement of some collective property can be used. This in turn often requires a theoretical treatment not just for estimating surface charge or potentials of the particles but even for obtaining single-particle properties as a previous step.

In this chapter, focus will be on the available techniques for the electrokinetic evaluation of concentrated suspensions when ac electric or pressure fields are applied. But let us consider as a first stage in our study their response to constant fields. We assume a suspension of spheres (radius a) with finite volume fraction of solids, ϕ, dispersed in an electrolyte solution for which we can define the ionic composition outside the electric double layers (EDLs) of the particles. Let us call n_i^∞ the number concentration of the ith ionic species in that external solution; $z_i e$ will be their respective charge. A constant, homogeneous electric field $\mathbf{E}$ is applied to the suspension, and we seek the value of the electrophoretic mobility u_e (i.e., the electrophoretic velocity v_e per unit field strength) of the particles as a function of ϕ and to investigate to what extent does this mobility differ from that of an isolated particle, u_e^∞.

The works of Zukoski and Saville (1987, 1989) are particularly interesting in the sense that they managed to measure the (micro)electrophoretic mobility of particles dispersed in concentrated suspensions by mixing intact red blood cells (colored because of their hemoglobin contents) and "ghost" erythrocytes in which osmotic lysis forced the cell contents out into the solution, leaving a transparent sack formed by the membrane and hemoglobin-free solution. In this way, the suspensions appeared optically dilute although in fact they were concentrated. From these significant studies, the authors concluded that electric and hydrodynamic interactions between particles cancelled each other, leaving a simple $u_e(\phi) = u_e^\infty(1-\phi)$ dependence between the mobility and the volume fraction. Such a dependence is, of course, the manifestation of the fluid counterflow generated by the electrophoresis of the particles: since each particle displaces back a volume of fluid equal to the particle volume when advancing under the action of the field, the contributions of the particles to this backflow add up, so that any individual particle moves against a liquid moving at a velocity $-u_e^\infty\phi$ (Ahualli 2006a). In general, as long as the EDLs are sufficiently thin in comparison with the particle radius, no effect of the volume fraction on the mobility beyond the $(1-\phi)$ correction is expected, according to these authors.

If we think of another kind of experiment typically performed in constant external fields with concentrated systems, streaming potential in porous plugs can be a clear example in which the volume fractions approach (or even exceed, in case of polydisperse systems) the maximum packing fraction of spheres. One could expect that the effects of volume fraction would be extreme in this case. However, this is not the case in many instances; in fact (Matijevic 1974), if

(a) the EDL thickness, or Debye length, κ^{-1}, given by

$$\kappa^{-1} = \left(\frac{\varepsilon_m \varepsilon_0 k_B T}{\sum_{i=1}^{N} n_i^\infty e^2 z_i^2} \right)^{1/2} \tag{14.1}$$

(ε_m is the relative permittivity of the liquid medium, ε_0 is the permittivity of vacuum, and $k_B T$ is the thermal energy) is much smaller than the radius of curvature of the pore surface at any point, $\kappa a \gg 1$, with a being the average particle radius in the plug. This is sometimes called (although not rigorously) the thin double layer approximation (Delgado et al. 2007a), and

(b) the minimum linear dimension of the pore is much larger than κ^{-1}, the simple Smoluchowski equation relating the observed response to the electrokinetic or zeta potential (ζ) remains valid, despite the complexity of the liquid paths inside the plug.

We must consider the most general situations in which the simplifying hypotheses are not valid anymore and in which the external fields are not stationary but alternating at some given frequency ω. This is the main aim of this chapter. It will be organized as follows: In Section 14.2, the methods available for the evaluation of electrokinetic properties in the systems will be briefly described in the case of rigid spherical particles, and we show some experimental results. In Section 14.3, we describe some models not based on the most used cell approach. The behavior of particles coated by a soft (deformable, charged, and permeable to the liquid) polymer layer will be considered, either for homogeneous or for brush-like structures of the layer in Section 14.4. Finally, Section 14.5 will focus on non-spherical particles, another situation in which the complications associated to finite volume fractions superimpose to those coming from spheroidal or planar geometries. Section 14.6 will summarize the contents of the chapter.

14.2   THE SPHERICAL PARTICLE CASE

14.2.1   General Features

The first approximate calculation of the electrophoretic mobility of a swarm of spherical particles was performed in the mid-20th century by Moeller et al. (1961a,b), but rigorous evaluations were first presented by Levine and Neale (1974a,b). They used so-called cell models as a way to account for hydrodynamic and electrical inter-actions between particles in concentrated suspensions, particularly (but not only) in electrophoresis. The former interactions had been considered in Happel and Brenner (1957) and Kuwabara (1959). The merit of Levine and Neale was to extend these ideas to charged particles under the action of an electric field. The cell model assumes that the interactions can be effectively considered by solving the problem of a single sphere in a "cell" of liquid medium, which is concentric with it and with radius b, fulfilling two conditions (Figure 14.1):

(a) The volume fraction of solids in the cell equals that in the real system.

$$\left(\frac{a}{b}\right)^3 = \phi \qquad (14.2)$$

(b) The interactions between particles are taken into account by selecting the boundary conditions for the velocity of the liquid, the electric potential, and the concentration of ions on the cell surface, $r = b$.

Although the use of this approach is not free of limitations (Batchelor 1972), most authors working on concentrated suspensions have used it as the best, if not the only, possible approach. We do not intend to be exhaustive (Stoilov and Stoimenova 2007), but some milestones, in addition to the above-mentioned pioneering contribu-tions, should be mentioned. The work of Ohshima (1997) described an analytical formula for the mobility in the case of low zeta potentials, and overlapping double layers, restrictions later eliminated in Hsu et al. (2000), Lee et al. (1999), and Lin et al. (2002). Suggestions by Dukhin et al. (1999a) concerning some inconsistencies in the Levine–Neale boundary conditions led a number of authors (Carrique et al. 2001a,b, 2002; Hsu et al. 2002) to consider a different approach based on the Shilov–Zharkikh model (Shilov et al. 1981).

As long as a mean field approach is acceptable, the solution to the problem requires to evaluate at every position $\mathbf{r}$ (the position vector has its origin at the particle center) the electric potential $\Psi(\mathbf{r})$, the fluid velocity $\mathbf{v}(\mathbf{r})$, and the concentrations and veloci-ties $n_i(\mathbf{r})$, $v_i(\mathbf{r})$ of each ionic species. This needs (DeLacey and White 1981; O'Brien and White 1978; Ohshima et al. 1984) the solution of the following set of equations (because an electric field of frequency ω, $\mathbf{E} \exp(-i\omega t)$, is assumed to be applied, an $\exp(-i\omega t)$ is implicit in all quantities):

(a) Poisson–Boltzmann equation:

$$\nabla^2 \Psi(\mathbf{r}) = -\frac{\rho_{el}(\mathbf{r})}{\varepsilon_0 \varepsilon_m}, \qquad (14.3)$$

with $\rho_{el}(\mathbf{r})$ being the charge density $\rho_{el}(\mathbf{r}) = \sum_{i=1}^{N} z_i e n_i(\mathbf{r})$.

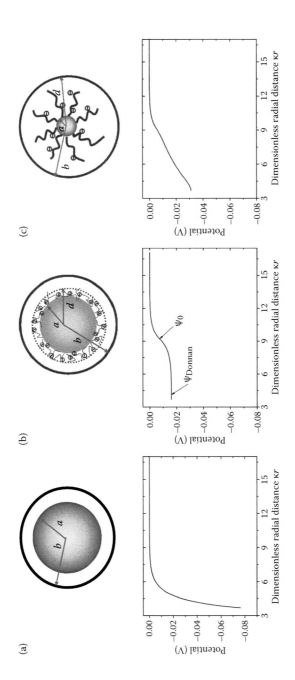

FIGURE 14.1 Top: Schematic representation of the ideal cell (radius *b*) used for the evaluation of electrokinetics in concentrated suspensions of spheres. (a) Rigid particles; (b) soft particles with a homogeneous polyelectrolyte layer of thickness $d - a$; (c) polyelectrolyte brushes on a rigid sphere. Bottom: Corresponding electric potential profiles.

(b) Navier–Stokes equations for an incompressible fluid:

$$\eta_m \nabla^2 \mathbf{v}(\mathbf{r}) - \nabla p(\mathbf{r}) - \rho_{el} \nabla \Psi(\mathbf{r}) = \rho_m \frac{\partial}{\partial t}[\mathbf{v}(\mathbf{r}) + \mathbf{v}_e]$$

(14.4)

$$\nabla \cdot \mathbf{v}(\mathbf{r}) = 0$$

where it has been taken into account that $\mathbf{v}$ is the fluid velocity with respect to the particle and that the particle moves with electrophoretic velocity $\mathbf{v}_e \exp(-i\omega t)$ in the presence of the electric field. In Equation 14.4, η_m, ρ_m are, respectively, the viscosity and density of the liquid medium.

(c) Ion conservation and Nernst–Planck equations:

$$\nabla \cdot [n_i(\mathbf{r})\mathbf{v}_i(\mathbf{r})] = -\frac{\partial}{\partial t}[n_i(\mathbf{r})]$$

$$\mathbf{v}_i(\mathbf{r}) = \mathbf{v}(\mathbf{r}) - \frac{D_i}{k_B T} \nabla \mu_i(\mathbf{r})$$

(14.5)

$$\mu_i(\mathbf{r}) = \mu_i^\infty + z_i e \Psi(\mathbf{r}) + k_B T \ln n_i(\mathbf{r})$$

where $\mu_i(\mathbf{r})$ is the electrochemical potential of the ith ions and μ_i^∞ is the corresponding standard value.

The problem can be solved if the non-equilibrium quantities are written in terms of equilibrium ones (superscript "0") and field-induced perturbations, assumed to be linear with the field and with the same frequency dependence (except for a phase term, i.e., all quantities will be complex, see Ohshima 1997, 1999):

$$\mu_i(\mathbf{r}) = \mu_i^0(r) + \delta\mu_i(\mathbf{r}) = \mu_i^0(r) - z_i e \phi_i(r) E \cos\theta \quad (i = 1, \ldots, N)$$

(14.6)

$$\Psi(\mathbf{r}) = \Psi^0(r) + \delta\Psi(\mathbf{r}) = \Psi^0(r) - \Gamma(r) E \cos\theta$$

(14.7)

$$\mathbf{v}(\mathbf{r}) = \left\{ -\frac{2}{r} h(r) E \cos\theta, \quad \frac{1}{r}\frac{d}{dr}[rh(r)]E \sin\theta, \quad 0 \right\},$$

(14.8)

where we note that the equilibrium quantities depend only on the radial spherical coordinate ($\mathbf{r}$) and axial symmetry around the field direction (θ is the angle with the field direction) allows us to leave the r-dependent functions $h(r)$, $\phi_i(r)$, and $\Gamma(r)$ as unknowns.

In terms of these new quantities, the partial differential equations of Poisson, Navier–Stokes, and ion conservation read, respectively:

$$L\left\{(L+\gamma^2)\left[h(r)\right]\right\} = -\frac{e}{\eta_m r}\frac{dy}{dr}\sum_{i=1}^{N}n_i^{\infty}z_i^2\exp(-z_iy)\phi_i(r) \tag{14.9}$$

$$L\phi_i(r)+\gamma_j^2\left[\Gamma(r)-\phi_j(r)\right]=\frac{dy}{dr}\left[z_i\frac{d\phi_i}{dr}-\frac{2D_i}{k_BTe}\frac{h(r)}{r}\right],\quad i=1,2,...,N \tag{14.10}$$

$$L\Gamma(r)=\frac{1}{\varepsilon_m\varepsilon_0 k_BT}\sum_{i=1}^{N}z_i^2e^2n_i^0(r)\left[\Gamma(r)-\phi_i(r)\right], \tag{14.11}$$

with $y = e\Psi^0(r)/k_BT$ and n_i^{∞} is the bulk number concentration of the ith ionic species. The second-order differential operator L is found to be

$$L\equiv\frac{d^2}{dr^2}+\frac{d}{dr}-\frac{2}{r^2} \tag{14.12}$$

and

$$\gamma^2\equiv\frac{i\omega\rho_m}{\kappa^2\eta_m}$$
$$\gamma_j^2\equiv\frac{i\omega D_j}{\kappa^2 k_B^2T^2}. \tag{14.13}$$

Considering Equations 14.9 through 14.12 together with Poisson–Boltzmann for the equilibrium potential distribution, it will be clear that we need $4 + 2N + 2 + 2 = 2N + 8$ boundary conditions. Two are immediate, and come from the definition of the (electrokinetic) surface charge density on the particle (total charge Q; surface charge density σ), and from the assumption that, in equilibrium, the cell is electroneutral:

$$\varepsilon_m\varepsilon_0\left.\frac{d\Psi^0}{dr}\right|_{r=a}=-\sigma=-\frac{Q}{4\pi a^2} \tag{14.14}$$

$$\left.\frac{d\Psi^0}{dr}\right|_{r=b}=0 \tag{14.15}$$

In terms of the electrokinetic or zeta potential (ζ), Equation 14.14 must be written:

$$\Psi^0 (r = a) = \zeta. \tag{14.14a}$$

This is all one can say about the equilibrium conditions. The next step is considering what happens in this respect when the ac field is applied. To begin with, the stagnancy of the liquid on the particle surface and the fact that the latter is impenetrable to ions lead to ($\hat{\mathbf{n}}$ is the outward unit vector normal to the surface):

$$\left.\mathbf{v}(\mathbf{r})\right|_{r=a} = 0 \rightarrow \begin{cases} h(a) = 0 \\ \left.\dfrac{dh}{dr}\right|_{r=a} = 0 \end{cases}$$

$$\left.\mathbf{v}_j \cdot \hat{\mathbf{n}}\right|_{r=a} = 0 \rightarrow \left.\dfrac{d\phi_j}{dr}\right|_{r=a} = 0, \quad j = 1,...,N. \tag{14.16}$$

In addition, the continuity of the potential and displacement perturbations on the particle surface can be written as

$$\left. \begin{array}{r} \delta\Psi(a^+) = \delta\Psi(a^-) \\[4pt] \varepsilon_p \left.\dfrac{\partial\delta\Psi}{\partial r}\right|_{a^-} = \varepsilon_m \left.\dfrac{\partial\delta\Psi}{\partial r}\right|_{a^+} \end{array} \right\} \rightarrow \left.\dfrac{d\Gamma}{dr}\right|_a - \dfrac{\varepsilon_p}{\varepsilon_m a}\Gamma(a) = 0, \tag{14.17}$$

where ε_p is the particle's relative permittivity.

The conditions on the cell surface are the key point of the model. We still need one condition for Γ, one for each of the ϕ_j functions, and two further conditions for h. The problem is still open and sometimes there are no clear criteria for taking an option. A thorough discussion can be found in Carrique et al. (2005), but we adhere here to the approach of Ahualli et al. (2006b): according to these authors, it suffices to establish a set of *general* physical statements on the cell dynamics. The first comes from the definition of the macroscopic electric field $\mathbf{E}$:

$$\mathbf{E} = \langle -\nabla\delta\Psi(\mathbf{r})\rangle \rightarrow \Gamma(b) = -b, \tag{14.18}$$

where the average $\langle\cdot\rangle$ must be evaluated throughout the whole cell volume. This Dirichlet-type condition is equivalent to the Shilov et al.'s proposal (Shilov et al. 1981) and has been used by our group (Carrique et al. 2002, 2003a) in all cases. The second condition is established on the average of the concentration perturbations:

$$\langle\nabla\delta n_j(\mathbf{r})\rangle = 0 \rightarrow \delta n_j(b) = 0 \rightarrow \phi_j(b) = b. \tag{14.19}$$

An alternative to this Dirichlet condition is a Neumann one, already used by other authors (Ding and Keh 2001; Ohshima 2000a):

$$\left.\frac{\partial \delta n_j}{\partial r}\right|_{r=b} = 0 \rightarrow \left.\frac{d\phi_j}{dr}\right|_{r=b} = -1 = \left.\frac{d\Gamma}{dr}\right|_{r=b}. \tag{14.20}$$

We will only use this condition when dealing with particles coated with a brush-like polyelectrolyte layer (see Section 14.5.2).

We finish with conditions for the fluid velocity on the cell. We will avoid a full discussion, which can be found in Ahualli et al. (2006b), for example. Again, general statements can be used. First, we assume that the average of the pressure perturbation (δp) gradient in the cell volume can be neglected, as no external pressure is imposed on the system:

$$\langle \nabla \delta p(\mathbf{r}) \rangle = 0 \rightarrow \left.\frac{d}{dr}\left[r(L + \gamma^2)h\right]\right|_{r=b} - 2h(b)\gamma^2 - \frac{\sum\limits_{i=1}^{N} z_i e n_i^0(b)}{\eta_m}, \tag{14.21}$$

with $n_i^0(b)$ being the equilibrium concentration of the ith species on the cell boundary. An additional condition, which will be essential in the determination of the dynamic electrophoretic mobility, stems from the condition of zero average for the fluid velocity, which happens to be equivalent to Kuwabara's condition that the radial velocity of the fluid on the cell coincides (except for a change of sign) with the component of the electrophoretic mobility in the same direction:

$$\left.\begin{array}{c} v_r(b) = -u_e E \cos\theta \\[1ex] \text{or} \\[1ex] \langle \mathbf{v}(\mathbf{r}) \rangle = 0 \end{array}\right\} \rightarrow u_e = \frac{2h(b)}{b}. \tag{14.22}$$

Finally, we must add the equation for the force balance on the cell (Ohshima 2000b):

$$\oint_{S(r=b)} \overset{\leftrightarrow}{\tau} \cdot \hat{\mathbf{n}}\, dS = -i\omega\left[\frac{4\pi}{3} a^3 \rho_p u_e \mathbf{E} + \rho_m \int_{V_m}\left[\mathbf{v}(\mathbf{r}) + \mathbf{v}_e\right] dV\right] \rightarrow$$

$$\rightarrow \left.L[h(r)]\right|_{r=b} = \gamma^2 \phi\left(\frac{\rho_p - \rho_m}{\rho_m}\right)h(b), \tag{14.23}$$

where ρ_p is the particle density and $\overset{\leftrightarrow}{\tau}$ is the total stress tensor on the surface of the cell ($r = b$), although the electroneutrality of the latter implies that the tensor only contains hydrodynamic contributions.

14.2.2 How Can the Permittivity and Dynamic Mobility Be Obtained?

The evaluation of the dynamic mobility or the electric permittivity requires performing a full solution of the differential equations and boundary conditions system described above. In the case of the mobility, this is immediate: once the radial dependence of $h(r)$ is known, it suffices to calculate its value at $r = b$ and use Equation 14.22. This must be repeated for every frequency of the applied field.

The calculation of the permittivity is more elaborate, as it can be obtained from the complex conductivity of the suspension, which is, in turn, based on the numerical calculation of the average current through the cell. However, it is still of interest, since the method can be applied to concentrated systems, covering much the same range as dynamic mobility determinations, but with the additional advantage of covering a wider frequency interval. Efforts in this direction are worth consideration, since both the frequency dependence of the permittivity and the amplitude of the relaxations eventually found are extremely sensitive to such properties as particle surface charge, size and shape, colloidal stability, surface nanostructure, and so on. (Delgado 2002).

It is fair to mention some disadvantages of this technique, related to its increased experimental difficulties, and the absence of a commercial device, resulting to most workers using their own designs. These features have probably prevented a more general use of the measurement of the frequency dependence of the permittivity of nanoparticle suspensions, the so-called low-frequency dielectric dispersion or LFDD. Nevertheless, methods have been proposed in the literature to minimize such unwanted effects as electrode polarization (Grosse and Tirado 1996; Jimenez et al. 2002; Kijlstra et al. 1993; Myers and Saville 1989; Rosen and Saville 1991; Tirado et al. 2000) or to improve theoretical treatments in order to take into account the possible presence of EDL overlap or non-zero conductance in the inner part of the EDL, a phenomenon generally known as SLC or stagnant-layer conductance (Arroyo et al. 1999a,b; Kijlstra et al. 1992; Rosen et al. 1993).

For the calculation of the (complex) relative permittivity of the suspension, $\varepsilon^*(\omega)$, we start from the expression for the average current density through the cell, and from it, for the complex conductivity of the system $K^*(\omega)$. Recall that they are related:

$$\langle \mathbf{j} \rangle = K^* \mathbf{E}$$
$$K^*(\omega) = K^*(\omega = 0) - i\omega\varepsilon^* \varepsilon_0. \tag{14.24}$$

It is usual to distinguish between the real and imaginary components of ε^*:

$$\varepsilon^*(\omega) = \varepsilon'(\omega) - i\varepsilon''(\omega) \tag{14.25}$$

and also to use the permittivity increment $\Delta\varepsilon^*(\omega)$ and its components:

$$\Delta\varepsilon^*(\omega) = \varepsilon^*(\omega) - \varepsilon_m = \Delta\varepsilon'(\omega) - i\Delta\varepsilon''(\omega). \tag{14.26}$$

Note that $\Delta\varepsilon^*$ will be linearly dependent on ϕ for dilute suspensions, while this will not be true in concentrated ones. In Carrique et al. (2003b), it was shown that

$$K^*(\omega) = \sum_{i=1}^{N} \left[\frac{z_i^2 e^2 n_i^{\infty} k_B T}{D_i} \left(1 - \frac{3\phi}{a^3} \Omega_i \right) - u_e z_i n_i^{\infty} \right]$$

$$\times \exp\left(-\frac{z_i e \Psi^0(b)}{k_B T} \right) - i\omega\varepsilon_m\varepsilon_0 \left(1 - \frac{3\phi}{a^3} \Omega \right) \tag{14.27}$$

where

$$\Omega_i \equiv -\frac{b^2}{3} \left(r \frac{d\phi_i}{dr} - \phi_i \right)_{r=b}, \quad i = 1,...,N$$

$$\Omega \equiv -\frac{b^2}{3} \left(\Gamma - r \frac{d\Gamma}{dr} \right)_{r=b}. \tag{14.28}$$

As observed, it is necessary again to solve the whole problem (in addition to the equilibrium one) in order to get the permittivity, since one requires the knowledge of the functions ϕ_i, Γ, and their derivatives at $r = b$.

14.2.3   SOME RESULTS

14.2.3.1   Dielectric Spectroscopy

We will now show which are the general features of the electric permittivity of the systems described. Figure 14.2 is a clear illustration of how the relative permittivity changes with frequency (the LFDD phenomenon). Note that, in all cases, double-layer polarization is very noticeable: at low frequencies, the permittivity can attain very high values and decreases to values below those of the solution if the frequency is large enough; in these conditions, the only observable effect is the polarization of the interface because of the small value of the permittivity of the solids as compared to that of aqueous solutions. The high values of the low-frequency permittivity (Delgado 2002; Dukhin and Shilov 1974; Grosse 2002; Shilov et al. 2000, 2002) are the manifestation of the existence of a gradient of neutral electrolyte concentration around the particle when an electric field is applied (concentration polarization). The characteristic size of such cloud is comparable to the particle radius: it will hence need a time of the order of

$$\tau_\alpha \simeq \frac{(a + \kappa^{-1})^2}{2D}$$

$$D = \frac{D_1 D_2 \left(|z_1| + |z_2| \right)}{(D_1 + D_2)} \tag{14.29}$$

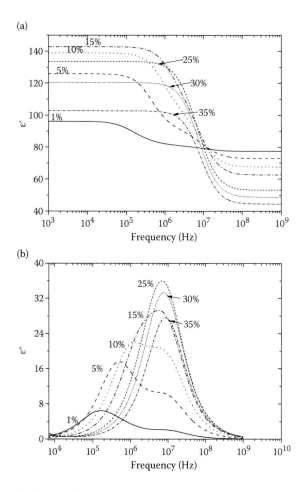

FIGURE 14.2 Real (a) and imaginary (b) parts of the relative permittivity of the suspensions of spherical particles ($\varepsilon_p = 2$) with 50 nm radius in 0.5 mM KCl solutions. The charge of the particles is 1×10^{-15} C.

to be established (D is the average diffusion coefficient of cations and anions in a solution containing only two types of ions). If the frequency of the field is below the so-called α-relaxation frequency $\omega_\alpha = 1/\tau_\alpha$, the electrolyte clouds can build up and exchange (between the two poles of the particle) upon inversion of the ac field. The existence of the concentration gradients induces changes in the dipole moment of the interface that manifest in the giant permittivity at low frequencies. At high frequencies, though, the clouds do not have time to form and the permittivity relaxes. At still higher frequencies, the Maxwell–Wagner–O'Konski (MWO) relaxation can be observed, although its amplitude is much lower and it does not have the importance of the alpha process. Figure 14.3 is an illustration of this. As observed, at low or moderate volume fractions ($\phi < \sim 15\%$), there is indication of the existence of two separate relaxations: the one at lower frequencies (left part of the curves) is associated

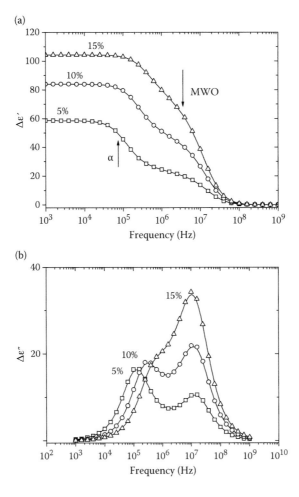

FIGURE 14.3 Real and imaginary components of the dielectric increment as a function of frequency for the indicated volume fractions. Particle radius: 100 nm; 10^{-4} mol/L KCl. Particle charge: 10^{-14} C.

with concentration polarization; at higher frequencies, the MWO relaxation can be observed. This is the passage from the situation in which counterions in the EDL are alternatively accumulated in excess on one side of the particle and are depleted on the other, to that in which the spherical symmetry of the charge distribution cannot be altered by the rapidly oscillating field. When the volume fractions are increased above the value previously cited, the two relaxations are not distinguished, as the frequency of the α-relaxation is higher the larger ϕ is, and the MWO frequency behaves the opposite way, as noted below (Dukhin and Shilov 1974):

$$\omega_{MWO} = \frac{K_m}{\varepsilon_m \varepsilon_0} \frac{2(1-\phi)Du + 2 + \phi}{2 + \phi}, \qquad (14.30)$$

where Du, the Dukhin number, is a dimensionless ratio between the surface conductivity K^σ (excess conductivity of the EDL) and the liquid medium conductivity K_m ($Du = 2K^\sigma/K_m a$) (Lyklema 1995).

Returning now to the results concerning the volume fraction effects on the dielectric dispersion, it is useful to consider in separate plots how the amplitude of the relaxation ($\Delta\varepsilon'(0)$) and its characteristic frequency ω_α evolve with volume fraction. This can be analyzed with the data in Figure 14.4. Note that the former quantity goes through a maximum: as the volume fraction is increased from the dilute region, the relaxation amplitude also increases, a manifestation of the larger number of particles involved in the polarization mechanisms. At the same time, the relaxation frequency also increases. Note however that $\Delta\varepsilon'(0)$ starts to decrease for volume fractions above

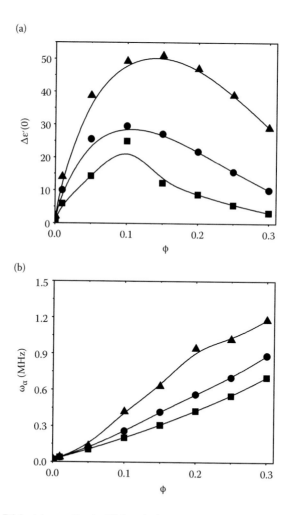

FIGURE 14.4   Dielectric amplitude (dielectric increment at zero frequency) (a) and alpha relaxation frequency (b) plotted as a function of volume fraction, for $\zeta = 75$ mV (squares), 100 mV (circles), and 130 mV (triangles). Particle radius: 100 nm; 10^{-4} mol/L KCl.

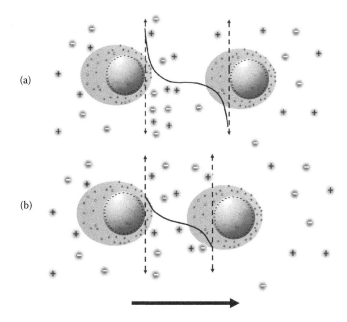

FIGURE 14.5 **(See color insert.)** Schematic representation of the polarization clouds around negatively charged spherical particles for dilute (a) and concentrated (b) suspensions.

10% or so: the concentration clouds of neighbor particles partially compensate each other, while the distance between them decreases as a matter of fact, this is also the explanation of the raise in alpha-frequency: the characteristic size of the cloud is not the particle size anymore, but it is rather controlled by the interparticle distance, shorter for larger volume fractions. Figure 14.5 is a scheme of the phenomenon. As a final comment, let us point out that Figures 14.4 and 14.5 clearly indicate the large errors that we could make when a theory designed for dilute systems, such as the classical one by Delacey and White (1981), is used with concentrated suspensions. Even for volume fractions as low as a few percent, the difference between the linear dependence predicted for dilute suspensions and the true volume fraction dependences amounts to approximately 10%.

14.2.3.2 Dynamic Mobility

Figure 14.6 provides an illustration of the prediction of cell models for the dynamic mobility of suspensions of spheres. First of all, it can be observed that the alpha relaxation, so extremely noticeable in the case of dielectric relaxation, is almost absent in this plot: only a small reduction in the mobility is observed in the kilohertz region (typical of alpha processes) if the zeta potential is high. On the contrary, a rise in the real part of u_e (and a corresponding minimum in its imaginary component) is clearly found in the 10^5–10^6 Hz range, followed by a sharp decrease (maximum in $Im(u_e)$) at frequencies of several megahertz. This kind of behavior is typical and this justifies the frequency range of the commercial instruments. The raise mentioned is the manifestation of MWO relaxation in the induced dipole moment (Arroyo et al. 2004),

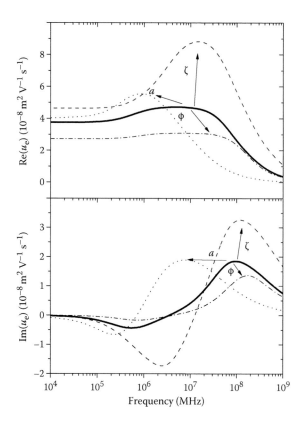

FIGURE 14.6 Real and imaginary parts of the dynamic mobility of suspensions of spheres (ρ_p = 2.2 g/cm³, ε_p = 2) in 5 × 10⁻⁵ mol/L KCl solutions plotted as a function of the field frequency for ζ = 100 mV (solid lines) and 180 mV (dashed lines), volume fraction ϕ = 2% (solid lines) and 8% (dashed dotted lines), and particle radii a = 50 nm (solid lines) and 250 nm (dotted lines).

and the final decrease is the inertial decay due to the impossibility of liquid flows and particle motions for following the rapid field oscillations (Landau and Lifshitz 2000). The characteristic frequency ω_{in} of this process is largely dependent on particle size, and it can be used for carrying out an independent estimation of this quantity:

$$\omega_{in} = \frac{2\eta_m}{\rho_m a^2} \tag{14.31}$$

(In the range 1–18 MHz, this corresponds to approximate radii between 100 and 500 nm in aqueous solution.) The very significant effect of increasing a, from, for instance, 50 to 250 nm, is very clearly seen in Figure 14.6.

Concerning the reason for mobility elevation when the MWO relaxation takes place, it is convenient to take into consideration that the electrophoretic mobility roughly depends on $(1 - C)$, if C is the induced dipole coefficient (proportional to the induced dipole strength: $\mathbf{d} = 3V\varepsilon_0\varepsilon_m C\mathbf{E}$, with V being the particle volume, see Shilov et al. 2000). At high Dukhin numbers, C is positive and thus $(1 - C)$ tends to reduce the mobility; if the process is not possible, at sufficiently high frequencies, the braking effect on the mobility disappears and $\text{Re}(u_e)$ increases. The effect is more important the larger the ζ potential is, in accordance with these arguments. The position of the relaxation is in agreement with the predictions of Equation 14.30: higher potential means larger frequencies.

Finally, the effect of volume fraction is the expected one: u_e is a decreasing function of ϕ owing to particle–particle interactions, whereas the inertial relaxation is also affected, since the characteristic length starts to be dominated by the interparticle distance instead of the particle size.

14.2.3.3 Comparison with Experiment

Despite the already mentioned limitations inherent to cell models, there are sufficient experimental data confirming its general validity as a tool in the electrokinetic characterization of concentrated suspensions. Although some data have been published in the evaluation of the permittivity of concentrates (Delgado et al. 1998, 2007b), the easy access to electroacoustic techniques has opened a wide range of possibilities in dynamic mobility determinations. The examples are very numerous.

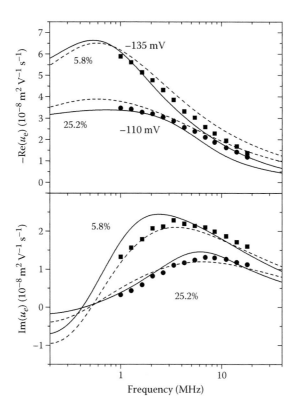

FIGURE 14.7 Real and imaginary components of the dynamic mobility of spherical silica particles of 525 nm diameter in 5×10^{-5} mol/L KCl solution. The zeta potential, the only fitting parameter, is indicated. Solid lines: cell model; dashed lines: O'Brien et al.'s formula.

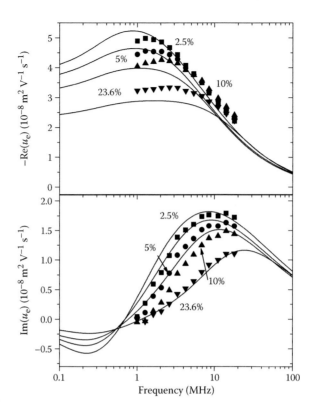

FIGURE 14.8 Real and imaginary components of the dynamic mobility of alumina particles 330 nm in diameter in 5×10^{-5} mol/L KCl. The lines are the results of the best fit of the data to the cell model calculations, using a common value of the zeta potential (-100 mV).

We will only show some examples in which the cell model was used in the evaluation of u_e data. Note first of all that the predicted mobility decrease when the volume fraction increases is in fact observed experimentally. In addition, the MWO mobility increase can be observed for the lowest electrolyte concentrations, and the inertia relaxation frequency increases with ϕ, a result equally predicted. The goodness of the model can be also confirmed by the fact that, except perhaps for the highest volume fractions, a single zeta potential value can be used to fit data corresponding to different ϕ values (Figures 14.7 and 14.8).

14.3 MODELS NOT BASED ON THE CELL APPROACH

To our knowledge, no rigorous treatment, other than those described in the cell model approach, has been developed for the evaluation of the permittivity of concentrated suspensions. It is worth mentioning the semiqualitative approximation proposed by Delgado et al. (1998), essentially based on the evaluation of the characteristic dimension L_D of the electrolyte cloud in concentration polarization, in relation to the volume fraction of solids in suspension. Specifically, these authors find that for a disperse system containing spheres of radius a:

$$L_D = a \left(1 + \frac{1}{(\phi^{-1/3} - 1)^2} \right)^{-\frac{1}{2}} \tag{14.32}$$

and from this, the alpha relaxation frequency in the dilute case, $\omega_{\alpha d}$, can be corrected for the presence of finite volume fractions as follows:

$$\omega_\alpha = \omega_{\alpha d} \left(1 + \frac{1}{(\phi^{-1/3} - 1)^2} \right). \tag{14.33}$$

A very significant effect is equally predicted on the relaxation amplitude of the concentration polarization (the subscript "d" refers again to the dilute systems):

$$\Delta\varepsilon'(0) = \Delta\varepsilon'_d(0) \left(1 + \frac{1}{(\phi^{-1/3} - 1)^2} \right)^{-\frac{3}{2}}. \tag{14.34}$$

Because of their simplicity, it may be of interest to consider the goodness of these approximate relationships in comparison with the numerical calculations reported above. This was done in Delgado et al. (2007a), where it was found that the agreement is reasonable for the permittivity increment and less accurate for the frequency, particularly for volume fractions above 10%. The fact that these formulas are so simple may still justify their use in situations where extreme precision is not needed.

The situation is better in the case of dynamic mobility models for concentrates. Two approaches can be cited in this case: O'Brien et al. (2003) and Rider and O'Brien (1993) used a "first-principles" kind of calculation, based on accounting for interactions between neighbor particles. We will not reproduce their formula here: the reader is referred to the original contributions, and to the work of Delgado et al. (2007b), but it is important to state the range of application, since it is again an analytical expression where only the Perkus–Yevick formula (Percus and Yevick 1958) is additionally required to evaluate the pair correlation function. It is also important to mention that, prior to any comparison between our calculations and O'Brien's (and in fact also for the comparison with experimental data), it is necessary to take into account necessary corrections. According to Dukhin et al. (1999b), the dynamic mobility deduced from any cell model $(u_e)_c$, does not correspond to that experimentally measured. This is because the experimental mobility is obtained from electroacoustics subjected to the condition of a reference system in which the macroscopic momentum per unit volume is zero (O'Brien et al. 2003). This means that the cell model calculations (based on a reference system fixed to the particle) must be corrected as follows (u_e would be the mobility actually measured):

$$u_e = (u_e)_c \frac{1}{1 + \dfrac{\rho_p - \rho_m}{\rho_m} \phi}. \tag{14.35}$$

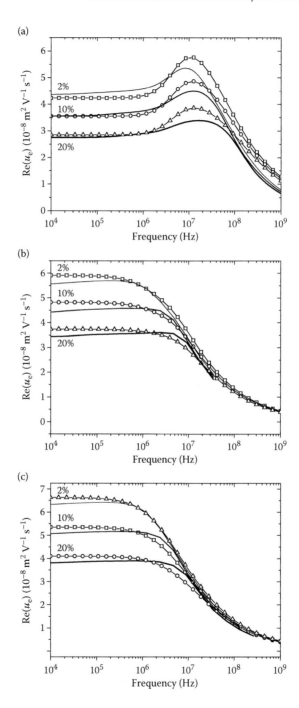

FIGURE 14.9 Real part of the dynamic mobility of suspensions of spheres with 300 nm radius and the indicated volume fraction of solids. (a) $\kappa a = 10$; (b) $\kappa a = 31$; (c) $\kappa a = 69$. Symbols: O'Brien et al.'s formula; lines: numerical calculations.

With all these precautions, it is possible to have a coherent description of the mobility spectrum of a concentrated suspension by either method, as shown in Figure 14.9. As expected, the analytical formula is less accurate for low values of κa, since it is based on the assumption of quasi-flat double layers.

A different focus on the electroacoustics of suspensions was originally developed by Dukhin et al. (1999b, 2002), through application of the so-called coupled-phase model (Dukhin et al. 1996) to the experimental signal of colloid vibration current (CVI) or colloid vibration potential (CVP): in this technique, the sample is set into vibration by an oscillating pressure wave, and the current (CVI) or potential (CVP) generated is measured. In this approach, the cell model is used, together with Shilov–Zharkikh's electrostatic boundary conditions. An interesting idea deals with considering CVI as a sedimentation current and emphasizing the aims of obtaining the zeta potential of the particles and their size distribution, with lesser emphasis on dynamic mobility (Dukhin et al. 2002).

14.4 THE CASE OF SOFT AND BRUSH PARTICLES

14.4.1 SOFT PARTICLES

We will briefly mention here the main modifications to the general cell model required to take into account the situation depicted in Figure 14.1b: the spherical particle, of radius a, is surrounded by a shell of polyelectrolyte with thickness $(d - a)$, and the whole unit is in the cell of electrolyte with radius b. The volume charge density of the polyelectrolyte layer is ρ_{pol} (total charge Q_{pol}). In the previous sections of this chapter, we have described different aspects of the electrokinetics of rigid, non-deformable particles. This is a reasonable model, given the small stresses applied to the particles, except if one wishes to study suspensions of fluid drops, or (as in the present section) when the particles are constituted by a truly rigid core and a deformable coating. The coating will be penetrable to the fluid, with a thickness that can change with the external conditions or with the charge of the layer itself.

This section is devoted to the investigation of such systems: the core particle is assumed to be covered by a layer of polyelectrolyte (this structure is denominated a *soft particle*). This confers special properties to the particles (Cohen Stuart et al. 2005), and the versatility of the layer structure, thickness, and charge allows a wide control of the properties of the suspension, including stability, rheology, electrokinetics, and so forth. For soft particles, the zeta potential (i.e., the potential at the particle core surface) becomes less important and, for most cases, loses its physical meaning. Instead, the following two potentials play an essential role in electrokinetics of soft particles, that is, the Donnan potential associated to the existence of fixed volume charges in the coating membrane Ψ_{DON} and the potential Ψ_0 (which we call the *surface potential* of a soft particle) at the boundary between the polyelectrolyte layer and the surrounding electrolyte solution (Figure 14.1b). The configuration of the coating depends to a large extent on the ionic strength of the medium and its pH. It is worth mentioning, without going into details, that a potential difference is established between the polyelectrolyte region and the external medium, namely, the Donnan potential: counterions of the charged polymer chains will be attracted toward it, while coions will be expelled. Because of the difference in electrolyte concentration in the external medium and in the coating volume, diffusive fluxes will be produced, but counterions and coions will not flow at the same rate because of the attraction or repulsion (respectively) of the chains: this will give rise to a potential difference, precisely the Donnan potential. In this section, we first describe the electrokinetic model that has been elaborated concerning the electrokinetics of soft particles and some theoretical predictions of the model. Our study includes both the dynamic mobility and the low-frequency dielectric relaxation of concentrated suspensions.

14.4.1.1 Model

Ohshima has been one of the most significant contributors to the electrokinetic theory of suspensions of soft particles, either dilute or concentrated, based again on cell models with Kuwabara conditions (Ohshima 1994, 1995a,b, 1996). Our description of the ac electrokinetics of these systems will be based on his ideas, although the use of numerical routines allows to extend the range of applicability of the results. The hydrodynamics of the polymer layer is modeled assuming a uniform distribution of point resistance centers, following the Debye–Bueche model (Ohshima 1994; Saville 2000), in such way that the Navier–Stokes equation (Equation 14.4) includes an extra term, compared to the case of rigid particle, in the region of polyelectrolyte layer:

$$
\rho_m \frac{\partial(\mathbf{v}+\mathbf{v}_e)}{\partial t} =
\begin{cases}
-\nabla P + \eta_m \nabla^2 \mathbf{v} - \sum_{j=1}^{N} ez_j n_j \nabla \Psi & \text{if } d < r < b \\[2em]
-\nabla P + \eta_m \nabla^2 \mathbf{v} - \sum_{j=1}^{N} ez_j n_j \nabla \Psi - \gamma \mathbf{v} & \text{if } a < r < d,
\end{cases}
\tag{14.36}
$$

where γ is the friction coefficient of the liquid in the polymer layer related to the force per unit volume as $6\pi\eta R N \mathbf{v} = \gamma \mathbf{v}$, where N is the number of polymer segments per unit of volume. Although this will be confirmed below, let us mention from the beginning that several authors (Levine et al. 1983; Wunderlich 1982) have shown that distributing charge throughout the layer may produce a mobility in excess of that when the same charge is confined to the rigid core.

Another necessary change affects the Poisson–Boltzmann equation (Equation 14.3), as the presence of the charged groups along the polymer chains is needed to be taken into account:

$$
\nabla^2 \Psi(\mathbf{r}) =
\begin{cases}
-\sum_{j=1}^{N} \dfrac{ez_j n_j(\mathbf{r})}{\varepsilon_m \varepsilon_0} & \text{if } d < r < b \\[2em]
-\sum_{j=1}^{N} \dfrac{ez_j n_j(\mathbf{r})}{\varepsilon_m \varepsilon_0} - \dfrac{\rho_{pol}}{\varepsilon_m \varepsilon_0} & \text{if } a < r < d.
\end{cases}
\tag{14.37}
$$

The procedure used to solve the resulting system of equations is similar to that used for hard particles, and it will not be repeated here. It must be mentioned, however, that new boundary conditions are required at the membrane/solution interface. These refer to the continuity of

- Electric potential and normal displacement:

$$\Gamma(d^-) = \Gamma(d^+)$$

$$\left.\frac{d\Gamma}{dr}\right|_{r=d^-} = \left.\frac{d\Gamma}{dr}\right|_{r=d^+} \tag{14.38}$$

- Fluid velocity:

$$h(d^-) = h(d^+)$$

$$\left.\frac{dh}{dr}\right|_{r=d^-} = \left.\frac{dh}{dr}\right|_{r=d^+} \tag{14.39}$$

- Vorticity:

$$\nabla \times \mathbf{v}\big|_{r=d^-} = \nabla \times \mathbf{v}\big|_{r=d^+} \tag{14.40}$$

- Concentrations and velocities of ionic species ($j = 1,2,\ldots N$):

$$\phi_j(d^-) = \phi_j(d^+)$$

$$\left.\frac{d\phi_j}{dr}\right|_{r=d^-} = \left.\frac{d\phi_j}{dr}\right|_{r=d^+} \tag{14.41}$$

14.4.1.2 Representative Results

The effect of volume fraction (for different values of the dimensionless friction parameter λa, where λ is defined as $(\gamma/\eta_m)^{1/2}$) on the dynamic mobility is plotted in Figure 14.10. Note that increasing the friction coefficient implies a strong reduction of the mobility, although if λa is not too large, it is confirmed that the electrophoretic velocity is higher if the charge is distributed than if it sits on a compact layer on the rigid particle. Nevertheless, whatever the range of values considered for λa, it is a general result that increasing the volume fraction of solids brings about a decrease in the mobility and an increase in the inertial relaxation frequency, which is better appreciated in the imaginary part of the dynamic mobility.

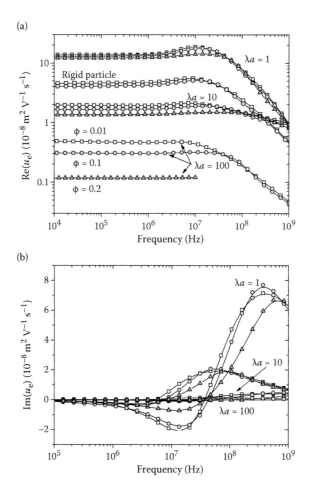

FIGURE 14.10 Real (a) and imaginary (b) components of the dynamic mobility of suspensions of spheres with radius $a = 100$ nm in 1 mmol/L KCl solutions. The particles have $\varepsilon_p = 4.5$, $\rho_p = 2200$ kg/m^3, $Q = 1.67 \times 10^{-15}$ C. The polyelectrolyte layer thickness is 150 nm, and $Q_{pol} = 1.4 \times 10^{-15}$ C. Dimensionless friction parameter, λa, as indicated. Volume fractions: 0.01 (squares), 0.1 (circles), and 0.2 (triangles). (Reprinted with permission from Ahualli, S. et al. 2009, 1896–1997. Copyright 2009 American Chemical Society.)

Similar plots for the relative permittivity are displayed in Figure 14.11. It is clearly dominated by the changes in the volume fraction of solids and, to a lesser extent, by the friction parameter: increasing this number reduces the amplitude of the alpha relaxation, with negligible effect on the alpha relaxation frequency. This is a manifestation of the hindered flow of counterions because of increased friction. It is also interesting to note that the permittivity spectra of suspensions of rigid particles lay between those of soft colloids with low and high friction. Two counteracting phenomena can be mentioned for the explanation of this behavior: the counterion concentration is enhanced by the presence of the membrane and spreads over a region (the thickness of the latter) wider than the EDL of the rigid particle. This will increase the permittivity, but, on

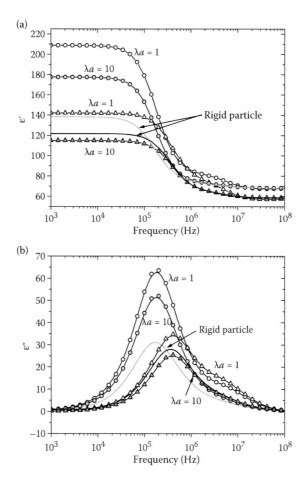

FIGURE 14.11 Real (a) and imaginary (b) components of the relative permittivity of suspensions of spheres with radius a = 100 nm in 1 mmol/L KCl solutions. The particles have ε_p = 4.5, ρ_p = 2200 kg/m³, Q = 1.67 × 10⁻¹⁵ C. The polyelectrolyte layer thickness is 150 nm, and Q_{pol} = 1.4 × 10⁻¹⁵ C. Dimensionless friction parameter, λa, as indicated. Volume fractions: 0.01 (squares), 0.1 (circles), and 0.2 (triangles). (Reprinted with permission from Ahualli, S. et al. 2009, 1896–1997. Copyright 2009 American Chemical Society.)

the other hand, as mentioned before, the friction tends to interfere with ion migration, and this is a negative contribution to the low-frequency permittivity.

Without intending to be exhaustive, it appears of interest to also consider how the charges on the particle core, Q, and in the polymer layer, Q_{pol}, affect the permittivity and mobility spectra. Figure 14.12 illustrates the main features. It is first of all clear that the electric permittivity of the suspensions is strongly affected by both the polymer and the particle charges: only when the signs of Q and Q_{pol} are opposite can we observe a reduction of the permittivity over that of a rigid particle. Such a reduction ceases as expected when the differences between the two charges are large enough. The dynamic mobility, in turn, shows similar behavior, although it is noticeable how

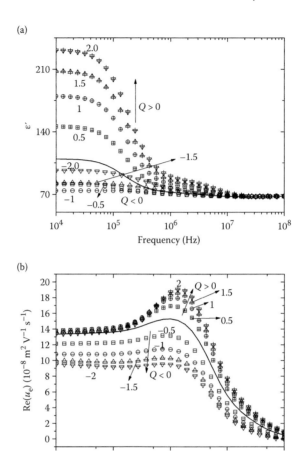

FIGURE 14.12   Effect of the particle charge Q (a and b; $Q_{pol} = 1.5 \times 10^{-15}$ C) and of the polymer charge Q_{pol} (c and d; $Q = 1.7 \times 10^{-15}$ C) on the real parts of the relative permittivity (a and c) and the dynamic mobility (b and d) of suspensions of 100 nm particles (ε_p and ρ_p as in Figure 14.10), with $\phi = 1\%$. The labels on the data are the corresponding charges (in 10^{-15} C). In all cases: 1 mmol/L KCl and 150 nm coating thickness. Solid lines: uncharged core (a,b), or uncharged layer (c,d). (Reprinted with permission from Ahualli, S. et al. 2009, 1896–1997. Copyright 2009 American Chemical Society.)

$Re(u_e)$ is more strongly dependent on the polymer than on the core charge: even the mobility sign is that of Q_{pol}.

14.4.2   Polyelectrolyte Brushes

A related but surprisingly different situation concerns the so-called *(spherical) polyelectrolyte brushes* or SPBs, complex particles consisting of a rigid core on which a polyelectrolyte layer is densely grafted, in such a way that the polymer chains emerge radially from the surface and arrange radially, as schematically shown in

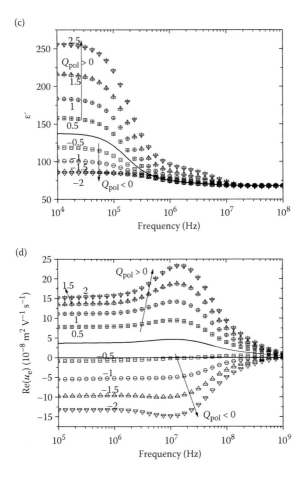

FIGURE 14.12 (Continued) Effect of the particle charge Q (a and b; Q_{pol} = 1.5 × 10^{-15} C) and of the polymer charge Q_{pol} (c and d; Q = 1.7 × 10^{-15} C) on the real parts of the relative permittivity (a and c) and the dynamic mobility (b and d) of suspensions of 100 nm particles (ε_p and ρ_p as in Figure 14.10), with ϕ = 1%. The labels on the data are the corresponding charges (in 10^{-15} C). In all cases: 1 mmol/L KCl and 150 nm coating thickness. Solid lines: uncharged core (a,b), or uncharged layer (c,d). (Reprinted with permission from Ahualli, S. et al. 2009, 1896–1997. Copyright 2009 American Chemical Society.)

Figure 14.1c (Hoffmann et al. 2009; Mei et al. 2006). If the ionic strength of the medium is low, the charged groups along the chains will keep the chains stretched. Shrinking of the chains will likely take place if the ionic concentration is raised or, as it is of interest to us here, if the concentrations of SPBs is also large.

Although the most usual technique of electrophoresis can be used for inferring structural changes in SPBs, the ac methods dealt with in the present chapter may be even more informative. In a recent contribution (Jiménez et al. 2011), it was shown that the dynamic mobility and permittivity spectra of SPB suspensions are hardly comparable to those of equivalent (as to size or surface charge) rigid or even soft particles. Figures 14.13 and 14.14 are a good illustration of this. Note, first of all, that the dynamic mobility of SPBs can be extremely high and was never reported for particles of any kind; in addition, the mobility increases with the volume fraction of particles, and this is against any results reported above (see Figure 14.2). This suggests existence of electro-osmotic flow associated to the counterions of the polymer charged groups. This occurs even though the mobility of monovalent counterions has been shown to be lower than that of free counterions (Hoffmann et al. 2009, Zimmermann et al. 2005).

As can be understood from our treatment above, this behavior should also manifest in the EDL dynamics as sensed by dielectric dispersion spectra. Figure 14.14 confirms that this is the case: the volume distributed charge in the polyelectrolyte layer produces an enhanced concentration polarization leading to extreme values of the low-frequency permittivity, far from the maximum permittivities achieved with polymer particles of comparable size and charge. Not only that, in Jiménez et al. (2011), it was demonstrated that the spectra of the relaxation increased in amplitude and characteristic frequency when the particle concentration increases.

All these features can be explained by using a modified cell model in which the friction parameter λa is assumed to decrease with the squared distance to the particle surface. The same applies to ρ_{pol}. A general treatment for the study of soft particles with non-homogeneous coating characteristics was developed by Duval and Ohshima (2006), but in our case, we deal with concentrated systems. The methods

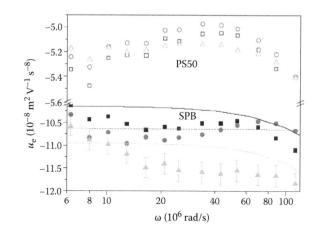

FIGURE 14.13 (**See color insert.**) Dynamic mobility of polyelectrolyte brushes (SPBs) and polystyrene spheres 50 nm in radius (PS50) as a function of the frequency of the applied electric field for different particle concentrations in 0.5 mM NaCl. Symbols denote the experimental results. Lines are the result of model predictions with the following parameters: particle charge: -5×10^{-16} C, polymer charge plus condensed counterions charge: -8.5×10^{-14} C, dimensionless friction parameter: $\lambda R_c = 110$ at the particle surface. Squares and solid line: 5 wt%. Circles and dashed line: 6 wt%. Triangles and dotted line: 7 wt%. The dispersion in these measurements is around 2%–3%. (Details in Jiménez et al. 2011. http://pubs.rsc.org/en/content/articlelanding/2011/sm/c0sm01544j. Reproduced by permission of the Royal Society of Chemistry.)

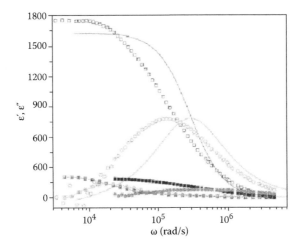

FIGURE 14.14 **(See color insert.)** Real (squares) and imaginary (circles) parts of the relative permittivity of suspensions of SPBs (open symbols), polystyrene particles 50 nm in radius, PS50 (filled symbols), and 168 nm in radius, PS168 (crossed symbols), as a function of the frequency of the applied field. Particle concentration: 4 wt%. Ionic strength: 0.5 mM (NaCl). Lines are the model predictions with the same parameters as in Figure 14.13, and particle concentration is 4 wt%. (Details in Jiménez et al. 2011. http://pubs.rsc.org/en/content/articlelanding/2011/sm/c0sm01544j. Reproduced by permission of the Royal Society of Chemistry.)

explained for soft particles in Section 14.4.1 are applicable to the present case, without any special modification of the general equations. The results found (lines in Figures 14.13 and 14.14) are an indication of the validity of our approach: a single set of parameters can explain the whole set of data, as indicated in the caption to Figure 14.13.

14.5 NON-SPHERICAL GEOMETRIES

14.5.1 MODELS

When particles with geometry different from the sphere (spheroids will be considered) are to be studied by ac methods, the polarization phenomena are considerably complicated by the presence of different length scales associated to the characteristic sizes of the particles. In this situation, the dipole coefficient becomes a tensor quantity, and the induced dipole moment is not necessarily parallel to the field:

$$\mathbf{d} = 3V\varepsilon_0\varepsilon_m \overset{\leftrightarrow}{\mathbf{C}} \cdot \mathbf{E}. \tag{14.42}$$

Fortunately, because of the linearity of the problem, we can consider separately every component, and the response of the whole system will be obtained from an appropriate average of them. Furthermore, many colloidal particles found in practical situations can be well described by assuming spheroidal geometry (Lee et al. 2011), thus allowing for analytical treatments. Thus, in the frame of reference defined by the symmetry axes of the spheroid, the only non-zero components of the resulting diagonal tensor are $C_\parallel$ and $C_\perp$, respectively, in directions parallel and perpendicular to the symmetry axis. We can write:

$$d_i = 4\pi\varepsilon_0\varepsilon_m ab^2 C_i E_i, \; i = \parallel, \perp \qquad (14.43)$$

where d_i and E_i are the components of the induced dipole and the field in either parallel or perpendicular directions to the symmetry axis of the spheroid, and $2a$ and $2b$ are the symmetry axis and the diameter of the spheroid, respectively. From this, we also distinguish between the parallel and perpendicular components of the dynamic mobility $u_{e,i}(\omega)$ and the relative electric permittivity $\varepsilon_i(\omega)$. Their macroscopic value is estimated from appropriate averages over the whole suspension. When the intensity of the applied field E is not very high, the interaction energy between the induced dipole moment and the field, W, is not enough to disturb the random orientation ensured by Brownian motion (i.e., $W \simeq \varepsilon_0\varepsilon_m V E^2 \ll k_B T$). In this case, the macroscopic values of the dynamic mobility and electric permittivity are estimated as the following averages over a homogeneous suspension:

$$\left\langle u_e(\omega) \right\rangle = \frac{u_{e,\parallel}(\omega) + 2u_{e,\perp}(\omega)}{3}$$

$$\left\langle \varepsilon(\omega) \right\rangle = \frac{\varepsilon_\parallel(\omega) + 2\varepsilon_\perp(\omega)}{3}. \qquad (14.44)$$

The dielectric and mobility spectra of non-spherical particles show the same relaxations already discussed in the case of spherical particles, namely, alpha and MWO, plus the inertia relaxation observed in electroacoustics. However, they can be complicated by the presence of different characteristic sizes. In particular, the strong dependence of the alpha relaxation on particle size manifests in the appearance of multiple relaxation processes in the low-frequency part of the spectrum, associated to the different characteristic sizes of the particles. In spheroidal particles, two alpha relaxations are expected, at characteristic frequencies $\omega_{\alpha,i} = 2D/L_{D,i}^2$, where $L_{D,i}$ are the two diffusion lengths of the particles oriented parallel or perpendicular to the applied field, which are of the order of the semiaxis of the particles, a and b. If these parameters are different enough, like in the case of highly elongated particles, the two alpha processes can be resolved in the dielectric spectrum, obtaining valuable information about the geometry of the suspended particles and the polarization state of the solid/liquid interface.

Let us now analyze how we can calculate the different components of the electrophoretic mobility and electric permittivity of moderately concentrated suspensions of spheroidal particles, to be used together with the averages given in Equation 14.44 to estimate the frequency response of the whole system. We do not introduce any specific EDL model, and we express these quantities as a function of the dipole coefficient of dilute suspensions of spheroidal particles, which can be calculated by well-established procedures (Dukhin and Shilov 1980; Fixman 2006; Grosse et al. 1999). The spectrum of the dynamic electrophoretic mobility of a concentrated suspension of spheroids can be found through a version of the Helmholtz–Smoluchowski equation, where additional effects are accounted for by multiplying it by three functions (Rica et al. 2009):

$$u_{e,i}(\omega) = \frac{\varepsilon_0 \varepsilon_m}{\eta_m} \zeta f_i^1(\omega) f_i^2(\omega) f_i^3(\omega). \tag{14.45}$$

Here, $f_i^1(\omega)$ is the inertia function, determining the cutoff frequency (inertia relaxation frequency, ω_{in}) above which the dynamic mobility falls to zero. This frequency can be estimated as $\omega_{in} = \eta_m / \rho_m l^2$, with l being the minimum dimension of the particle in the direction perpendicular to its motion. $f_i^2(\omega)$ considers the EDL polarization, establishing the dependence with the dipole coefficient. It was first evaluated by Loewenberg and O'Brien (1992) for $\kappa l_{min} \ll 1$, with l_{min} being the minimum dimension of the spheroid:

$$f_i^2(\omega) = 1 - L_i - 3L_i(1 - L_i)C_i(\omega), \tag{14.46}$$

where L_i are the depolarization factors, carrying information on the axes of the spheroid, and whose expression can be found elsewhere (Rica et al. 2012). Finally, $f_i^3(\omega)$ accounts for the interactions between particles. A simple way to estimate their effects is through the following expression:

$$f_i^3(\omega) = \frac{1 - \phi}{1 - \phi C_i(\omega)} \left(1 + \phi \frac{\rho_p - \rho_m}{\rho_m} \right). \tag{14.47}$$

With this, the components of the electric permittivity along each direction are estimated as (Rica et al. 2010):

$$\varepsilon_i(\omega) = \varepsilon_m + \Delta\varepsilon_i(\omega) = \varepsilon_m [1 + 3\phi C_i(\omega)] - j \frac{3\phi K_m}{\omega \varepsilon_0} [C_i(\omega) - C_i(0)]. \tag{14.48}$$

The treatment still requires consideration of finite volume fractions of solids. A modified version of Equations 14.33 and 14.34 can be used with this purpose, introducing a certain degree of approximation, recalling the semiqualitative origin of such equations:

$$\Delta\varepsilon'(0) = \Delta\varepsilon_d'(0) \left(1 + \frac{1}{(\phi_{eff}^{-1/3} - 1)^2} \right)^{-\frac{3}{2}}$$

$$\omega_\alpha = \omega_{\alpha d} \left(1 + \frac{1}{(\phi_{eff}^{-1/3} - 1)^2} \right) \tag{14.49}$$

where we have introduced an *effective volume fraction* $\phi_{eff} = f\phi$ needed to properly fit our experimental results (Rica et al. 2011). Here, f is a factor including information on the size and shape of the concentration polarization clouds and therefore on the particle geometry. Note that f will also be different for each orientation of the particles with respect to the applied field.

14.5.2 REPRESENTATIVE RESULTS

Figure 14.15 illustrates how powerful is the LFDD technique for analyzing the polarization state of concentrated suspensions of non-spherical particles. For convenience in the correction for electrode polarization, we focus in the logarithmic derivative of the $\varepsilon'(\omega)$ spectrum (Jimenez et al. 2002):

$$\varepsilon_D''(\omega) = -\frac{1}{2\pi} \frac{\partial \varepsilon'(\omega)}{\partial \ln \omega} \qquad (14.50)$$

very similar, in fact, to the spectrum of the imaginary component of the relative permittivity. The derivative spectra show clearly two alpha processes at frequencies 5 kHz and 100 kHz, which are better resolved the larger the particle concentration is, owing to their different tendencies. The low-frequency (LF) relaxation process is associated to the large semiaxis of the spheroids, while the high frequency (HF) is associated to the small semiaxis (Rica et al. 2011). It is interesting to note that while the HF process describes the same dependence with the volume fraction of solids already discussed in the case of spherical particles (the amplitude of the dielectric increment goes through a maximum and then decreases, while its characteristic frequency $\omega_{\alpha,\perp}$ increases continuously with ϕ), the LF one has an anomalous behavior: the dielectric increment reaches a plateau at high ϕ, while the characteristic frequency of the alpha process does not depend on the volume fraction in the studied range of particle concentration. The origin of this anomalous behavior has not been fully clarified yet, and available studies justify it in terms of the presence of a nematic phase or

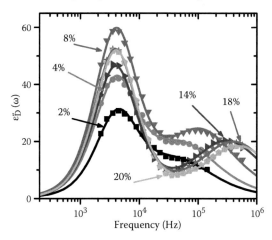

FIGURE 14.15 **(See color insert.)** Spectra of the logarithmic derivative of the relative electric permittivity of suspensions of elongated hematite particles (semiaxes $a = 276 \pm 18$ nm, $b = 45 \pm 6$ nm) in a 0.5 mM KCl solution at pH 4, for the indicated values of the volume fraction of solids (ϕ). Symbols: experimental data. Lines: best fits to the logarithmic derivative of a combination of two Cole–Cole relaxation functions. (Details in Rica et al. 2011. http://pubs.rsc.org/en/content/articlelanding/2011/sm/c1sm05153a. Reproduced by permission of The Royal Society of Chemistry.)

the asymmetric geometry of the concentration polarization clouds around elongated particles. Whatever the origin of this behavior, this experiment illustrates the large amount of information provided by LFDD. The joint analysis of LFDD and electroacoustic data gives access to the most relevant part of the spectra of colloidal suspensions, as it allows us to identify the alpha and MWO dielectric relaxations together with the inertia decay. An example of this comparison is given in Figure 14.16.

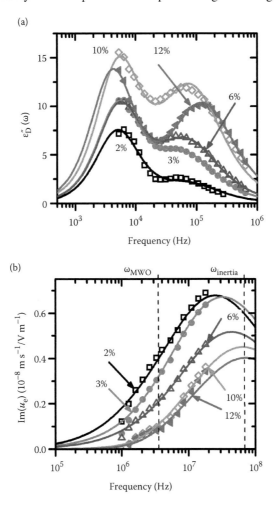

FIGURE 14.16 **(See color insert.)** Spectra of (a) the logarithmic derivative of the relative electric permittivity and (b) the imaginary part of the dynamic electrophoretic mobility of suspensions of elongated hematite particles (semiaxes $a = 276 \pm 18$ nm, $b = 45 \pm 6$ nm) in a 0.5 mM KCl solution at pH 5.8, for the indicated values of the volume fraction of the solids (ϕ). Symbols: experimental data. Lines: best fits to the models. Dashed horizontal lines in panel (b) indicate the approximate position of the MWO relaxation and inertia decay in the electroacoustic spectrum of the suspension at 10% concentration. (Details in Rica et al. 2012. http://pubs.rsc.org/en/content/articlelanding/2012/sm/c2sm07365j. Reproduced by permission of the Royal Society of Chemistry.)

The dynamic mobility spectra, displayed on the panel (b) of Figure 14.16, are coherent with the mobility reduction associated to volume fraction increase. Both the MWO relaxations and the beginning of the inertia decay are observable, mainly at the highest volume fractions, because the MWO relaxation is largely unaffected by volume fraction whereas the inertia decay declines in amplitude when ϕ is increased. Finally, let us mention that the observation of the MWO process in the megahertz region of electroacoustics confirms our interpretation of the presence of two alpha relaxations in the LFDD experiments.

14.6 SUMMARY AND CONCLUSIONS

In this chapter, we have tried to give an overview of the main electrokinetic phenomena in concentrated systems under the action of alternating fields. We have justified their importance as methods largely suited for the characterization of concentrated slurries, mostly interesting in many different applications. Our theoretical approach has been mainly based on cell models, which have been tested against experimental data in many different cases and are conceptually simple. However, other approaches exist, particularly in the case of dynamic mobility, which have also been briefly described. We can find applicability of the described techniques and, most importantly, of the theoretical approaches available even in situations where we encounter particles far from the ideal rigid sphere (soft coated spheres, spheroids), but in approximate terms in some cases.

ACKNOWLEDGMENTS

Financial support by Project P08-FQM-3993 (Junta de Andalucía and FEDER funds) and MICINN, Spain (project FIS2010-19493), is greatly appreciated. One of us (M.L.J.) also acknowledges support from Ministerio de Economía y Competitividad, Spain, for her "Ramón and Cajal" Contract.

REFERENCES

Ahualli, S., A.V. Delgado, and C. Grosse, A simple model of the high-frequency dynamic mobility in concentrated suspensions. *Journal of Colloid and Interface Science*, 2006a. 301(2): pp. 660–667.

Ahualli, S. et al., Dynamic electrophoretic mobility of concentrated dispersions of spherical colloidal particles. On the consistent use of the cell model. *Langmuir*, 2006b. 22(16): pp. 7041–7051.

Alejo, B. and A. Barrientos, Model for yield stress of quartz pulps and copper tailings. *International Journal of Mineral Processing*, 2009. 93(3–4): pp. 213–219.

Arroyo, F.J. et al., Dielectric dispersion of colloidal suspensions in the presence of stern layer conductance: Particle size effects. *Journal of Colloid and Interface Science*, 1999a. 210(1): pp. 194–199.

Arroyo, F.J., F. Carrique, and A.V. Delgado, Effects of temperature and polydispersity on the dielectric relaxation of dilute ethylcellulose suspensions. *Journal of Colloid and Interface Science*, 1999b. 217(2): pp. 411–416.

Arroyo, F.J. et al., Dynamic mobility of concentrated suspensions. Comparison between different calculations. *Physical Chemistry Chemical Physics*, 2004. 6(7): pp. 1446–1452.

Batchelor, G.K., Sedimentation in a dilute dispersion of spheres. *Journal of Fluid Mechanics*, 1972. 52(Mar 28): pp. 245–250.

Beirowski, J. et al., Freeze-drying of nanosuspensions, part 3: Investigation of factors compromising storage stability of highly concentrated drug nanosuspensions. *Journal of Pharmaceutical Sciences*, 2012. 101(1): pp. 354–362.

Carrique, F., F.J. Arroyo, and A.V. Delgado, Sedimentation velocity and potential in a concentrated colloidal suspension—Effect of a dynamic Stern layer. *Colloids and Surfaces A—Physicochemical and Engineering Aspects*, 2001a. 195(1–3): pp. 157–169.

Carrique, F., F.J. Arroyo, and A.V. Delgado, Electrokinetics of concentrated suspensions of spherical colloidal particles: Effect of a dynamic Stern layer on electrophoresis and DC conductivity. *Journal of Colloid and Interface Science*, 2001b. 243(2): pp. 351–361.

Carrique, F., F.J. Arroyo, and A.V. Delgado, Electrokinetics of concentrated suspensions of spherical colloidal particles with surface conductance, arbitrary zeta potential, and double-layer thickness in static electric fields. *Journal of Colloid and Interface Science*, 2002. 252(1): pp. 126–137.

Carrique, F. et al., Dielectric response of concentrated colloidal suspensions. *Journal of Chemical Physics*, 2003a. 118(4): pp. 1945–1956.

Carrique, F. et al., Influence of double-layer overlap on the electrophoretic mobility and DC conductivity of a concentrated suspension of spherical particles. *Journal of Physical Chemistry B*, 2003b. 107(14): pp. 3199–3206.

Carrique, F. et al., Influence of cell-model boundary conditions on the conductivity and electrophoretic mobility of concentrated suspensions. *Advances in Colloid and Interface Science*, 2005. 118(1–3): pp. 43–50.

Cohen Stuart, M., R. de Vries, and J. Lyklema, Polyelectrolytes. In *Fundamentals of Interface and Colloid Science*, J. Lyklema, editor. 2005, Amsterdam: Elsevier.

Delacey, E.H.B. and L.R. White, Dielectric response and conductivity of dilute suspensions of colloidal particles. *Journal of the Chemical Society—Faraday Transactions II*, 1981. 77: pp. 2007–2039.

Delgado, A.V. et al., The effect of the concentration of dispersed particles on the mechanisms of low-frequency dielectric dispersion (LFDD) in colloidal suspensions. *Colloids and Surfaces A—Physicochemical and Engineering Aspects*, 1998. 140(1–3): pp. 139–149.

Delgado, A.V., ed. *Interfacial Electrokinetics and Electrophoresis*. Surfactant Science. Vol. 106. 2002, New York: Marcel Dekker.

Delgado, A.V. et al., Measurement and interpretation of electrokinetic phenomena. *Journal of Colloid and Interface Science*, 2007a. 309(2): pp. 194–224.

Delgado, A.V. et al., Electrokinetics of concentrated colloidal dispersions. In *Molecular and Colloidal Electro-Optics*. 2007b. pp. 149–191, Boca Raton, FL: CRC Press.

Derkach, S.R., Rheology of emulsions. *Advances in Colloid and Interface Science*, 2009. 151(1–2): pp. 1–23.

Ding, J.M. and H.J. Keh, The electrophoretic mobility and electric conductivity of a concentrated suspension of colloidal spheres with arbitrary double-layer thickness. *Journal of Colloid and Interface Science*, 2001. 236(1): pp. 180–193.

Dukhin, S.S. and V.N. Shilov, *Dielectric Phenomena and the Double Layer in Disperse Systems and Polyelectrolytes*. 1974, New York: John Wiley & Sons.

Dukhin, S.S. and V.N. Shilov, Kinetic aspects of electrochemistry of disperse systems. 2. Induced dipole-moment and the nonequilibrium double-layer of a colloid particle. *Advances in Colloid and Interface Science*, 1980. 13(1–2): pp. 153–195.

Dukhin, A.S., P.J. Goetz, and C.W. Hamlet, Acoustic spectroscopy for concentrated polydisperse colloids with low density contrast. *Langmuir*, 1996. 12(21): pp. 4998–5003.

Dukhin, A.S., V. Shilov, and Y. Borkovskaya, Dynamic electrophoretic mobility in concentrated dispersed systems. Cell model. *Langmuir*, 1999a. 15(10): pp. 3452–3457.

Dukhin, A.S. et al., Electroacoustics for concentrated dispersions. *Langmuir*, 1999b. 15(10): pp. 3445–3451.

Dukhin, A.S. et al., Electroacoustic phenomena in concentrated dispersions: Theory, experiment, applications. In *Interfacial Electrokinetics and Electrophoresis*, A.V. Delgado, Editor. 2002, New York: Marcel Dekker.

Duval, J.F.L. and H. Ohshima, Electrophoresis of diffuse soft particles. *Langmuir*, 2006. 22(8): pp. 3533–3546.

Fixman, M., A macroion electrokinetics algorithm. *Journal of Chemical Physics*, 2006. 124(21): 214506, pp. 1–19.

Grosse, C., Relaxation mechanisms of homogeneous particles and cells suspended in aqueous electrolyte solutions. In *Interfacial Electrokinetics and Electrophoresis*, A.V. Delgado, editor. 2002, New York: Marcel Dekker.

Grosse, C. and M.C. Tirado, Measurement of the dielectric properties of polystyrene particles in electrolyte solution. In *Microwave Processing of Materials V*. 1996. pp. 287–293.

Grosse, C., S. Pedrosa, and V.N. Shilov, Calculation of the dielectric increment and characteristic time of the LFDD in colloidal suspensions of spheroidal particles. *Journal of Colloid and Interface Science*, 1999. 220(1): pp. 31–41.

Happel, J. and H. Brenner, Viscous flow in multiparticle systems—Motion of spheres and a fluid in a cylindrical tube. *AIChE Journal*, 1957. 3(4): pp. 506–513.

Hobbie, E.K., Shear rheology of carbon nanotube suspensions. *Rheologica Acta*, 2010. 49(4): pp. 323–334.

Hoffmann, M. et al., Surface potential of spherical polyelectrolyte brushes in the presence of trivalent counterions. *Journal of Colloid and Interface Science*, 2009. 338(2): pp. 566–572.

Hsu, J.P., E. Lee, and F.Y. Yen, Electrophoresis of concentrated spherical particles with a charge-regulated surface. *Journal of Chemical Physics*, 2000. 112(14): pp. 6404–6410.

Hsu, J.P., E. Lee, and F.Y. Yen, Dynamic electrophoretic mobility in electroacoustic phenomenon: Concentrated dispersions at arbitrary potentials. *Journal of Physical Chemistry B*, 2002. 106(18): pp. 4789–4798.

Hunter, R.J., *Foundations of Colloid Science*. Vol. 1. 1987, Oxford: Oxford University Press.

Jiménez, M.L. et al., Analysis of the dielectric permittivity of suspensions by means of the logarithmic derivative of its real part. *Journal of Colloid and Interface Science*, 2002. 249(2): pp. 327–335.

Jiménez, M.L. et al., Giant permittivity and dynamic mobility observed for spherical polyelectrolyte brushes. *Soft Matter*, 2011. 7(8): pp. 3758–3762.

Johnson, S.B. et al., Surface chemistry–rheology relationships in concentrated mineral suspensions. *International Journal of Mineral Processing*, 2000. 58(1–4): pp. 267–304.

Kijlstra, J., H.P. Vanleeuwen, and J. Lyklema, Effects of surface conduction on the electrokinetic properties of colloids. *Journal of the Chemical Society—Faraday Transactions*, 1992. 88(23): pp. 3441–3449.

Kijlstra, J., H.P. Vanleeuwen, and J. Lyklema, Low-frequency dielectric-relaxation of hematite and silica sols. *Langmuir*, 1993. 9(7): pp. 1625–1633.

Kuwabara, S., The forces experienced by randomly distributed parallel circular cylinders or spheres in a viscous flow at small Reynolds numbers. *Journal of the Physical Society of Japan*, 1959. 14(4): pp. 527–532.

Landau, L.D. and E.M. Lifshitz, *Fluid Mechanics*. 2nd ed. Course of Theoretical Physics. 2000, Oxford: Butterworth-Heinemann.

Lee, E., J.W. Chu, and J.P. Hsu, Electrophoretic mobility of a concentrated suspension of spherical particles. *Journal of Colloid and Interface Science*, 1999. 209(1): pp. 240–246.

Lee, K.J., J. Yoon, and J. Lahann, Recent advances with anisotropic particles. *Current Opinion in Colloid and Interface Science*, 2011. 16(3): pp. 195–202.

Levine, S. and G. Neale, Electrophoretic mobility of multiparticle systems. *Journal of Colloid and Interface Science*, 1974a. 49(2): pp. 330–332.

Levine, S. and G.H. Neale, Prediction of electrokinetic phenomena within multiparticle systems. 1. Electrophoresis and electroosmosis. *Journal of Colloid and Interface Science*, 1974b. 47(2): pp. 520–529.

Levine, S. et al., Theory of the electrokinetic behavior of human-erythrocytes. *Biophysical Journal*, 1983. 42(2): pp. 127–135.

Lin, W.H., E. Lee, and J.P. Hsu, Electrophoresis of a concentrated spherical dispersion at arbitrary electrical potentials. *Journal of Colloid and Interface Science*, 2002. 248(2): pp. 398–403.

Loewenberg, M. and R.W. Obrien, The dynamic mobility of nonspherical particles. *Journal of Colloid and Interface Science*, 1992. 150(1): pp. 158–168.

Lyklema, J., *Fundamentals of Interface and Colloid Science*. Vol. 2. 1995, London: Academic Press.

Matijevic, E., ed. *Electrokinetic Phenomena*. Surface and Colloid Science. Vol. 7. 1974, New York: John Wiley & Sons.

Mei, Y. et al., Collapse of spherical polyelectrolyte brushes in the presence of multivalent counterions. *Physical Review Letters*, 2006. 97(15): 158301, pp. 1–4.

Melito, H.S. and C.R. Daubert, Rheological innovations for characterizing food material properties. In *Annual Review of Food Science and Technology*, Vol. 2, M.P. Doyle and T.R. Klaenhammer, Editors. 2011. pp. 153–179, Palo Alto: Annual Reviews.

Moeller, W.J., J.T. Overbeek, and G.A.J. Vanos, Interpretation of conductance and transference of bovine serum albumin solutions. *Transactions of the Faraday Society*, 1961a. 57(2): p. 325.

Moeller, W.J., G.A.J. Vanos, and J.T. Overbeek, Electric conductivity and transference of alkali albuminates. *Transactions of the Faraday Society*, 1961b. 57(2): p. 312.

Myers, D.F. and D.A. Saville, Dielectric-spectroscopy of colloidal suspensions. 1. The dielectric spectrometer. *Journal of Colloid and Interface Science*, 1989. 131(2): pp. 448–460.

O'Brien, R.W. and L.R. White, Electrophoretic mobility of a spherical colloidal particle. *Journal of the Chemical Society—Faraday Transactions II*, 1978. 74p: pp. 1607–1626.

O'Brien, R.W., A. Jones, and W.N. Rowlands, A new formula for the dynamic mobility in a concentrated colloid. *Colloids and Surfaces A—Physicochemical and Engineering Aspects*, 2003. 218(1–3): pp. 89–101.

Ohshima, H., Electrophoretic mobility of soft particles. *Journal of Colloid and Interface Science*, 1994. 163(2): pp. 474–483.

Ohshima, H., Electrophoresis of soft particles. *Advances in Colloid and Interface Science*, 1995a. 62(2–3): pp. 189–235.

Ohshima, H., Electrophoretic mobility of soft particles. *Colloids and Surfaces A—Physicochemical and Engineering Aspects*, 1995b. 103(3): pp. 249–255.

Ohshima, H., Electrostatic interaction between two parallel cylinders. *Colloid and Polymer Science*, 1996. 274(12): pp. 1176–1182.

Ohshima, H., Electrophoretic mobility of spherical colloidal particles in concentrated suspensions. *Journal of Colloid and Interface Science*, 1997. 188(2): pp. 481–485.

Ohshima, H., Electrical conductivity of a concentrated suspension of spherical colloidal particles. *Journal of Colloid and Interface Science*, 1999. 212(2): pp. 443–448.

Ohshima, H., Sedimentation potential and velocity in a concentrated suspension of soft particles. *Journal of Colloid and Interface Science*, 2000a. 229(1): pp. 140–147.

Ohshima, H., Cell model calculation for electrokinetic phenomena in concentrated suspensions: an Onsager relation between sedimentation potential and electrophoretic mobility. *Advances in Colloid and Interface Science*, 2000b. 88(1–2): pp. 1–18.

Ohshima, H., *Theory of Colloid and Interfacial Electric Phenomena*. Interface Science and Technology. Vol. 12. 2006, London: Academic Press.

Ohshima, H. et al., Sedimentation-velocity and potential in a dilute suspension of charged spherical colloidal particles. *Journal of the Chemical Society—Faraday Transactions II*, 1984. 80: pp. 1299–1317.

Percus, J.K. and G.J. Yevick, Analysis of classical statistical mechanics by means of collective coordinates. *Physical Review*, 1958. 110(1): pp. 1–13.

Rica, R.A., M.L. Jimenez, and A.V. Delgado, Dynamic mobility of rodlike goethite particles. *Langmuir*, 2009. 25(18): pp. 10587–10594.

Rica, R.A., M.L. Jimenez, and A.V. Delgado, Electric permittivity of concentrated suspensions of elongated goethite particles. *Journal of Colloid and Interface Science*, 2010. 343(2): pp. 564–573.

Rica, R.A., M.L. Jimenez, and A.V. Delgado, Effect of the volume fraction of solids on the concentration polarization around spheroidal hematite particles. *Soft Matter*, 2011. 7(7): pp. 3286–3289.

Rica, R.A., M.L. Jimenez, and A.V. Delgado, Electrokinetics of concentrated suspensions of spheroidal hematite nanoparticles. *Soft Matter*, 2012. 8(13): pp. 3596–3607.

Rider, P.F. and R.W. Obrien, The dynamic mobility of particles in a non-dilute suspension. *Journal of Fluid Mechanics*, 1993. 257: pp. 607–636.

Rosen, L.A. and D.A. Saville, Dielectric-spectroscopy of colloidal dispersions—Comparisons between experiment and theory. *Langmuir*, 1991. 7(1): pp. 36–42.

Rosen, L.A., J.C. Baygents, and D.A. Saville, The interpretation of dielectric response measurements on colloidal dispersions using the dynamic stern layer model. *Journal of Chemical Physics*, 1993. 98(5): pp. 4183–4194.

Saville, D.A., Electrokinetic properties of fuzzy colloidal particles. *Journal of Colloid and Interface Science*, 2000. 222(1): pp. 137–145.

Shilov, V.N., N.I. Zharkikh, and Y.B. Borkovskaya, Theory of non-equilibrium electrosurface phenomena in concentrated disperse systems. 1. Application of non-equilibrium thermodynamics to cell model of concentrated dispersions. *Colloid Journal of the USSR*, 1981. 43(3): pp. 434–438.

Shilov, V.N. et al., Polarization of the electrical double layer. Time evolution after application of an electric field. *Journal of Colloid and Interface Science*, 2000. 232(1): pp. 141–148.

Shilov, V.N. et al., Suspensions in alternating external electric field: Dielectric and electrorotation spectroscopies. In *Interfacial Electrokinetics and Electrophoresis*, A.V. Delgado, Editor. 2002, New York: Marcel Dekker.

Siebenburger, M., M. Fuchs, and M. Ballauff, Core-shell microgels as model colloids for rheological studies. *Soft Matter*, 2012. 8(15): pp. 4014–4024.

Stoilov, S.P. and M.V. Stoimenova, eds. *Molecular and Colloidal Electro-Optics*. Surfactant Science Series. Vol. 134. 2007, Boca Raton: Taylor and Francis.

Tirado, M.C. et al., Measurement of the low-frequency dielectric properties of colloidal suspensions: Comparison between different methods. *Journal of Colloid and Interface Science*, 2000. 227(1): pp. 141–146.

Vauthier, C., B. Cabane, and D. Labarre, How to concentrate nanoparticles and avoid aggregation? *European Journal of Pharmaceutics and Biopharmaceutics*, 2008. 69(2): pp. 466–475.

Wall, S., The history of electrokinetic phenomena. *Current Opinion in Colloid & Interface Science*, 2010. 15(3): pp. 119–124.

Wunderlich, R.W., The effects of surface-structure on the electrophoretic mobilities of large particles. *Journal of Colloid and Interface Science*, 1982. 88(2): pp. 385–397.

Zimmermann, R. et al., Electrokinetic characterization of poly(acrylic acid) and poly(ethylene oxide) brushes in aqueous electrolyte solutions. *Langmuir*, 2005. 21(11): pp. 5108–5114.

Zukoski, C.F. and D.A. Saville, Electrokinetic properties of particles in concentrated suspensions. *Journal of Colloid and Interface Science*, 1987. 115(2): pp. 422–436.

Zukoski, C.F. and D.A. Saville, Electrokinetic properties of particles in concentrated suspensions—heterogeneous systems. *Journal of Colloid and Interface Science*, 1989. 132(1): pp. 220–229.

15 Interfacial Rheology of Viscoelastic Surfactant–Polymer Layers

Theodor D. Gurkov, Boryana Nenova,
Elena K. Kostova, and Wolfgang Gaschler

CONTENTS

15.1 INTRODUCTION

The two-dimensional rheology of fluid interfaces has been a subject of numerous studies, because of its link with the stability of foams and emulsions (Langevin 2000; Wilde 2000). Basically, when the stress response to deformation is stronger, this immobilizes the surfaces and the thin films and prevents them from being disturbed too much, and the dispersion is stabilized. Experimental measurements with various systems, containing surfactants, proteins, polymers, and so on, have revealed that the interfacial rheological behavior is often of the viscoelastic type (Sagis 2011).

The dilatational rheology relies on widely used experimental methods that are based on small harmonic (sinusoidal) deformations—waves on a flat surface or pulsating expansion/compression of deformed or spherical drops and bubbles (Miller et al. 2010; Mucic et al. 2011). Two moduli are measured directly—storage, E', and loss, E''. They are commonly regarded as characteristics of elasticity and viscous dissipation, respectively. However, the exact relation between E', E'', and physical coefficients of elasticity and viscosity (G, η) depends on the rheological model or, in general, on the physical processes underlying the rheological response or how the material actually behaves.

A case of great practical importance is when surfactant can be exchanged between the interface and the volume phase in which it is soluble. Then, surface expansion (or contraction) will be accompanied by adsorption (or desorption) and diffusion of molecules from (or toward) the bulk interior. This leads to complicated dependence of the rheological moduli upon the oscillation frequency. Theoretical analysis of this scenario was carried out in comprehensive details (Horozov et al. 1997; Kotsmar et al. 2009; Lucassen and van den Tempel 1972).

Freer et al. (2004) applied the diffusion theory for analysis of storage and loss moduli of β-casein. They reached the conclusion that the Lucassen–van den Tempel framework should be supplemented with a static modulus $\left(E'_\infty \right)$ of irreversibly adsorbed protein molecules. The latter quantity could be obtained as the limit of the elasticity at zero frequency (Freer et al. 2004). For interpretation of our data in this work, we also need such an elastic modulus at very slow deformation—see G_2 in Section 15.3.2. It takes into account the contribution of adsorbed molecules that cannot be exchanged with the bulk or subsurface.

It has been recognized that the diffusion is not the only possible relaxation mechanism that leads to effective viscous dissipation in the 2D rheology. For example, proteins may undergo reorientation after adsorption, internal reconformation, molecular shrinking under increased surface pressure, and so on (Benjamins et al. 2006). With simpler molecules, a feasible scenario is a reversible exchange between the adsorbed layer and the adjacent subsurface (Boury et al. 1995; Liggieri and Miller 2010; Wantke et al. 2005). In the present work, we elaborate on this mechanism and derive an explicit equation that connects the apparent viscosity with the mass transfer coefficient.

Surfaces that exhibit linear viscoelasticity are often described in terms of the Maxwell rheological model; a number of literature citations for this are listed in the review by Sagis (2011). With an additional elastic element, responsible for the insoluble molecules, attached in parallel to the Maxwell model, one obtains the Zener model (Boury et al. 1995); the latter turns out to be adequate for our needs. In Section 15.3.2, we discuss the physical relevance of the parameters that take part in the Maxwell and Zener models, in the case when the relaxation is due to out-of-plane mass transport.

In this work, we investigate the layer response to deformation whose time dependence is not sinusoidal, but has a triangle-shaped waveform. Correspondingly, the strain is represented as a Fourier series. When the constitutive equation for the rheological model is solved, it predicts the engendered stress, again as a Fourier series. Here, we demonstrate that this theoretical development is suitable for fitting experimental data, collected from Langmuir trough measurements with a mixed layer of surfactant and polymer. Fourier transform rheology has recently been proposed by Hilles et al. (2006), but they used harmonic disturbances with high amplitude (in the non-linear regime) and studied only insoluble layers.

15.2   EXPERIMENTAL MEASUREMENTS

15.2.1   MATERIALS

The main surfactant is a mixture of acids (sodium salts) from wood resin. We use the commercial product Dynakoll VS 50 FS (CAS No. 68201-59-2), supplied by Akzo Nobel, which contains 50% surface-active ingredients—the so-called resin acids. The abietic and levopimaric acids are among the predominant chemical substances; their structure is shown in Figure 15.1a and b. Further information about these and other similar components in the resin can be found in Peng and Roberts (2000). For us, the most important property of the molecules in Dynakoll is that they adsorb readily on the air/water interface and cause a significant decrease of the surface tension.

We investigate adsorbed layers of resin acids in the presence of a cationic polymer. Solvitose BPN (CAS No. 56780-58-6), from Avebe GmbH (Germany), represents

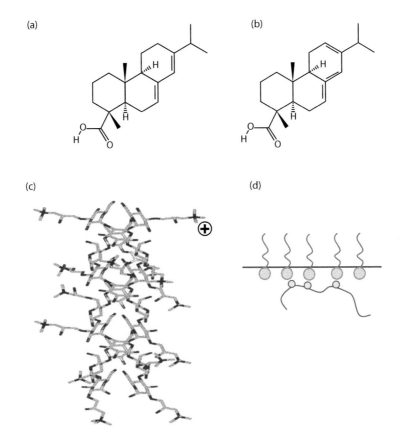

FIGURE 15.1   Substances used in this work: (a) abietic acid; (b) levopimaric acid; (c) cationized starch, with ~4.0%–4.5% trimethylammonium groups; (d) possible attachment of polymer to adsorbed surfactant at the air/water boundary.

a derivative of potato starch, namely, hydroxypropyl-trimethylammonium chloride ether. This is a hydrophilic polymer, whose degree of substitution with cationic groups is approximately 4.0–4.5 mol%. The chemical structure is sketched in Figure 15.1c. The Solvitose itself does not possess surface activity; there is no adsorption from solutions of Solvitose alone.

The subject of our study is the mixed system of 0.01 wt% Dynakoll and 0.1 wt% Solvitose. All solutions were prepared with deionized water from a Milli-Q Organex purification system (Millipore, USA).

15.2.2 METHODS

The dilatational rheology of adsorbed layers on A/W boundary is studied by means of a Langmuir trough with a traditional design, sketched in Figure 15.2. The model of the apparatus is 302 LL/DI, manufactured by Nima Technology Ltd., UK. The area of the trough is varied with two parallel Teflon barriers that move symmetrically; their speed of linear translation is constant and can be set by the software. The surface tension, σ, is measured with a Wilhelmy plate, made of chromatographic paper. The choice of paper ensures complete wetting and also prevents contamination by impurities (a new piece is used for each experiment). The Wilhelmy plate is positioned exactly in the middle between the two barriers. It is oriented in parallel direction to the barriers. As far as our layers are fluid-like, the orientation actually does not matter (the surface tension is isotropic). The measurements are performed at 40.0°C of the aqueous solution. The setup is equipped with thermostating jacket, contacting with the bottom of the Teflon trough from below.

Data acquisition is performed continuously; the apparatus records the area between the barriers, $A(t)$, and the surface pressure, $\Pi(t)$, as functions of time. By definition, Π is the decrease of the surface tension caused by the presence of surfactant: $\Pi = \sigma_0 - \sigma$, where σ_0 refers to the bare air/water interface. At 40°C, $\sigma_0 = 69.6$ mN/m.

Initially, the solution is loaded in the Langmuir trough with open barriers, at $A = 150$ cm^2. Some time is allowed for equilibration of the layer (typically, it is left at rest

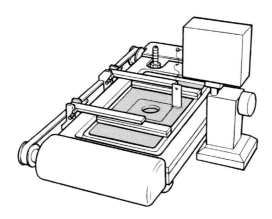

FIGURE 15.2 Sketch of the Langmuir trough; symmetric deformation is created by the two barriers, and the Wilhelmy plate sensor is positioned at the midpoint.

for approximately 15 min). The first stage of adsorption is very fast; the earliest possible measurement with the Wilhelmy plate, after placing it properly, and so on, gives $\Pi \sim 16$ mN/m. Next, shrinking is applied with the barriers until the desired area is reached (it is often 80 or 100 cm^2). After another ~4–5 min for relaxation, the cyclic compression/expansion starts. Figure 15.3 provides an example of raw experimental data for $A(t)$ and $\Pi(t)$.

In order to achieve better characterization of the mixed system, we performed some measurements with oscillating pendant drops. Those were made on a DSA 100 automated instrument for surface and interfacial tension determination (Krüss GmbH, Germany); the setup was complemented with the special ODM/EDM module dedicated to oscillations. The drop shape analysis technique was employed to extract information from the shape of pendant drops, deformed by gravity, on which harmonic surface perturbation was imposed. A sinusoidal variation of the drop surface area, with defined angular frequency, led to the due response of oscillatory change in the surface tension. The method is described in detail by Russev et al. (2008). We obtained values for the storage modulus, E' (representative for the surface elasticity). The solution of 0.01 wt% Dynakoll gave $E' = 54$ mN/m, while the mixed solution of 0.01 wt% Dynakoll and 0.1 wt% Solvitose showed $E' = 78$ mN/m (the oscillation period, T, was 10 s). For different periods T in the interval 5–20 s, E' was considerably greater in the presence of Solvitose, as compared to the case of Dynakoll alone. We interpret this fact as a strong evidence that the cationic polymer is engaged in the interfacial layer and influences its properties substantially. Our hypothesis for the molecular structure is depicted in Figure 15.1d. The low-molecular-weight surfactant is adsorbed, and its polar heads attract some polymer segments, so the chains of the starch are attached to the surface from below (at certain points). The interaction is most probably of electrostatic origin, because the resin acids carry partial negative

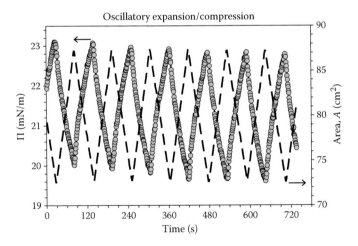

FIGURE 15.3 Example of an experiment with cyclic deformation; the dashed lines correspond to the surface area, $A(t)$, and the circles represent the measured surface pressure, $\Pi(t)$. The barrier speed gives $dA/dt = 15.91$ cm^2/min.

charge at the carboxylic group, while the Solvitose contains 4.0%–4.5% cationic groups.

15.3 RHEOLOGICAL INTERPRETATION OF THE DATA FROM PERIODIC DEFORMATION

15.3.1 THEORETICAL DESCRIPTION OF THE STRESS RESPONSE IN TERMS OF A RHEOLOGICAL MODEL

The results presented in Figure 15.3 indicate that the layer behaves as a viscoelastic material (pure elasticity would have given a strictly linear $\Pi(t)$ dependence, as far as $A(t)$ is linear). We attempt to explain the data using the known Zener model (Boury et al. 1995; Ouis 2003), also known as the standard viscoelastic body. Its mechanistic depiction is shown in Figure 15.4; a linear spring G_2 is coupled in parallel with a Maxwell element (that consists of elastic and viscous parts, G_1 and η). The constitutive relation between the stress τ and the strain γ reads:

$$\left(1+\frac{G_2}{G_1}\right)\frac{d\gamma}{dt}+\frac{G_2}{\eta}\gamma=\frac{1}{G_1}\frac{d\tau}{dt}+\frac{1}{\eta}\tau. \tag{15.1}$$

In our case, the deformation $\gamma(t)$ is defined by the experimental setup. Then, Equation 15.1 allows one to calculate the theoretical response of the system, $\tau(t)$, by solving the differential equation for τ. The constants G_1, G_2, and η will naturally stand as model parameters and can be used to fit measured data for the stress. In order to implement this strategy, we first represent the strain $\gamma(t)$ as an explicit function.

The expansion/compression of the area in the Langmuir trough, $A(t)$, is performed by translation of the barriers with constant speed. Figure 15.5 displays several cycles of such deformation. The strain, γ, is the integrated relative change of the surface area $(d\gamma = dA/A)$

$$\int_0^\gamma d\gamma = \int_{A_0}^A \frac{dA}{A} = \ln\frac{A}{A_0} = \gamma \tag{15.2}$$

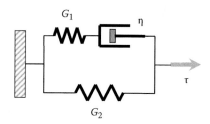

FIGURE 15.4   Scheme of the Zener model (standard linear viscoelastic body).

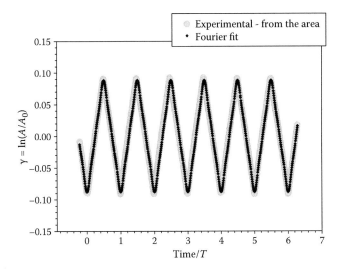

FIGURE 15.5 Illustration of the fit of the cyclic deformation, $\gamma(t)$, with Fourier series. Three terms in the right-hand side of Equation 15.3 were taken into account; $\gamma_{\text{ampl}} = 0.1885$; $T = 112.95$ s; $\ln(A_0, \text{cm}^2) = 4.3785$.

and the reference state A_0, where $\gamma = 0$, is the mean area. The whole curve in Figure 15.5 can be described by the following Fourier series:

$$\gamma(t) = -\gamma_{\text{ampl}} \frac{4}{\pi^2} \left\{ \cos\left(\frac{2\pi}{T}t\right) + \frac{1}{9}\cos\left(\frac{6\pi}{T}t\right) + \frac{1}{25}\cos\left(\frac{10\pi}{T}t\right) + ... \right\}. \quad (15.3)$$

The amplitude γ_{ampl} is the difference between the maximum and the minimum of γ; T is the period of the oscillations. In principle, these two quantities are known from the software of the trough, when a given deformation is set. Some fine adjustment of γ_{ampl} and T is still made in Equation 15.3, in order to match the actual $A(t)$ data. We take three terms in the right-hand side of Equation 15.3, but of course, the series can be truncated at a different length, according to the needs for precision.

Next, the formula (Equation 15.3) is substituted into Equation 15.1, and the resulting differential equation for τ is solved analytically; τ is obtained in the form:

$$\tau = p\cos\left(\frac{2\pi}{T}t\right) + q\sin\left(\frac{2\pi}{T}t\right) + u\cos\left(\frac{6\pi}{T}t\right) + v\sin\left(\frac{6\pi}{T}t\right)$$
$$+ r\cos\left(\frac{10\pi}{T}t\right) + s\sin\left(\frac{10\pi}{T}t\right) + ... \quad (15.4)$$

Here, the constant coefficients p, q, u, v, r, and s are related to the material parameters of the model, G_1, G_2, η, and γ_{ampl}. For the sake of completeness, we list these coefficients below:

$$p = -\frac{4\gamma_{ampl}}{\pi^2}\frac{BD+AC\omega^2}{D^2+C^2\omega^2}, \qquad q = -\frac{4\gamma_{ampl}}{\pi^2}\frac{BC-AD}{D^2+C^2\omega^2}\omega,$$

$$u = -\frac{4\gamma_{ampl}}{9\pi^2}\frac{BD+AC(3\omega)^2}{D^2+C^2(3\omega)^2}, \qquad v = -\frac{4\gamma_{ampl}}{3\pi^2}\frac{BC-AD}{D^2+C^2(3\omega)^2}\omega, \qquad (15.5)$$

$$r = -\frac{4\gamma_{ampl}}{25\pi^2}\frac{BD+AC(5\omega)^2}{D^2+C^2(5\omega)^2}, \qquad s = -\frac{4\gamma_{ampl}}{5\pi^2}\frac{BC-AD}{D^2+C^2(5\omega)^2}\omega$$

where

$$A = 1+\frac{G_2}{G_1}, \quad B = \frac{G_2}{\eta}, \quad C = \frac{1}{G_1}, \quad D = \frac{1}{\eta}, \quad \omega = 2\pi/T.$$

The jth and the $(j+1)$st terms in Equation 15.4 read:

$$\tau = ... - \frac{4\gamma_{ampl}}{j^2\pi^2}\frac{BD+AC(j\omega)^2}{D^2+C^2(j\omega)^2}\cos\left(\frac{2j\pi}{T}t\right) - \frac{4\gamma_{ampl}}{j\pi^2}\frac{BC-AD}{D^2+C^2(j\omega)^2}\omega\sin\left(\frac{2j\pi}{T}t\right) - ...$$

$$(15.4a)$$

where $j = 1, 3, 5, 7, ...$ (odd integer numbers).

Thus, Equation 15.4 represents the theoretical prediction for the layer response to a deformation of triangular shape, such as that depicted in Figure 15.5. We use Equation 15.4 for fitting of experimental results for $\tau(t)$; the rheological characteristics G_1, G_2, and η serve as three adjustable parameters to be varied and determined from the best fit.

15.3.2 Physical Relevance of the Material Constants

According to the Zener model (Figure 15.4 and Equation 15.1), at very fast deformation, the strain on the viscous element approaches zero and the system will become purely elastic, with a modulus $G_1 + G_2$; specifically, $d\tau = (G_1 + G_2)d\gamma$. In the opposite case of very slow deformation, the viscous element will fully relax to zero stress and only the element G_2 will deform. The system will be again elastic, but with a modulus G_2. This behavior suggests that one can attribute the viscous dissipation (η) to a certain exchange of molecules from the interface with the subsurface or the bulk phase. Such an exchange should have a characteristic timescale and will happen only when the deformation is sufficiently slow.

A plausible physical picture might be that two types of molecules are present in the interfacial layer: (I) Irreversibly adsorbed ones, which are associated with the elasticity G_2. Such species are commonly called "insoluble surfactant" in the literature. (II) Molecules that go to the interface reversibly; they can be exchanged with the bulk phase or the subsurface: adsorption will take place upon expansion, and desorption will happen on compression. If the change of surface area is made in a quasi-static way (infinitely slowly), these molecules (II) will have equilibrium adsorption, $\Gamma_{r, eq}$, and will not bring about any deviation in the surface tension, σ (hence, there will be no contribution to the stress, τ). Here, we assume that σ depends only on the instantaneous number of molecules per unit area at the interface, which is denoted by Γ_r for the reversibly adsorbed species and by Γ_{ir} for the insoluble ones. In other words, $\sigma = \sigma(\Gamma_r(t), \Gamma_{ir}(t))$. Any effects of interfacial reconfiguration, gradual reorganization, and so on, are discarded.

In general,

$$N_r = A\Gamma_r, \tag{15.6}$$

where N_r is the number of reversibly adsorbed molecules on the whole area A. When a change δ is applied because of deformation, Equation 15.6 yields a differential expression that can be cast into the following convenient form:

$$\delta \ln A = \delta(-\ln \Gamma_r) + \delta \ln N_r. \tag{15.7}$$

Let us now consider the Maxwell section of the rheological model in Figure 15.4 (that consists of G_1 and η connected in series). One writes the total strain $\delta\gamma$ as a sum of two contributions, on G_1 and on η:

$$\delta\gamma = \delta\gamma_{1elastic} + \delta\gamma_{viscous}. \tag{15.8}$$

Comparing Equations 15.7 and 15.8, we can identify the corresponding terms; it is already set that $\delta\gamma = \delta\ln A$, Equation 15.2, and the dissipation is supposedly associated with mass exchange to or from the interface—reversible adsorption/desorption—and the concomitant variation of N_r.

$$\delta\gamma_{1elastic} = -\delta \ln \Gamma_r; \; \delta\gamma_{viscous} = \delta \ln N_r \tag{15.9}$$

The full stress $\delta\tau$, according to the Zener model, is

$$\delta\tau = \delta\tau_1 + \delta\tau_2 = G_1\delta\gamma_{1elastic} + G_2\delta\gamma. \tag{15.10}$$

We wish to reveal the physical meaning of the elasticities G_1 and G_2. For this purpose, it should be specified how the adsorbed surfactant molecules influence the surface tension, σ. Under the restriction of small deviations, the following expansion holds:

$$\delta\sigma = \left(\frac{\partial\sigma}{\partial \ln \Gamma_r}\right)_{\Gamma_{ir}} \delta \ln \Gamma_r + \left(\frac{\partial\sigma}{\partial \ln \Gamma_{ir}}\right)_{\Gamma_r} \delta \ln \Gamma_{ir}. \tag{15.11}$$

This equation takes into account the independent effects from the reversibly adsorbed species, Γ_r, and the insoluble ones, Γ_{ir}. The number of irreversibly adsorbed molecules, $N_{ir} = A\Gamma_{ir}$, should remain constant; therefore, $\delta\ln\Gamma_{ir} = -\delta\ln A = -\delta\gamma$.

The stress is in fact the change of σ, so that $\delta\tau = \delta\sigma$; now, one can compare Equations 15.10 and 15.11, in view of the first equation of Equation 15.9. The result reads

$$G_1 = -\frac{\partial\sigma}{\partial\ln\Gamma_r} = E_G. \tag{15.12}$$

Hence, the modulus G_1 coincides with the Gibbs elasticity of the soluble surfactant, E_G; this is the physical meaning of G_1. It is known that E_G is a thermodynamic quantity that characterizes the adsorption layer; it may be found from the equation of state. If an alternative elasticity is defined as $[d\sigma/d(\ln A)]$, the latter will be influenced by the surfactant transfer rate and the rate of strain (see, e.g., Liggieri and Miller 2010).

Similarly to the above calculation, from Equations 15.10 and 15.11, we deduce

$$G_2 = \left(\frac{\partial\sigma}{\partial\ln A}\right)_{\Gamma_r=\text{const.}} = -\left(\frac{\partial\sigma}{\partial\ln\Gamma_{ir}}\right)_{\Gamma_r}. \tag{15.13}$$

It is confirmed that G_2 is the mechanical elasticity of the layer at very slow deformation (when Γ_r stays constant). G_2 is due to the presence of molecules that cannot be transferred between the interface and the bulk or subsurface, at least not during our experiments of cyclic expansion/compression.

In general, both G_1 and G_2 are expected to depend on the density of the surfactant-laden interface, that is, on the particular values of Γ_r and Γ_{ir}. This implies a possibility that G_1 and G_2 may exhibit a trend when the layer is subjected to different degrees of compression or with the increase of the average surface pressure $\langle\Pi\rangle$.

For the dissipative component of the rheological model, the stress is determined according to the usual constitutive relation for (apparent) viscosity and the second equation of Equation 15.9:

$$\delta\tau_1 = \eta\frac{d}{dt}(\delta\gamma_{\text{viscous}}) = \eta\frac{d}{dt}(\delta\ln N_r) = \eta\frac{1}{N_r}\frac{dN_r}{dt}. \tag{15.14}$$

The deviation δN_r is assumed to be with respect to some reference state (e.g., at the mean area of the surface during the deformation cycles), and the latter state is essentially independent of time—it may be equilibrium, or at least it should change much more slowly, as compared to the expansions/compressions. This conjecture has led to the last equality in Equation 15.14.

The kinetics of mass exchange between the interface and the adjacent bulk region can be described in a usual macroscopic way, as flux proportional to the driving force:

$$\frac{1}{A}\frac{dN_r}{dt} = -K_r\Gamma_r\frac{\delta\mu}{k_BT_K}, \text{ or } \frac{1}{N_r}\frac{dN_r}{dt} = -K_r\frac{\delta\mu}{k_BT_K}. \qquad (15.15)$$

Here, the driving force is the change of the chemical potential of the reversibly adsorbed molecules, $\delta\mu$, when the layer is deformed with respect to the reference (equilibrium) state. $\delta\mu > 0$ would correspond to desorption, since the molecules will have lower chemical potential in the bulk, while $\delta\mu < 0$ would lead to adsorption. In Equation 15.15, K_r is a kinetic coefficient of mass transfer, whose dimension is time^{-1}; k_BT_K denotes the thermal energy (k_B is the Boltzmann constant and T_K is the temperature in Kelvin).

The surface chemical potential, μ, is a function of the adsorption, Γ_r; the type of this function depends on the specific equation of state (the isotherm) for the given system. In a general form, we will write $\delta\mu$ as follows:

$$\delta\mu = k_BT_K f(\Gamma_r)\,\delta\ln\Gamma_r, \qquad (15.16)$$

where the dimensionless function $f(\Gamma_r)$ pertains to a particular isotherm. The $\mu(\Gamma_r)$ relations, for a number of different widely used equations of state, are listed in the work of Kralchevsky et al. (2008). For instance, the well-known Langmuir isotherm gives $f(\Gamma_r) = (1 - \Gamma_r/\Gamma_\infty)^{-1}$, where Γ_∞ is the maximum attainable value of Γ_r.

Now, we can find a connection between the apparent viscosity, in the frames of the Zener model, and the transfer kinetics of surfactant from/to the interface, represented by the coefficient K_r. The stress component $\delta\tau_1$ is the same on the elastic (G_1) and the viscous (η) elements in Figure 15.4, whence

$$\delta\tau_1 = -E_G\,\delta\ln\Gamma_r \qquad (15.17)$$

from Equations 15.9, 15.10, and 15.12, and

$$\delta\tau_1 = -\eta K_r f(\Gamma_r)\,\delta\ln\Gamma_r \qquad (15.18)$$

from Equations 15.14 through 15.16. The combination of Equations 15.17 and 15.18 easily yields the desired relation:

$$K_r = \frac{E_G}{\eta f(\Gamma_r)} \qquad (15.19)$$

It is seen that Maxwell's relaxation time, η/E_G, can be of the order of the characteristic time of mass exchange from/to the surface, K_r^{-1}; however, the two quantities are not identical. The macroscopic viscosity η is influenced by the equation of state of the surfactant layer. Equation 15.19 suggests that larger values of the apparent viscosity η correspond to slower transfer of molecules between the interface and its bulk surroundings. On the other hand, in the limiting case of very fast mass exchange, Γ_r will not significantly deviate from equilibrium, and $\eta \rightarrow 0$.

Existing previous studies of other authors, which address the role of surface–bulk transfer, describe the process in the framework of diffusion. Theories were developed to account for the frequency dependence of the elastic and viscous moduli by solving the diffusion problem (Horozov et al. 1997; Lucassen and van den Tempel 1972; Lucassen-Reynders et al. 2001). However, one can encounter physical scenarios in which it is more important what happens locally, in the immediate vicinity of the interface, rather than how the concentration disturbance propagates further away to the bulk. Thus, the diffusion is not the only possible mechanism to interfere with the distribution of material in and around a phase boundary that undergoes deformation. For example, reversible out-of-plane escape of molecules, polymer fragments, aggregates, and so on, may affect the surface tension considerably. In this context, it seems feasible that some segments of proteins or other polymers may be expelled, because of steric repulsion within the plane of the interface, and after subsequent expansion, these segments can adsorb back. There may be no time for diffusion, or no freedom to leave the interfacial zone completely. Such cases are envisaged in this work, where exchange with the subsurface is only considered (Equation 15.15). A similar idea was followed by Wantke et al. (2005) and Boury et al. (1995), who found that the interface/subsurface transfer of surfactant molecules and protein segments can be important in different systems.

15.4 DISCUSSION OF MEASURED DATA AND THE RESULTING RHEOLOGICAL PARAMETERS

The rheological response of the adsorbed layer is studied by keeping track of the changes in the surface tension, σ, during oscillatory deformation. The measured stress, τ, is defined as the difference between the running value of σ and the average, $\langle\sigma\rangle$, from several full cycles:

$$\tau = \sigma - \langle\sigma\rangle = \langle\Pi\rangle - \Pi. \tag{15.20}$$

Equation 15.20 gives τ also in terms of the surface pressure, Π; the relation $d\Pi = -d\sigma$ always holds. We present the raw data in the scale of stress, τ, as a function of the strain, γ (see Figure 15.6). The experimental points that are selected for analysis span about three complete compression/expansion cycles, with the time interval being centered at around ~550 s from the start of the deformation (cf. Figure 15.3). There is an initial slight decrease of $\langle\Pi\rangle$ for a few oscillations (Figure 15.3), which we would like to avoid.

The results for the stress are fitted with the theoretical function for its time dependence, $\tau(t)$ (Equation 15.4). The adjustable parameters G_1, G_2, and η are varied until the standard error of the regression, RMSE, is minimized.

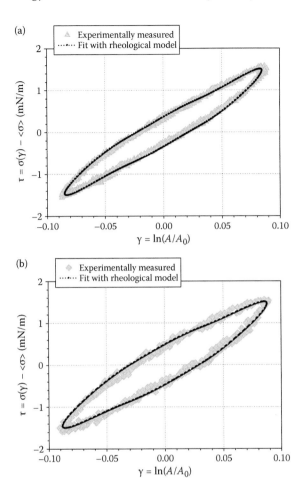

FIGURE 15.6 Measured data for the stress response, τ (gray symbols), fitted with the Zener model (Equation 15.4, black dotted curves). The time is excluded from $\tau(t)$ and $\gamma(t)$. (a) System 4 from Table 15.1; (b) system 2 from Table 15.1.

$$\text{RMSE} = \sqrt{\sum_{i=1}^{n}[\tau_{\text{measured},i}(t) - \tau_{\text{fit},i}(t)]^2 / \text{residual degrees of freedom}}$$

The error of the fits in Figure 15.6 is 0.077 mN/m for Figure 15.6a and 0.092 mN/m for Figure 15.6b, respectively. This RMSE is below the experimental uncertainty in τ, which proves that the model is adequate. In order to draw the plots in Figure 15.6, the time was eliminated from the experimental and theoretical sets of $\tau(t)$ and $\gamma(t)$ data.

One notices that Figure 15.6 contains results for several consecutive compressions and expansions of the interface, which lie on the same curve; moreover, both the stage of compression and that of expansion are described by one and the same model, whose rheological constants G_1, G_2, and η are determined from the fit of all points. These facts show definitely that the physical processes during the cyclic deformation are fully reversible (at least for ~3 oscillations and to the extent that $\langle \Pi \rangle$ does not change). Such a reversible dissipation (apparent viscosity) could be due to expulsion and readsorption of some surfactant molecules and polymer segments, which are exchanged between the planar interface and the immediately adjacent subsurface.

The values of the material parameters G_1, G_2, and η are listed in Table 15.1. The columns are labeled in direction of increasing $\langle \Pi \rangle$; the rate of strain, $d\gamma/dt$, and the overall extent of layer compression [$\langle A \rangle$] are different. We observe a clear trend that G_2 rises with $\langle \Pi \rangle$. This behavior is illustrated in Figure 15.7 and can be attributed to the higher density of the layer (or greater Γ_{ir}) at higher $\langle \Pi \rangle$. The slope in Figure 15.7 is approximately 2.0. It seems physically plausible to anticipate such a trend: The work by Boury et al. (1995) reports increasing elasticity (E, corresponding to our $G_1 + G_2$) of bovine serum albumin with growing density (Γ) and surface pressure (Π) of the adsorbed layer, with $E(\Pi)$ being linear. Other examples can be found in the article of Benjamins et al. (2006), with proteins on oil/water and air/water boundaries.

There is no particular dependence of G_1 and η on $\langle \Pi \rangle$ (Table 15.1). Perhaps the differences in $\langle \Pi \rangle$, and in the layer density, between experimental runs 1–4 are too small to affect these two rheological properties. As far as the rate of strain, $d\gamma/dt$ does indeed have an influence on the results in Table 15.1, the mean adsorption $\langle \Gamma_r \rangle$ is not expected to be in full equilibrium with the bulk. Still, the deviations from the true $\Gamma_{r,eq}$ seem to be modest in the studied range of conditions.

The Maxwell characteristic time for the reversibly exchangeable molecules, $t_M = \eta/G_1$, is confined in the interval 15.5–18.9 s for the data in Table 15.1. This t_M is shorter in comparison with the oscillation period, T (the latter varies between 99.5 and 191.1 s, see Table 15.1). Hence, the mass transfer is relatively fast. Nevertheless,

---

TABLE 15.1

Values of the Material Parameters, Determined from the Best Fit of $\tau(\gamma)$ Data for Different Deformation Rates and Degrees of Layer Compression [$\langle A \rangle$]

	System #			
	1	**2**	**3**	**4**
G_1 (mN/m)	13.90	13.22	10.40	12.76
G_2 (mN/m)	7.62	9.10	9.88	13.59
η (Pa s m) (or $\times 10^3$ sP)	0.2149	0.2421	0.1968	0.2064
$\langle \Pi \rangle$ (mN/m)	**20.47**	**21.22**	**22.27**	**23.38**
$\langle A \rangle$ (cm²)	99.93	79.72	99.68	79.72
$d\gamma/dt$ (s⁻¹)	3.51×10^{-3}	3.40×10^{-3}	1.89×10^{-3}	1.85×10^{-3}
T (s)	99.51	112.95	170.51	191.05

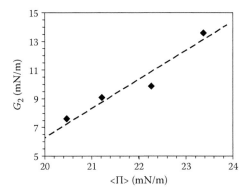

FIGURE 15.7   Results for the elasticity modulus owing to irreversibly adsorbed molecules, G_2, plotted as a function of the average surface pressure. The points correspond to separate independent experiments (see the values marked with shade and boldface in Table 15.1).

it causes a rather significant effect of viscous dissipation, manifested as the "loop" in the graphs of Figure 15.6.

15.5   CONCLUSIONS

This work reports an analysis of the rheological behavior of a surface layer that is subjected to cyclic expansion/compression in a Langmuir trough, with a triangle-shaped waveform versus time. We derive an exact solution for the stress response, represented as a Fourier series (Equation 15.4a), in the case when the Zener model is applicable. The same methodology can be used in combination with other rheological models of choice.

The physical meaning of the material parameters (two elasticities and one viscosity) is discussed in view of the effects that influence the surface tension upon deformation. It is shown that one elastic modulus (G_2) can be ascribed to irreversibly adsorbed molecules, equivalent to insoluble in the bulk subphase. The second modulus (G_1), which is part of the Maxwell element in the Zener model, coincides with the Gibbs elasticity (E_G) of adsorbed molecules capable of reversible exchange with the subsurface or the volume phase. The Gibbs elasticity is a thermodynamic quantity, related to the equation of state; it is independent from the non-equilibrium effects in the surfactant distribution. The apparent viscosity (η) is connected with the mass transfer coefficient of the exchange between the interface and its bulk surroundings. A formula is proposed for this connection, Equation 15.19, which follows from the premise that the transport flux is proportional to the deviations in the chemical potential.

The theory is employed to fit experimental results for a mixed layer consisting of a low-molecular-weight surfactant and a polymer. Their structure suggests that electrostatic attraction may be operative between the species and can cause attachment of the polymer to the surfactant-laden A/W boundary. The measured data are in good agreement with the model; the obtained values of the material parameters are discussed in relation to the average density of the adsorption layer.

ACKNOWLEDGMENTS

This work was funded by BASF SE. T. Gurkov also wishes to acknowledge partial financial support from the project DCVP 02/2-2009 with the Bulgarian Science Fund (National Centre for Advanced Materials "UNION": Module 1, Centre for Advanced Materials).

REFERENCES

Benjamins, J., J. Lyklema and E. H. Lucassen-Reynders. 2006. Compression/expansion rheology of oil/water interfaces with adsorbed proteins. Comparison with the air/water surface. *Langmuir* 22: 6181–88.

Boury, F., Tz. Ivanova, I. Panaiotov, J. E. Proust, A. Bois and J. Richou. 1995. Dilatational properties of adsorbed poly(D,L-lactide) and bovine serum albumin monolayers at the DCM/water interface. *Langmuir* 11: 1636–44.

Freer, E. M., K. S. Yim, G. G. Fuller and C. J. Radke. 2004. Shear and dilatational relaxation mechanisms of globular and flexible proteins at the hexadecane/water interface. *Langmuir* 20: 10159–67.

Hilles, H., F. Monroy, L. J. Bonales, F. Ortega and R. G. Rubio. 2006. Fourier-transform rheology of polymer Langmuir monolayers: Analysis of the non-linear and plastic behaviors. *Adv. Colloid Interface Sci.* 122: 67–77.

Horozov, T. S., P. A. Kralchevsky, K. D. Danov and I. B. Ivanov. 1997. Interfacial rheology and kinetics of adsorption from surfactant solutions. *J. Dispersion Sci. Technol.* 18: 593–607.

Kotsmar, Cs., V. Pradines, V. S. Alahverdjieva, E. V. Aksenenko, V. B. Fainerman, V. I. Kovalchuk, J. Krägel, M. E. Leser, B. A. Noskov and R. Miller. 2009. Thermodynamics, adsorption kinetics and rheology of mixed protein–surfactant interfacial layers. *Adv. Colloid Interface Sci.* 150: 41–54.

Kralchevsky, P. A., K. D. Danov and N. D. Denkov. 2008. Chemical physics of colloid systems and interfaces. In *Handbook of Surface and Colloid Chemistry*, ed. K. S. Birdi, 3rd updated edition, 197–377. Boca Raton, FL: CRC Press.

Langevin, D. 2000. Influence of interfacial rheology on foam and emulsion properties. *Adv. Colloid Interface Sci.* 88: 209–222.

Liggieri, L. and R. Miller. 2010. Relaxation of surfactants adsorption layers at liquid interfaces. *Curr. Opin. Colloid Interface Sci.* 15: 256–63.

Lucassen, J. and M. van den Tempel. 1972. Dynamic measurements of dilational properties of a liquid interface. *Chem. Eng. Sci.* 27: 1283–91.

Lucassen-Reynders, E. H., A. Cagna and J. Lucassen. 2001. Gibbs elasticity, surface dilational modulus and diffusional relaxation in nonionic surfactant monolayers. *Colloids Surf. A Physicochem. Eng. Aspects* 186: 63–72.

Miller, R., J. K. Ferri, A. Javadi, J. Krägel, N. Mucic and R. Wüstneck. 2010. Rheology of interfacial layers. *Colloid Polym. Sci.* 288: 937–50.

Mucic, N., A. Javadi, N. M. Kovalchuk, E. V. Aksenenko and R. Miller. 2011. Dynamics of interfacial layers—Experimental feasibilities of adsorption kinetics and dilational rheology. *Adv. Colloid Interface Sci.* 168: 167–78.

Ouis, D. 2003. Combination of a standard viscoelastic model and fractional derivate calculus to the characterization of polymers. *Mater. Res. Innovat.* 7: 42–6.

Peng, G. and J. C. Roberts. 2000. Solubility and toxicity of resin acids. *Water Res.* 34: 2779–85.

Russev, S. C., N. Alexandrov, K. G. Marinova, K. D. Danov, N. D. Denkov, L. Lyutov, V. Vulchev and C. Bilke-Krause. 2008. Instrument and methods for surface dilatational rheology measurements. *Rev. Sci. Instrum.* 79: 104102 (1–10).

Sagis, L. M. C. 2011. Dynamic properties of interfaces in soft matter: Experiments and theory. *Rev. Mod. Phys.* 83: 1367–403.

Wantke, K.-D., J. Örtegren, H. Fruhner, A. Andersen and H. Motschmann. 2005. The influence of the sublayer on the surface dilatational modulus. *Colloids Surf. A Physicochem. Eng. Aspects* 261: 75–83.

Wilde, P. J. 2000. Interfaces: Their role in foam and emulsion behaviour. *Curr. Opin. Colloid Interface Sci.* 5: 176–81.

16 Hofmeister Effect in Ion-Selective Electrodes from the Fluid–Fluid Interface Perspective

Kamil Wojciechowski

CONTENTS

16.1 ION-SELECTIVE ELECTRODES

The past decades have witnessed an intense development of chemical sensors for a variety of analytes. Electrochemical sensors, especially potentiometric ones, which provide an electrical signal that is relatively straightforward for further processing, have dominated many aspects of clinical diagnostics and environmental and process monitoring. Everyday clinical analyses of K^+, Na^+, Ca^{2+}, Cl^-, and so on are based on the use of potentiometric sensors, and a pH-sensitive glass electrode is probably the most popular chemical sensor.

Potentiometric sensors rely on the generation of an electrical potential difference between an analyte solution and an ion-selective membrane, the latter representing a receptor (sensing) part of a sensor. Changes in the activity of the ions in the analyte solution are assumed to affect only the ion-selective membrane potential, while all the other contributions to the overall potential difference are believed to be independent of the sample composition. Depending on a transducer part (i.e., responsible for signal detection), two types of potentiometric sensors can be distinguished. The "classic" ion-selective electrodes (ISEs) use typically an Ag/AgCl wire immersed in a KCl

369

solution (*internal solution*) as the signal transducer. An alternative approach, introduced in the early 1970s, takes advantage of field effect transistors (FETs). The first sensors of this type were developed as pH-sensitive devices, because of an inherent pH sensitivity of the gate of FET, typically made of SiO_2, that is, the same material used in glass electrodes. Further development in this area resulted in fabrication of chemically modified FET, where an ion-selective membrane (e.g., the same one used for ISE) is deposited on the FET's gate. Although ISE and ISFET are different from a metrological point of view, the mechanism of electrical potential generation seems to be the same; hence, in this chapter, the whole discussion will focus on ISE. The potential changes of the ion-selective membrane are measured against a reference electrode (e.g., Ag/AgCl) immersed in the sample. The resulting electrical potential difference between the indicator electrode and the reference electrode (electromotive force [EMF]) provides an analytical signal of potentiometric sensors (Figure 16.1).

The ion-selective membranes of potentiometric sensors use solid crystalline, glass, liquid, or polymeric matrices. The most popular polymeric ion-selective membranes consist of a polymer matrix (e.g., plasticized poly(vinyl chloride) [PVC]) and at least one electroactive additive, the so-called *lipophilic salt*. These membranes, displaying a characteristic selectivity pattern (*Hofmeister series*) will be the main subject of the following discussion. For specific applications, where the *Hofmeister selectivity* does not allow for reliable analysis of a given ion, the second electroactive additive—ionophore—can be added. The latter interacts specifically with given ionic

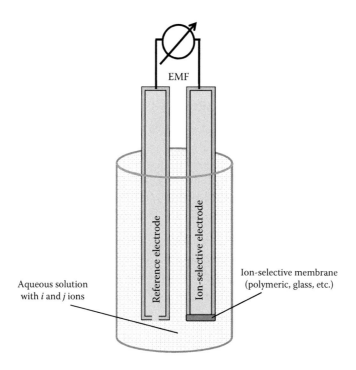

FIGURE 16.1 Setup for EMF measurements in ISE.

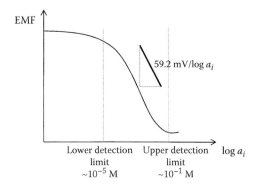

FIGURE 16.2   Typical calibration curve for ISE selective to singly charged anionic species.

species, providing the selectivity of the sensor to these species. The lipophilic salts and ionophore (if needed) are incorporated into the bulk of an organic polymeric membrane. The most commonly used polymeric matrix in membranes of ISE—plasticized PVC—consists typically of one-third of the polymer and two-thirds of the plasticizer. The latter can be considered as the solvent of the membrane components and its physicochemical properties (lipophilicity, polarity) to a large extent will determine the membrane's physicochemical properties. The PVC ion-selective membranes are prepared by dissolving all the components in a volatile organic solvent (e.g., tetrahydrofuran); the resulting solution is then poured into a glass mold or applied directly on the surface of the transducer, and the solvent is allowed to slowly evaporate.

As in any other analytical technique, a calibration curve is the basis for quantitative analysis with ISE. The curve is determined by recording the changes in EMF with the variation of activity of the target ions (Figure 16.2). The slope of the linear range of the calibration curve represents the sensitivity of the sensor and is typically equal to $59.2/z$ mV/decade (ΔE versus $\log a_i$) at $25°C$ (*Nernstian* slope), where z is the charge of the ion. The measuring range of the ISE is limited by the upper and, especially, lower detection limit, where deviations of the electrode response from the theoretical (Nernstian) slope become significant. The lower detection limit of membrane ISE is usually located between 10^{-5} and 10^{-6} M, provided that no external control of ion fluxes through the membrane is exerted (Bakker and Pretsch 2007).

16.2   NERNST EQUATION

The main feature of potentiometry, clearly distinguishing it from other electroanalytical techniques, is the absence of an electrical current in the circuit. Instead, an electrical potential difference is measured in a cell consisting of a reference electrode (e.g., Ag/AgCl immersed in 3 M KCl internal solution electrolyte) and an ISE (Figure 16.1). For this reason, in potentiometry, only the difference in electrical potential (EMF)—never its absolute value—is measurable.

The potential difference across an ion-selective membrane can be divided into three parts: two boundary potentials and a diffusion potential. The boundary potential at the internal solution/membrane interface is governed by the composition of the solution and is essentially independent of the sample composition. The difference in the ion activity on both sides of the membrane drives the diffusion of ions through the membrane. If the mobilities of positively and negatively charged species are different, the diffusion potential arises. However, if the concentration of free ions in the membrane phase is low, the latter can be neglected. Therefore, the analytical signal of ISE is generated at the interface between an aqueous solution of the analyte and the membrane, owing to an asymmetric electrical charge distribution across the membrane/aqueous interface.

In a classical view, often referred to as the *phase boundary model*, bulk partitioning equilibria are responsible for this charge distribution (Bakker, Bühlmann et al. 2004). The basic equation of this model is the Nernst equation, usually derived from the equality of electrochemical potentials in the membrane and in the aqueous (analyte) solution:

$$\tilde{\mu}_i^a \equiv \mu_i^{0,a} + RT \ln a_i^a + z_i F \phi^a = \tilde{\mu}_i^m \equiv \mu_i^{0,m} + RT \ln a_i^m + z_i F \phi^m, \qquad (16.1)$$

where μ_i^0 is the standard chemical potential of species i, a_i is its activity, and R, T, and F have their usual meanings. The superscripts a and m refer to the aqueous and membrane phases, respectively.

Each ion is then characterized by its standard Gibbs transfer free energy from the aqueous phase (aq) to the organic one (org), $\Delta G_{tr,i}^{0,a \to m}$, equal to the difference in the standard chemical potential of the ion $\left(\tilde{\mu}_i^0\right)$ in both phases. After rearrangement, the Nernst equation is obtained:

$$\Delta_m^a \phi \equiv \phi^a - \phi^m = \Delta_m^a \phi^0 + \frac{RT}{z_i F} \ln\left(\frac{a_i^m}{a_i^a}\right) = \frac{\Delta G_{tr,i}^{0,a \to m}}{z_i F} + \frac{RT}{z_i F} \ln\left(\frac{a_i^m}{a_i^a}\right). \qquad (16.2)$$

16.3   ISE SELECTIVITY

Depending on the composition of an ion-sensitive membrane, different values of EMF will be generated at the membrane/analyte interface in response to the same ion present in the contacting aqueous solution (analyte). Thus, the selectivity of a potentiometric sensor can be tuned by a proper choice of the membrane composition. In the simplest case, which is the main subject of this chapter, an asymmetric salt can be added to introduce cationic or anionic sites into the membrane. These salts (lipophilic salts) consist of a large lipophilic cation (e.g., quaternary ammonium) or anion (e.g., tetraphenylborate) in combination with a small hydrophilic counterion, providing an anion or a cation selectivity, respectively, and giving rise to the characteristic *Hofmeister selectivity pattern*. When an ionophore is present, the selectivity of the potentiometric sensor is imposed by the ion recognition process. The driving force of this (reversible) process can be electrostatic, ion–dipole, dipole–dipole, and hydrogen-bonding interactions or, in certain cases, the formation of covalent bonds.

The recognition of the guest molecules sometimes involves an appropriate arrangement of the binding sites (functional groups) of the receptor (*key-lock configuration*).

The problem of the ISE selectivity becomes especially important when more than one species are present in solution, which is very common in practice. Since only an overall potential difference can be measured in potentiometry, its decomposition into contributions of individual ions becomes crucial. In the absence of solid theoretical basis, the selectivity of ISE is usually described and discussed in the framework provided by a semiempirical Nikolsky–Eisenman equation:

$$\text{EMF} = \text{const} + \frac{2.303RT}{z_i F} \log\left(a_i + \sum_j K_{i,j} a_j^{z_i/z_j} \right), \tag{16.3}$$

where EMF is the measured electrical potential difference (EMF), a is activity of the ion, z is its charge, and $K_{i,j}$ is the potentiometric selectivity coefficient. The subscripts i and j refer to the *primary* and *interfering* ions, respectively. The other symbols have their usual meaning.

The primary ion is defined as the one for which the sensor is supposed to be selective, while all other ions are defined as interfering ones. Equation 16.3 defines the potentiometric selectivity coefficient $K_{i,j}$ as a measure of the effect of a given interfering ion on the overall response of the sensor. The selectivity coefficient is a relative number and describes the relative weight of the contribution of the interfering ions, j, with respect to the primary ions, i. Negative values of log $K_{i,j}$ indicate the preference for the primary ions relative to the interfering ions. It should be stressed at this point that the Nikolsky–Eisenman equation was obtained by "intuitive" modification of the Nernst equation for the potential difference across the membrane, but its derivation is not strict. Consequently, the Nikolsky–Eisenman selectivity coefficients have no clear thermodynamic meaning.

Several methods have been proposed for the determination of the potentiometric selectivity coefficients, although three of them are the most frequently used: separate solution method (SSM), fixed interference method (FIM), and matched potential method (MPM). SSM and MPM require recording of the calibration curves of the sensors in solutions containing separately the primary and interfering ions. In FIM, the calibration curve is measured for the primary ions in a constant background of interfering ions. For SSM and FIM, the selectivity coefficients are calculated from the transformed Nikolsky–Eisenman equation, while MPM allows only for the determination of conditional selectivities. All the methods require that the calibration curves toward primary and interfering ions have Nernstian slopes in order to calculate reliable values of selectivity coefficients.

16.4   HOFMEISTER EFFECT

The terms *ion-specific interactions* and *Hofmeister effect* refer to interactions and observations that cannot be explained by the classic theory of electrolytes based on the Poisson–Boltzmann equation (Lo Nostro and Ninham 2012). The latter involves only electrostatic terms, which are exactly the same for ions of the same charge (sign and value). In fact, ion-specific interactions are most pronounced when the electrostatic interactions are either weak (low ionic strength; Enami, Mishra et al. 2012; Wojciechowski, Bitner et al. 2011; Wojciechowski and Linek 2012) or screened to a large extent (high ionic strength; Kunz 2010). For the latter reason, these effects play key roles in biology and biochemistry under physiological conditions.

Ion-specific interactions have been observed in numerous phenomena in colloid and surface science: from protein solubility (Boström, Williams et al. 2003; Moreira, Boström et al. 2006) to shape and size of micelles (Moreira and Firoozabadi 2010), to hydrogel swelling (Swann, Bras et al. 2010), to interfacial water structure (Chen, Yang et al. 2007), to zeolite synthesis (Leontidis 2002; Li and Shantz 2010; Parsons and Ninham 2010), to crystal nucleation (Lee, Sanstead et al. 2010), to wetting properties (Silbert, Klein et al. 2010). The first reports on ion specificity date back to the 19th century when F. Hofmeister studied the effect of electrolytes on precipitation of hen egg proteins (Hofmeister 1888). It should be stressed that despite the fact that ion specificity is a universal phenomenon, and for obvious reasons cannot be decomposed into contributions from cations and anions, the effect of the latter is clearly more pronounced. Although there is no unique Hofmeister series for all types of surfaces, typical ordering of singly charged anions in the Hofmeister series follows the following order:

$$ClO_4^-, \ SCN^-, \ I^-, \ NO_3^-, \ Br^-, \ Cl^-, \ HCO_3^-, \ CH_3CO_2^-, \ F^-, \ H_2PO_4^-.$$

Traditionally, the ions from the left-hand side are called *chaotropes* (water structure breakers), and those from the right-hand side are called *kosmotropes* (water structure makers). Until only recently, the Hofmeister effect was associated with structuring of the water molecules adjacent to the given ion (Nucci and Vanderkooi 2008). Depending on the strength of interaction, some ions were suggested to enhance the structure of water (kosmotropes), while others would rather break it (chaotropes) (Marcus 2009). Nevertheless, several recent spectroscopic studies suggest that the ions present in bulk solution have only little effect on water structure (Omta, Kropman et al. 2003; Smith, Saykally et al. 2007). As it was recently pointed out by Marcus, the Hofmeister effect is inherently linked to the presence of a phase boundary (e.g., a protein's surface in biochemistry, or macroscopic phase boundaries in colloid and interface science) and not the bulk phase (Marcus 2009). Extensive surface tension (Gilányi, Varga et al. 2004; Manev, Sazdanova et al. 2008; Para, Jarek et al. 2006) and spectroscopic (Hua, Chen et al. 2011; Knock and Bain 2000; Padmanabhan, Daillant et al. 2007; Turshatov, Zaitsev et al. 2008; Wang and Morgner 2011) studies clearly show that counterions are indeed selectively co-adsorbed within the Stern layer of an oppositely charged adsorbed layer at several interfaces (Ao, Liu et al. 2011; Dos Santos, Diehl et al. 2010; Lyklema 2009; Parsons, Boström et al. 2011; Petrache, Zemb et al. 2006).

The origin of the ion specificity is still the subject of an intense debate (Chen, Yang et al. 2007; Kunz 2010; López-León, Santander-Ortega et al. 2008). Collins et al. proposed a concept of *matching water affinities*, which explains the observed Hofmeister series by short-range ion–water interactions (Collins, Neilson et al. 2007;

Jagoda-Cwiklik, Vácha et al. 2007; Vlachy, Jagoda-Cwiklik et al. 2009). Others focus more on the role of interfacial interactions (Baer, Kuo et al. 2009; Chen, Yang et al. 2007; López-León, Santander-Ortega et al. 2008; Marcus 2009), with polarizability playing a central role in the ions' affinity to the interface (Dang 2002; Jungwirth and Tobias 2002). Manciu and Ruckenstein proposed a simple extension to the mean-field Poisson–Boltzmann approach by inclusion of ion hydration effects, which are significantly altered by the environment (bulk vs. surface) (Manciu and Ruckenstein 2003). Pegram and Record in a series of papers (Pegram and Record 2007, 2008a,b, 2009a,b) introduced a new concept of solute partitioning model with the corresponding partitioning coefficients derived from the surface tension concentration increments. Their approach provides a direct link between the Hofmeister effect and surface phenomena, for example, bubble coalescence (Henry and Craig 2010).

16.5 HOFMEISTER EFFECT IN ISE

In potentiometry, the so-called *potentiometric Hofmeister effect* has been known since the beginning of the era of liquid membranes with ion exchangers. The term *Hofmeister series* refers to a characteristic ordering of potentiometric selectivity coefficients ($K_{i,j}$) for ISEs with polymeric membranes containing long-chain ion exchangers (lipophilic salts) as sole electroactive components. This ordering is usually the same as that reported originally by Hofmeister. In line with other manifestations of the Hofmeister effect, in potentiometry, it is most pronounced for anions when salts with the same cation are used.

The term migrated to potentiometry probably in the 1970s (Koryta 1975) from the ion exchangers literature (Neihof and Sollner 1956), where the ion-exchange selectivity correlated with differences in the Gibbs free hydration energies of ions (Wegmann, Weiss et al. 1984). The major argument in favor of this classic approach is a relatively good correlation between the standard Gibbs free energies of hydration of ions, $\Delta G_{tr,i}^{0,a\to m}$, and the potentiometric selectivity coefficients obtained experimentally. While this interpretation was acceptable at that time, it is far from the contemporary interpretation of the Hofmeister effect in biochemistry, colloid, and interface science. Despite this discrepancy, many authors in the potentiometric literature continue to use the terms *lipophilic* or *lyotropic series* and assign the effect to differences in solubility of ions in the lipophilic membrane of ISE (Bakker and Pretsch 2007).

The Hofmeister selectivity pattern for anion-selective potentiometric sensors is usually achieved by the addition of quaternary ammonium salts (QASs) to the plasticized PVC membrane (Legin, Makarychev-Mikhailov et al. 2004). The selectivity pattern of such ISE does not significantly depend on the chemical structure of QASs. Nevertheless, some authors have reported significant deviations from the Hofmeister selectivity pattern (Braven, Ebdon et al. 2003; Hara, Ohkubo et al. 1993; Ozawa, Miyagi et al. 1996; Schwake, Cammann et al. 1999; Sutton, Braven et al. 1999). Ozawa et al. have shown that the membranes containing asymmetric QASs (e.g., methyltrialkylammonium salts with an alkyl chain length of more than 14 methylene units) show a non-Hofmeister pattern (Ozawa, Miyagi et al. 1996). They display enhanced selectivity to chloride in comparison to lipophilic and hydrophilic organic anions. Asymmetric QASs have been shown to promote higher selectivity to doubly charged anions, as compared to singly charged anions (Egorov, Rakhman'ko et al. 2004; Legin, Makarychev-Mikhailov et al. 2004; Smirnova, Tarasevitch et al. 1994). Schwake et al. (Schwake, Cammann et al. 1999) and Wroblewski et al. (Legin, Makarychev-Mikhailov et al. 2004; Wróblewski, Chudy et al. 2000) have observed that increasing the number of methylene units in the alkyl chains enhances the electrode selectivity to hydrophilic anions. Legin et al. (Egorov, Rakhman'ko et al. 2004; Legin, Makarychev-Mikhailov et al. 2004; Smirnova, Tarasevitch et al. 1994) proposed that the selectivity pattern depends on the degree of QAS dissociation in the membrane, which in turn depends on steric accessibility, the concentration of QAS in the membrane, and its polarity.

16.6 PARTITIONING-BASED APPROACH

In the framework of the phase boundary model, many authors try to explain the potentiometric response with a rather vague concept of a *boundary layer* of an unknown, yet small thickness (Bakker, Buhlmann et al. 1997). This "quasi-interfacial" layer should, however, possess bulk-like properties, and all potential-determining partitioning equilibria would have to be restricted to this very thin phase boundary region. An interesting alternative to the simple phase boundary model has been presented by Sokalski, Lewenstam et al. (Lingenfelter, Bedlechowicz-Sliwakowska et al. 2006; Sokalski, Lingenfelter et al. 2003), who pointed to the role of the mass and charge transport phenomena in the buildup of the potential difference across the membrane. They performed a series of simulations using the Nernst–Planck–Poisson set of equations in order to numerically calculate both temporal and spatial distributions of ionic concentrations and electrical potentials in the membrane.

The classic theory of ISE focuses on macroscopic partitioning of ions between the aqueous phase and the membrane, which is a natural consequence of the use of the Nernst equation. Below, this classic view is presented using a more general approach commonly used in voltammetry at liquid/liquid interfaces, termed hereafter ITIES (interface between two immiscible electrolyte solutions). In the framework of the Nernst equation (or more generally, ITIES), the response of membrane-based potentiometric sensors is described by means of ion partitioning equilibria, derived from Guggenheim's concept of equality of electrochemical potentials $\left(\tilde{\mu}_i^a \right)$ (Guggenheim 1929) in both the membrane and the aqueous phase (Equation 16.1).

In the case of a fully dissociated single electrolyte A_1B_1 in partitioning equilibrium between the two immiscible phases (water and membrane), the Nernst equation for each species A_1^+ and B_1^- (e.g., CTA$^+$ and Br$^-$) can be written in the form:

$$A_1B_1, \text{water} \mid \text{membrane}$$

$$\Delta_m^a \phi = \Delta_m^a \phi_{A_1^+}^0 + \frac{RT}{F} \ln\left(\frac{a_{A_1^+}^m}{a_{A_1^+}^a} \right) = \Delta_m^a \phi_{A_1^+}^0 + \frac{RT}{F} \ln\left(\frac{c_{A_1^+}^m}{c_{A_1^+}^a} \right) + \frac{RT}{F} \ln\left(\frac{\gamma_{A_1^+}^m}{\gamma_{A_1^+}^a} \right) \quad (16.4)$$

$$\Delta_m^a\phi = \Delta_m^a\phi_{B_1^-}^0 - \frac{RT}{F}\ln\left(\frac{a_{B_1^-}^m}{a_{B_1^-}^a}\right) = \Delta_m^a\phi_{B_1^-}^0 - \frac{RT}{F}\ln\left(\frac{c_{B_1^-}^m}{c_{B_1^-}^a}\right) - \frac{RT}{F}\ln\left(\frac{\gamma_{B_1^-}^m}{\gamma_{B_1^-}^a}\right), \quad (16.5)$$

where $\Delta_m^a\phi$ is the Galvani potential difference, $\Delta_m^a\phi_i^0$ is the standard transfer potential of the ion i, a_i is its activity, and γ_i is its activity coefficient. The superscripts a and m refer to the aqueous and membrane phases, respectively.

Given that $c_{A_1^+}^m = c_{B_1^-}^m$ and $c_{A_1^+}^a = c_{B_1^-}^a$, the phase boundary potential for the A_1B_1 electrolyte is equal to the partitioning potential:

$$\Delta_m^a\phi = \frac{\Delta_m^a\phi_{A_1^+}^0 + \Delta_m^a\phi_{B_1^-}^0}{2} + \frac{RT}{2F}\ln\left(\frac{\gamma_{A_1^+}^m \gamma_{B_1^-}^a}{\gamma_{A_1^+}^a \gamma_{B_1^-}^m}\right). \quad (16.6)$$

For singly charged ions at ionic strength < 0.01 M, the activity coefficients are close to 1; therefore, the last term in Equation 16.6 containing the activity coefficients can be omitted in many situations encountered in potentiometry. Therefore, in the case of a single electrolyte, the measured potential difference is constant, independent of its concentration. Changing the nature of the organic phase would only alter the absolute value of $\Delta_m^a\phi$.

The partitioning potential can only depend on concentration if more than two ions are present in the system. When the second electrolyte (A_2B_2) is added to the aqueous solution together with A_1B_1, the phase boundary in the present experimental setup is of the type:

$$A_1B_1, A_2B_2, \text{water} \mid \text{membrane}$$

for which $\Delta_m^a\phi$ can be calculated by solving the equation derived first by Hung for a general case of β ionic components (Hung 1980):

$$\sum_{i=1}^{\beta} \frac{z_i c_i^{a,0}}{1 + \frac{\gamma_i^a}{\gamma_i^m}e_i r} + \sum_{i=1}^{\beta} \frac{r z_i c_i^{m,0}}{1 + \frac{\gamma_i^a}{\gamma_i^m}e_i r} = 0, \quad (16.7)$$

where $e_i = \exp\left[\frac{z_i F}{RT}\left(\Delta_m^a\phi - \Delta_m^a\phi_i^0\right)\right]$, r is the volume ratio of the organic and aqueous phases ($r = V^m/V^a$), and $c_i^{a,0}, c_i^{m,0}$ refer to the initial concentrations of i in the aqueous and membrane phases, respectively.

Kakiuchi (1996) developed an asymptotic solution of the Hung equation for $z:z$-type electrolytes for the special case of $r \to 0$, which is especially relevant for ISE (typically, $r \approx 2 \times 10^{-4}$). The correct version of the equation reads:

$$\Delta_m^a\phi = \frac{RT}{2zF}\ln\left[-\frac{\sum_B\left(\sum_{k\neq B}z_k c_k^{a,0}\right)e_B}{\sum_A\left(\sum_{k\neq A}z_k c_k^{a,0}\right)/e_A}\right] \quad (16.8)$$

where the subscripts A and B refer to cations and anions, respectively, and
$e_x = \exp\left[\dfrac{F}{RT}\Delta_m^a\phi_x^0\right]$. The summations for A and B are for all anionic and cationic
species, respectively. The summations for $k \neq A$ and $k \neq B$ are done for all ionic species, except for A and B, respectively.

In order to experimentally verify the applicability of this approach to explain the mechanism of electrical potential generation and selectivity of ISE, the membranes consisting only of the polymer matrix (plasticized PVC) and containing neither ion exchanger nor ionophore (*blank* membrane) were prepared. Plasticizers of low (bis(2-ethylhexyl)sebacate [DOS]) and high (*o*-nitrophenyl octyl ether [*o*-NPOE]) dielectric constant were chosen. For practical reasons, a water-soluble QAS, cetyl-trimethylammonium bromide (CTAB), was used. Chemically, CTAB is similar to the lipophilic salts used in anion-selective membranes displaying the characteristic Hofmeister selectivity pattern. The solubility of CTAB in water greatly simplified the experimental procedure of varying the concentration of the A_1B_1 salt (see Equation 16.6): the EMF of the cell consisting of a reference electrode (Ag/AgCl) and an ISE with the blank membrane was monitored during a stepwise addition of CTAB. If the membrane-soluble QASs were used, each QAS concentration would require the preparation of a new electrode, which would certainly reduce the reliability of the data. From the point of view of the adsorbed layer, the initial location of the surface-active species (aqueous or membrane phase) is of secondary importance; hence, the fact of having a water-soluble QAS instead of the membrane-soluble one should not influence the conclusions.

In the framework of ITIES, according to Equation 16.6, the Galvani potential difference $\left(\Delta_m^a\phi\right)$ should not depend on the CTAB concentration, since it is the sole electrolyte in the system. The experimental results for the concentration dependence of the potential in blank ISE (Figure 16.3) do not, however, follow these predictions. This result is not likely to be an artifact caused by the presence of anionic impurities in the membrane, since an analogous (anionic) response was observed for the same membranes immersed in solutions of an anionic surfactant, sodium dodecyl sulfate solution. If the response of ISE with blank membranes was linked to the presence of ionic impurities in the membrane matrix, the latter would then have to contain both anionic and cation sites in appreciable amounts. For later comparison with an alternative "adsorption-based" approach (see below), a concentration dependence of the Stern layer potential ($\Delta\Psi_S$) calculated from the interfacial tension data using the surface quasi two-dimensional electrolyte (STDE) model is also presented.

Despite the ITIES approach failing completely to explain the ISE behavior in pure ionic surfactant solutions, the applicability of this approach was tested for a more close-to-real-ISE CTAB/NaX system (A_1B_1/A_2B_2), with X = Br⁻, NO_3^-, and F⁻ (Wojciechowski, Kucharek et al. 2010). In order to calculate the theoretical dependence of the Galvani potential difference, $\Delta_m^a\phi$, on electrolyte concentration for the system CTAB/NaX, the standard transfer potential for CTA⁺ ion, $\Delta_{NPOE}^a\phi_{CTA^+}^0$, was estimated to be −0.324 V. Other standard potential values for the calculations were taken as the average from all the data available for NPOE in the Girault

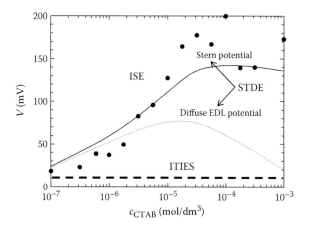

FIGURE 16.3 Dependence of the EMF measured with blank membranes on the bulk concentration of CTAB in the aqueous solution (●), compared with the predictions from the ITIES (dashed line: $\Delta_m^a \phi$, Equation 16.8) and STDE models (solid line: Stern potential, Ψ_S, Equation 16.13; dotted line: diffuse layer potential, Ψ_d, Equation 16.14).

ElectroChemical DataBase (Girault): $\Delta_{NPOE}^a \phi_{Na^+}^0 = 0.376$ V, $\Delta_{NPOE}^a \phi_{Br^-}^0 = -0.389$ V, $\Delta_{NPOE}^a \phi_{NO_3^-}^0 = -0.299$ V, or from Quentel, Mirčeski et al. (2008) $\Delta_{NPOE}^a \phi_{F^-}^0 = -0.725$ V.

The comparison between the ITIES predictions (Hung and Kakiuchi equations [Equations 16.7 and 16.8] both gave the same results) and the experimental results from the blank membrane ISE for CTAB/NaBr, CTAB/NaNO$_3$, and CTAB/NaF mixtures is displayed in Figure 16.4. For comparison, the predictions from the adsorption-based approach are also shown and will be discussed later. It should be noted here that for metrological reasons described above, the EMFs obtained from the ISE measurements are not the absolute values of potential difference, $\Delta_m^a \phi$. Therefore, in order to facilitate comparison with the calculated ITIES potentials, the potentiometric response curves shown in Figure 16.4 were shifted vertically to align with the latter. The comparison shows that the ITIES approach does not properly predict the effect of the presence of inorganic electrolytes (NaBr, NaF, or NaNO$_3$), even when the estimated value of $\Delta_{NPOE}^a \phi_{CTA^+}^0$ was allowed to change during data fitting (not shown). A similar disagreement with the experimental data was obtained while using the standard potential values for the transfer from the aqueous to the nitrobenzene phase, instead of those for the water/ NPOE [$\Delta_{NB}^a \phi_{Na^+}^0 = 0.331$ V, $\Delta_{NB}^a \phi_{Br^-}^0 = -0.345$ V, $\Delta_{NB}^a \phi_{NO_3^-}^0 = -0.263$ V (Girault), and $\Delta_{NB}^a \phi_{F^-}^0 = -0.591$ V (Charreteur, Quentel et al. 2008)]. The standard transfer potential for CTA$^+$ was estimated, $\Delta_{NB}^a \phi_{CTA^+}^0 = -0.370$ V, by using a simple formula (Jensen, Devaud et al. 2002) for the partition coefficient between the aqueous and nitrobenzene phase, $K_p^{a,NB} = 1.62$ (Blank 1966):

$$K_p^{NB,a} = \frac{c_{CTA^+Br^-}^{NB}}{c_{CTA^+Br^-}^a} = \exp\left(-\frac{F}{2RT}\left(\Delta_{NB}^a \phi_{CTA^+}^0 - \Delta_{NB}^a \phi_{Br^-}^0\right)\right). \tag{16.9}$$

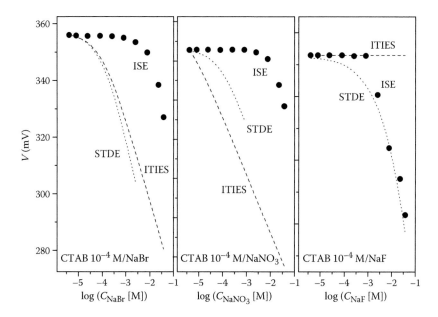

FIGURE 16.4 Dependence of the EMF measured with blank membranes on the bulk concentration of NaX (X = Br⁻, NO₃⁻, F⁻) in the aqueous solution at fixed (10^{-4} M) concentration of CTAB (●), compared with the predictions from the ITIES (dashed line: $\Delta_m^a \phi$, Equation 16.8) and STDE models (dotted line: Stern potential, Ψ_S, Equation 16.13).

The ITIES approach, representing the generalized Nernst equation, failed to predict the ISE response to the water-soluble QAS, CTAB, in the absence and presence of Br⁻, NO₃⁻, and F⁻. This casts some doubts on the applicability of the Nernst "partitioning-based" approach to interpret the ISE response mechanism, including the origin of the Hofmeister effect. Another argument against the use of this approach comes from the comparison of the timescales of ion partitioning and the experimentally observed ISE response times. In practice, it is hardly possible to establish an equilibrium between the two bulk phases (of which one is highly viscous) on the timescale of typical response time of the ISE. For a typical thickness and diffusion coefficient of the plasticized PVC membrane (d = 0.2 mm, D = 10^{-8} cm²/s), the characteristic diffusion time $\left(\tau = \dfrac{d^2}{2D} \right)$ would be of the order of 10^4 s. On the other hand, the pioneering experiments already performed in the 1980s by the group of Pungor revealed that the electrical potential difference is generated even within 2×10^{-2} s, much shorter than the time required for penetration of ions into a PVC membrane (Kellner, Fischbock et al. 1985).

An additional argument questioning the applicability of the Nernst equation in interpreting the ISE response mechanism comes from considerations of the changes in hydration of ions during their transfer through the interface. Ion partitioning between the polar aqueous and low-to-medium polarity membrane would require an almost complete stripping of the ion's hydration shell, which is energetically very expensive, especially for highly hydrated ions. For example, the hydration shell of halide anion consists of typically six to eight water molecules (Marcus 2009), of which only about one to two remain upon transfer even to a solvent as highly polar as nitrobenzene (Osakai, Ogata et al. 1997). Avoiding this costly process, for example, by adsorbing instead of partitioning, would very likely be energetically favorable. Below, an alternative approach, which focuses on adsorption rather than on partitioning equilibria and shifts the central point of the discussion on the Hofmeister effect in ISE from the bulk phases toward the interface between them, is described.

16.7   ADSORPTION-BASED APPROACH

While the Nernst equation has proven extremely useful for practical purposes, mostly for its mathematical simplicity, it is certainly not a good starting point for discussing the mechanism of electrical potential generation in ISE. In fact, there is no single experimental proof that the actual Nernst equation is indeed obeyed in potentiometry with polymeric membranes. It is a common practice in many electrochemistry handbooks to write an equation for the total potential difference between the two electrodes (half-cells) in the form (neglecting any corrections due to selectivity, see Equation 16.3):

$$\text{EMF} = \text{const} + \frac{2.303RT}{z_i F} \log(a), \tag{16.10}$$

where const gathers all the (mostly unknown) constant terms, including the standard Gibbs free energies of transfer, $\Delta G_{\text{tr},i}^{0,\text{a}\rightarrow\text{m}}$, from Equation 16.2. This value is, however, not measurable and there is no possibility to verify whether it indeed contains the $\Delta G_{\text{tr},i}^{0,\text{a}\rightarrow\text{m}}$ terms. The latter are crucial from the point of view of the ISE response mechanism, and especially the potentiometric Hofmeister effect. Therefore, upon closer inspection, it becomes obvious that, in fact the only experimentally available quantity is the slope of the ISE calibration curves. This is indeed usually close to the Nernstian value of 59.2 mV/log a at room temperature. In other words, the equation that characterizes the experimental ISE response is not Equation 16.10, but, in fact, its differential form:

$$\frac{d\text{EMF}}{d\log(a)} = 59.2[\text{mV/log}\,a]. \tag{16.11}$$

The const terms in Equation 16.10 will always depend on the experimental setup. Although the difference between these terms is used in practice for the determination of potentiometric selectivity coefficients and for real ISE measurements, their physical interpretation is rather vague.

As it will be shown below, Equation 16.11 can also be obtained using the electrical double layer (EDL) theory of Gouy–Chapman, which conveniently relates the electrical potential (Ψ) with the surface charge density (σ) through the Grahame equation:

$$\Psi = \frac{2kT}{z_i e} \sinh^{-1}\left(\frac{\sigma}{(8RT\varepsilon\varepsilon_0 c_i)^{1/2}}\right), \tag{16.12}$$

where e is the elementary charge, k is the Boltzmann constant, c_i is the bulk concentration, z_i is the ionic charge, and ε and ε_0 are the permittivity of the medium (relative) and vacuum, respectively. R and T have their usual meaning.

The surface charge density is provided by the adsorption of the surface-active cation or anion of the lipophilic salt. In the case of anion-selective ISE, it is reasonable to assume that adsorption of quaternary ammonium cation (QA^+) from the lipophilic salt charges the membrane surface positively. At high electrical potentials ($\Psi > 100$ mV) and constant surface charge, Equation 16.12 for singly charged anions reads $\dfrac{d\Psi}{d\log(c)} = -2.3kt$, which is equal to -59.2 mV/log a at room temperature, that is, exactly the Nernstian slope (Zubrowska, Wróblewski et al. 2011). Note that this Nernstian slope is not obtained from the Nernst equation here.

In view of the above, a simple requirement for the Nernstian slope of ISE would be the constancy of the surface charge and sufficiently high value of the surface potential, without the need for the actual Nernst equation. Although the condition of high surface potential cannot be experimentally verified (only the potential difference can be measured), it is likely that Ψ indeed exceeds 100 mV given the fact that the potentiometric sensors usually respond over five to six orders of magnitude with the slope of 59.2 mV/log a. Provided that a sufficient amount of counterions is present in the aqueous solution, their surface concentration can closely follow that of the surface-active ions of the lipophilic salt (e.g., QA^+), maintaining the surface charge density constant. Therefore, the Nernstian slope is not unique to the Nernst equation and, upon closer inspection, does not provide a strong proof of the validity of the Nernst equation in potentiometry.

Given the arguments provided above and an amphiphilic nature of the lipophilic salts used in potentiometry, the origin of ion specificity in potentiometric sensors should not necessarily be sought in the bulk partitioning, but at the interface between the aqueous phase and the membrane. The importance of interfacial phenomena in potentiometric signal generation has already been signaled by several authors during the 1980s and 1990s, for example, by the groups of Pungor (Pungor 1997), Umezawa (Tohda, Umezawa et al. 1995), and Ren (Ren 2000). The surface-sensitive spectroscopy (Fourier transform infrared attenuated total reflectance and second harmonic generation) proved that, indeed, the charge separation takes place across the layer of only a few molecules thick (Kellner, Fischbock et al. 1985; Tohda, Umezawa et al. 1995). In this respect, the structure of an outermost membrane layer seems to be of crucial importance for the response mechanism of polymeric ISE. Following the pioneering experiments of Eisenman et al. (Szabo, Eisenman et al. 1969), Umezawa et al. (Minami, Sato et al. 1991) showed that the potentiometric signal can be generated even with the use of the black-film bilayer membranes, with thicknesses of only a few nanometers. Although the aim of their studies was different, these results confirm that bulk partitioning is not necessary for the ISE signal generation. The importance of specific anion adsorption in pH measurements with glass electrodes has also been raised by Salis et al. (Salis, Cristina Pinna et al. 2006).

In order to test the hypothesis of the non-negligible role of interfacial phenomena in the generation of electrical potential at the ISE membrane/aqueous interface and in determining the Hofmeister selectivity, the actual membrane/aqueous interface of ISE was mimicked in several model systems, as described below.

The first indication that the adsorption and electrical potential generation might be related comes from simultaneous measurements of both interfacial tension (as a measure of adsorption) and EMF of the cell (as a measure of electrical potential difference across the interface). For this purpose, a new device combining a drop-shape analysis tensiometer and a high-input impedance voltmeter was used, as described in detail in Wojciechowski (in press). Briefly, the key element of the setup is a specially designed cell that enables both optical and electrical measurements. The organic phase was placed in a 1 cm optical path spectrophotometric cuvette. One reference electrode was immersed in a pool of an aqueous phase placed at the bottom of the cuvette, and the second (pseudo)reference electrode (platinum wire) was melted into the syringe used to produce the aqueous drops for interfacial tension measurements. The simplest model that can be used to mimic the polymeric membrane of ISE is an organic solvent with properties resembling those of the plasticizer, which normally constitutes about two-thirds of the membrane matrix. The two most popular membrane plasticizers, DOS and o-NPOE, possess intermediate polarity ($\varepsilon = 4.2$ and 21, respectively). When mixed with PVC at a 2:1 ratio, the dielectric constants change to 4.8 and 14, respectively (Armstrong, Covington et al. 1988). The less polar plasticizer (DOS) was used as an organic phase filling the cuvette and the drop was formed from the aqueous solution of CTAB. The dynamic interfacial tension was obtained by fitting the shape of the drop to the Gauss–Laplace equation. Upon formation of each new drop of CTAB solution, the interfacial tension immediately started to decrease, while EMF began to increase, both as a result of accumulation of the positive charge of CTA$^+$ ion at the interface.

Although the plasticized PVC used as a typical ISE membrane matrix has some appreciable polarity, the use of simple non-polar solvents is preferable for studying adsorption phenomena. The high polarity of the organic phase would promote the competitive processes of ion partitioning; hence, for the initial assessment, the non-polar toluene ($\varepsilon = 2.4$) was chosen to verify whether adsorption may play any role in the Hofmeister effect in ISE. In terms of EDL properties, the toluene/water system is similar to that reported in the pioneering potentiometric experiments of the group of Eisenman (Ciani, Eisenman et al. 1969) and Umezawa (Sato, Wakabayashi et al. 1997). They used black lipid membranes and bilayer lipid membranes, with the EDL thickness (Debye length) in the non-aqueous phase being large compared to the overall thickness of the membrane. In our case, because of the very low solubility of ionic species in toluene, the thickness of EDL in the non-aqueous phase is also large.

Surface tension measurements are a sensitive tool to study ion-specific adsorption at the water/air interface (Gilányi, Varga et al. 2004; Ivanov, Marinova et al. 2007; Kralchevsky, Danov et al. 1999; Para, Jarek et al. 2006). Since surface tension data can, in principle, provide information on surface concentrations of the charged species (e.g., QA$^+$ and inorganic anions present in solution, X$^-$) adsorbed in the surface region, consequently the corresponding surface charge densities can be obtained. The latter provide a direct access to calculating the electrical potential through the Gouy–Chapman EDL theory. Among several adsorption models available in the literature, only few are capable of describing ion specificity (Gilányi, Varga et al. 2004; Kalinin, Radke 1996; Kralchevsky, Danov et al. 1999; Warszynski, Lunkenheimer et al. 2002). The STDE model of Warszynski et al. (Warszynski, Lunkenheimer et al. 2002) was chosen to interpret the interfacial tension data for CTAB at the water/toluene interface, mainly for the highest number of documented experimental results from the water/air interface consistent with the model. The main features of the model, useful in the context of the Hofmeister selectivity mechanism, are outlined below:

- The surface-active ions of an ionic surfactant (e.g., CTA$^+$) together with their counterions $\left(X^- \equiv Br^-, NO_3^-, F^-, \text{etc.}\right)$ are accumulating in the Stern layer (Figure 16.5).
- The counterions may originate both from dissociation of the ionic surfactant and from an added electrolyte (if present).
- The total charge in the Stern layer is a sum of positive charges of the adsorbed surfactant headgroups (CTA$^+$) and negative charges of counterions (X$^-$).
- Accumulation of the surfactant co-ions (e.g., Na$^+$) in the Stern layer is negligible owing to a strong electrostatic repulsion.

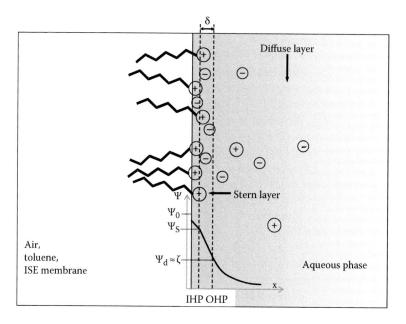

FIGURE 16.5 Scheme of the model of the fluid/aqueous interface with the corresponding EDL potentials (Stern, Ψ_S, and diffuse layer, Ψ_d) and electrokinetic (ζ) potential. IHP and OHP indicate the location of inner and outer Helmholtz planes, respectively.

- The surface charge of the Stern layer is compensated by the opposite charge of the ions accumulated in the diffuse part of the EDL, which renders the interface electrically neutral.
- Ions adsorbed in the Stern layer preserve their freedom of motion; the Stern layer is considered a quasi two-dimensional electrolyte, which does not fulfill the electroneutrality condition.
- Degree of penetration of the Stern layer by a given counterion depends on its effective size and interactions in this layer, which is taken into account by considering a finite size of the headgroups and counterions as well as the lateral electric interactions between them.
- Ion pairing between the adsorbed ions is not considered.

The underlying assumptions of the model and its applicability to the CTAB adsorbed layers were verified experimentally using the total reflection x-ray fluorescence (TRXF). With a penetration depth of the exciting x-ray beam comparable to the Debye length of the EDL, TRXF enabled the determination of the bromide ion concentration in the Stern and diffuse layers (Wojciechowski, Gutberlet et al. in press). Upon addition of inorganic salts to the solution of CTAB (10^{-4} M), the Br x-ray fluorescence intensity diminished, confirming the replacement of Br$^-$ in the interfacial region with anions introduced with the inorganic salt. The ordering of anions closely followed that of the Hofmeister series, with the exception of phosphate ions, which is probably related to the partial dissociation of $H_2PO_4^-$.

To gain a quantitative insight into the relation between adsorption of ionic species and electrical potential generation, a series of interfacial tension measurements were performed with either pure CTAB or its mixtures with inorganic salts containing different anions (Wojciechowski, Kucharek et al. 2009, 2010) in the water/toluene system. For the latter case, in order to mimic the situation encountered during the operation of ISE, the concentration of CTAB was made constant, while the concentration of inorganic salts added to the aqueous phase was varied. Using these data, surface concentrations (Γ) of the ionic species present in solution and the corresponding surface charge densities (σ) were calculated. The electric potential of the Stern layer was found from

$$\Psi_S = \Psi_d + \frac{\sigma \delta}{\varepsilon_0 \varepsilon_S}, \tag{16.13}$$

where the diffuse layer potential, Ψ_d, at the boundary between the Stern layer and the diffuse part of EDL can be determined from the formula:

$$\Psi_d = \frac{2kT}{e} \sinh^{-1} \left(\frac{\sigma e}{2\varepsilon_0 \varepsilon kT \kappa} \right) \tag{16.14}$$

where e is the elementary charge, k is the Boltzmann constant, ε_0 is the vacuum dielectric permittivity, ε is the dielectric constant of the solution, κ is the reciprocal Debye length, δ is the thickness of the Stern layer, and ε_S is the dielectric constant in the Stern layer.

The EDL potentials calculated using the STDE model were compared to those experimentally obtained for the blank membrane–based ISE, as described earlier in this chapter for testing the ITIES approach. The concentration dependence of the electrical potential of the membrane-based ISEs is very similar to that of the Stern potential (Ψ_s) for pure CTAB solutions (Figure 16.3). It is worth recalling that the partitioning-based approach (ITIES) predicted a constant value of electrical potential at the membrane, in strong disagreement with the experimental results. In the case of mixed CTAB/NaX solutions (Figure 16.4), the STDE model fits the data comparably to the ITIES, although for F⁻ (extreme right-hand side of the Hofmeister series), the adsorption-based model (STDE) is much better.

Some efforts to extend the study of interfacial tension to a more polar organic phase were also taken (Wojciechowski and Kucharek 2009). Replacing the non-polar toluene by, for example, nitrobenzene provides a non-aqueous phase, where the lipophilic QASs can be dissolved, as in real ISE. On the other hand, however, ion partitioning may also take place in the system, because of the high affinity of small counterions toward the aqueous phase. Four lipophilic salts (tetraoctylammonium bromide [TOAB], tetradodecylammonium bromide [TDDAB], tetrahexadecyl-ammonium bromide [THDAB], and tetraoctadecylammonium bromide) with chain lengths varying in the range from $4 \times C_8H_{17}$ to $C_{18}H_{37}$, that is, between 32 and 72 total carbon atoms, were dissolved in nitrobenzene and their dynamic interfacial tension against pure water was recorded. Unfortunately, instead of an expected decay of dynamic interfacial tension, typically observed during adsorption of CTAB or any other surfactant from the aqueous phase, an oscillatory behavior was observed for all salts. An exemplary "apparent" dynamic interfacial tension profile for THDAB at a concentration of 10^{-4} M is shown in Figure 16.6.

Even though typically the amplitude of oscillation did not exceed a few millinewtons per meter, in some cases, it was reaching up to 50 mN/m. Interestingly,

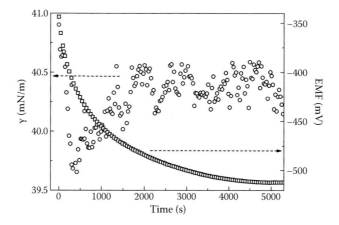

FIGURE 16.6 Apparent oscillations of dynamic interfacial tension for THDAB (10^{-4} M) at the nitrobenzene/water interface (left axis), together with the corresponding time course of EMF (right axis).

the frequency and amplitude of oscillations did not depend critically on the length of the alkyl chains in the tetraalkylammonium salt. This suggests that the solubility of tetraalkylammonium bromide in the aqueous phase does not play a major role in the process. Interestingly, the apparent dynamic interfacial tension oscillations were always accompanied by clouding in the nitrobenzene phase, which points to a spontaneous emulsification as the origin of the observed oscillations. As such, they do not have to be related to any real changes of adsorbed amount at the interface, but might just be an experimental artifact owing to the optical method employed for determination of interfacial tension. This was indeed confirmed using the setup described above for simultaneous measurements of dynamic interfacial tension and EMF. As shown in Figure 16.6, in the system THDAB–nitrobenzene/water, the oscillations of interfacial tension are not accompanied by any oscillations of electrical potential. Instead, a slow decay could be seen, suggesting that the oscillations do not involve any periodic exchange of ionic species. Moreover, the microscopic observation of the nitrobenzene phase in the proximity of an aqueous drop indeed confirmed the emulsion formation. Therefore, the most probable mechanism of oscillations involves a transfer of water necessary for hydration of bromide ions transferring through the interfacial region from the nitrobenzene to the aqueous phase. Unfortunately, this phenomenon precluded any reliable measurements of interfacial tension in the system polar organic solvent/water.

Although the Stern layer (inner Helmholtz plane [IHP]; see Figure 16.5) would be the best location in EDL to probe the ion-specific effects, unfortunately no direct way exists to measure the Stern potential, Ψ_s (Dynarowicz-Latka, Dhanabalan et al. 2001). On the other hand, the electrical potential at the slip plane (electrokinetic or ζ-potential) is easily obtainable experimentally, for example, from streaming current/potential or electrophoretic mobility measurements (Delgado, González-Caballero et al. 2005). Even though the ζ-potential is not a well-defined EDL parameter, its value is believed to best approximate the diffuse layer potential, Ψ_d (i.e., electrical potential at the outer Helmholtz plane [OHP]; see Figure 16.5) (Delgado, González-Caballero et al. 2005; Lyklema 2011). So far, only few reports on electrokinetic studies of ion specificity are available in the literature (Das, Borah et al. 2010; Djerdjev and Beattie 2008; Wojciechowski, Bitner et al. 2011; Zimmermann, Dukhin et al. 2001; Zimmermann, Rein et al. 2009), mostly due to a rather low sensitivity of the ζ-potential to adsorption processes taking place relatively far away from the diffuse layer location (see Figure 16.5). Some authors even question the existence of any measurable ion specificity in ζ-potential; see, for example, Creux, Lachaise et al. (2007) and Franks, Djerdjev et al. (2005).

If ion-specific adsorption of counterions is indeed responsible for the selectivity of ISE, the electrokinetic potential should also carry some information on this ion specificity. To test this hypothesis, the electrokinetic mobility of a series of toluene-in-water emulsions (0.2% v/v) was measured. The emulsions were stabilized by adsorption of the positively charged CTA^+ ions from CTAB, and the total charge on the surface of the aqueous droplets was modulated by the presence of different anions added as an inorganic salt NaX (with $X = Br^-$, NO_3^-, F^-, ClO_4^-) (Wojciechowski, Bitner et al. 2011). Given that a large κa limit holds for the emulsions used in the study (where κ is the Debye reciprocal length and a is the radius of emulsion drop), the electrokinetic mobility was converted into ζ-potentials using the simple Helmholtz–Smoluchowski equation. It should be, however, borne in mind that this theory has been developed for rigid particles, and not oil drops, where the notion of shear plane is not clearly defined.

The CTAB concentration for this and subsequent studies (10^{-4} M) was chosen on the basis of the adsorption isotherms at water/air and water/toluene interfaces (Para, Jarek et al. 2006; Wojciechowski, Kucharek et al. 2009). At this concentration, the adsorbed layer is already densely packed, but still sufficiently far from the saturation (surface coverage, $\theta \approx 30\%$) to be sensitive to small changes induced by the addition of anions. The inorganic salt concentration of NaBr, NaNO$_3$, and NaF was varied in the range 10^{-5} to 10^{-2} M. In the presence of even relatively small concentrations of NaClO$_4$, the precipitate of CTA-ClO$_4$ was formed by metathesis; hence, the effect of this salt could only be studied up to the concentration of 3×10^{-5} M. The effect of added anions varies with their concentration, with an exception of fluoride, which does not have any significant effect on the zeta potential, which is consistent with its low affinity to the interface (Figure 16.7). In general, the current zeta potentials follow the STDE predictions and reproduce reasonably well the expected small differences between different anions, although a fully quantitative interpretation of the zeta potential at the toluene–water interface is complicated by its diffuse nature (Djerdjev and Beattie 2008).

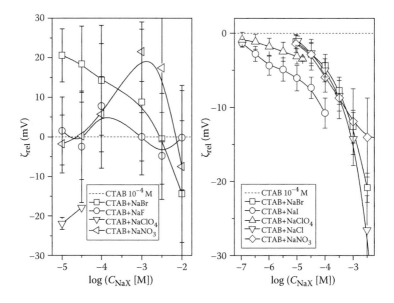

FIGURE 16.7 Variation of electrokinetic potential of toluene-in-water emulsions (left) and at the PVC membrane interface (right) with concentration of sodium salts, NaX, at fixed CTAB concentration (10^{-4} M). The data are presented as a difference between ζ-potential in the given solution and that in bare CTAB solution (without added inorganic electrolyte), ζ_{rel}.

The ion specificity in electrokinetic potential was also assessed in a more realistic membrane/aqueous system (Wojciechowski and Linek 2012). The blank (i.e., with no added lipophilic salts or ionophore) polymeric membranes of the same composition as those typically employed for potentiometric measurements for ISE (Bakker, Buhlmann et al. 1997) were used throughout this study. They consisted of PVC plasticized with DOS in 1:2 w/w ratio.

The ζ-potential was obtained by measuring the applied pressure dependence of the streaming current ($\Delta I_{str}/\Delta P$) during the flow of the aqueous solution of CTAB/NaX through the channel, whose walls were covered with the DOS-plasticized PVC membranes:

$$\frac{\Delta I_{str}}{\Delta P} = \zeta(I_{str})\frac{\varepsilon_0\varepsilon_r A_c}{\eta L}, \tag{16.15}$$

where ε_0 is the electric permittivity of vacuum, ε_r and η are the relative permittivity and dynamic viscosity of the liquid medium, A_c is the channel cross section, and L is its length.

At fixed concentration of CTAB (10^{-4} M), the two most left hand–sided *Hofmeister anions* (ClO_4^- and I^-) are capable of decreasing the ζ-potential even at very low ($10^{-7} \div 10^{-6}$ M) concentrations (Figure 16.7). Unfortunately, for the reasons described earlier, the measurements above the concentrations of 2×10^{-5} and 10^{-4} M, for ClO_4^- and I^-, respectively, were not possible. Nevertheless, even at these relatively low concentrations, a significant reduction of the ζ-potential was noticed. Other salts employed in the study (NaCl, NaBr, NaF, NaNO$_3$) do not pose any solubility problems under the same conditions, and their effect on ζ-potential was studied in the concentration range of 10^{-5} to 3×10^{-3} M (Figure 16.7). Unexpectedly, fluoride ions, despite being one of the most right hand–sided Hofmeister anions, exert strong effect on the measured ζ-potential, even higher than, for example, bromide ions. The effect of anions on ζ-potential shown in Figure 16.7 in general again follows the anion ordering in the Hofmeister series, with F^- being the only exception.

The hypothesis of ion adsorption at the ISE membrane/aqueous interface as a driving force for the electrical potential development was finally tested in real ISE setup. For this purpose, a series of ISE with membranes containing different QASs were prepared: TOAB, TDDAB, tridodecylmethylammonium chloride (TDMAC), and dimethyldioctadecylammonium chloride (DODMAC) (see Figure 16.8). The QAS employed contained 32 (4×8), 48 (4×12), 37 ($[3 \times 12] + 1$), and 38 ($[2 \times 1] + [2 \times 18]$) methylene units, respectively. Hence, they should show a similar affinity to the organic phase (lipophilicity) but should differ in surface affinity (amphiphilicity). For the preliminary experiments, the QAS concentration in the membrane was maintained at the level of typical ISE, that is, 1% (Legin, Makarychev-Mikhailov et al. 2004; Wróblewski, Chudy et al. 2000). Since the QAS-based ISEs are typically used as nitrate-selective sensors, the results are presented as response curves in solutions of increasing concentrations of nitrate ions (FIM) or as potentiometric selectivity coefficients for a given ion with respect to nitrate, $K_{NO3-,\ X-}$ (SSM).

The electrodes based on symmetric QASs with 4×8 and 4×12 methylene units in the alkyl chain (TOAB, TDDAB) as well as an asymmetric salt TDMAC

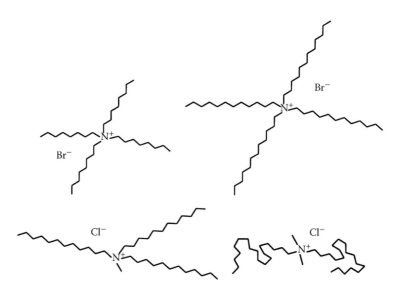

FIGURE 16.8 Structures of tetraalkylammonium salts employed typically in ISE membranes exhibiting the Hofmeister selectivity: TOAB, TDDAB, TDMAC, and DODMAC.

($[3 \times 12] + 1$ methylene units) exhibited the highest slopes of the calibration curves in a broad concentration range (3×10^{-5} to 1.9×10^{-2} M) (Zubrowska, Wróblewski et al. 2011). In full agreement with the conclusions of previous studies (Legin, Makarychev-Mikhailov et al. 2004; Ozawa, Miyagi et al. 1996; Schwake, Cammann et al. 1999; Wróblewski, Chudy et al. 2000), the symmetry of the QAS structure is not a decisive factor determining the performance of ISE. Interestingly, however, the two salts with a comparable number of methylene units but with different structures (TDMAC and DODMAC: 37 vs. 38 methylene units, respectively) show significant differences in their potentiometric behavior. In this case, the asymmetry of TDMAC seems to favor the near-Nernstian behavior of the corresponding ISE in a broad nitrate concentration range. This observation fully supports the hypothesis of the importance of amphiphilicity of the QAS in its ability to modulate a potential response of the ISE. A similar conclusion can be drawn from the potentiometric selectivity coefficients determined using the SSM method. The most pronounced differences in selectivity coefficients for different anions were found for electrodes with membranes containing TDDAB and TDMAC, and the least, for DODMAC (Zubrowska, Wróblewski et al. 2011).

In the second step, the effect of lipophilic salt concentration on the ISE performance was tested in the range 1% to 10^{-4}% (corresponding to 2×10^{-2} to 2×10^{-6} mol kg^{-1} of the membrane plasticizer) for TDDAB-based membranes. A near-Nernstian sensitivity in the range 10^{-1}% to 1% was observed. At 10^{-2}% of QAS in the membrane phase, a significant worsening of nitrate selectivity was noticed for these electrodes, and practically no response was observed below 10^{-3}% (Figure 16.9).

A concentration dependence of the potentiometric selectivity coefficient determined using the SSM method for potentiometric electrodes with membranes

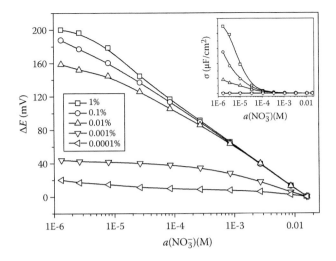

FIGURE 16.9 The NO_3^- response curves for ISE with membranes containing different amounts of TDDAB. The curves were recorded in 0.1 M Na_2SO_4. All the curves were intentionally shifted vertically to merge at the highest $a(NO_3^-)$ value (1.9×10^{-2} M), assigned $\Delta E = 0$. The inset shows the surface charge densities calculated from the ISE responses using Equation 16.19.

containing TDDAB in the concentration range $10^{-4}\%$ to 1% is shown in Table 16.1. The SSM selectivity pattern observed at high QAS concentrations is consistent with the Hofmeister series, with F^- being the most discriminated among the right hand–side anions. With decreasing QAS content in the membrane, the right hand–side anions of the Hofmeister series become less and less discriminated with respect to nitrate and the left hand–side anions. On the other hand, the response to the latter is not significantly altered with decreasing QAS concentration.

The potentiometric results described above cannot be explained using a simple Nernst equation, which does not predict any effect on the ISE performance by either

TABLE 16.1

SSM Selectivity Coefficients ($\log K_{NO_3^-, j}$) of ISE with Membranes Containing Different Concentrations of TDDAB for Different Interfering Ions, $j = F^-, Cl^-, Br^-, ClO_4^-$ and $H_2PO_4^-$

TDDAB	$\log K_{NO_3^-, j}$				
Concentration	F^-	Cl^-	Br^-	ClO_4^-	$H_2PO_4^-$
1%	−5.0	−2.1	−0.9	3.2	−3.1
$10^{-1}\%$	−1.3	−2.2	−0.8	3.0	−1.5
$10^{-2}\%$	−0.6	−1.6	−0.5	2.8	−1.4
$10^{-3}\%$	0.9	−0.3	0.5	2.7	−1.3
$10^{-4}\%$	1.4	0.1	0.4	1.5	−1.0

composition or concentration of the lipophilic salt. Below, an attempt to explain them in the framework of the adsorption-based mechanism is described. The major assumption in the discussion, following the arguments presented in the preceding parts, is that the electrical potential difference in QAS-based ISE is generated at the interface, mostly because of a preferential adsorption of the surface-active QA^+ (from the membrane side) accompanied by co-adsorption of inorganic anions (from both the membrane and aqueous sides). These two processes lead to an incomplete charge neutralization within the Stern layer and give rise to a net electrical potential at the aqueous/membrane interface.

The electrical potential, Ψ, at the membrane surface originating from the partially screened positive charge of the adsorbed layer of QA^+ can be easily related to the surface charge density, σ, in the Stern layer (see Figure 16.5) within the framework of the Gouy–Chapman theory:

$$\sigma = -\varepsilon_0 \varepsilon_r \left(\frac{d\psi}{dx} \right), \tag{16.16}$$

where ε_0 is the vacuum dielectric permittivity, ε_r is the dielectric constant of the solution, and x is the distance from the interface (see Figure 16.5).

For simplicity, the EDL is assumed to extend only toward the aqueous phase. In view of many experimental results (Samec 1988), this is not fully realistic, but it is sufficient for the purpose of semiquantitative discussion presented below. The electrical potential gradient $\left(\dfrac{d\psi}{dx} \right)$ can be found from the Poisson–Boltzmann equation:

$$\frac{d^2\psi}{dx^2} = -\frac{e}{\varepsilon_0 \varepsilon_r} \sum z_i n_i^\infty \exp\left(\frac{-z_i e\psi}{kT} \right) \tag{16.17}$$

which, after rearrangement and transformation to the molar concentration scale, gives

$$\frac{d\psi}{dx} = \left[\frac{1.204 \times 10^{27} kT}{\varepsilon_0 \varepsilon_r} \sum z_i c_i^\infty \left(\exp\left(\frac{-z_i e\psi}{kT} \right) - 1 \right) \right]^{1/2} \tag{16.18}$$

where e is the elementary charge, k is the Boltzmann constant, T is the absolute temperature, z_i is the ionic charge of ions i, n_i^∞ is the number density of ions i in bulk, and c_i^∞ is their bulk concentration in molar.

For a mixture of 1:1 and 1:2 electrolytes used for recording the FIM response curves ($NaNO_3$ and Na_2SO_4), a combination of Equations 16.16 and 16.18 gives the surface charge density in the following form of the Grahame equation:

$$\sigma = 0.118 \sinh\left(\frac{e\psi}{2kT} \right) \left[c_1^\infty + c_2^\infty \left(\exp\left(\frac{e\psi}{kT} \right) + 2 \right) \right]^{1/2}. \tag{16.19}$$

The surface charge density in this case originates from the presence of all adsorbed ionic species: $\sigma = F(\Gamma_{QA} - 2\Gamma_{SO42} - \Gamma_{NO3})$, where F is the Faraday constant and Γ_i is the surface concentration of ionic species i (QA^+, SO_4^{2-}, and NO_3^-, respectively). It should be noted here that during the measurements, the anions of the QASs (Cl^-, Br^-) in the surface layer are most likely immediately exchanged with the aqueous phase anions (NO_3^- in the present study). For this reason, the original QAS counterions (Cl^- or Br^-) are not included in the present considerations. Since the bulk concentration of Na_2SO_4 in the aqueous solution during FIM measurements is practically constant, the surface concentration of SO_4^{2-} (Γ_{SO42}) can also be assumed constant. Moreover, Na_2SO_4 is regarded as the most excluded salt from the macroscopic aqueous/air interface (Para and Warszynski 2007; Pegram and Record 2008a,b); thus, the SO_4^{2-} surface concentration is not only constant but also probably negligible. Hence, the major contribution to σ in the present setup would be QA^+ (from the membrane) and NO_3^- (both from the membrane and from the sample).

The surface charge densities as a function of nitrate ions concentration in the sample calculated from the FIM curves using the Grahame equation (Equation 16.19) for different TDDAB concentrations in the ISE membrane are shown in Figure 16.9. It should be reminded here that the absolute values of the electrical potential of ISE (E) are not experimentally available. As a consequence, the absolute values of the surface charge density cannot be calculated from the experimental ISE data with Equation 16.19 as well (hence, no numerical values of σ were provided in the figure). As an effort to normalize the ISE results, the original responses were intentionally shifted vertically in order to merge at $\Delta E = 0$ for the highest NO_3^- activity. Nevertheless, the values presented in Figure 16.9 should in no case be treated as absolute ones; they will only be discussed semiquantitatively.

The results (Figure 16.9) clearly show that the surface charge decreases nonlinearly with the anion concentrations both in the membrane (Br^- or Cl^-, varying in line with QAS concentration) and in the sample (NO_3^- varying independently from QAS in the membrane). This provides a comprehensive explanation of the dependence of the ISE performance on the concentration and structure of QAS. In the framework of the Gouy–Chapman theory (Grahame equation), the anionic response of QAS-based ISE requires a positive charge of the membrane/aqueous interface, which is assumed by a high surface concentration of QA^+ (Γ_{QA}). The latter can be achieved by a combination of high surface activity and high concentration of QAS in the membrane. When the QAS concentration in the membrane is low and its surface activity is low, the positive charge density of the membrane is also low (Figure 16.9). In that case, the Stern layer potential is dominated by the (weak) spontaneous adsorption of anions, and the differences in potentiometric selectivities are minor. Only when the positive charge density in the Stern layer develops with increasing QA^+ adsorption are the left hand–side anions getting more and more preferably recognized with respect to the right-hand ones. This is because the major driving force for anion adsorption is still an electrostatic interaction of an anion with an interface (Moreira, Boström et al. 2006). A combination of several non-electrostatic interactions, whose concerted action is commonly known as the Hofmeister effect, can only fine-tune the ion distribution at the ISE membrane/aqueous interface, consequently regulating the value of the net surface charge. The latter then influences the electrical potential measured in

ISE. As long as the interface maintains its ability to keep the surface charge approximately constant (and non-zero), the Nernstian slope is observed (which of course does not imply that the Nernst equation is obeyed, see above).

16.8 CONCLUSIONS

Electrical phenomena at fluid/aqueous interfaces are often coupled with ion partitioning between the two phases or with adsorption from one or both phases onto these interfaces. If the non-aqueous phase is polar, usually the partitioning dominates and the electrical potential difference develops across the interface following the predictions of the Nernst equation (Reymond, Fermín et al. 2000). On the other hand, the electrical potential at the interface between the aqueous and the non-polar organic phase or air is driven by adsorption or orientation of ions and dipoles at the interface. The polymeric membranes of ISEs represent an example of an intermediate polarity medium (Armstrong, Covington et al. 1988), where probably both phenomena (i.e., partitioning and adsorption) may co-exist. As a result, the electrical potential generation and selectivity of ISE can be interpreted from two different points of view. One, so far dominating (partitioning based), is centered on the Nernst equation and assumes that the electrical potential is generated through the bulk partitioning of ions between the aqueous phase (analyte) and the membrane. This implies that the consequent selectivity is also governed by selective ion partitioning. In the case of membranes with lipophilic salts but without deliberately added ionophores, the characteristic selectivity pattern (Hofmeister series) would then stem from differences in free Gibbs energies of ion transfer $\left(\Delta G_{\mathrm{tr},i}^{0,\mathrm{a}\rightarrow\mathrm{m}}\right)$ from the aqueous phase to the membrane. This interpretation, although very simple, both mathematically (Nernst equation) and conceptually, is not consistent with the current views on the origin of ion specificity in biochemistry and in colloid and interface science.

An alternative view (adsorption based), presented in this chapter focuses on interfacial processes, rather than bulk partitioning. The electrical potential difference measured in ISE is assumed to originate from specific adsorption of ions at the membrane/aqueous interface. The latter is electrically charged owing to adsorption of a surface-active ion of the lipophilic salt (e.g., quaternary ammonium cation, QA^+, for anion-selective membranes). For anion-selective ISE, the membrane/aqueous interface is then positively charged and attracts counterions (anions) electrostatically. However, because of the specific conditions of the interface, this purely electrostatic attraction is fine-tuned by specific non-electrostatic interactions, commonly described as the Hofmeister effect, eventually giving rise to the potentiometric Hofmeister selectivity pattern. From the perspective of interfacial phenomena, as long as the surface charge density remains constant (and non-zero), the electrical potential at the interface varies linearly with the logarithm of the counterion concentration, giving the so-called Nernstian slope. Any deviations from this slope (theoretically excluded in the framework of Nernst equation, although very often observed experimentally) would then originate from a loss of the ability to maintain the total surface charge constant, for example, because of insufficiently high surface concentration of QA^+ in the membrane or of the anions in the aqueous solution. It should be stressed that the Nernstian slope is not necessarily a proof of the validity of

Nernst equation, as often stated in the literature. Using the "adsorption" approach to the ISE response and selectivity, some other experimental observations can be easily explained as well. For example, the detection limit, which for most ISEs lies around 10^{-5} M and is not predicted by the Nernst equation, can be justified by realizing that a certain minimum concentration of counterions is necessary to maintain the surface potential constant. More importantly, however, ion specificity can be easily linked to different surface charge densities in the presence of different counterions and related to their affinity to the oppositely charged interface.

Despite the fact that the new adsorption-based approach to the selectivity of ISE offers some advantages over the simple partitioning-based one, it is still not clear which of the two reflects better the real situation at the ISE membrane/analyte interface. In practice, partitioning and adsorption phenomena at fluid/fluid interfaces cannot be separated, and most likely both participate in the generation of the electrical potential difference in ISE. Therefore, the distinction between adsorption and ion partitioning is somehow arbitrary. Mathematically, both give practically the same dependence on logarithm of concentration, with a characteristic Nernstian slope. In the future, perhaps a combination of partitioning and adsorption equilibria will provide the most realistic description of the Hofmeister effect in potentiometry with polymeric membranes.

ACKNOWLEDGMENTS

This work was financially supported by the Warsaw University of Technology, Poland.

REFERENCES

Ao, Z., Liu, G. and Zhang, G., 2011. Ion specificity at low salt concentrations investigated with total internal reflection microscopy. *Journal of Physical Chemistry C*, 115(5), pp. 2284–2289.

Armstrong, R.D., Covington, A.K. and Proud, W.G., 1988. Solvent properties of PVC membranes. *Journal of Electroanalytical Chemistry*, 257(1–2), pp. 155–160.

Baer, M.D., Kuo, I.-W., Bluhm, H. and Ghosal, S., 2009. Interfacial behavior of perchlorate versus chloride ions in aqueous solutions. *Journal of Physical Chemistry B*, 113(48), pp. 15843–15850.

Bakker, E. and Pretsch, E., 2007. Modern potentiometry. *Angewandte Chemie—International Edition*, 46(30), pp. 5660–5668.

Bakker, E., Buhlmann, P. and Pretsch, E., 1997. Carrier-based ion-selective electrodes and bulk optodes. 1. General characteristics. *Chemical Reviews*, 97(8), pp. 3083–3132.

Bakker, E., Bühlmann, P. and Pretsch, E., 2004. The phase-boundary potential model. *Talanta*, 63(1), pp. 3–20.

Blank, M., 1966. Some effects due to the flow of current across a water–nitrobenzene interface. *Journal of Colloid and Interface Science*, 22(1), pp. 51–57.

Boström, M., Williams, D.R.M. and Ninham, B.W., 2003. Specific ion effects: Why the properties of lysozyme in salt solutions follow a Hofmeister series. *Biophysical Journal*, 85(2), pp. 686–694

Braven, J., Ebdon, L., Frampton, N.C., Le Goff, T., Scholefield, D. and Sutton, P.G., 2003. Mechanistic aspects of nitrate-selective electrodes with immobilised ion exchangers in a rubbery membrane. *Analyst*, 128(8), pp. 1067–1072.

Charreteur, K., Quentel, F., Elleouet, C. and L'her, M., 2008. Transfer of highly hydrophilic ions from water to nitrobenzene, studied by three-phase and thin-film modified electrodes. *Analytical Chemistry*, 80(13), pp. 5065–5070.

Chen, X., Yang, T., Kataoka, S. and Cremer, P.S., 2007. Specific ion effects on interfacial water structure near macromolecules. *Journal of the American Chemical Society*, 129(40), pp. 12272–12279.

Ciani, S., Eisenman, G. and Szabo, G., 1969. A theory for the effects of neutral carriers such as the macrotetralide actin antibiotics on the electric properties of bilayer membranes. *Journal of Membrane Biology*, 1(1), pp. 1–36.

Collins, K.D., Neilson, G.W. and Enderby, J.E., 2007. Ions in water: Characterizing the forces that control chemical processes and biological structure. *Biophysical Chemistry*, 128(2–3), pp. 95–104.

Creux, P., Lachaise, J., Graciaa, A. and Beattie, J.K., 2007. Specific cation effects at the hydroxide-charged air/water interface. *Journal of Physical Chemistry C*, 111(9), pp. 3753–3755.

Dang, L.X., 2002. Computational study of ion binding to the liquid interface of water. *Journal of Physical Chemistry B*, 106(40), pp. 10388–10394.

Das, M.R., Borah, J.M., Kunz, W., Ninham, B.W. and Mahiuddin, S., 2010. Ion specificity of the zeta potential of α-alumina, and of the adsorption of *p*-hydroxybenzoate at the α-alumina–water interface. *Journal of Colloid and Interface Science*, 344(2), pp. 482–491.

Delgado, A.V., González-Caballero, F., Hunter, R.J., Koopal, L.K. and Lyklema, J., 2005. Measurement and interpretation of electrokinetic phenomena: (IUPAC technical report). *Pure and Applied Chemistry*, 77(10), pp. 1753–1805.

Djerdjev, A.M. and Beattie, J.K., 2008. Electroacoustic and ultrasonic attenuation measurements of droplet size and ζ-potential of alkane-in-water emulsions: Effects of oil solubility and composition. *Physical Chemistry Chemical Physics*, 10(32), pp. 4843–4852.

Dos Santos, A.P., Diehl, A. and Levin, Y., 2010. Surface tensions, surface potentials, and the Hofmeister series of electrolyte solutions. *Langmuir*, 26(13), pp. 10778–10783.

Dynarowicz-Latka, P., Dhanabalan, A. and Oliveira, O.N., 2001. Modern physicochemical research on Langmuir monolayers. *Advances in Colloid and Interface Science*, 91(2), pp. 221–293.

Egorov, V., Rakhman'ko, E., Okaev, E., Nazarov, V., Pomelyenok, E. and Pavlova, T., 2004. Novel anion exchangers for electrodes with improved selectivity to divalent anions. *Electroanalysis*, 16(17), pp. 1459–1462.

Enami, S., Mishra, H., Hoffmann, M.R. and Colussi, A.J., 2012. Hofmeister effects in micromolar electrolyte solutions. *Journal of Chemical Physics*, 136(15), pp. 154707_1–154707_5.

Franks, G.V., Djerdjev, A.M. and Beattie, J.K., 2005. Absence of specific cation or anion effects at low salt concentrations on the charge at the oil/water interface. *Langmuir*, 21(19), pp. 8670–8674.

Gilányi, T., Varga, I. and Mészáros, R., 2004. Specific counterion effect on the adsorption of alkali decyl sulfate surfactants at air/solution interface. *Physical Chemistry Chemical Physics*, 6(17), pp. 4338–4346.

Girault, H.H., The ElectroChemical DataBase. Available at http://serveur-isic.epfl.ch/labo/girault/cgi/DB/InterrDB.pl.

Guggenheim, E.A., 1929. The conceptions of electrical potential difference between two phases and the individual activities of ions. *The Journal of Physical Chemistry*, 33(6), pp. 842–849.

Hara, H., Ohkubo, H. and Sawai, K., 1993. Nitrate ion-selective coated-wire electrode based on tetraoctadecylammonium nitrate in solid solvents and the effect of additives on its selectivity. *The Analyst*, 118(5), pp. 549–552.

Henry, C.L. and Craig, V.S.J., 2010. The link between ion specific bubble coalescence and Hofmeister effects is the partitioning of ions within the interface. *Langmuir*, 26(9), pp. 6478–6483.

Hofmeister, F.E., 1888. Zur Lehre von der Wirkung der Salze. *Naunyn-Schmiedeberg's Archives of Pharmacology*, 24(4), pp. 247–260.

Hua, W., Chen, X. and Allen, H.C., 2011. Phase-sensitive sum frequency revealing accommodation of bicarbonate ions, and charge separation of sodium and carbonate ions within the air/water interface. *Journal of Physical Chemistry A*, 115(23), pp. 6233–6238.

Hung, L.Q., 1980. Electrochemical properties of the interface between two immiscible electrolyte solutions. Part I. Equilibrium situation and Galvani potential difference. *Journal of Electroanalytical Chemistry and Interfacial Electrochemistry*, 115(2), pp. 159–174.

Ivanov, I.B., Marinova, K.G., Danov, K.D., Dimitrova, D., Ananthapadmanabhan, K.P. and Lips, A., 2007. Role of the counterions on the adsorption of ionic surfactants. *Advances in Colloid and Interface Science*, 134–135, pp. 105–124.

Jagoda-Cwiklik, B., Vácha, R., Lund, M., Srebro, M. and Jungwirth, P., 2007. Ion pairing as a possible clue for discriminating between sodium and potassium in biological and other complex environments. *Journal of Physical Chemistry B*, 111(51), pp. 14077–14079.

Jensen, H., Devaud, V., Josserand, J. and Girault, H.H., 2002. Contact Galvani potential differences at liquid-liquid interfaces. *Journal of Electroanalytical Chemistry*, 537(1–2), p. 77.

Jungwirth, P. and Tobias, D.J., 2002. Ions at the air/water interface. *Journal of Physical Chemistry B*, 106(25), pp. 6361–6373.

Kakiuchi, T., 1996. Limiting behavior in equilibrium partitioning of ionic components in liquid-liquid two-phase systems. *Analytical Chemistry*, 68(20), pp. 3658–3664.

Kalinin, V.V. and Radke, C.J., 1996. An ion-binding model for ionic surfactant adsorption at aqueous–fluid interfaces. *Colloids and Surfaces A: Physicochemical and Engineering Aspects*, 114, pp. 337–350.

Kellner, R., Fischbock, G., Gotzinger, G., Pungor, E., Toth, K., Polos, L. and Lindner, E., 1985. FTIR-ATR spectroscopic analysis of bis-crown-ether based PVC-membrane surfaces. *Fresenius' Journal of Analytical Chemistry*, 322(2), pp. 151–156.

Knock, M.M. and Bain, C.D., 2000. Effect of counterion on monolayers of hexadecyltrimethylammonium halides at the air–water interface. *Langmuir*, 16(6), pp. 2857–2865.

Koryta, J., 1975. *Ion-Selective Electrodes*. Cambridge: Cambridge University Press.

Kralchevsky, P.A., Danov, K.D., Broze, G. and Mehreteab, A., 1999. Thermodynamics of ionic surfactant adsorption with account for the counterion binding: Effect of salts of various valency. *Langmuir*, 15(7), pp. 2351–2365.

Kunz, W., 2010. Specific ion effects in colloidal and biological systems. *Current Opinion in Colloid and Interface Science*, 15(1–2), pp. 34–39.

Lee, S., Sanstead, P.J., Wiener, J.M., Bebawee, R. and Hilario, A.G., 2010. Effect of specific anion on templated crystal nucleation at the liquid–liquid interface. *Langmuir*, 26(12), pp. 9556–9564.

Legini, A., Makarychev-Mikhailov, S., Kirsanov, D., Mortensen, J. and Vlasov, Y., 2004. Solvent polymeric membranes based on tridodecylmethylammonium chloride studied by potentiometry and electrochemical impedance spectroscopy. *Analytica Chimica Acta*, 514(1), pp. 107–113.

Leontidis, E., 2002. Hofmeister anion effects on surfactant self-assembly and the formation of mesoporous solids. *Current Opinion in Colloid and Interface Science*, 7(1–2), pp. 81–91.

Li, X. and Shantz, D.F., 2010. Specific ion effects on nanoparticle stability and organocation–particle interactions in tetraalkylammonium–silica mixtures. *Langmuir*, 26(23), pp. 18459–18467.

Lingenfelter, P., Bedlechowicz-Sliwakowska, I., Sokalski, T., Maj-Zurawska, M. and Lewenstam, A., 2006. Time-dependent phenomena in the potential response of ion-selective electrodes treated by the Nernst–Planck–Poisson model. 1. Intramembrane processes and selectivity. *Analytical Chemistry*, 78(19), pp. 6783–6791.

Lo Nostro, P. and Ninham, B.W., 2012. Hofmeister phenomena: An update on ion specificity in biology. *Chemical Reviews*, 112(4), pp. 2286–2322.

López-León, T., Santander-Ortega, M.J., Ortega-Vinuesa, J.L. and Bastos-González, D., 2008. Hofmeister effects in colloidal systems: Influence of the surface nature. *Journal of Physical Chemistry C*, 112(41), pp. 16060–16069.

Lyklema, J., 2009. Simple Hofmeister series. *Chemical Physics Letters*, 467(4–6), pp. 217–222.

Lyklema, J., 2011. Surface charges and electrokinetic charges: Distinctions and juxtapositionings. *Colloids and Surfaces A: Physicochemical and Engineering Aspects*, 376(1–3), pp. 2–8.

Manciu, M. and Ruckenstein, E., 2003. Specific ion effects via ion hydration: I. Surface tension. *Advances in Colloid and Interface Science*, 105(1–3), pp. 63–101.

Manev, E.D., Sazdanova, S.V., Tsekov, R., Karakashev, S.I. and Nguyen, A.V., 2008. Adsorption of ionic surfactants. *Colloids and Surfaces A: Physicochemical and Engineering Aspects*, 319(1–3), pp. 29–33.

Marcus, Y., 2009. Effect of ions on the structure of water: Structure making and breaking. *Chemical Reviews*, 109(3), pp. 1346–1370.

Minami, H., Sato, N., Sugawara, M. and Umezawa, Y., 1991. Comparative study on the potentiometric responses between a valinomycin-based bilayer lipid membrane and a solvent polymeric membrane. *Analytical Sciences*, 7(6), p. 853.

Moreira, L. and Firoozabadi, A., 2010. Molecular thermodynamic modeling of specific ion effects on micellization of ionic surfactants. *Langmuir*, 26(19), pp. 15177–15191.

Moreira, L.A., Boström, M., Ninham, B.W., Biscaia, E.C. and Tavares, F.W., 2006. Hofmeister effects: Why protein charge, pH titration and protein precipitation depend on the choice of background salt solution. *Colloids and Surfaces A: Physicochemical and Engineering Aspects*, 282–283, pp. 457–463.

Neihof, R. and Sollner, K., 1956. The physical chemistry of the differential rates of permeation of ions across porous membranes. *Discussions of the Faraday Society*, 21, pp. 94–101.

Nucci, N.V. and Vanderkooi, J.M., 2008. Effects of salts of the Hofmeister series on the hydrogen bond network of water. *Journal of Molecular Liquids*, 143(2–3), pp. 160–170.

Omta, A.W., Kropman, M.F., Woutersen, S. and Bakker, H.J., 2003. Negligible effect of ions on the hydrogen-bond structure in liquid water. *Science*, 301(5631), pp. 347–349.

Osakai, T., Ogata, A. and Ebina, K., 1997. Hydration of ions in organic solvent and its significance in the Gibbs energy of ion transfer between two immiscible liquids. *Journal of Physical Chemistry B*, 101(41), pp. 8341–8348.

Ozawa, S., Miyagi, H., Shibata, Y., Oki, N., Kunitake, T. and Keller, W.E., 1996. Anion-selective electrodes based on long-chain methyltrialkylammonium salts. *Analytical Chemistry*, 68(23), pp. 4149–4152.

Padmanabhan, V., Daillant, J., Belloni, L., Mora, S., Alba, M. and Konovalov, O., 2007. Specific ion adsorption and short-range interactions at the air aqueous solution interface. *Physical Review Letters*, 99(8), p. 086105.

Para, G. and Warszynski, P., 2007. Cationic surfactant adsorption in the presence of divalent ions. *Colloids and Surfaces A: Physicochemical and Engineering Aspects*, 300(3 Spec. Issue), pp. 346–352.

Para, G., Jarek, E. and Warszynski, P., 2006. The Hofmeister series effect in adsorption of cationic surfactants—Theoretical description and experimental results. *Advances in Colloid and Interface Science*, 122(1–3), pp. 39–55.

Parsons, D.F. and Ninham, B.W., 2010. Charge reversal of surfaces in divalent electrolytes: The role of ionic dispersion interactions. *Langmuir*, 26(9), pp. 6430–6436.

Parsons, D.F., Boström, M., Nostro, P.L. and Ninham, B.W., 2011. Hofmeister effects: Interplay of hydration, nonelectrostatic potentials, and ion size. *Physical Chemistry Chemical Physics*, 13(27), pp. 12352–12367.

Pegram, L.M. and Record, M.T., 2007. Hofmeister salt effects on surface tension arise from partitioning of anions and cations between bulk water and the air–water interface. *The Journal of Physical Chemistry B*, 111(19), pp. 5411–5417.

Pegram, L.M. and Record Jr., M.T., 2008a. Quantifying accumulation or exclusion of H^+, HO^-, and Hofmeister salt ions near interfaces. *Chemical Physics Letters*, 467(1–3), pp. 1–8.

Pegram, L.M. and Record Jr., M.T., 2008b. Thermodynamic origin of Hofmeister ion effects. *Journal of Physical Chemistry B*, 112(31), pp. 9428–9436.

Pegram, L.M. and Record Jr., M.T., 2009a. Quantifying the roles of water and solutes (denaturants, osmolytes, and Hofmeister salts) in protein and model processes using the solute partitioning model. *Methods in Molecular Biology (Clifton, N.J.)*, 490, pp. 179–193.

Pegram, L.M. and Record Jr., M.T., 2009b. Using surface tension data to predict differences in surface and bulk concentrations of nonelectrolytes in water. *Journal of Physical Chemistry C*, 113(6), pp. 2171–2174.

Petrache, H.I., Zemb, T., Belloni, L. and Parsegian, V.A., 2006. Salt screening and specific ion adsorption determine neutral-lipid membrane interactions. *Proceedings of the National Academy of Sciences of the United States of America*, 103(21), pp. 7982–7987.

Pungor, E., 1997. How to understand the response mechanism of ion-selective electrodes. *Talanta*, 44(9), pp. 1505–1508.

Quentel, F., Mirčeski, V., Elleouet, C. and L'her, M., 2008. Studying the thermodynamics and kinetics of ion transfers across water-2-nitrophenyloctyl ether interface by means of organic-solution-modified electrodes. *The Journal of Physical Chemistry C*, 112(39), pp. 15553–15561.

Ren, K., 2000. The ion adsorption effect on selectivity of liquid state, O,O'-didecylodithiophosphate chelate based ion-selective electrodes. *Talanta*, 52(6), pp. 1157–1170.

Reymond, F., Fermín, D., Lee, H.J. and Girault, H.H., 2000. Electrochemistry at liquid/liquid interfaces: Methodology and potential applications. *Electrochimica Acta*, 45(15–16), pp. 2647–2662.

Salis, A., Cristina Pinna, M., Bilanicǒva, D., Monduzzi, M., Lo Nostro, P. and Ninham, B.W., 2006. Specific anion effects on glass electrode pH measurements of buffer solutions: Bulk and surface phenomena. *The Journal of Physical Chemistry B*, 110(6), pp. 2949–2956.

Samec, Z., 1988. Electrical double layer at the interface between two immiscible electrolyte solutions. *Chemical Reviews*, 88(4), pp. 617–632.

Sato, H., Wakabayashi, M., Iro, T., Sugawara, M. and Umezawa, Y., 1997. Potentiometric responses of ionophore-incorporated bilayer lipid membranes with and without added anionic sites. *Analytical Sciences*, 13(3), pp. 437–446.

Schwake, A., Cammann, K., Smirnova, A.L., Levitchev, S.S., Khitrova, V.L., Grekovich, A.L. and Vlasov, Y., 1999. Potentiometric properties and impedance spectroscopic data of poly(vinyl chloride) membranes containing quaternary ammonium salts of different chemical structure. *Analytica Chimica Acta*, 393(1–3), pp. 19–28.

Silbert, G., Klein, J. and Perkin, S., 2010. The effect of counterions on surfactant-hydrophobized surfaces. *Faraday Discussions*, 146, pp. 309–324.

Smirnova, A.L., Tarasevitch, V.N. and Rakhman'ko, E.M., 1994. Some properties of sulfate ion-selective PVC membranes based on a neutral carrier and ion-exchanger. *Sensors and Actuators: B. Chemical*, 19(1–3), pp. 392–395.

Smith, J.D., Saykally, R.J. and Geissler, P.L., 2007. The effects of dissolved halide anions on hydrogen bonding in liquid water. *Journal of the American Chemical Society*, 129(45), pp. 13847–13856.

Sokalski, T., Lingenfelter, P. and Lewenstam, A., 2003. Numerical solution of the coupled Nernst–Planck and Poisson equations for liquid junction and ion selective membrane potentials. *Journal of Physical Chemistry B*, 107(11), pp. 2443–2452.

Sutton, P.G., Braven, J., Ebdon, L. and Scholefield, D., 1999. Development of a sensitive nitrate-selective electrode for on-site use in fresh waters. *Analyst*, 124(6), pp. 877–882.

Swann, J.M.G., Bras, W., Topham, P.D., Howse, J.R. and Ryan, A.J., 2010. Effect of the Hofmeister anions upon the swelling of a self-assembled pH-responsive hydrogel. *Langmuir*, 26(12), pp. 10191–10197.

Szabo, G., Eisenman, G. and Ciani, S., 1969. The effects of the macrotetralide actin antibiotics on the electrical properties of phospholipid bilayer membranes. *Journal of Membrane Biology*, 1(1), pp. 346–382.

Tohda, K., Umezawa, Y., Yoshiyagawa, S., Hashimoto, S. and Kawasaki, M., 1995. Cation permselectivity at the phase boundary of ionophore-incorporated solvent polymeric membranes as studied by optical second harmonic generation. *Analytical Chemistry*, 67(3), pp. 570–577.

Turshatov, A.A., Zaitsev, S.Y., Sazonov, S.K., Vedernikov, A.I., Gromov, S.P., Alfimov, M.V. and Möbius, D., 2008. Anion effects on monolayers of a new amphiphilic styryl-pyridinium dye at the air–water interface. *Colloids and Surfaces A: Physicochemical and Engineering Aspects*, 329(1–2), pp. 18–23.

Vlachy, N., Jagoda-Cwiklik, B., Vácha, R., Touraud, D., Jungwirth, P. and Kunz, W., 2009. Hofmeister series and specific interactions of charged headgroups with aqueous ions. *Advances in Colloid and Interface Science*, 146(1–2), pp. 42–47.

Wang, C. and Morgner, H., 2011. The competitive adsorption of counter-ions at the surface of anionic surfactants solution. *Physical Chemistry Chemical Physics*, 13(9), pp. 3881–3885.

Warszynski, P., Lunkenheimer, K. and Czichocki, G., 2002. Effect of counterions on the adsorption of ionic surfactants at fluid–fluid interfaces. *Langmuir*, 18(7), pp. 2506–2514.

Wegmann, D., Weiss, H., Ammann, D., Morf, W.E., Pretsch, E., Sugahara, K. and Simon, W., 1984. Anion-selective liquid membrane electrodes based on lipophilic quaternary ammonium compounds. *Microchimica Acta*, 84(1), pp. 1–16.

Wojciechowski, K., 2012. A new device for simultaneous measurements of dynamic interfacial tension and electrical potential difference. *Chemistry Letters*, 41(10), pp. 1099–1100.

Wojciechowski, K. and Kucharek, M., 2009. Interfacial tension oscillations without surfactant transfer. *Journal of Physical Chemistry B*, 113(41), pp. 13457–13461.

Wojciechowski, K. and Linek, K., 2012. Anion selectivity at the aqueous/polymeric membrane interface: A streaming current study of potentiometric Hofmeister effect. *Electrochimica Acta*, 71, pp. 159–165.

Wojciechowskii, K., Kucharek, M., Wróblewski, W. and Warszyński, P., 2009. The double layer potentials at the toluene–aqueous interface in the presence of CTAB/NaBr. Implications for ion-selective electrodes. *Colloids and Surfaces A: Physicochemical and Engineering Aspects*, 343(1–3), pp. 83–88.

Wojciechowski, K., Kucharek, M., Wróblewski, W. and Warszyński, P., 2010. On the origin of the Hofmeister effect in anion-selective potentiometric electrodes with tetraalkylammonium salts. *Journal of Electroanalytical Chemistry*, 638(2), pp. 204–211.

Wojciechowski, K., Bitner, A., Warszyński, P. and Zübrowska, M., 2011. The Hofmeister effect in zeta potentials of CTAB-stabilised toluene-in-water emulsions. *Colloids and Surfaces A: Physicochemical and Engineering Aspects*, 376(1–3), pp. 122–126.

Wojciechowski, K., Gutberlet, T. and Konovalov, O., 2012. Anion-specificity at water-air interface probed by total reflection X-ray fluorescence (TRXF). *Colloids and Surfaces A*, 413, pp. 184–190.

Wróblewski, W., Chudy, M. and Dybko, A., 2000. Nitrate-selective chemically modified field effect transistors for flow-cell applications. *Analytica Chimica Acta*, 416(1), pp. 97–104.

Zimmermann, R., Dukhin, S. and Werner, C., 2001. Electrokinetic measurements reveal interracial charge at polymer films caused by simple electrolyte ions. *Journal of Physical Chemistry B*, 105(36), pp. 8544–8549.

Zimmermann, R., Rein, N. and Werner, C., 2009. Water ion adsorption dominates charging at nonpolar polymer surfaces in multivalent electrolytes. *Physical Chemistry Chemical Physics*, 11(21), pp. 4360–4364.

Zubrowska, M., Wróblewski, W. and Wojciechowski, K., 2011. The effect of lipophilic salts on surface charge in polymeric ion-selective electrodes. *Electrochimica Acta*, 56(17), pp. 6114–6122.

Section III

Interfaces and Nanocolloidal Dispersions

17 Human Serum Albumin Adsorption on Solid Substrates
Electrokinetic Studies

*Zbigniew Adamczyk, Maria Dąbkowska,
Marta Kujda, and Kamila Sofińska*

CONTENTS

17.1 INTRODUCTION

Adsorption of proteins at solid interfaces is involved in thrombosis, artificial organ failure, plaque formation, fouling of contact lenses and heat exchangers, ultrafiltration, and membrane filtration units. On the other hand, controlled protein deposition on various surfaces is a prerequisite of their efficient separation and purification by chromatography, filtration, biosensing, bioreactors, immunological assays, and so on.

One of the most extensively studied proteins is the human serum albumin (HSA) and the analogous bovine serum albumin (BSA), abundant at a high level, ca. 4%–5% in blood plasmas (Carter and Ho 1994; Peters 1985).

HSA synthesized in the liver is a single non-glycosylated, α-chain protein consisting of 585 amino acids. The most abundant are listed in Table 17.1 together with their nominal pK_a values. Its molecular mass calculated from this amino acid composition is 66,439 kDa (Peters 1985) and the crystalline structure shown in Table 17.2 (RCSB Protein Data Bank 2012) consists of consists of 70% α-helix (Haynes and Norde

405

TABLE 17.1

Amino Acid Composition of HSA with Their Nominal pK_a Values

Amino Acids	Content (%)	Nominal pK_a Value		
		Carboxyl Group	Ammonium Ion	Side-Chain Group
Asp	6.15	1.88	9.60	3.65
Asn	2.90	2.02	8.80	–
Thr	4.70	2.09	9.10	–
Ser	4.10	2.21	9.15	–
Glu	10.59	2.19	9.67	4.25
Gln	3.41	2.17	9.13	–
Pro	4.10	1.99	10.60	–
Gly	2.0	2.34	9.60	–
Ala	10.59	2.34	9.69	–
Val	7.0	2.32	9.62	–
Cys	5.98	1.96	10.28	8.18
Met	1.02	2.28	9.21	–
Ile	1.36	2.36	9.60	–
Leu	10.42	2.36	9.60	–
Tyr	3.07	2.20	9.11	10.07
Phe	5.29	1.83	9.13	–
Lys	10.07	2.18	8.95	10.53
His	2.70	1.82	9.17	6.00
Trp	0.17	2.82	9.39	–
Arg	4.10	2.17	9.04	12.48

1994; Norde 1986; RCSB Protein Data Bank 2012). The BSA molecule exhibits quite an analogous chemical structure and composition as HSA.

Albumins (HSA or BSA) are mainly responsible for osmotic pressure regulation and, because of their ligand binding capacity, for transport of numerous compounds such as fatty acids, drugs, metals, and hormones. Fatty acids are carried by albumins from the intestines through the liver, and muscles to and from adipose tissue (Peters 1995). For many hormones and vitamins, HSA is not only a transporter but also an important reservoir.

In addition to the transport role, HSA is involved in the inactivation of a group of compounds, for example, several exogenous toxins. HSA binds also many therapeutic drugs and controls their active concentration (Nicholson et al. 2000; Peters 1995).

Albumin has antioxidant properties and plays an important role in scavenging of oxygen free radicals (reactive oxygen species), which have been implicated in the pathogenesis of inflammatory diseases. For example, HSA shows protection against α1-antiproteinase inactivation by hypochlorous acid (HOCl), a product of myeloperoxidase enzyme released by activated neutrophils in inflammation state. The free radical-trapping properties of albumin may be related to the one reduced residue (Cys34) that is able to scavenge hydroxyl radicals. Also, six HSA methionine

TABLE 17.2

Physicochemical Properties and Molecular Shape of HSA

Property (Unit)	Value	References, Remarks
Molar mass (kDa)	66,439	From composition (Peters 1985)
	67,120	Dissolved (Peters 1985)
	69,000	From physical data (Peters 1985)
Specific density (g·cm^{-3})	1.36	(Peters 1985)
Specific volume in crystalline state (nm^3)	88.3	Calculated from atomic coordinates (He and Carter 1992)
Diffusion coefficient (cm^2 s^{-1})	6.1×10^{-7}	Measured in Dąbkowska and Adamczyk (2012a) by DLS for $T = 293$ K, pH = 3.5, $I = 1.3 \times 10^{-3}$ to 0.15 M NaCl
Hydrodynamic diameter (nm)	7.0	Calculated from Equation 17.3
Geometrical dimensions, spheroid (nm)	$9.5 \times 5 \times 5$	(Jachimska et al. 2008)
Geometrical cross-section area (nm^2)	37	Calculated from geometry
Molecular shape		Monomer (RCSB Protein Data Bank)

residues are susceptible to oxidation from a wide variety of oxidants, which leads to the production of methionine sulfoxide (Nicholson et al. 2000; Roche et al. 2008).

In addition to its important physiological role, HSA is often used for drug delivery, cell therapy products, and medical device coating. For therapeutic active peptide and drugs, albumin is one of the most important carriers used in the treatment and diagnosis of inflammatory metabolic and viral diseases.

Albumins are also widely used as anti-adherent coatings against many bacteria (Brokke et al. 1991; Hogt et al. 1985; Reynolds and Wong 1983) platelets, fibrinogen and IgG adsorption in hemodialyzer membranes and pacemakers, and postoperative infections on orthopedic titanium implants (An et al. 1996; Kottke-Marchant et al. 1989; Zdanowski et al. 1993).

For medical and biopharmaceutical applications, it is important to use high-quality, contaminant-free albumin. This demand fulfills recombinant human serum albumin (rHSA), characterized by hypoallergenic properties, stability, absence of aggregates (dimers), and homogeneity.

Although the molecular shape of albumins is rather irregular, characterized by no symmetry axis (Table 17.2), one can approximate it by a spheroid having dimensions of $9.5 \times 5 \times 5$ nm (Jachimska et al. 2008; Rezwan et al. 2005) with the effective cross-section area in the side-on orientation equal to 37 nm^2. This spheroidal model is useful for the interpretation of experimental results concerning HSA adsorption kinetics and streaming potential measurements.

In electrolyte solutions, albumin molecules exhibit amphoteric properties; that is, their effective (electrokinetic) charge can be varied by pH being on average positive for pH < 5 and negative otherwise. Hence, the isoelectric point (pH value where the net electrokinetic charge vanishes) for fatty acids free of HSA is 5.1 (Jachimska et al. 2008). The slightly lower value of 4.7 reported in the literature (Parmelee et al. 1978; Peters 1985) is characteristic for fatty acids containing samples of HSA. This can also be due to the various buffers used in these studies (Jachimska et al. 2008). The negative charge distributed heterogeneously over the molecule is mostly due to the glutamic and aspartic acid residues, whereas the positive charge is due to cysteine (Cys), tyrosine (Tyr), lysine (Lys), and arginine (Arg) amino acid residues (Peters 1985). This can be deduced from pK_a values of the side-chain groups as shown in Table 17.1. As discussed in Haynes and Norde (1994) and Norde (1986), the electrokinetic (uncompensated) charge of the HSA molecule is always much smaller in absolute terms than the proton charge derived from titration.

Because of its significance, especially for immunological assays, HSA adsorption on various surfaces was widely studied by a multitude of techniques (Elgersma et al. 1992; Kurrat et al. 1994, 1997; Norde 1986; Norde and Lyklema 1978; Malmsten 1994; Ortega-Vinuesa et al. 1998; Van Dulm and Norde 1983) in both kinetic and equilibrium aspects. For example, Norde and Lyklema (1978) determined adsorption isotherms of HSA at negatively charged polystyrene latexes. The role of pH (4.0–7.0) and the temperature (5°C–37°C) was systematically studied. Interestingly, for the highly charged latex and a temperature of 22°C, the protein coverage in the limit of zero bulk concentration approached 1 mg m^{-2}, which indicates a high degree of irreversibility of HSA adsorption. For higher bulk concentration range (10–500 per mg L^{-1}), the HSA coverage increased to ca. 2 mg m^{-2}.

Kinetic measurements of HSA adsorption on glass slides at pH 4–7.4 and I = 0.001–0.1 M (protein bulk concentration, 38 mg L^{-1} were performed by Van Dulm and Norde (1983) using the radioisotope techniques. The maximum coverage was ca. 1 mg m^{-2} for pH 7.4 and I = 0.1 M and ca. 2 mg m^{-2} for pH 4.0 and I = 0.1 M.

Interesting kinetic measurements of HSA adsorption on polystyrene-covered silicon wafers and glass slides were also performed by Elgersma et al. (1992) using reflectometry and streaming-potential measurements. The maximum coverage of the protein was determined as the function of pH and protein bulk concentration (1 to 100 mg L^{-1}). It was demonstrated that for pH 4.0, the amount of adsorbed protein equal to 0.8 mg m^{-2} was independent of the bulk concentration of the protein.

Malmsten (1994) determined the kinetics of HSA adsorption on methylated silica for physiological conditions, that is, pH 7.4 and I = 0.15 M using ellipsometry. The maximum coverage of the protein found in these measurements varied between 0.5 and 1 mg m^{-2} for HSA bulk concentrations changed between 50 and 1000 mg L^{-1}.

Kurrat et al. (1994, 1997) studied the kinetics of HSA adsorption on hydrated Si and Ti oxides (optical waveguides) using the integrated optics technique (pH 7.4). The maximum coverage of irreversibly bound protein determined in these experiments was 1.7 mg m^{-2}.

Systematic measurements of HSA adsorption were performed by Ortega-Vinuesa et al. (1998), who measured the thickness of the protein layer on silicon plates by ellipsometry as a function of the solution pH and ionic strength for the bulk concentration of 20 mg L^{-1}. For pH 4.0, the coverage of HSA varied between 2 and 2.7 mg m^{-2} depending on the ionic strength, whereas for pH close to the isoelectric point (4.7), the maximum coverage of HSA approached 3 mg m^{-2}, fairly independent of the ionic strength.

Similar values were obtained in precise kinetic measurements of BSA adsorption on silicon and modified glass surfaces forming parallel-plate channels using the *in situ* total internal reflection fluorescence technique (Wertz and Santore 2001). The maximum coverage varied between 2.5 and 3.5 mg m^{-2} for the bulk concentration of BSA changed between 10 and 50 mg L^{-1}.

Analogous behavior was observed in the case of BSA adsorption on polystyrene latexes (Fair and Jamieson 1980; Revilla et al. 1996) and colloidal alumina (Rezwan et al. 2004, 2005).

As can be noticed, a significant spread in the maximum coverage of albumin was reported in these works with the lowest and highest value equal to 0.5 and 3.5 mg m^{-2}, respectively.

In Dąbkowska and Adamczyk (2012a), systematic experiments were performed aimed at elucidating HSA adsorption mechanisms on mica, a model hydrophilic substrate of well-defined and reproducible surface properties. Using the *in situ* streaming-potential measurements, the adsorption and desorption kinetics of HSA was determined for various ionic strengths and pH 3.5. It was shown that the amount of irreversibly adsorbed albumin increased systematically with the ionic strength attaining 1.4 mg m^{-2} for $I = 0.15$ NaCl. However, a significant amount of reversibly adsorbed protein was revealed, which could be removed upon prolonged washing.

In Dąbkowska et al. (2013), additional experiments were performed using the x-ray photoelectron spectroscopy (XPS) method, which allowed one to precisely determine the maximum coverage of HSA on mica, confirming previous results obtained by the streaming-potential method.

In a recent publication (Dąbkowska and Adamczyk 2012b), the influence of pH on HSA adsorption and the stability of monolayers were systematically studied using the combined atomic force microscopy (AFM), streaming-potential, and XPS methods. An unusual adsorption of HSA on mica for pH above 5.1 (isoelectric point), where the molecule was on average negatively charged, was explained in terms of a heterogeneous charge distribution.

Therefore, the goal of this work is to discuss these recent results obtained in model systems, which can be used as useful reference data for interpretation of more complicated mechanisms appearing for HSA adsorption on colloid carrier particles. The latter problem is of vital practical significance for efficiently performing various immunological assays.

17.2   MATERIALS AND METHODS

In our studies, two samples of fatty acid-free HSA were used: (i) the crystallized and lyophilized powder 99% (Sigma), with the nominal fatty acid content given by the supplier being 0.02%, and (ii) the recombinant HSA, hereafter referred to as rHSA. The purity of the albumin samples was checked by dynamic surface tension measurements carried out using the pendant drop shape method. No measurable changes in surface tension of supernatants acquired by ultrafiltration or dialysis of HSA solutions were observed over a prolonged period reaching 12 h.

The purity of the samples (especially the presence of high-molecular-weight components) of both HSA solutions was also determined via sodium dodecyl sulfate polyacrylamide gel electrophoresis (SDS-PAGE) in the Laemmli system (Laemmli 1970), using non-reducing sample buffer and 12% polyacrylamide gel. The appropriate amount of albumin was applied to each well of the polyacrylamide gel and molecular protein markers were added in one well. The gel electrophoresis was run at room temperature in 0.025 M Tris, 0.192 M glycine, and 0.1% SDS. After the run, the gel was fixed and silver stained (Blum et al. 1987).

No species having a molecular mass below 60 kDa were detected for both HSA samples. However, in the case of the HSA of Sigma, there appeared a band at 130 kDa, which was attributed to the HSA dimer commonly appearing in commercial HSA samples (Hunter and Carta 2001). The amount of the dimer was quantitatively evaluated via the special software Quantity One program (Bio-Rad) from the intensity of the bands. The average value taken from six measurements was 13.5% (by weight), which corresponds to 6.7% by number.

The bulk concentration of HSA after dissolving in phosphate-buffered saline solution at pH 7.4 was determined spectrophotometrically using a bicinchoninic acid (BCA) protein assay (Smith et al. 1985). HSA samples were incubated 30 min in 37°C with BCA protein assay, and the purple color was measured at 562 nm using a Shimadzu spectrophotometer. The concentration of HSA was calculated using a prepared calibration curve with known concentrations of the dissolved standard protein (BSA) and correlated with spectrophotometric absorbance at 280 nm.

Additionally, the bulk concentration of HSA after dissolving the powder in appropriate electrolyte solutions and filtration was determined using a high-precision densitometer (Anton Paar, type DMA 5000 M). The specific density of the HSA solutions of a nominal weight concentration 500–2000 mg L^{-1} was measured as well as the supernatant solution acquired by membrane ultrafiltration using regenerated cellulose filters (Millipore nominal molecular weight limit: 30 kDa).

These concentrated stock solutions of HSA of known concentrations were diluted to a desired bulk concentration (usually 0.5–10 mg L^{-1}) prior to each experiment without any filtration procedure. Ruby muscovite mica obtained from Continental Trade was used as a substrate for albumin adsorption measurements. The solid pieces of mica were freshly cleaved into thin sheets prior to every experiment. Water was purified using a Millipore Elix 5 apparatus.

Additionally, in albumin adsorption experiments, negatively charged polystyrene latex particles were used. The latex suspension was produced in emulsion polymerization of styrene using a persulfate initiator according to the standard method of Goodwin et al. (1974). The reaction mixture was purified by steam distillation and prolonged membrane filtration. In this way, a stock suspension of the latex, hereafter referred to as L800, containing 10% solid was obtained (Adamczyk et al. 2012a). In albumin adsorption experiments, dilute latex suspensions typically having a solid content of 40–100 mg L^{-1} (0.004% to 0.01%) were used.

Chemical reagents (sodium chloride, hydrochloric acid) were commercial products of Sigma-Aldrich and used without further purification. The experimental temperature was kept constant at 293 ± 0.1 K.

The diffusion coefficient of HSA was determined by dynamic light scattering (DLS), using the Zetasizer Nano ZS Malvern instrument for protein concentrations varying between 200 and 2000 mg L^{-1}. The microelectrophoretic mobility of the protein solution and the bare and protein-covered latex was measured via the laser Doppler velocimetry technique using the same Malvern device.

The streaming potential of bare and HSA-covered mica was determined using a homemade apparatus previously described (Adamczyk et al. 2010b; Wasilewska and Adamczyk 2011). A laminar flow of the electrolyte (or the protein suspension) was forced by a regulated hydrostatic pressure difference ΔP through the parallel-plate channel of dimensions $2b_c \times 2c_c \times L = 0.027 \times 0.29 \times 6.2$ cm, formed by mica sheets separated by a perfluoroethylene spacer. The resulting streaming potential E_s was measured by a pair of AgCl electrodes for various pressures to obtain the slope of the E_s versus ΔP dependence. The cell electric resistance R_e was determined using another pair of platinum electrodes to account for the surface conductivity effect. Knowing the slope of the E_s versus ΔP dependence, the zeta potential of substrate surface (ζ_i) can be calculated from the Smoluchowski relationship

$$\zeta_i = \frac{\eta L}{4\varepsilon b_c c_c R_e}\left(\frac{\Delta E_s}{\Delta P}\right) = \frac{\eta K_e}{\varepsilon}\left(\frac{\Delta E_s}{\Delta P}\right),$$

(17.1)

where ε is the dielectric permittivity of water and K_e is the specific conductivity of the cell, which is connected with the electric resistance via the constitutive relationship

$$R_e = \frac{L}{\Delta S_c K_e} = \frac{L}{\Delta S_c\left(K_e' + K_s\right)}$$

(17.2)

$\Delta S_c = 2b_c \times 2c_c$ is the channel cross-section area, K_e' is the specific conductivity due to electrolyte, and K_s is the surface conductivity, depending in general on the channel shape.

It should be mentioned that the correction caused by surface conductivity was practically negligible for the ionic strength of the electrolyte above 10^{-3} M used in our experiments.

The procedure for determining the zeta potential of protein-covered mica consisted of three stages:

- Measuring the reference zeta potential of bare mica in pure electrolyte
- *In situ* formation of the HSA monolayer of controlled coverage under the diffusion-controlled deposition conditions by filling the cell with the suspension of appropriate concentration for an appropriate period of time
- Washing the cell with pure electrolyte and measuring streaming potentials for HSA-covered mica

The surface concentration of albumins in monolayers formed on mica was determined by AFM imaging in air using the NT-MDT Solver BIO device with the SMENA SFC050L scanning head. All measurements were performed in semi-contact mode by using high-resolution silicon probes (NT-MDT ETALON probes, HA NC series, polysilicon cantilevers with resonance frequencies 240 kHz ± 10% or 140 kHz ± 10%, force constants 9.5 N/m ± 20% or 4.4 N/m ± 20%). The number of adsorbed HSA molecules was determined using scan areas of 0.5 μm × 0.5 μm. About 500 molecules were calculated over a few randomly chosen areas over the mica sheet, which ensures a relative precision of these measurements better than 5%.

The XPS method was also used to quantify HSA adsorption kinetics on mica, especially the maximum coverage. The XPS measurements were carried out on a hemispherical spectrometer (SES R4000, Gammadata Scienta) using MgK_α (1253.6 eV, 11 kV, 17 mA) as an x-ray source. The photoelectrons were analyzed at a takeoff angle of 90°. The base pressure in the analysis chamber was about 8×10^{-9} Pa and about 3×10^{-8} Pa during measurements. The system was calibrated according to ISO 15472:2001. The energy resolution of the system, measured as a full width at half maximum for Ag $3d_{5/2}$ excitation line, was 0.9 eV. The analysis area of the samples (rectangle plates of mica) was approximately 4.8 mm². All survey scans were collected at pass energy of 200 eV. The mica surfaces were immersed in the protein solution containing 10 mg L^{-1} HSA for 1–20 min at 0.15 M NaCl (pH 3.5) in the channel. After incubation, the mica samples were rinsed by ultrapure water for 1 minute to remove any non-adsorbed molecules in the fluid film and air dried prior to analyzing. High-resolution spectra for N1s were obtained. Elemental atomic percentages of nitrogen were determined.

Since the AFM and XPS measurements were carried out under air or vacuum conditions, much attention was devoted to elaborate an appropriate experimental procedure of drying the samples. This was facilitated by the fact that HSA monolayers on mica were hydrophilic, so the water film could be evaporated under controlled humidity and temperature in a continuous manner without forming drops. This minimized the action of meniscus forces and possible distortion of monolayers.

17.3 RESULTS AND DISCUSSION

17.3.1 BULK CHARACTERISTICS OF HSA

The diffusion coefficient of HSA (denoted by D) was measured in solutions of NaCl for concentration range 10^{-3} to 0.15 M and for pH range 3.5 to 11 regulated by an addition of HCl or NaOH ($T = 293$ K) using the DLS method. It was determined that D was practically constant within this ionic strength and pH range, equal to 6.1×10^{-7} cm² s^{-1} (Table 17.2) (Dąbkowska and Adamczyk 2012a). Similar values were previously reported in Jachimska et al. (2008) and Peters (1985).

Knowing the diffusion coefficient, the hydrodynamic diameter of the protein was calculated using the Stokes–Einstein dependence

$$d_H = \frac{kT}{3\pi\eta D}, \tag{17.3}$$

where d_H is the hydrodynamic radius, k is the Boltzmann constant, T is the absolute temperature, and η is the dynamic viscosity of water.

The advantage of using d_H is that it is independent of the temperature and viscosity of the suspending medium.

Using Equation 17.3, it was calculated that $d_H = 7.0$ nm for $I = 10^{-3}$ to 10^{-2} M and pH 3.5. Hence, the "hydrodynamic" size of HSA is equal to 7 nm, which can be treated as a good estimate of its geometrical dimension, given the structural stability of the protein and its compact shape (Jachimska et al. 2008).

Another parameter that characterizes the electrokinetic (uncompensated) charge on the protein is electrophoretic mobility μ_e, defined as the average translation velocity of the protein U under given electric field E; that is, $\mu_e = U/E$. This quantity can be directly measured using the microelectrophoretic method. Values of μ_e measured for various ionic strengths and pH values are collected in Table 17.3. As can be noticed, the electrophoretic mobility of HSA at pH 3.0 remains positive, indicating that the molecule acquired a net positive charge. For $I = 10^{-3}$ M NaCl, μ_e attained the maximum value of 2.78 μm cm s^{-1} V^{-1}, and for $I = 0.15$, the minimum value of $\mu_e = 1.39$ μm cm s^{-1} V^{-1}.

However, for the physiological pH 7.4, the electrophoretic mobility of albumin becomes negative. Thus, for $I = 10^{-3}$ M NaCl, μ_e was -2.78 μm cm s^{-1} V^{-1}, and for $I = 0.15$, $\mu_e = -0.84$ μm cm s^{-1} V^{-1}. Even more negative values were measured for pH 9.5 (Table 17.3).

Knowing the electrophoretic mobility, one can calculate the average number of free (electrokinetic) charges per molecule from the Lorenz–Stokes relationship (Jachimska et al. 2008; Wasilewska and Adamczyk 2011)

$$N_c = \frac{30\pi\eta}{1.602} d_H \mu_e, \tag{17.4}$$

TABLE 17.3
Electrophoretic Mobility, Electrokinetic Charge, and Zeta Potential of HSA for Various Ionic Strengths and pH Values

pH	NaCl Concentration (M)	Electrophoretic Mobility (μm cm s^{-1} V^{-1})	Number of Electrokinetic Charges (e)	Electrokinetic Surface Charge (e nm^{-2})	Zeta Potential (mV) (Henry Model)
3.5	10^{-3}	2.78	11.4	0.31	53
	10^{-2}	2.10	8.7	0.24	40
	0.15	1.39	5.7	0.15	27
7.4	10^{-3}	−2.78	−11.4	−0.31	−53
	10^{-2}	−1.37	−5.6	−0.15	−26
	0.15	−0.84	−3.5	−0.095	−16
9.5	10^{-3}	−3.16	−13	−0.35	−60
	10^{-2}	−2.00	−8.2	−0.22	−38
	0.15	−1.26	−5.2	−0.14	−24

where d_H is expressed in nanometer and N_c is expressed as the number of elementary charges (e) per molecule. It should be noted that $|e| = 1.602 \times 10^{-19}$ C.

Using Equation 17.4, one can calculate that for pH 3.5, $N_c = 11.4$ for $I = 10^{-3}$ M NaCl and $N_c = 5.7$ for $I = 0.15$ M NaCl. For pH 7.4, $N_c = -11.4$ for $I = 10^{-3}$ M NaCl and $N_c = -3.5$ for $I = 0.15$ M NaCl (for the sake of convenience, values of N_c calculated for various ionic strengths and pH values are collected in Table 17.3).

However, for a quantitative interpretation of streaming-potential measurements of HSA-covered mica, it is necessary to also know the zeta potential ζ_p of the protein in the bulk. Knowing the electrophoretic mobility, this quantity can be calculated from the well-known Henry relationship

$$\zeta = \frac{\eta}{\varepsilon f(\kappa a)} \mu_e, \qquad (17.5)$$

where $f(\kappa a)$ is the dimensionless Henry function, a is the characteristic protein dimension, for example, the hydrodynamic radius $1/2\ d_H$, and $\kappa^{-1} = (\varepsilon kT/2e^2 I)^{1/2}$ is the double-layer thickness, where $I = \frac{1}{2}\left(\sum_i c_i z_i^2\right)$ is the ionic strength c_i are the ion concentrations, z_i their valences. For thin double layers, where $\kappa a \gg 1$, $f(\kappa a)$ approaches 3/2, and for thick double layers, where $\kappa a < 1$, $f(\kappa a)$ approaches unity.

The dependence of ζ_p on pH for 10^{-3}, 10^{-2}, and 0.15 M NaCl is graphically presented in Figure 17.1a.

As can be seen, for 10^{-3} M NaCl, ζ_p decreased from 53 mV at pH = 3.5 to –60 mV for pH 9.5. For 0.15 M NaCl, ζ_p was 27 mV at pH = 3.5 and –24 mV for pH 9.5. For all ionic strengths, ζ_p vanished at pH 5.1 (Figure 17.1a), which can be, therefore, treated as the isoelectric point of the protein. Slightly smaller isoelectric point values (4.7) reported in the literature (Parmelee et al. 1978; Peters 1985) are probably due to the use of various buffers, such as the MES or MES-TRIS buffers.

The zeta potential of mica, denoted by ζ_i, was determined via the streaming-potential measurements according to the above-described procedure. The dependence of ζ_i on pH for various ionic strengths is also shown in Figure 17.1b.

As seen, the zeta potential of mica was negative for the entire range of ionic strength, varying between –60 mV for $I = 10^{-3}$ M NaCl and –30 mV for $I = 0.15$ (pH 3.5). For pH 7.4, $\zeta_i = -101$ mV for $I = 10^{-3}$ M NaCl and –58 mV for $I = 0.15$ M NaCl.

Knowing the zeta potential, one can calculate the electrokinetic (uncompensated) charge of mica using the Gouy–Chapman (GC) relationship valid for a symmetric 1:1 electrolyte (Jachimska et al. 2008; Wasilewska and Adamczyk 2011).

$$\sigma_0 = \frac{(8\varepsilon kTn_b)^{\frac{1}{2}}}{0.1602} \sinh\left(\frac{e\zeta}{2kT}\right), \qquad (17.6)$$

where σ_0 is the electrokinetic charge density expressed in e nm^{-2} (for the sake of convenience, $e = |e|$ is used hereafter) and n_b is the number concentration of ions expressed in m^{-3}.

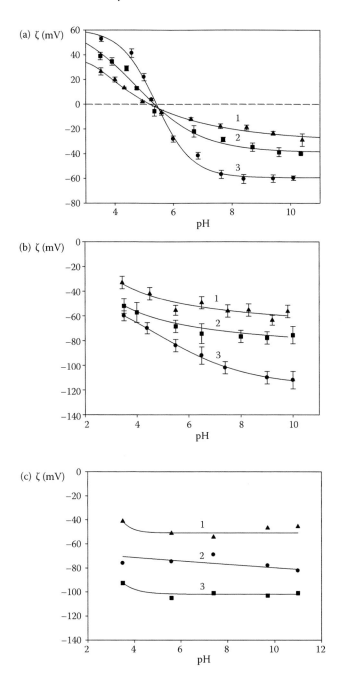

FIGURE 17.1 Dependence of zeta potentials on pH. (a) HSA zeta potential determined by microelectrophoresis: 1, (▲) 0.15 M NaCl; 2, (■) 10^{-2} M NaCl; 3, (•) 10^{-3} M NaCl. (b) Mica zeta potential determined by streaming-potential measurements: 1, (▲) 0.15 M NaCl; 2, (■) 10^{-2} M NaCl; 3, (•) 10^{-3} M NaCl. (c) Latex zeta potential determined by microelectrophoresis: 1, (▲) 0.15 M NaCl; 2, (•) 10^{-3} M NaCl; 3, (■) 10^{-2} M NaCl. The solid lines denote non-linear fits.

Using the above zeta potential values, one can calculate from Equation 17.6 that $\sigma_0 = -0.042$ e nm^{-2} for 10^{-3} M NaCl, and $\sigma_0 = -0.18$ e nm^{-2} for 0.15 M (pH 3.5). For pH 7.4, $\sigma_0 = -0.085$ e nm^{-2} for 10^{-3} M NaCl, and $\sigma_0 = -0.375$ e nm^{-2} for 0.15 M (for the sake of convenience, these values are collected in Table 17.4).

As can be noticed, the electrokinetic charge density on mica decreased significantly with the ionic strength. It should be observed, however, that even the lowest value of -0.415 e nm^{-2} (pH 9.5, 0.15 M NaCl) remains much higher than the lattice charge of the basal plane of mica, equal to -2.1 e nm^{-2} (Rojas 2002). This effect is caused by the specific adsorption of cations, including H$^+$, which is often referred to as the ion condensation effect (Norde and Lyklema 1978; Norde 1986) extensively studied by Scales et al. (1990, 1992).

Bulk characteristics of the latex particles used as substrates for albumin adsorption were also carried out (Adamczyk et al. 2012a). From the DLS measurements, the diffusion coefficient of latex was measured for various ionic strengths and pH values. From these data, the hydrodynamic diameter of latex d_L was calculated using Equation 17.3. As shown in Adamczyk et al. (2012a), for high ionic strength $I >$ 10^{-2} M, the hydrodynamic diameter of latex particles was 810 nm, independently of pH. However, for lower ionic strength $I = 10^{-3}$ M NaCl, the hydrodynamic diameter increased slightly, attaining 850 nm for pH 3.5 and 860 nm for pH 7.4. As discussed in Adamczyk et al. (2012a), this decrease in the hydrodynamic diameter with the ionic strength is caused by the presence of the residue polymeric moieties forming loops or chains. Since these polymeric moieties are charged, an increase in the ionic strength decreases electrostatic repulsion among them, leading to their collapse on the latex core and the decrease in the apparent size of latex particles.

TABLE 17.4

Zeta Potential and the Electrokinetic Charge Density of Mica for Various Ionic Strengths and pH Values

pH	NaCl Concentration (M)	Zeta Potential, ζ_i (mV)	Electrokinetic Surface Charge σ_0 (e nm^{-2})
3.5	$I = 10^{-3}$	-60 ± 4	-0.042
	$I = 10^{-2}$	-52 ± 5	-0.090
	$I = 0.15$	-30 ± 3	-0.18
5.1	$I = 10^{-3}$	-78 ± 8	-0.052
	$I = 10^{-2}$	-64 ± 5	-0.120
	$I = 0.15$	-45 ± 3	-0.305
7.4	$I = 10^{-3}$	-101 ± 7	-0.085
	$I = 10^{-2}$	-72 ± 5	-0.145
	$I = 0.15$	-58 ± 3	-0.375
9.5	$I = 10^{-3}$	-112 ± 4	-0.106
	$I = 10^{-2}$	-76 ± 5	-0.158
	$I = 0.15$	-59 ± 3	-0.415
–	Crystallographic, lattice charge	–	-2.1 (Rojas et al. 2002)

This hypothesis was confirmed by measurements of the electrophoretic mobility of latex particles. Knowing the mobility, the zeta potential of latex can be determined using Equation 17.5. The dependence of zeta potential of the latex particles on pH in solutions of 10^{-3}, 10^{-2}, and 0.15 M NaCl is shown in Figure 17.1c. As seen, for a fixed ionic strength, the zeta potential of latex was negative and independent of pH, which confirms the presence of strong acidic groups on the surface of the latex with a pK_a value well below pH 3.0. Interestingly, however, the increase in ionic strength from 10^{-3} to 10^{-2} M NaCl causes the decrease of zeta potential of latex from −76 to −92 mV for pH 3.5. This indicates that the negative charge on the latex surface increased (in absolute terms) because of the collapse of polymeric chains as previously mentioned. However, a further increase in the ionic strength reduced the negative zeta potential to −41 mV for pH 3.5 and $I = 0.15$ M. Knowing the zeta potential, the electrokinetic charge density was calculated using Equation 17.6. Accordingly, for pH 3.5, $\sigma_0 = -0.068\ e\ \mathrm{nm}^{-2}$ for 10^{-3} M NaCl, and $\sigma_0 = -0.41\ e\ \mathrm{nm}^{-2}$ for $I = 0.15$ M. For pH 7.4, $\sigma_0 = -0.052\ e\ \mathrm{nm}^{-2}$ for 10^{-3} NaCl, and $\sigma_0 = -0.40\ e\ \mathrm{nm}^{-2}$ for $I = 0.15$ M. As can be noticed, the electrokinetic charge density on latex particles decreases significantly with ionic strength, analogously as for mica. However, for $I = 0.15$ M and pH 3.5, the negative charge density on latex is more than two times higher (in absolute terms) than for mica.

These physicochemical characteristics of HSA, mica, and latex substrates were exploited for a quantitative calculation of the coverage of the protein via streaming-potential measurements.

17.3.2 HSA Adsorption on Mica

As mentioned, albumin adsorption was studied by various methods, most of them of an indirect nature, providing only average information about protein monolayers, expressed most often as mass per unit area of the substrate surface. Few attempts were reported in the literature aimed at imaging adsorbed HSA molecules at solid substrates. In Dąbkowska and Adamczyk (2012a), a direct AFM imaging was used to quantitatively determine HSA adsorption kinetics on mica. Albumin monolayers were produced for low bulk concentration over a desired period of time under diffusion transport in a thermostatic cell. Afterward, a series of AFM micrographs was obtained by semi-contact imaging in air.

A typical micrograph obtained for pH 3.5 and 0.15 M NaCl is shown in the inset of Figure 17.2. As observed, HSA molecules appear as isolated entities, the enumeration of which is enabled by an image-analyzing software. In this way, the number of molecules per unit (surface concentration, denoted by N) can be directly determined by taking averages from various surface areas randomly chosen over the mica substrate. For the sake of convenience, N is expressed hereafter as the number of molecules per square micrometer; hence, its unit is $\mu\mathrm{m}^{-2}$. It should be mentioned that by determining the surface concentration in this way, no information about protein size, shape, and is required, which makes the above approach simple and reliable. By determining N as a function of adsorption time, one obtains experimental kinetic dependencies, which can be used to validate various theoretical approaches. For example, it was theoretically predicted (Adamczyk 2000) that for an irreversible

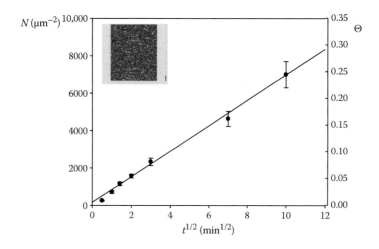

FIGURE 17.2 Dependence of the surface concentration of HSA, N (μm^{-2}) on the square root of adsorption time $t^{1/2}$ ($min^{1/2}$). The points denote experimental results obtained by a direct AFM enumeration of adsorbed HSA molecules (pH 3.5, $I = 0.15$ M NaCl, bulk protein concentration $c_b = 1$ mg L^{-1}). The solid line shows exact theoretical results obtained by numerical solution of the diffusion equation using the random sequential adsorption model. The inset shows the HSA monolayer on mica for $N = 4.64 \times 10^3$ μm^{-2} ($\Theta = 0.15$).

adsorption of proteins under diffusion-controlled transport, the initial adsorption kinetics, for not too high surface concentration, is described by

$$N = 2\left(\frac{D}{\pi}\right)^{\frac{1}{2}} t^{\frac{1}{2}} n_b, \tag{17.7}$$

where n_b is the bulk concentration of protein expressed in m^{-3}.

One can alternatively formulate Equation 17.7 in terms of the protein concentration expressed in mg L^{-1} (denoted by c_b) and N expressed in μm^{-2}

$$N = 2\times10^{-12}\frac{A_v}{M_w}\left(\frac{D}{\pi}\right)^{\frac{1}{2}} t^{\frac{1}{2}} c_b, \tag{17.8}$$

where A_v is Avogadro's constant and M_w is the molar mass of the protein. It is interesting to mention that the commonly used protein coverage unit Γ, expressed in mg m^{-2}, is connected with N in μm^{-2} through the linear relationship

$$\Gamma = 10^{15}\frac{M_w}{A_v} N. \tag{17.9}$$

However, a disadvantage of using Γ compared to the surface concentration N is that its calculation requires the knowledge of the effective molecular mass of a protein, which depends on the unknown degree of hydration.

Often, to facilitate comparisons of experimental data with theoretical approaches, it is convenient to express the albumin surface concentration in terms of the dimensionless coverage (two-dimensional [2D] density) defined as

$$\Theta = 10^{-6} N S_g, \tag{17.10}$$

where S_g is expressed in nm^2.

As seen in Figure 17.2, the experimental data are well reflected by theoretical results calculated from Equation 17.8 for a broad range of N up to 6000 μm^{-2}, which corresponds to $\Theta = 0.22$. However, for higher coverage, a direct AFM enumeration of adsorbed albumin molecules became inaccurate (Dąbkowska and Adamczyk 2012a).

The results shown in Figure 17.2 have interesting practical implications, indicating that one can determine in a convenient way an unknown bulk concentration of albumin (or other proteins) via the AFM measurements of their adsorption kinetics. Using the experimental dependence of N on the square root of adsorption time, $t^{1/2}$, one can calculate the slope $s_a = \Delta N / \Delta t^{1/2}$. Knowing this parameter, one can calculate from Equation 17.8 the bulk concentration of the protein from the relationship

$$c_b = \frac{M_w}{A_v} \times 10^{13} \left(\frac{\pi}{2.4 D} \right)^{1/2} s_a, \tag{17.11}$$

where c_b is expressed in mg L^{-1} and s_a is expressed in $\mu m^{-2} min^{-1/2}$.

Using the data pertinent to albumin (Table 17.2), one can express Equation 17.11 in the simple form

$$c_b = 0.00164 \, s_a. \tag{17.12}$$

From Equation 17.12, one can directly determine the bulk concentration of albumin at a level of 0.1 mg L^{-1} and less, which is rather difficult using other methods.

The agreement of experimental and theoretical results observed in Figure 17.2 also suggests that albumin adsorption was irreversible for pH 3.5 and the ionic strength of 0.15 M NaCl. In this respect, its adsorption proceeds analogously to colloid particle deposition, for example, gold nanoparticles (Kooij et al. 2002) silver nanoparticles (Oćwieja et al. 2011), hematite nanoparticles (Oćwieja et al. 2012), or dendrimers (Pericet-Camara et al. 2007).

In order to study this issue in more detail, a series of thorough experiments was performed in Dąbkowska and Adamczyk (2012a) aimed at determining the role of ionic strength on the adsorption and desorption kinetics of albumin. Except for the above-described AFM enumeration, streaming-potential and XPS measurements were used to evaluate albumin adsorption at a higher coverage range.

The streaming-potential measurements were carried out according to the procedure describe above. The parallel-plate cell was filled up with the HSA solution of a prescribed concentration for various periods ranging from 5 to 1600 min.

Afterward, the cell was flushed with pure electrolyte and the streaming potential of HSA-covered mica was determined as a function of time, which allowed one also to determine the albumin desorption kinetics.

The measured streaming potential was converted to the zeta potential of HSA monolayer on mica ζ using Equation 17.1. In this way, the dependencies of ζ on time were obtained for various ionic strengths, which served as primary data for assessing the extent of HSA desorption.

Typical results obtained in these experiments are shown in Figure 17.3. The square root of the adsorption time $t^{1/2}$ was used there as the independent variable. As seen, the experimental dependence of ζ on $t^{1/2}$ is characterized by a steep, quasi-linear increase for $t^{1/2} < 10$ min$^{1/2}$ (100 min). For longer times, an inversion of the negative zeta potential occurs. For adsorption time exceeding 400 min, a saturation value of ζ is asymptotically attained, being markedly lower than the corresponding bulk value of the zeta potential of HSA. Analogous results were previously reported for the PEI/mica system (Adamczyk et al. 2007), for the positive polystyrene latex particles/mica system (Adamczyk et al. 2010b), and for the fibrinogen/mica system at pH 3.5 (Wasilewska and Adamczyk 2011). The desorption run is also depicted in Figure 17.3 by hollow points. As observed, the variations in the zeta potential of HSA monolayers with the flushing time were limited to 1–2 mV for longer adsorption times. This suggests that HSA desorption was negligible. Similar series of experiments were also performed for other ionic strengths.

A quantitative interpretation of these experiments is feasible because the albumin coverage can be determined, as above described, via AFM enumeration and, at a higher coverage range, via the XPS measurements. In this way, dependencies of ζ on the dimensionless coverage Θ can be obtained for various experimental conditions. In Figure 17.4, typical results are shown, obtained for $I = 0.15$ M NaCl and pH 3.5.

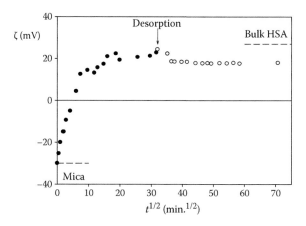

FIGURE 17.3 HSA adsorption/desorption on mica expressed as the dependence of the zeta potential ζ on the square root of the adsorption time $t^{1/2}$. The full points denote experimental results obtained from the streaming-potential measurements for pH 3.5, $I = 0.15$ M NaCl, $c_b = 2$ mg L^{-1}. The arrow and the hollow points show desorption runs, where the cell was flushed with a pure electrolyte.

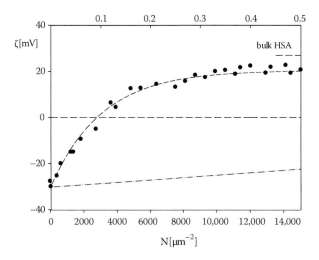

FIGURE 17.4 Dependence of the zeta potential of mica ζ on the coverage of HSA, Θ. The points denote experimental results obtained from the streaming-potential measurements for pH 3.5 and $I = 0.15$ M NaCl. The solid line denotes exact theoretical results calculated from the 3D adsorption model (Equations 17.15 and 17.18), and the dashed/dotted line denotes the theoretical results calculated from the GC model (Equation 17.14).

These experimental data were interpreted in Dąbkowska and Adamczyk (2012a) in terms of the two electrokinetic models:

(i) The mean-field adsorption model, based on the GC concept, where the adsorbate (HSA) is treated as a flat (2D) object having a uniform charge distribution

(ii) The irreversible, side-on adsorption model, considering a heterogeneous three-dimensional (3D) charge distribution over the adsorbed molecules or particles (Adamczyk et al. 2010a; Sadlej et al. 2009), applied before for colloid particles and polyelectrolytes

Accordingly, in the case of (i), because charge additivity is assumed, the net charge density of mica after HSA adsorption is calculated from the charge balance equation (Dąbkowska et al. 2013; Wasilewska and Adamczyk 2011).

$$\sigma = \sigma_0 + \Theta \, \sigma_{HSA}, \tag{17.13}$$

where $\sigma_{HSA} = N_c/S_g$ is the electrokinetic surface charge of HSA (Table 17.3).

Since all parameters appearing in Equation 17.13 are experimentally accessible, one can unequivocally determine the zeta potential of the interface covered by HSA using a relationship obtained by inversion of the GC relationship (Equation 17.6),

$$\zeta = \pm \frac{2kT}{e} \ln \frac{|\bar{\sigma}| + (\bar{\sigma}^2 + 4)^{\frac{1}{2}}}{2}, \tag{17.14}$$

where the plus and minus sign corresponds to the positive and negative sign of σ and $\bar{\sigma} = \sigma/(2\varepsilon k T n_b)^{1/2}$ is the dimensionless electrokinetic surface charge of the substrate covered by protein.

It should be mentioned that according to the GC model, the entire substrate surface covered by adsorbed species is treated as uniformly charged, characterized by an average (mean-field) value of the electrokinetic surface charge σ and the zeta potential ζ. Hence, the charge distribution is assumed to be strictly 2D and uniformly spread over the interface.

As can be seen in Figure 17.4, the GC model significantly deviates from experimental data, predicting results considerably lower. An analogous poor performance of the GC model was also observed for other ionic strength (Dąbkowska and Adamczyk 2012a).

It was, therefore, concluded that the GC model is not adequate. This suggests that the charge distribution in the adsorbed HSA monolayer is indeed 3D and heterogeneous rather than 2D. This fact can be accounted for by the second electrokinetic model, where adsorbed molecules are treated as isolated entities exhibiting a 3D charge distribution (Adamczyk et al. 2010a, 2011; Sadlej et al. 2009). Using this approach, the general expression for the streaming potential of an interface covered by particles (protein molecules) can be formulated as

$$\zeta(\Theta) = F_i(\Theta)\zeta_i + F_p(\Theta)\zeta_p, \tag{17.15}$$

where $\zeta(\Theta)$ is the zeta potential of protein-covered substrate and $F_i(\Theta)$, $F_p(\Theta)$ are dimensionless functions of the HSA coverage.

It should be mentioned that Equation 17.15 does not involve any fitting parameters because the functions $F_i(\Theta)$, $F_p(\Theta)$ were theoretically determined. For example, in the limit of low coverage and spherical particles, these functions become (Adamczyk et al. 2010a)

$$F_i(\Theta) = 1 - C_i\Theta \tag{17.16}$$

$$F_p(\Theta) = C_p\Theta,$$

where C_i and C_p are functions of the κa parameter alone. They were numerically calculated in Adamczyk et al. (2010a) for $0.1 < \kappa a < 100$. As shown, these functions become practically constant for $\kappa a > 1$ (thin double layers) approaching the limiting values of $C_i^0 = 10.2$ and $C_p^0 = 6.51$ for $\kappa a \gg 1$.

Thus, in the limit of $\kappa a > 1$ and low coverage of protein, Equation 17.15 assumes the linear form

$$\zeta(\Theta) = \zeta_i + (C_p\zeta_p - C_i\zeta_i)\Theta. \tag{17.17}$$

Theoretical results were also reported in Adamczyk et al. (2010a) and Sadlej et al. (2009), which allowed one to determine the $F_i(\Theta)$, $F_p(\Theta)$ functions for the entire range of coverage in the limit of thin double layers. These results were obtained by numerically evaluating the flow field in the vicinity of adsorbed particles using the multipole expansion method. The exact numerical results can be approximated by the analytical interpolation functions, with a precision better than 1% (Wasilewska and Adamczyk 2011).

$$F_i(\Theta) = e^{-C_i\theta}$$

$$F_p(\Theta) = \frac{1}{\sqrt{2}}\left(1 - e^{-\sqrt{2}C_p\theta}\right) \tag{17.18}$$

Results calculated using Equations 17.15 and 17.18 assuming a uniform 3D charge distribution over adsorbed HSA molecules are plotted in Figure 17.4 (solid line). As seen, they properly reflect the experimental data for the entire coverage range. This suggests that the heterogeneity of charge distribution was rather minor, because the main acidic amino acids (Table 17.1) became neutral since they exhibit pK_a above 3.5, whereas all basic amino acids remained positively charged. This resulted in a quasi-uniform distribution of positive charge over the entire molecule.

It is interesting to observe that these results obtained for HSA for high ionic strength are analogous to results previously obtained for positive latex particle deposition on bare mica (Adamczyk et al. 2010b), where the 3D electrokinetic model also proved adequate. In the latter case, the use of direct optical microscope enumeration of adsorbed particles under wet conditions enabled a unequivocal determination of the coverage. These facts confirm the hypothesis of a particle-like and irreversible adsorption of HSA at pH 3.5–driven and electrostatic interactions.

As seen in Figure 17.4, the sensitivity of the streaming-potential measurements in respect to albumin coverage is especially high for low $\Theta < 0.2$. This can be exploited for a convenient and sensitive determination of the coverage using the streaming-potential measurements, without relying on tedious AFM enumeration procedure. The dependence of Θ on ζ can be obtained by numerically inverting Equations 17.17–17.18 (Dąbkowska and Adamczyk 2012a). However, a useful analytical result can be derived by iteratively solving these equations, which yields the expression

$$\Theta = -\frac{1}{C_i} \ln X_1(\zeta), \tag{17.19}$$

where $X_1(\zeta)$ can be treated as the normalized zeta potential given by

$$X_1(\zeta) = \frac{\zeta - \zeta_p/\sqrt{2}}{\zeta_i - \zeta_p/\sqrt{2}} = \frac{\zeta - \zeta_\infty}{\zeta_i - \zeta_\infty}, \tag{17.20}$$

where $\zeta_\infty = \zeta_p/\sqrt{2}$.

Equation 17.19 is useful for practical applications because it enables a reliable determination of the unknown protein coverage under *in situ* conditions via direct streaming-potential measurements. In this way, one can efficiently study both adsorption and desorption kinetics of proteins exhibiting arbitrary zeta potential including a zero potential, where the protein net charge vanishes. If the bulk zeta potential of a protein is not known, ζ_∞ can be approximated by the value of the zeta potential obtained at a long adsorption time limit using the streaming-potential measurements. Additionally, by knowing the coverage and the time of adsorption, one can accurately determine the bulk concentration of proteins, if it is not known.

The validity of the streaming-potential measurements for determining protein coverage was confirmed in Dąbkowska et al. (2013) using complementary XPS measurements. The advantage of XPS measurements is that their precision increases with the albumin coverage, which was determined from the surface area of the nitrogen peak at 400 eV. The results obtained by XPS, by streaming potential (obtained via Equation 17.19), and by AFM imaging are shown in Figure 17.5. In this figure, the albumin coverage obtained by various methods is expressed as the square root of the adsorption time, which facilitates a quantitative comparison with theoretical predictions shown by the solid line. These theoretical predictions were obtained by numerical solution of the governing diffusion transport equation by exactly considering surface blocking effects (Adamczyk 2000). As seen, experimental data obtained by various methods agree with theoretical predictions, giving the maximum HSA coverage $\Theta_{mx} = 0.45$. This is less than the maximum monolayer coverage of 0.583 theoretically predicted using the random sequential adsorption (RSA) model for spheroidal particles having an aspect ratio of 2 (Adamczyk 2006).

It is interesting to mention that this value of Θ_{mx} corresponds to $\Gamma = 1.4$ mg m^{-2} for a bare molecule. Assuming 20% hydration, the corrected value of Γ is equal to 1.65 mg

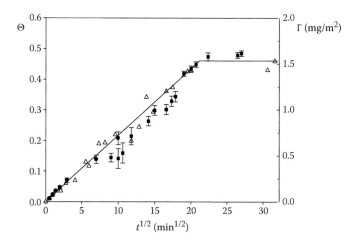

FIGURE 17.5 Comparison of experimental data characterizing HSA adsorption on mica (pH 3.5, $I = 0.15$ M NaCl) shown as the dependence of Θ and Γ (mg/m^2) on the square root of the adsorption time $t^{1/2}$. The points denote experimental values obtained by various methods: (•) AFM, (■) XPS (normalized), and (Δ) streaming potential. The solid line denotes exact theoretical results calculated by solving the diffusion transport equation with the RSA blocking function.

m^{-2}. This agrees with values reported by Kurrat et al. (1994, 1997) for oxidized silicon and titanium.

The deviation of the maximum coverage from results pertinent to hard (non-interacting) particles is quite analogous to that previously observed for colloid particles, interpreted in terms of electrostatic repulsive lateral interactions (Adamczyk 2006). In order to verify if this holds for albumin adsorption, a series of thorough experiments was performed in Dąbkowska and Adamczyk (2012a) where the effect of ionic strength was studied. In these experiments, also carried out using the streaming potential method, the amount of irreversibly adsorbed albumin was determined for a fixed pH of 3.5 and NaCl concentration varied between 10^{-3} and 0.15 M. It was determined that maximum coverage of irreversibly bound HSA increased systematically with the ionic strength from 0.14 ($\Gamma = 0.42$ mg m^{-2}) for $I = 10^{-3}$ M NaCl to 0.45 (1.4 mg m^{-2}) for 0.15 M. It is interesting to observe that these results agree with some previous measurements reported in the literature (Elgersma et al. 1992; Norde and Lyklema 1978; Yoon et al. 1996) obtained using other indirect techniques, for example, reflectometry.

As seen, the maximum coverage of irreversibly bound albumin monotonically increases with the ionic strength, that is, with the $\kappa d_H/2$ parameter. This correlation suggests that this effect is caused by the lateral electrostatic interactions among adsorbed HSA molecules, whose range decreases proportionally to $1/(\kappa d_H/2)^{1/2}$ (Adamczyk 2006, Adamczyk et al. 2010a). This was also observed for colloid particles (Adamczyk et al. 2010a), dendrimers (Pericet-Camara et al. 2007), gold (Kooij et al. 2002), and silver (Oćwieja et al. 2011) nanoparticles.

As shown in Adamczyk (2006), this effect can be interpreted in terms of the effective hard particle concept. According to this approach, the net electrostatic interaction energy between particles (referred to as the soft interaction potential) is replaced by the hard particle potential. This is tantamount to postulating that at the distance $2h^*$ between particle surfaces, the interaction energy tends to infinity, and zero otherwise. Physically, this means that the geometrical cross-section area of the particle is increased by the factor $(1+ 2h^*/d_H)^2$. Accordingly, in the case of HSA adsorption, the following expression for the maximum coverage of HSA can be formulated:

$$\theta_{mx} = \theta_{\infty} \frac{1}{(1+2h^*/d_H)^2},$$ (17.21)

where Θ_{∞} is the maximum coverage for hard (non-interacting) particles (Adamczyk 2006).

The effective interaction range, characterizing the repulsive double-layer interaction particles, can be calculated from the formula (Adamczyk 2006)

$$2h^*/d_H = \frac{1}{\kappa d_H} \left\{ \ln \frac{\phi_0}{2\phi_{ch}} - \ln \left[1 + \frac{1}{\kappa d_H} \ln \frac{\phi_0}{2\phi_{ch}} \right] \right\},$$ (17.22)

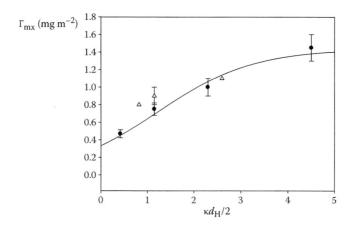

FIGURE 17.6 Dependence of the maximum coverage of irreversibly bound HSA, Γ_{mx} (mg m^{-2}) on the $\kappa d_H/2$ parameter. The full points denote experimental results obtained via the streaming-potential measurements for pH 3.5, Dąbkowska and Adamczyk 2012a and the hollow triangles denote the literature data reported in Norde and Lyklema (1978) and Elgersma et al. (1992). The solid line shows the theoretical results calculated from Equations 17.21 and 17.22 using the effective hard particle concept assuming no hydration.

where $\phi_0 = 4\pi\varepsilon d_H \left(\dfrac{kT}{e}\right)^2 \tanh^2\left(\dfrac{\zeta_p e}{4kT}\right)$ is the characteristic interaction energy of HSA molecules, ζ_p is the zeta potential of has, and ϕ_{ch} is the scaling interaction energy, close to the kT unit (Adamczyk 2006).

The theoretical results calculated from Equations 17.21 and 17.22 are shown in Figure 17.6. As can be noticed, the agreement between theoretical predictions and the experimental data is quite satisfactory for the entire range of ionic strength occurring in experiments, corresponding to the $\kappa d_H/2$ parameter, varying between 0.42 and 4.5.

Results discussed in this section indicate that HSA adsorption on mica at pH 3.5 is fully analogous to solid and polymeric nanoparticle adsorption, governed by electrostatic interactions, which can be accounted for by classical approaches.

However, a more complicated situation arises in the case of other pH values, where the HSA charge heterogeneity effects play a pronounced role. These results are discussed in the next section.

17.3.3   Role of pH in HSA Adsorption on Mica

Because the streaming-potential measurements are carried out under wet, *in situ* conditions, which eliminated conformation changes and desorption of protein upon drying, they can efficiently be used to study albumin adsorption for a wide range of pH values. This is vital because no coherent mechanism was proposed to explain anomalous albumin adsorption at negatively charged surfaces for pH equal to or above the isoelectric point (pH 5.1). As shown in Figure 17.1a, the average (mean-field) zeta potential of albumin is either zero or negative for this pH range, which prohibits a natural explanation of its adsorption in terms of the classical DLVO theory.

In order to elucidate this discrepancy, thorough studies were performed in a publication from Dąbkowska and Adamczyk (2012b) using the streaming-potential measurements, where the role of pH, which varied from 3.5 to 10.5, was determined. Results obtained in these measurements, carried out at a NaCl concentration of 0.15 M, are shown in Figure 17.7 as the dependence of the zeta potential of mica on the nominal coverage of albumin Θ. For comparison, the previous results obtained for pH 3.5 are also shown as reference data. As can be seen, for pH 3.5, 5.1 and 7.4, the zeta potential of mica increased abruptly with Θ, attaining saturation values close to the bulk zeta potentials of albumin, determined by electrophoresis. For pH 9.5, the zeta potential of mica increased only slightly from the initial value of −58 mV to ca. −40 mV (bulk value of albumin equal to −30 mV), and for pH 10.5, there was practically no change in zeta potential of mica. Qualitatively, these results indicate that there was a significant adsorption of albumin for pH 3.5, 5.1 and 7.4, less adsorption for pH 9.5, and no adsorption for pH 10.5. These predictions are confirmed by interpreting the experimental data in terms of the above-defined 3D electrokinetic model (Equations 17.15 and 17.18). As can be seen in Figure 17.7, theoretical results calculated from this model adequately reflect the experimental data for pH 3.5, 5.1 and pH 7.4. However, the effective zeta potential of HSA, which fitted well the experimental data, were −8 mV for pH 5.1 (compared to the bulk value close to zero) and −20 mV for pH 7.4 (compared to the bulk value of −16 mV). It is interesting to mention that an analogous behavior was previously reported for fibrinogen adsorption on mica for pH 7.4, where the effective zeta potential was much lower than the bulk zeta potential (Wasilewska and Adamczyk 2011). To further confirm the validity of the 3D electrokinetic model, the primary results shown in Figure 17.7 were transformed using Equations 17.19 and 17.20 to the universal relationship Θ versus $-\dfrac{1}{C_i}\ln X_1(\zeta)$.

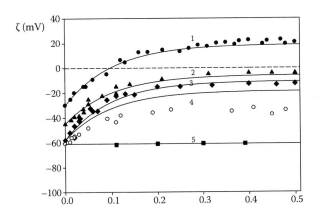

FIGURE 17.7 Dependence of the zeta potential of mica ζ on the coverage of HSA, Θ. The points denote experimental results obtained from the streaming-potential measurements for $I = 0.15$ M and various pH values: 1, (•) pH 3.5; 2, (▲) pH 5.1; 3, (♦) pH 7.4; 4, (○) pH 9.5; 5, (■) pH 10.5. The solid lines denote exact theoretical results calculated from the 3D electrokinetic model (Equations 17.15 and 17.18).

As noticed (Figure 17.8), this resulted in a straight-line dependence valid for a wide range of albumin coverage (up to of 0.3), which can have practical implications. The agreement indicates that one can sensitively determine an unknown coverage of albumin via streaming potential measurements using Equation 17.19.

The agreement of the experimental data with theoretical predictions shown in Figures 17.7 and 17.8 indicates quite unequivocally that for pH 5.1 and 7.4, albumin adsorbed efficiently on negatively charged mica surface. This was confirmed more directly using the AFM enumeration of adsorbed molecules (which worked efficiently for coverage up to 0.2) and the XPS measurements for higher coverage range.

This anomalous HSA adsorption on negatively charged surfaces can be explained analogously as done for fibrinogen (Adamczyk et al. 2012b; Wasilewska and Adamczyk 2011) in terms of a heterogeneous charge distribution. Thus, at pH 5.1 and 7.4, there exist positively charged patches on HSA molecules, despite that it is on average neutral or negatively charged. Hence, the albumin molecule effectively behaves as a dipole. The positively charged areas of the molecule are directed to the mica surface, which creates attractive electrostatic interactions ensuring efficient attachment of albumin molecules. Given the high charge density of mica at $I = 0.15$ M and pH 5.1–7.4 (Table 17.4), one can estimate that the presence of a few positive charges on albumin would suffice to produce the energy minimum depth of ca. $-17\ kT$. Combined with the van der Waals attraction amounting to a few kT units, this ensures irreversible albumin adsorption on mica for this pH range. Because the positively charged part of the albumin molecule is directed to mica, the negatively charged part is exposed to the electrolyte solution (flow), which can explain why the effective zeta potential of HSA on mica is more negative than in the bulk.

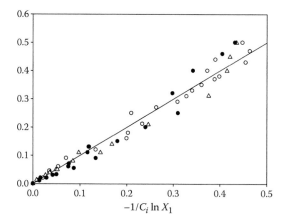

FIGURE 17.8 Universal graph showing the dependence of the HSA coverage on $-(1/C_i)\ln \dfrac{\zeta - \zeta_\infty}{\zeta_i - \zeta_\infty}$ $I = 0.15$ M and various pH values: (o) pH 3.5, (Δ) pH 5.1, and (•) pH 7.4. The points denote experimental results from using the streaming-potential measurements. The solid line denotes exact theoretical results calculated from the 3D electrokinetic model (Equations 17.19 and 17.20).

It should be mentioned that the existence of similar dipolar charge distribution in the case of chymotrypsin was reported by Vasina and Dejardin (2004), who used analogous streaming-potential measurements.

A decisive role of electrostatic interactions is further confirmed by the fact that for a higher pH of 9.5, albumin adsorbs less efficiently, and for pH 10.5, its adsorption vanishes. The latter pH value well correlates with the pK_a values of the most abundant basic amino acids in the molecule; that is, lysine pH 10.5 (Table 17.1). Otherwise, if albumin adsorption was governed by van der Waals interactions or hydrogen bonding, the change in its adsorption efficiency with pH would be negligible or much less abrupt.

The above results obtained for planar, homogeneous, and smooth interfaces can be used as convenient reference data for more complicated situations of albumin adsorption on colloid carrier particles, widely studied in the literature (Norde and Lyklema 1978).

17.3.4 ALBUMIN ADSORPTION ON COLLOID CARRIER PARTICLES

Albumin (rHSA) adsorption on latex particles was determined by measuring changes in its electrophoretic mobility (zeta potential) induced by this process (Adamczyk and Sofińska, in preparation) The experimental procedure was analogous to that previously used for fibrinogen (Adamczyk et al. 2012a) and consisted of three main steps:

(i) Measuring the reference zeta potential of bare latex particles in suspensions of the concentration c_1 changed between 40 and 100 mg L^{-1}
(ii) *In situ* formation of rHSA monolayers on latex particles under diffusion-controlled transport conditions using albumin solutions of an appropriate concentration c_a. ranging between 0.1 and 10 mg L^{-1} in 10 min
(iii) Washing the latex suspension using a membrane filtration unit with pure electrolyte and determining residual concentration in the supernatant

This procedure proved reliable, enabling a direct determination of the zeta potential variations as a function of the bulk concentration of albumin.

It is interesting to mention that the relaxation time of albumin monolayer formation under these conditions varied between 1.5 s for $c_1 = 40$ mg L^{-1} and 0.7 s for $c_1 = 100$ mg L^{-1} (Adamczyk and Sofińska, in preparation). This is much shorter than the experimental adsorption time of 600 s, which suggests that all albumin molecules could adsorb on latex during this period.

In order to prove the irreversibility of albumin adsorption on latex, additional series of experiments were performed, where the stability of the zeta potential of covered latex in time was measured. It was shown that there were no changes in zeta potential of latex covered by various degrees of albumin over a prolonged period up to 48 h (for pH 3.5 and NaCl concentration varied between 10^{-3} and 0.15 M).

This observation confirms that the albumin coverage can be directly calculated from the mass balance equation, that is, assuming that all albumins initially present in solution were irreversibly adsorbed on latex particle surfaces:

$$\Gamma = c_a/S_1, \tag{17.23}$$

where Γ is the coverage of albumin on latex particles expressed in mg m^{-2} and S_1 is the specific surface area of latex expressed in m^2 L^{-1} given by

$$S_1 = 6 \times 10^{-6} c_1/(d_1\rho_1), \tag{17.24}$$

where d_1 is the latex particle diameter in meters (m) and ρ_1 is the specific density of the latex (kg m^{-3}).

For the latex of the bulk concentration of 40 mg L^{-1}, the specific surface area S_1 is equal to 2.86×10^{-1} m^2 L^{-1}; for $c_1 = 60$ mg L^{-1}, $S_1 = 4.29 \times 10^{-1}$ m^2 g^{-1}; and for $c_1 = 100$ mg L^{-1}, $S_1 = 7.14 \times 10^{-1}$ m^2 L^{-1}.

By knowing Γ, one can calculate the surface concentration of albumin molecules on latex particles N (in µm^{-2}), which has a direct physical interpretation. This quantity is given by

$$N = 10^{-15} \Gamma A_v/M_w. \tag{17.25}$$

Accordingly, the coverage of albumin can be calculated from Equation 17.10 as $\Theta = 10^{-6} N S_g$.

By knowing the coverage, one can quantitatively interpret the results obtained for albumin adsorption on latex in terms of the 3D electrokinetic and the GC models previously applied. These results obtained for pH 3.5 and NaCl concentrations of 0.15 and 10^{-3} M NaCl are shown in Figure 17.9a and b, respectively. As can be observed, all results obtained for various bulk latex concentrations (40, 60, 80, and 100 mg L^{-1}) are transformed to one universal dependence, which supports the hypothesis of irreversible albumin adsorption mechanism.

As can be seen in Figure 17.9a, for an ionic strength of 0.15 M, pertinent to physiological conditions, the GC model (depicted by the dashed/dotted line) proved completely inadequate, similarly as previously observed for albumin adsorption on mica (see Figure 17.4). The 3D electrokinetic model (solid line in Figure 17.9a) performed much better, although it underestimated the experimental results for the intermediate coverage range below 0.3. This behavior shown in Figure 17.9a for albumin is fully analogous to that previously observed for fibrinogen adsorption on the same latex sample (Adamczyk et al. 2012a). Considering that the latex particles bear a fuzzy layer whose thickness decreases with the ionic strength, a two-stage mechanism of fibrinogen adsorption was postulated. Thus, at initial stages of adsorption, protein molecules fill up local cavities, partially penetrating the fuzzy layer on the latex particles. In the second stage, because of the lack of favorable adsorption spots, the molecules adsorb side-on, exposing its entire surface area to the bulk electrolyte solution. It seems that a quite analogous mechanism is also valid in the presently studied case of albumin adsorption.

In order to confirm the validity of these adsorption mechanisms, further experiments were performed by Adamczyk and Sofińska (in preparation) with the aim of evaluating the role of ionic strength in albumin adsorption on latex particles. In Figure 17.9b, results of these experiments obtained for 10^{-3} NaCl are shown. As

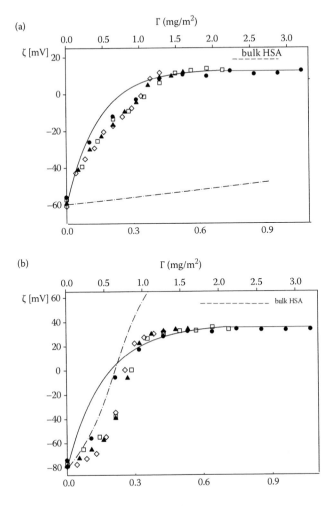

FIGURE 17.9 Dependencies of the zeta potential of latex ζ on the coverage of rHSA. The points denote experimental results obtained from the microelectrophoretic measurements for pH 3.5 and various bulk concentrations of latex, 40–100 mg L^{-1}. (a) 0.15 M NaCl. (b) 10^{-3} M NaCl. The solid lines denote theoretical results calculated from the corrected 3D adsorption model (Equations 17.15 and 17.18) and the dashed/dotted lines denote the theoretical results calculated from the GC model (Equation 17.14).

observed, in this case, the GC model performs better, giving a quite reasonable estimate of the experimental results for coverage range below 0.3. On the other hand, the 3D model was rather inadequate, overestimating the experimental data at the lower coverage range and underestimating the data at the higher coverage range. A better performance of the GC model can be interpreted in terms of increased thickness of the fuzzy layer for such a low ionic strength, which considerably exceeds the albumin molecular dimensions. In consequence, the molecules adsorb inside the fuzzy layer, that is, well below the slip plane. Thus, albumin adsorption leads to the increase of the net electrokinetic charge of latex, described by Equation 17.13, and does not affect the flow pattern in the vicinity of the latex particles.

As can be seen in Figure 17.9a and b, albumin adsorption results in an abrupt increase in the zeta potential of latex, whose sign is inverted for Θ close to 0.3. This finding has practical significance, because one can determine the coverage of albumin from electrophoretic mobility measurements. In this way, an unknown bulk concentration of albumin can be accurately determined from Equation 17.23 at a level of 0.1 mg L^{-1}, which is impractical using chemical-type methods.

However, as noticed in Figure 17.9a, because of a significant decrease in slopes of the ζ on Θ dependencies, it is rather difficult to precisely determine the maximum coverage of albumin on latex Θ_{mx}. It can be estimated that for 0.15 M NaCl, Θ_{mx} was close to 0.45–0.5 ($\Gamma = 1.4$–1.5 mg m^{-2}), which agrees with a previous value determined for albumin adsorption on mica. However, for 10^{-3} M NaCl, Θ_{mx} was close to 0.3 ($\Gamma = 0.9$ mg m^{-2}), which exceeds the previous value determined for albumin adsorption on mica, where $\Theta_{mx} = 0.14$ ($\Gamma = 0.42$ mg m^{-2}). This difference can be attributed to the above-suggested possibility of albumin penetration into the fuzzy layer on latex, which screens the lateral electrostatic repulsion among the protein molecules.

A more accurate determination of the maximum albumin coverage was acquired from the determination of the residual albumin concentration in the solution after contacting the latex particle suspension. This method was efficiently used to determine the maximum coverage in the case of fibrinogen (Adamczyk et al. 2012a). For low initial albumin concentrations, all molecules are irreversibly adsorbed on latex particles, leaving no residue concentration in the suspension. However, at higher bulk concentration, when a saturated albumin monolayer is formed, there should appear free albumin molecules in the supernatant solution at a level that is usually in the range of 0.1–1 mg L^{-1}. This residual concentration was determined as described Adamczyk et al. (2012a) by a direct AFM enumeration of the albumin molecules adsorbed on mica after a fixed adsorption time (typically 600 s).

Using the above determined value of $c_a = 0.85$, one can calculate from Equation 17.9 that the breakthrough (maximum) coverage of albumin was 0.5 (1.5 mg m^{-2}) for 0.15 M NaCl. Analogously, for 10^{-3} M NaCl, $\Theta_{mx} = 0.35$ ($\Gamma = 1$ mg m^{-2}). As seen, both values determined by AFM agree with the maximum coverage determined by electrophoretic measurements.

It is interesting to mention that our results obtained for albumin are analogous to those previously obtained by Serra et al. (1992) who measured adsorption of immunoglobulin (rabbit IgG, polyclonal) on polystyrene latex particles (diameter 297 nm). The amount of adsorbed protein was determined by direct UV spectrometry measurements at a wavelength of 280 nm and the mobility of the latex/protein complex was determined by microelectrophoresis. They observed that the μ_e of the protein/latex complex was almost a linear function of the coverage of IgG, up to the value of 2.0 mg m^{-2}, with the electrophoretic mobility varying between –4.0 and 0.5 μm cm (V s)$^{-1}$.

Similar linear dependencies of the zeta potential on the coverage were reported in the above-mentioned work of Kalasin and Santore (2009) for the fibrinogen/silica particle system.

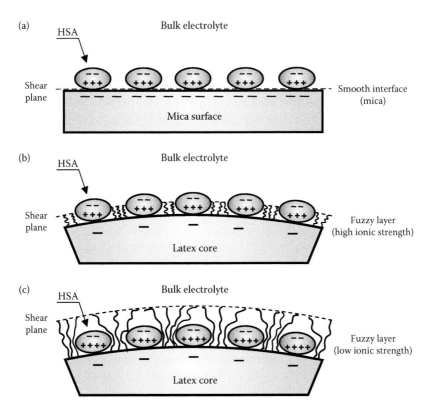

FIGURE 17.10 Mechanisms of HSA adsorption on various interfaces, a schematic view. (a) Adsorption on smooth surfaces (3D electrokinetic model). (b) Adsorption on rough surfaces (corrected 3D electrokinetic model). (c) Adsorption on polymer-covered surfaces (GC model).

On the basis of experimental evidences discussed in this chapter, three distinct adsorption mechanisms of albumin on various surfaces were proposed, shown schematically in Figure 17.10.

17.4 CONCLUSIONS

In situ streaming-potential measurements combined with AFM and XPS measurements of protein coverage furnished reliable clues on HSA adsorption mechanisms on mica, a model hydrophilic substrate.

These experiments, performed for NaCl concentrations varying between 10^{-3} and 0.15 M and a broad range of pH (3.5 to 10.5), were quantitatively interpreted in terms of the theoretical model postulating a 3D adsorption of HSA molecules as discrete particles. The GC model, based on a patch-like 2D adsorption model, proved inadequate.

The streaming-potential measurements combined with this 3D model enabled one to determine the amount of irreversibly adsorbed HSA as a function of ionic strength and pH. For pH 3.5, where HSA molecules are positively charged, the coverage of irreversibly adsorbed HSA increased from 0.42 mg m^{-2} for 10^{-3} M NaCl to 1.4 mg m^{-2} for $I = 0.15$ M (pH 3.5). This significant role of ionic strength was attributed to the lateral electrostatic repulsion among adsorbed HSA molecules. This effect was quantitatively interpreted in terms of the effective hard particle concept previously used for colloid particles.

Anomalous adsorption of HSA on mica observed for pH > 5.1 (isoelectric point), where the molecules bear a net negative charge, was explained in terms of heterogeneous charge distribution. Hence, the positively charged patches are contacting the negatively charged mica surface, whereas the more negative patches are pointing out to the solution. This hypothesis was confirmed by the streaming-potential measurements. On the other hand, for pH 10.5, no adsorption of HSA on mica was observed. These experimental observations revealed a dominant role of electrostatic interactions in HSA adsorption.

The results obtained for homogeneous and smooth interfaces were used as convenient reference data for more complicated situations of albumin adsorption on colloid carrier particles (polystyrene latex particles) studied by microelectrophoretic measurements. Abrupt changes in zeta potentials of latex particles with albumin coverage were observed. For high ionic strength (0.15 M), this effect was reasonably accounted for by the 3D electrokinetic model, whereas for low ionic strength (10^{-3} M), the GC model proved more adequate. Variation in HSA adsorption mechanisms with ionic strength were interpreted as due to the penetration of the protein into the fuzzy layer on the latex particle, whose thickness decreases with ionic strength. These results can be exploited for a precise determination of the albumin coverage on colloid carrier particles via simple measurements of electrophoretic mobility.

The experimental results confirmed that monolayers of irreversibly bound HSA of a well-controlled coverage can be produced both on mica and on latex particles by adjusting the ionic strength of the suspension.

ACKNOWLEDGMENTS

This work was financially supported by the COST D43 Special Grant of MNiSZW and by a Ministry of Science and Higher Education (MNiSzW) grant (N N204 026438).

REFERENCES

Adamczyk, Z., 2000. Kinetics of diffusion-controlled adsorption of colloid particles and proteins. *J. Colloid Interface Sci.* 229:477–489.

Adamczyk, Z., 2006. *Particles at Interfaces: Interactions, Deposition, Structure.* Elsevier/Academic Press, Amsterdam.

Adamczyk, Z., Bratek-Skicki, A., Dąbrowska, P., Nattich-Rak, M., 2012a. Mechanisms of fibrinogen adsorption on latex particles determined by zeta potential and AFM measurements. *Langmuir* 28:474–485.

Adamczyk, Z., Cichocki, B., Egiel-Jeżewska, M., Słowicka, A., Wajnryb, E., Wasilewska, M., 2012b. Fibrinogen conformations and charge in electrolyte solutions derived from DLS and dynamic viscosity measurements. *J. Colloid Interface Sci* 385:244–57.

Adamczyk and Sofińska in preparation.

Adamczyk, Z., Nattich, M., Wasilewska, M., Zaucha, M., 2011. Colloid particle and protein deposition—Electrokinetic studies. *Adv. Colloid Interface Sci.* 168:3–28.

Adamczyk, Z., Sadlej, K., Wajnryb, E., Nattich, M., Ekiel-Jeżewska, M.L., Bławzdziewicz, J., 2010a. Streaming potential studies of colloid, polyelectrolyte and protein deposition. *Adv. Colloid Interface Sci.* 153:1–29.

Adamczyk, Z., Zaucha, M., Zembala, M., 2010b. Zeta potential of mica covered by colloid particles: A streaming potential study. *Langmuir* 26:9368–9377.

Adamczyk, Z., Michna, A., Szaraniec, M., Bratek, A., Barbasz, J., 2007. Characterization of poly(ethylene imine) layers on mica by the streaming potential and particle deposition methods. *J. Colloid Interface Sci.* 313:86–96.

An, Y., Stuart, G.W., McDowell, S.J., McDaniel, S.E., Kang, Q., Friedman, R.J., 1996. Prevention of bacterial adherence to implant surfaces with a crosslinked albumin coating in vitro. *J. Orthop. Res.* 14:846–849.

Blum, H., Beier, H., Gross, H.J., 1987. Improved silver staining of plant proteins, RNA and DNA in polyacrylamide gels. *Electrophoresis* 8:93–99.

Brokke, P., Dankert, J., Carballo, J., Feijen, J., 1991. Adherence of coagulase-negative Staphylococci onto polyethylene catheters in vitro and in vivo: a study on the influence of various plasma proteins. *J. Biomater. Appl.* 5:204–226.

Carter, D.C., Ho, J.X., 1994. Structure of serum albumin. *Adv. Protein Chem.* 45:153–203.

Dąbkowska, M., Adamczyk, Z., Kujda, M., 2013. Mechanism of HSA adsorption on mica determined by streaming potential, AFM and XPS measurements. *Colloids Surf., B.* 101:442–449.

Dąbkowska, M., Adamczyk, Z., 2012a. Ionic strength effect in HSA adsorption on mica determined by streaming potential measurements. *J. Colloid Interface Sci.* 366:105–113.

Dąbkowska, M., Adamczyk, Z., 2012b. Human serum albumin monolayers on mica: electrokinetic characterictics. *Langmuir* 28:15663–15673.

Elgersma, A.V., Zsom, R.L.J., Lyklema, J., Norde, W., 1992. Kinetics of single and competitive protein adsorption studied by reflectometry and streaming potential measurements. *Colloids Surf.* 65:17–28.

Fair, B.D., Jamieson, A.M., 1980. Studies of protein adsorption on polystyrene latex surfaces. *J. Colloid Interface Sci.* 77:525–534.

Goodwin, J.W., Hearn, J., Ho, C.C., Ottewill, R.H., 1974. Studies on the preparation and characterisation of monodisperse polystyrene lattices. *Colloid Polym. Sci.* 252:464–471.

Haynes, C.A., Norde, W., 1994. Globular proteins at solid/liquid interfaces. *Colloids Surf., B* 2:517–566.

He, X.M., Carter, D.C. 1992. Atomic structure and chemistry of human serum albumin. *Nature* 358:209–215.

Hogt, A.H., Dankert, J., Feijen, J., 1985. Adhesion of *Staphylococcus epidermidis* and *Staphylococcus saprophyticus* to a hydrophobic biomaterial. *J. Gen. Microbiol.* 131:2485–2491.

Hunter, A.K., Carta, G., 2001. Effects of bovine serum albumin heterogeneity on frontal analysis with anion-exchange media. *J. Chromatogr., A* 937:13–19.

Jachimska, B., Wasilewska, M., Adamczyk, Z., 2008. Characterization of globular protein solutions by dynamic light scattering, electrophoretic mobility, and viscosity measurements. *Langmuir* 24:6866–6872.

Kalasin, S., Santore, M.M., 2009. Non-specific adhesion on biomaterial surfaces driven by small amounts of protein adsorption. *Colloids Surf., B* 73:229–236.

Kooij, E.S., Brouwer, E.A.M., Wormeester, H., Poelsema, B., 2002. Ionic strength mediated self-organization of gold nanocrystals: An AFM study. *Langmuir* 18:7677–7682.

Kottke-Marchant, K., Anderson, J.M., Umemura, Y., Marchant, R.E., 1989. Effect of albumin coating on the in vitro blood compatibility of Dacron® arterial prostheses. *Biomaterials* 10:147–155.

Kurrat, R., Ramsden, J.J., Prenosil, J.E., 1994. Kinetic model for serum albumin adsorption: experimental verification. *J. Chem. Soc., Faraday Trans.* 90:587–590.

Kurrat, R., Prenosil, J.E., Ramsden, J.J., 1997. Kinetics of human and bovine serum albumin adsorption at silica–titania surfaces. *J. Colloid Interface Sci.* 185:1–8.

Laemmli, U.K., 1970. Cleavage of structural proteins during the assembly of the head of bacteriophage T4. *Nature* 227:680–685.

Malmsten, M., 1994. Ellipsometry studies of protein layers adsorbed at hydrophobic surfaces. *J. Colloid Interface Sci.* 166:333–342.

Nicholson, J.P., Wolmarans, M.R., Park, G.R., 2000. The role of albumin in critical illness. *Br. J. Anaesth.* 85:599–610.

Norde, W., 1986. Adsorption of proteins from solution at the solid–liquid interface. *Adv. Colloid Interface Sci.* 25:267–340.

Norde, W., Lyklema, J., 1978. The adsorption of human plasma albumin and bovine pancreas ribonuclease at negatively charged polystyrene surfaces: I. Adsorption isotherms. Effects of charge, ionic strength, and temperature. *J. Colloid Interface Sci.* 66:257–265.

Oćwieja, M., Adamczyk, Z., Morga, M., Michna, A., 2011. High density silver nanoparticle monolayers produced by colloid self-assembly on polyelectrolyte supporting layers. *J. Colloid Interface Sci.* 364:39–48.

Oćwieja, M., Adamczyk, Z., Morga, M., 2012. Hematite nanoparticle monolayers on mica preparation by controlled self-assembly. *J. Colloid Interface Sci.* DOI: 10.1016/j.jcis.2012.06.056

Ortega-Vinuesa, J.L., Tengvall, P., Lundström, I., 1998. Molecular packing of HSA, IgG, and fibrinogen adsorbed on silicon by AFM imaging. *Thin Solid Films* 324:257–273.

Parmelee, D.C., Evenson, M.A., Deutsch, H.F., 1978. The presence of fatty acids in human α-fetoprotein. *J. Biol. Chem.* 253:2114–2119.

Pericet-Camara, R., Cahill, B.P., Papastavrou, G., Borkovec, M., 2007. Nano-patterning of solid substrates by adsorbed dendrimers. *Chem. Commun. (Cambridge, U. K.)*:266–268.

Peters, Jr., T. 1985. Serum albumin. *Adv. Protein Chem.* 37:161–245.

Peters, Jr., T., 1995. *All About Albumin: Biochemistry, Genetics, and Medical Applications.* Academic Press, California, USA.

RCSB Protein Data Bank, 2012. Available at http://www.pdb.org/.

Revilla, J., Elaïssari, A., Carriere, P., Pichot, C., 1996. Adsorption of bovine serum albumin onto polystyrene latex particles bearing saccharidic moieties. *J. Colloid Interface Sci.* 180:405–412.

Reynolds, E.C., Wong, A., 1983. Effect of adsorbed protein on hydroxyapatite zeta potential and Streptococcus mutans adherence. *Infect. Immun.* 39:1285–1290.

Rezwan, K., Meier, L.P., Rezwan, M., Vörös, J., Textor, M., Gauckler, L.J., 2004. Bovine serum albumin adsorption onto colloidal Al_2O_3 particles: a new model based on zeta potential and UV–Vis measurements. *Langmuir* 20:10055–10061.

Rezwan, K., Meier, L.P., Gauckler, L.J., 2005. A prediction method for the isoelectric point of binary protein mixtures of bovine serum albumin and lysozyme adsorbed on colloidal titania and alumina particles. *Langmuir* 21:3493–3497.

Roche, M., Rondeau, P., Singh, N.R., Tarnus, E., Bourdon, E., 2008. The antioxidant properties of serum albumin. *FEBS Lett.* 582:1783–1787.

Rojas, O.J., 2002. Adsorption of polyelectrolytes on mica. In: *Encyclopedia of Surface and Colloid Science,* 2nd edition. Taylor & Francis, Boca Raton, FL, pp. 517–535.

Sadlej, K., Wajnryb, E., Bławzdziewicz, J., Ekiel-Jeżewska, M.L., Adamczyk, Z., 2009. Streaming current and streaming potential for particle covered surfaces: Virial expansion and simulations. *J. Chem. Phys.* 130:144706–144711.

Scales, P.J., Grieser, F., Healy, T.W., 1990. Electrokinetics of the muscovite mica–aqueous solution interface. *Langmuir* 6:582–589.

Scales, P.J., Grieser, F., Healy, T.W., White, L.R., Chan, D.Y.C., 1992. Electrokinetics of the silica-solution interface: a flat plate streaming potential study. *Langmuir* 8:965–974.

Serra, J., Puig, J.E., Martín, A.M., Galisteo, F.C., Gálvez, M., Hidalgo-Álvarez, R., 1992. On the adsorption of IgG onto polystyrene particles: electrophoretic mobility and critical coagulation concentration. *Colloid Polym. Sci.* 270:574–583.

Smith, P.K., Krohn, R.I., Hermanson, G.T., Mallia, A.K., Gartner, F.H., Provenzano, M.D., Fujimoto, E.K., Goeke, N.M., Olson, B.J., Klenk, D.C., 1985. Measurement of protein using bicinchoninic acid. *Anal. Biochem.* 150:76–85.

Van Dulm, P., Norde, W., 1983. The adsorption of human plasma albumin on solid surfaces, with special attention to the kinetic aspects. *J. Colloid Interface Sci.* 91:248–255.

Vasina, E.N., Déjardin, P., 2004. Adsorption of α-chymotrypsin onto mica in laminar flow conditions. adsorption kinetic constant as a function of tris buffer concentration at pH 8.6. *Langmuir* 20:8699–8706.

Wasilewska, M., Adamczyk, Z., 2011. Fibrinogen adsorption on mica studied by AFM and in situ streaming potential measurements. *Langmuir* 27:686–696.

Wertz, C.F., Santore, M.M., 2001. Effect of surface hydrophobicity on adsorption and relaxation kinetics of albumin and fibrinogen: single-species and competitive behavior. *Langmuir* 17:3006–3016.

Yoon, J.Y., Park, H.Y., Kim J.H., Kim, W.S. 1996. Adsorption of BSA on highly carboxylated microspheres-quantitative effects of surface functional groups and interaction forces. *J. Colloid Interface Sci.* 177:613–620.

Zdanowski, Z., Ribbe, E., Schalén, C., 1993. Influence of some plasma proteins on in vitro bacterial adherence to PTFE and Dacron vascular prostheses. *APMIS* 101:926–932.

18 Co-Adsorption of the Proteins β-Casein and BSA in Relation to the Stability of Thin Liquid Films and Foams

Krastanka G. Marinova, Rumyana D. Stanimirova,
Mihail T. Georgiev, Nikola A. Alexandrov,
Elka S. Basheva, and Peter A. Kralchevsky

CONTENTS

18.1   INTRODUCTION

Proteins are primary natural stabilizers of numerous food dispersion products, for example, creams, cheese, spreads, mousses, souses, and so on (Dickinson 2001; McClements 2005). The practical applications usually involve the use of natural protein mixtures, for example, the mixtures of caseins and globular whey proteins from milk (Mackie et al. 2001; Sengupta and Damodaran 2000; Zhang et al. 2004) and egg protein mixtures (Foegeding et al. 2006). Competitive adsorption and the formation of mixed adsorption layers in these systems are expected to govern the foaming and emulsifying ability, as well as the subsequent product stability (Foegeding et al. 2006; Maldonado-Valderrama et al. 2007; Zhang et al. 2004). Co-adsorption has been investigated and used in foam fractionation applications (Merz et al. 2001).

Co-adsorption and structure of protein adsorption layers from various protein mixtures have been studied by using different techniques: atomic force microscopy, fluorescent labeling and imaging, radiolabeling, and radiotracing (Damodaran 2004; Mackie et al. 2001; Sengupta and Damodaran 2000). The dynamics of protein adsorption/desorption and the displacement of proteins by surfactants can be investigated by recording the adsorption by ellipsometry (Russev et al. 2000) and the surface tension by various methods, for example, by drop shape analysis (DSA) (Hoorfar and Neumann 2006; Kotsmar et al. 2009; Rotenberg et al. 1983). The latter method allows replacing the outer phase around a pendant drop or buoyant bubble with another protein solution and investigating the sequential adsorption of proteins (Svitova et al. 2003; Svitova and Radke 2005). The microfluidics techniques also allow exchanging the protein solution within a pendant drop by using two coaxial capillaries (Cabrerizo-Vilchez et al. 1999; Ferri et al. 2008, 2010).

The transfer of knowledge from the model experiments with single surfaces to explain the behavior of foams and emulsions is not always straightforward (Wierenga et al. 2009b). The foam longevity depends not only on the rate of surfactant or protein adsorption but also on the stability of the thin liquid films (TLFs) formed between the bubbles in the foam (Marinova et al. 2009; Saint-Jalmes et al. 2005). Another important factor is the permeability of the films to the transport of gases (that leads to Ostwald ripening), which can be blocked by densely packed adsorption layers (Danov et al. 2012; Tcholakova et al. 2011).

In this chapter, we combine experimental measurements with single bubbles, TLFs, and foams to obtain information on the adsorption behavior and interactions of two proteins at the air/water interface in relation to the stabilization of foams. We selected a pair of rather different proteins: the disordered protein β-casein and the globular bovine serum albumin (BSA). Their adsorption behavior has been studied separately (Bantchev and Schwartz 2004; Cascão-Pereira et al. 2003a; Dickinson et al. 1993; Engelhardt et al. 2012; Graham and Philips 1979). At the best of our knowledge, this chapter is the first systematic study on the co-adsorption of BSA and β-casein. Our goal is to obtain information for the structure of the mixed adsorption layers and the foam film stability by analysis of data obtained by different methods. For this goal, we employed the DSA technique with buoyant bubbles. This technique is upgraded with the oscillating bubble method and with an option for replacement of the liquid phase. The used device for TLFs also allows phase exchange. These methods yield the dynamic and equilibrium surface tension, the surface dilatation storage and loss moduli, and the film thickness at low and high applied pressures, and allow comparing data from parallel and sequential adsorption of the two proteins. The properties of foams from the mixed solutions are compared with foams from solutions of the separate components. The results give information about the most favorable structure of the compound protein adsorption layer in relation to foam stability.

18.2   MIXED PROTEIN SOLUTIONS—EXPERIMENTAL METHODS

18.2.1   BSA versus β-Casein

The BSA is a globular protein of molecular weight 66,382 g/mol; the used sample is a product of Sigma (A7511). The BSA molecule consists of 580 amino acid residues with 17 interchain disulfide bonds. The second used protein is the disordered β-casein from bovine milk of molecular weight 23,983 g/mol, also from Sigma (C6905). The β-casein consists of 290 amino acid residues without any disulfide bonds. Aqueous solutions of the two proteins were prepared at different concentrations, which are listed in Table 18.1. The first five solutions in the table correspond to molar fractions of β-casein, $x_{\beta CS}$, in the protein blend varying from 0 to 1. Solution no. 6 contains the same concentration of BSA as solution no. 1 plus the concentration of β-casein present in solution no. 5. The total protein weight concentration, c_{tot}, varies from 0.010 to 0.026 wt% (Table 18.1). This concentration range was chosen because it allows distinguishing between more and less stable liquid films and foams (Cascão-Pereira et al. 2003b; Mackie et al. 2001; Wierenga et al. 2009a). All used β-casein concentrations are above the critical micellization concentration (CMC) of this protein, which is 0.62 μM at 25°C (Portnaya et al. 2006).

Deionized water from Elix purification system (Millipore) was used for the preparation of all solutions. One millimolar NaCl (sodium chloride, Merck) was added in the solutions for the measurements with TLFs. All solutions were used overnight after the preparation. The acidity of the solutions was not adjusted—they were used at their natural pH = 6.1 ± 0.1. The experiments were carried out at a room temperature of 25 ± 1°C.

TABLE 18.1
Chemical Composition of the Investigated Protein Solutions

No.	Solution	$x_{\beta CS}$	c_{tot} (wt%)
1	1.5 μM BSA	0.00	0.010
2	1.4 μM BSA + 0.7 μM β-casein	0.33	0.011
3	1.2 μM BSA + 1.3 μM β-casein	0.52	0.011
4	0.75 μM BSA + 3.3 μM β-casein	0.82	0.013
5	6.7 μM β-casein	1.00	0.016
6	1.5 μM BSA + 6.7 μM β-casein	0.82	0.026

18.2.2 EXPERIMENTS WITH SINGLE INTERFACES

In these experiments, a buoyant bubble was formed at the tip of a U-shaped needle dipped in the aqueous phase. The bubble was observed by the instrument DSA100R (Krüss GmbH, Hamburg, Germany), and the surface tension, σ, was determined from the bubble instantaneous shape using the software DSA1 (Krüss GmbH). The decrease of surface tension was recorded during a period of 30 min after the bubble formation.

At a given moment (in our experiments—10 min from the beginning), sinusoidal oscillations of the bubble volume (and of its surface area) were applied to determine the surface dilatational storage and loss moduli, E' and E'', following a standard procedure (Benjamins et al. 1996; Lucassen and van den Tempel 1972). The periods of oscillations varied from 1 to 20 s; the data were processed as described by Russev et al. (2008).

For the needs of the phase-exchange experiments, the investigated solution was loaded in a rectangular cuvette of dimensions $35 \times 25 \times 55$ mm (Figure 18.1). The volume of the solution in the cuvette was 12–13 mL. To exchange the solution in the cuvette with a new one, two thin tubes were used, which were positioned in two diagonal corners of the cuvette. A cartridge pump (model no. 7523-27, 10–600 RPM, manufactured by Barnant Co., Cole Parmer Instrument Company, USA) was used to simultaneously supply the new solution and suck out the old one. To exchange the aqueous phase, we ran the pump for 1 min at a flow rate of 150 mL/min. Thus, the volume of liquid in the cuvette was exchanged 12 times for 1 min work of the pump. According to the estimate by Svitova et al. (2003), when the volume of the inserted new solution reaches 10 times the volume of the cuvette, the solution becomes practically identical with the newly supplied solution.

The experimental protocol for sequential adsorption was as follows: (i) A bubble is formed in a given solution where its surface tension is measured for 10 min, and then oscillations are applied to determine the moduli E' and E''. (ii) Phase exchange: the first solution is exchanged with a second solution for 1 min as described above (see Figure 18.1). (iii) The bubble surface tension is measured for 10 min in the

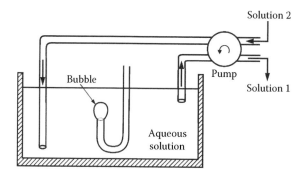

FIGURE 18.1 Sketch of the setup for exchange of the aqueous phase around a bubble formed at the tip of a U-shaped capillary. The cuvette with the solution is mounted on the table of the instrument DSA100R. The bubble is observed in transmitted light during the whole experiment.

second solution, and then oscillations are applied again to determine the new values of the moduli E' and E''. Furthermore, in the same way the second solution can be exchanged with a third solution, and so on.

In some experiments, we exchanged the protein solution with pure water ("rinsing" of the bubble). The aim of this experiment is to verify whether the protein desorbs from the bubble surface into the pure water, that is, whether the protein adsorption is irreversible. If the values of σ, E', and E'' remain the same after the exchange of the protein solution with pure water, this would indicate that the protein adsorption is irreversible.

18.2.3 EXPERIMENTS WITH THIN FOAM FILMS

We used two devices for experiments with thin foam films. The first is the Scheludko–Exerowa (SE) cell with one side capillary (Scheludko 1967; Scheludko and Exerowa 1959). This cell can be applied to investigate both foam and emulsion films stabilized by proteins (Basheva et al. 2006, 2011; Marinova et al. 1997).

The second setup, called the "flush cell" for brevity, is a modification of the SE cell with two capillaries, which allows exchange of the solutions within the cell with a new solution (Wierenga et al. 2009a). The flush cell allows to adsorb a given protein on the film surfaces and afterward to create a second adsorption layer from another protein (sequential adsorption).

The foam films were observed in reflected monochromatic light (564 nm, interference filter) by using a Jenavert microscope (Carl Zeiss, Jena) equipped with a long working distance objective (×20). The film thickness was determined by the interferometric method (Scheludko 1967). Substituting the refractive index of water, $n = 1.33$, in the basic formula of this method, we obtain the equivalent water thickness of the investigated films from the measured intensity of the reflected light.

Both the SE and flush cells can be used in the regimes of *closed* and *open* cell. In the case of "closed cell," the water vapors in the gas phase around the film are equilibrated with the aqueous (film) phase, so that there is no evaporation of water from the film. In this case, the equilibrium capillary pressure of the meniscus around the film, which is counterbalanced by the film's disjoining pressure, Π, can be estimated from the formula $P_c = 2\sigma R/(R^2 - r_c^2)$, where R is the inner radius of the capillary cell and r_c is the film radius (Nikolov and Wasan 1989). Using experimental parameter values, $R = 1.2$ mm, $r_c = 100$ μm, and $\sigma \approx 50$ mN/m, we estimate $P_c \approx 84$ Pa for the SE cell.

In the case of "open cell," the cover of the cell is removed so that evaporation of water from the film takes place. The evaporation drives a flux of water from the Plateau border around the film toward the center of the film. Under steady-state conditions, this hydrodynamic flux leads to a significant pressure difference, $P_c \approx 5 \times 10^5$ Pa, between the film's periphery and center (Basheva et al. 2011; Kralchevsky and Nagayama 2001). This pressure difference forces the two film surfaces against each other, which leads to overcoming of the soft electrostatic (double layer) repulsion between them. In this way, the film thickness becomes considerably smaller, approximately equal to the thickness of the two protein adsorption layers.

In our experiments with the SE cell, foam films of radius $r_c = 100$ µm were formed after 15 min aging of their surfaces and the film thinning was observed for 15 min in closed cell and for 1–2 min in open cell. Here, the SE cell was used to investigate the films in the case of *parallel* adsorption of the two proteins.

In the case of *sequential* adsorption of the two proteins, the experiments with foam films were carried out in the flush cell following the protocol given by Wierenga et al. (2009a). Again, films of radius $r_c = 100$ µm were formed after 15 min aging of their surfaces and the film thinning was observed for 15 min in closed cell and for 1–2 min in open cell.

18.2.4 EXPERIMENTS WITH FOAMS

Foams were formed and observed by means of the Dynamic Foam Analyzer DFA100 (Krüss GmbH). In this instrument, the foam is formed by bubbling of gas through a glass porous frit into the foaming solution. The frit is mounted at the base of a glass cylinder of 40 mm inner diameter and 250 mm height. The height of the foam and of the liquid beneath (the serum) has been detected and recorded automatically by means of a computer-controlled linear LED panel and a line sensor mounted along the column height. The foam height is calculated as a difference between the upper and lower foam levels. For the used cylinder, a foam height of 10 mm corresponds to a foam volume of 12.56 mL.

The used foaming procedure was as follows: 50 mL of the solution was loaded into the cylinder and the gas was bubbling for 10 s with a flow rate of 0.5 L/min through a frit of porosity G2 (40–100 µm pore size). The foam decay was recorded for 1 h after the foaming. The glass column with the foam was sealed with Parafilm-M at the top, after ceasing of the bubbling, to prevent foam destruction because of water evaporation.

18.3 CHARACTERIZATION OF THE PROTEIN ADSORPTION LAYERS

In this section, we present the experimental results. They are analyzed and discussed in Section 18.4.

18.3.1 SURFACE TENSION AND DILATATIONAL RHEOLOGY

Figure 18.2 shows data for the relaxation of surface tension with time, $\sigma(t)$, measured by the DSA method applied to buoyant bubbles (Figure 18.1). Measurements with solution nos. 1–5 in Table 18.1 have been carried out. The data in Figure 18.2 indicate that σ decreases faster with the rise of the β-casein concentration and molar fraction, $x_{\beta CS}$. This can be explained by the fact that β-casein is more surface active than BSA (Graham and Phillips 1979) and by the circumstance that the used β-casein concentrations are above the CMC of this protein (Portnaya et al. 2006).

At the longer times, the curves in Figure 18.2 follow the law $(\sigma - \sigma_{eq}) \propto t^{-1/2}$, which is the relaxation asymptotics in the case of adsorption under diffusion control (Sutherland 1952). The extrapolation of the σ versus $t^{-1/2}$ dependence at $t \to \infty$ yields

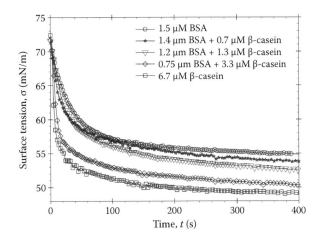

FIGURE 18.2 Relaxation of the surface tension, σ, of mixed aqueous solutions of BSA and β-casein at concentrations shown in the figure (see also Table 18.1)—experiments with buoyant bubbles by the DSA method.

the equilibrium surface tension σ_{eq} = 48 and 53 mN/m for β-casein and BSA, respectively. The lower equilibrium surface tension of the β-casein solutions confirms the higher surface activity of this protein.

In Figure 18.3, the surface tension data for all investigated solutions are plotted in terms of surface pressure, $\pi_s = \sigma_0 - \sigma_{eq}$, versus the molar fractions of β-casein, $x_{\beta CS}$ (see Table 18.1); σ_0 = 72 mN/m is the surface tension of pure water at 25°C. π_s gradually increases with the rise of $x_{\beta CS}$ from 0 to 1. Indications for synergism of the

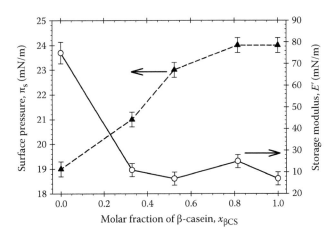

FIGURE 18.3 Plots of the measured equilibrium surface pressure, $\pi_s = \sigma_0 - \sigma_{eq}$, and surface dilatational storage modulus, E', versus the molar fraction of β-casein, $x_{\beta CS}$, in mixed aqueous solutions with BSA (see Table 18.1)—experiments with buoyant bubbles by the DSA method and applied oscillations to determine E'. The lines are guides to the eye.

two proteins (a local maximum in π_s) are missing. The π_s-versus-$x_{\beta CS}$ curve in Figure 18.3 indicates an increasing influence of the more surface active β-casein with the rise of its molar fraction.

Ten to fifteen minutes after the bubble formation, we applied oscillations of frequency 0.5 Hz to determine the surface dilatational storage and loss moduli, E' and E''. As seen in Figure 18.3, there is a considerable difference between the two proteins with respect to the storage modulus: $E' = 76$ mN/m for BSA versus 15 mN/m for β-casein. In other words, the adsorption layers from BSA are considerably more elastic. For the mixed solutions, $x_{\beta CS} = 0.33, 0.52$, and 0.82, E' is close to its value for β-casein alone. In other words, the β-casein dominates the surface elasticity in the case of simultaneous (parallel) adsorption of the two proteins. Concerning the loss modulus, E'', it gradually decreases with the rise of $x_{\beta CS}$, from 11 mN/m for BSA to 7 mN/m for β-casein.

Furthermore, we carried out phase-exchange experiments; the results are summarized in Table 18.2. In experiment no. 1, the values of σ, E', and E'' have been measured for the adsorption layer of 1.5 μM BSA solution. Next, the BSA solution was exchanged with pure water, and after 15 min, the values of σ, E', and E'' were measured again (experiment no. 2). The data in Table 18.2 show that the values of these three parameters before and after the exchange coincide in the framework of the experimental reproducibility, indicating that the adsorption of BSA at the air/water interface is irreversible. This result is in agreement with the conclusions by Svitova et al. (2003).

Likewise, experiment nos. 4 and 5 in Table 18.2 indicate that the adsorption of β-casein is also irreversible. Having once adsorbed at the air/water interface, the

---

TABLE 18.2

Surface Tension, σ, and Dilatational Storage and Loss Moduli, E' and E'', Measured with Buoyant Bubbles at a Frequency of 0.5 Hz

No.	Protein Solutions and Phase Exchanges (PhE)	σ (mN/m)	E' (mN/m)	E'' (mN/m)
1	BSA (1.5 μM)	55	76	11
2	BSA (1.5 μM) after PhE with water	54	74	15
3	BSA (1.5 μM) after PhE with β-casein (6.7 μM)	48.5	42	17
4	β-Casein (6.7 μM)	48.5	15	7
5	β-Casein (6.7 μM) after PhE with water	48.5	13	5
6	β-Casein (6.7 μM) after PhE with BSA (1.5 μM)	50	33	13
7	BSA (1.2 μM) + β-casein (1.3 μM)	51	16	8
8	BSA (1.4 μM) + β-casein (0.7 μM)	52	21	10

Note: The values of σ, E', and E'' are averaged over four runs; the standard deviations are $\Delta\sigma = 1$ mN/m, $\Delta E' = 5$ mN/m, and $\Delta E'' = 3$ mN/m.

β-casein cannot be washed out by flushing with pure water. An analogous result was obtained by MacRitchie (1998) (see also the review by Fainerman et al. 2006).

The practically equal values of the loss modulus E'' before and after the exchange of the protein solution with pure water deserves a special discussion. This indicates that the energy dissipation (giving rise to E'') is not related to diffusion transport of protein molecules to/from the interface but is due to interfacial processes. Such processes can be the attachment/detachment of protein molecules to surface aggregates, the energy dissipation accompanying breakage and restoration of adhesive contacts between the adsorbed proteins, or conformational changes in their molecules (Ivanov et al. 2005). In the case of BSA, the relatively great value of E'' is probably due to adhesive contacts between the molecules of this protein. An indication for the existence of such adhesive contacts is the fact that in the vicinity of the isoelectric point of BSA, where our experiments have been carried out, a *bilayer* (rather than monolayer) of BSA molecules is formed at the air/water interface (Engelhardt et al. 2012).

In another series of experiments, the bubble was first formed in a solution of BSA, which was subsequently exchanged with a β-casein solution (rather than with pure water) (compare experiment nos. 1 and 3 in Table 18.2). The surface tension σ drops from 55 mN/m for BSA to 48.5 mN/m after the phase exchange; the latter coincides with the value of σ for β-casein. However, the large storage modulus $E' = 42$ mN/m (approximately two times greater than that of β-casein) indicates that BSA molecules are still present at the interface after the phase exchange.

The above experiment was carried out also in the opposite order: first, β-casein was adsorbed on the bubble surface and then the β-casein solution was replaced with a BSA solution (compare experiment nos. 4 and 6 in Table 18.2). After the phase exchange, the values of both σ and E' are intermediate between the values for BSA and β-casein alone. In the two cases of *sequential* adsorption (experiment nos. 3 and 6 in Table 18.2), the dilatational loss modulus, E'', reaches its maximal values, indicating the formation of a thicker composite layer of the two proteins (see below).

Finally, the last two rows of Table 18.2 show data for the case of *parallel* adsorption of β-casein and BSA. In this case, σ is intermediate between its values for β-casein and BSA, whereas the values of the storage modulus E' are close to those for β-casein layers.

An additional illustration is given in Figure 18.4, where the solid line shows the surface tension relaxation in the case of parallel adsorption of the two proteins, whereas the symbols show the two cases of sequential adsorption: β-casein over a BSA layer and BSA over a β-casein layer. The contact of the BSA adsorption layer with a β-casein solution leads to a drop in the surface tension, which becomes close to that of the β-casein layer before the exchange (the lowest curve). In contrast, after the contact of a β-casein adsorption layer with a BSA solution, σ undergoes a small jump upward, but the higher value of σ for BSA alone has not been reached.

The above experiments on parallel and sequential adsorption of BSA and β-casein indicate the formation of mixed adsorption layers of the two proteins. The experiments with thin foam films (two interacting adsorption layers) bring additional information for the structure of these layers (see Section 18.3.2).

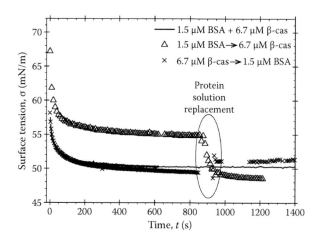

FIGURE 18.4 Comparison of surface tension, σ, versus time, t, curves in the cases of parallel and sequential adsorption from solutions of BSA and β-casein. In the cases of sequential adsorption, approximately 900 s after the beginning of the experiment, the solution of the first protein is replaced by a solution of the second protein using the device sketched in Figure 18.1.

18.3.2 Thickness and Stability of the Liquid Films

We carried out experiments with free foam films in the SE and flush cells at concentrations of BSA and β-casein corresponding to solution nos. 1, 3, 5, and 6 in Table 18.1 ($x_{\beta CS}$ = 0, 0.52, 0.82, and 1). In the closed cell, all films had a radius r_c = 100 μm, which was maintained by control of the applied pressure, P_c. [We recall the relation $P_c = 2\sigma R/(R^2 - r_c^2)$.] The thickness of these films (measured interferometrically) was greater than 100 nm for all investigated solutions. Such thick films have been observed in many experiments with protein solutions and explained with the long-range electrostatic (double layer) repulsion owing to the charge of the adsorbed protein molecules (Basheva et al. 2006; Dimitrova et al. 2004; Marinova et al. 1997).

To decrease the strong electrostatic repulsion in the investigated foam films, we added 1 mM NaCl to all solutions. This resulted in a decrease of the film thickness, h, from above 100 nm to approximately 80 nm for the films with β-casein and 41 nm for the films with BSA (see Table 18.3 and Figure 18.5, closed cell). The smaller thickness of the films from BSA evidences for a lower surface charge density. Note that for all investigated protein solutions with 1 mM NaCl, the Debye screening length is the same: κ^{-1} = 9.6 nm.

The soft character of the long-range double-layer repulsion leads to a sensitivity of the film thickness, h, to the applied pressure, P_c: The increase of P_c leads to smaller values of h. This is the reason for the different h values reported for films from BSA solutions with 1 mM NaCl, varying from 8 to 80 nm (Cascão-Pereira et al. 2003b; Wierenga et al. 2009a; Yampolskaya and Platikanov 2006), which most probably correspond to different P_c values. In our experiments, all measured values of the film thickness h correspond to a film radius r_c = 100 μm, which defines also

TABLE 18.3

Thickness, *h*, and Stability of Foam Films Formed from Solutions of BSA and β-Casein

No.	Solution	*h* (nm) in Closed Cell ($P_c \approx 85$ Pa)	*h* (nm) of Spot in Open Cell ($P_c \approx 5 \times 10^5$ Pa)
1	1.5 μM (0.01%) BSA	41	18 (film rupture)
2	6.7 μM (0.016%) β-casein	87	12 (stable film)
3	1.2 μM (0.01%) BSA + 1.3 μM (0.016%) β-casein	78	12 (stable film)
4	1.5 μM (0.01%) BSA after PhE with 6.7 μM (0.016%) β-casein	79	13 (stable film)
5	6.7 μM (0.016%) β-casein after PhE with 1.5 μM (0.01%) BSA	79	≥18 (film rupture)

Note: The solutions also contain 1 mM NaCl. In closed cell, all films are stable.

the value of the capillary pressure P_c at a given σ. Thus, in the closed cell, the range of applied pressures was $81 < P_c < 89$ Pa, the lowest limit being for β-casein alone, and the upper one, for BSA alone.

In the open cell, the pressure difference is much higher, $P_c \approx 5 \times 10^5$ Pa (see above). At such high pressures, the long-range electrostatic repulsion is overcome and the two protein adsorption layers come into a direct contact; that is, we are dealing with a steric stabilization. This is evidenced by the appearance of *dark spots*, that is, films of smaller thickness in the open cell—see the right column of photos in Figure 18.5 (observations in reflected light). In Figure 18.5b and h (BSA-dominated films), the spots have an equivalent water thickness of 18 nm, but the films ruptured soon after the spot appearance. In Figure 18.5d and f (β-casein-dominated films), the spots have a smaller thickness, $h = 12–13$ nm, and the films remained stable despite the spot expansion.

Each value of the film thickness h in Table 18.3 is the average from experiments with at least five films formed under the same conditions. The standard deviation of h, which is due to the reproducibility of the experiments, is ±3 nm. In Table 18.3, experiment nos. 1–3 have been carried out with the SE cell, whereas experiment nos. 4 and 5 have been carried out with the flush cell. Experiment no. 3 corresponds to parallel adsorption of the two proteins, whereas experiment nos. 4 and 5 correspond to sequential adsorption. The interpretation of these experiments is discussed in Section 18.4, together with the results on surface tension and dilatational rheology.

18.3.3 FOAMINESS AND FOAM STABILITY

We performed foam experiments with solution nos. 1, 4, 5, and 6 in Table 18.1. As mentioned above, the foaming process represents gas bubbling (0.5 L/min) through 50 mL of solution for 10 s. This resulted in the formation of foam with a height of

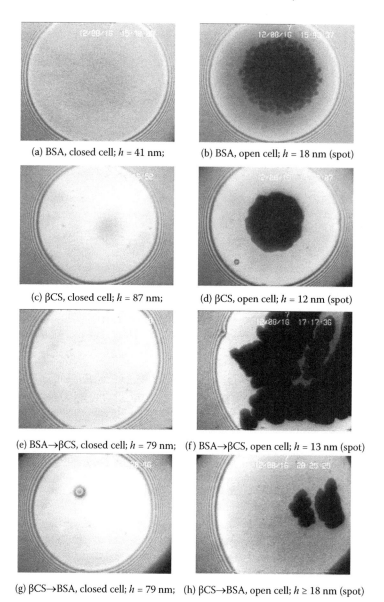

(a) BSA, closed cell; h = 41 nm;   (b) BSA, open cell; h = 18 nm (spot)

(c) βCS, closed cell; h = 87 nm;   (d) βCS, open cell; h = 12 nm (spot)

(e) BSA→βCS, closed cell; h = 79 nm;   (f) BSA→βCS, open cell; h = 13 nm (spot)

(g) βCS→BSA, closed cell; h = 79 nm;   (h) βCS→BSA, open cell; $h \geq$ 18 nm (spot)

FIGURE 18.5 Photographs of foam films formed from protein solutions. The four rows correspond to experiment nos. 1, 2, 4, and 5 in Table 18.3. The left column shows films in closed cell, whereas the right column shows films in open cell. All solutions contain 1 mM NaCl.

70–80 mm (Figure 18.6). The fact that the initial foam height is the same for all investigated solutions indicates that there is no bubble coalescence with the upper gas phase during the foaming.

At these relatively low protein concentrations (Table 18.1), all foams were unstable and disappeared within an hour. First, the foam volume decreases owing to the

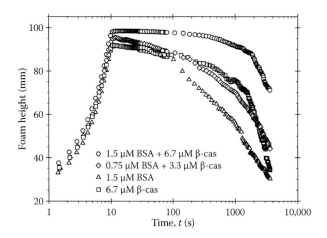

FIGURE 18.6 Height of the foam column versus time for protein solutions of concentrations shown in the figure. The foam is produced during the first 10 s by bubbling of air through the solution. Afterward, the foams gradually decay.

drainage of liquid from the foam. After that, intensive bubble coalescence begins, which is evidenced by the steeper regions of the experimental curves at longer times (Figure 18.6).

Most stable is the foam from the most concentrated solution (no. 6) (see the upper curve in Figure 18.6). Solution no. 4 represents the twice-diluted solution no. 6; this dilution leads to a considerably faster decay of the foam. This is probably due to the faster protein adsorption from the more concentrated solution, which leads to denser adsorption layers on the bubble surfaces and to a better protection of the bubbles against coalescence.

It is interesting that the foams from the less concentrated mixed solution no. 4 and from the more concentrated solution no. 5 (6.7 μM β-casein) drain almost in the same way (Figure 18.6). Hence, the mixing of the two proteins produces a stabilizing effect.

The foam from the solution of BSA alone exhibits the fastest decay, which correlates with the fact that, for this solution, the adsorption kinetics is the slowest among the investigated solutions and that the films rupture at the higher applied pressure (Table 18.3). It seems that the highest elasticity of the BSA adsorption layers (see the values of E' in Table 18.2) is insufficient to produce a stabilizing effect.

18.4   DISCUSSION ON THE STRUCTURE AND INTERACTIONS

18.4.1   ADSORPTION LAYERS FROM BSA

The shape of the BSA molecule is close to a prolate ellipsoid of length 14 nm and cross-sectional diameter 4 nm (Peters 1985; Wright and Thompson 1975). It has hydrophobic patches and pockets, which play the role of adhesion centers for the adsorption of BSA at air/water and oil/water interfaces, as well as in the interaction of BSA with surfactants (Díaz et al. 2003; Zhang et al. 2011). In the vicinity of the isoelectric point, the BSA adsorption layer represents a bilayer (Engelhardt et al. 2012). A foam film with BSA bilayers at its surfaces is sketched in Figure 18.7a. The equivalent water thickness of the black spots in the films from BSA solutions is $h = 18$ nm (Figure 18.5b and Table 18.3), which corresponds to a real thickness of $18 \times 1.33/1.45 \approx 16$ nm, where 1.45 is a typical value for the refractive index of proteins (Russev et al. 2000). The latter thickness corresponds to two bilayers (like those in Figure 18.7a) in contact: 4 layers × 4 nm = 16 nm. At 1 mM added NaCl, the appearance of BSA film of equivalent water thickness 18 nm leads to film destabilization and rupture. Similar results have been obtained by Cascão-Pereira et al. (2003b). At a higher salt concentration (25 mM NaCl), or at the lower pH = 5.2, the latter authors have observed stable films of thickness $h = 18$ nm, which thin down to $h = 8.6$ nm (approximately two monolayers) if the two film surfaces are pressed stronger against each other. In other words, the two inner layers of BSA (Figure 18.7a) are loosely attached and can be squeezed out of the film at higher pressures. The formation of a weakly attached sublayer has been observed also with β-casein at higher concentrations (Grigoriev et al. 2002).

The relatively high storage modulus, E', of the BSA adsorption layers (see experiment no. 1 in Table 18.2) can be attributed to the rigid packing of the polypeptide chain in the BSA molecules, which is stabilized by 17 disulfide bonds. In other

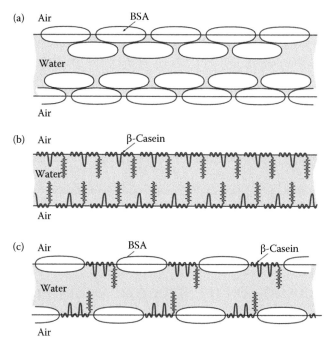

FIGURE 18.7 Models of foam films from solutions of (a) BSA, (b) β-casein, and (c) their mixture in the case of parallel adsorption, constructed on the basis of the data for σ, E', and h (Tables 18.2 and 18.3 and Figures 18.1 through 18.5).

words, the dilatational elasticity of the BSA adsorption layers (Figure 18.7a) is most probably due to the elastic response of the relatively rigid BSA molecules when subjected to compression or expansion.

18.4.2   Adsorption Layers from β-Casein

The β-casein molecule consists of a shorter hydrophilic part (≈20%) and a longer hydrophobic part (≈80%). The hydrophilic chain includes the amino acids at position from 1 to 43 at the N-terminus. This chain has 21 negative charges and 5 positive charges, so that its net charge is −16 (Farrell et al. 2001). If the length per amino acid residue is 0.3 nm (Horne 2006), then the length of the extended hydrophilic chain of β-casein is approximately 13 nm. However, in water, this chain forms an undulated tail of end-to-end distance that could be about twice shorter.

The remaining 80% of the β-casein molecule is very hydrophobic; its net charge is either +1 or +2, depending on the pH. The water is a poor solvent for this chain, so that the β-casein forms micelles of diameter ≈13 nm in aqueous solutions (Portnaya et al. 2006). Upon adsorption at the air/water interface, the hydrophobic chain of the β-casein molecule forms trains and loops, whereas the charged hydrophilic part protrudes as a tail in the water phase (Graham and Phillips 1979) (see Figure 18.7b). A compression of the β-casein adsorption layer results in the increase of the relative part of the loops with respect to those of the trains. The relatively low values of the storage modulus E' (see experiment nos. 4 and 5 in Table 18.2) means that these train-to-loop transformations are accompanied with relatively small changes in the elastic energy of the molecule (in comparison with the deformations of the more rigid BSA molecule). The 12 nm thickness of the thinnest films with β-casein (experiment no. 2 in Table 18.3) can be explained with the steric overlap repulsion (Israelachvili 2011), which is due to the protruding hydrophilic parts of the β-casein molecules (see Figure 18.7b).

18.4.3   Parallel Adsorption of BSA and β-Casein

In this case, the surface tension, σ, has intermediate values between those for BSA and β-casein, whereas the values of the storage modulus, E', are close to those for β-casein (see experiment nos. 7 and 8 in Table 18.2). In addition, the 12 nm film thickness at high pressure is identical with that for the case of β-casein alone (compare experiment nos. 2 and 3 in Table 18.3).

The above experimental facts can be explained with the attachment of both β-casein and BSA molecules at the air/water interface, as sketched in Figure 18.7c. This can explain the intermediate value of σ. Moreover, surface compressions and expansions would result in deformations of the adsorbed flexible chain of the β-casein (which undergoes train-to-loop transformations), whereas the more rigid BSA molecules are not deformed. This could explain why the values of E' are close to those for β-casein. Finally, the 12 nm thickness of the thinnest films with β-casein can be attributed to the steric overlap repulsion owing to the protruding hydrophilic portions of the β-casein molecules (compare Figure 18.7b and c).

18.4.4 SEQUENTIAL ADSORPTION OF BSA AND β-CASEIN

Let us begin with the case where the BSA adsorbs first, and then the aqueous phase is exchanged with a β-casein solution. After the exchange, the surface tension σ decreases from its value for BSA to that for β-casein, but E' remains considerably greater than for β-casein (see Figure 18.4 and experiment no. 3 in Table 18.2). At the higher pressure, the respective foam films are stable and have a thickness of 13 nm, almost the same as in the case of β-casein alone (compare experiment nos. 2 and 4 in Table 18.3).

These experimental results can be explained with the structural model sketched in Figure 18.8a. After the exchange of the aqueous phases, the β-casein molecules penetrate in the adsorption layer of BSA; their hydrophobic chains spread out and occupy the air/water interface. This can explain the lowering of the σ values. The BSA molecules can adhere by their hydrophobic spots to the trains of the spread β-casein molecules (Figure 18.8a), whereas the initially present second, loosely attached BSA layer (Figure 18.7a) is detached during the flushing with the β-casein solution. The presence of BSA in the adsorption layer can explain the registered E' values, which are markedly greater than those for β-casein. The 13 nm thickness of the thinnest films (experiment no. 4 in Table 18.3) can be explained with the steric overlap repulsion caused by the protruding hydrophilic portions of the β-casein molecules (compare Figures 18.7b and 18.8a).

Finally, let us consider the case where the β-casein adsorbs first, and then the aqueous phase is exchanged with a BSA solution. After the exchange, the surface tension σ is slightly greater than for β-casein solutions, but E' remains considerably

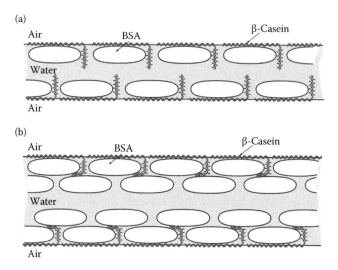

FIGURE 18.8 Structural models of foam films from BSA and β-casein in the case of sequential adsorption at the air/water interface. (a) The BSA is first adsorbed; the β-casein is adsorbed second; it penetrates between the BSA molecules and the interface, and spreads out its hydrophobic chains. (b) β-Casein is first adsorbed; BSA is adsorbed second and forms a bilayer over the β-casein layer.

greater than for β-casein (see Figure 18.4 and experiment no. 6 in Table 18.2). In addition, at the higher pressure, the respective foam films have a thickness close to that for BSA alone, and like them, they are unstable and rupture soon after the spot formation (compare experiment nos. 1 and 5 in Table 18.3).

These results can be explained with the structural model sketched in Figure 18.8b. After the exchange of the aqueous phases, the β-casein adsorption layer is brought into contact with a BSA solution. BSA molecules adsorb on the hydrophobic chains of β-casein, and afterward, a second BSA layer is formed just as in Figure 18.7a. Because the hydrophilic chains of the β-casein molecules are built in the BSA bilayer, the steric overlap repulsion caused by the β-casein chains is absent, and the film behavior is dominated by the BSA (Figure 18.8b), which leads to film rupture at the higher applied pressures as in the case sketched in Figure 18.7a.

18.5   SUMMARY AND CONCLUSIONS

In the present study, we combined surface tension and dilatational rheology measurements with experiments on TLFs to investigate the properties and structure of mixed adsorption layers from the proteins BSA and β-casein and the stability of the respective foam films. These two proteins, which find applications in aerated and emulsified foods, are rather different in structure and properties. The molecule of BSA is 2.8 times bigger than that of β-casein by mass and represents a rigid globule stabilized by 17 disulfide bonds. In contrast, the β-casein molecule has no disulfide bonds. It is a disordered protein consisting of 20% hydrophilic chain and 80% hydrophobic chain. For this reason, the β-casein is strongly amphiphilic: in the bulk, it forms micelles, whereas at the air/water interface, its hydrophobic part forms trains and loops, whereas its hydrophilic part protrudes as a tail into the water phase. Our goal is to investigate how these two quite different macromolecules interact in mixed adsorption layers and how the combination of their properties affects the stability of liquid films and foams.

Our experiments showed that in the investigated range of concentrations, the adsorption layers from β-casein have a lower surface tension and dilatational elasticity than the adsorption layers from BSA. At low applied pressures (in closed cell), the foam films from the aqueous solutions of both proteins are thick and electrostatically stabilized. However, at high applied pressures (in open cell), the films with β-casein are stable, whereas those with BSA are unstable.

In the case of mixed adsorption layers, we investigated the cases of parallel and sequential adsorption of the two proteins. The experiments show that adsorption layers and films of different properties and stability are obtained depending on the order of protein adsorption. The analysis of the whole set of data allowed us to draw conclusions about the possible structure of the formed adsorption layers. The experiments indicate that in the case of parallel adsorption, β-casein molecules are situated among the BSA molecules at the interface, thus lowering the surface tension and elasticity (Figure 18.7c). In the case of sequential adsorption, the situation is different depending on whether BSA or β-casein adsorbs first. If β-casein adsorbs first, the film behavior is dominated by the BSA, and the films are unstable under high applied pressures. In contrast, if BSA adsorbs first, the surface tension is relatively low, the surface elasticity is high, and the foam films are stable, which is the most favorable combination.

The results can be useful for the design and control of the properties of foams and emulsions stabilized by protein mixtures.

ACKNOWLEDGMENTS

The authors gratefully acknowledge the support from the National Science Fund of Bulgaria (grant no. DO-02-121/2009) and from the ESF COST Actions D43 and CM1101.

REFERENCES

Bantchev, G. B. and Schwartz, D. K. 2004. Structure of β-casein layers at the air/solution interface: atomic force microscopy studies of transferred layers. *Langmuir* 20:11692–7.

Basheva E. S., Gurkov, T. D., Christov, N. C. and Campbell, B. 2006. Interactions in oil/water/oil films stabilized by β-lactoglobulin; role of the surface charge. *Colloids Surf. A*, 282:99–108.

Basheva, E. S., Kralchevsky, P. A., Christov, N. C., Danov, K. D., Stoyanov, S. D., Blijdenstein, T. B. J. et al. 2011. Unique properties of bubbles and foam films stabilized by HFBII hydrophobin. *Langmuir* 27:2382–92.

Benjamins, J., Cagna, A. and Lucassen-Reynders, E. H. 1996. Viscoelastic properties of triacylglycerol/water interfaces covered by proteins. *Colloids Surf. A*, 114:245–54.

Cabrerizo-Vilchez, M. A., Wege, H. A., Holgado-Terriza, J. A. and Neumann, A. W. 1999. Axisymmetric drop shape analysis as penetration Langmuir balance. *Rev. Sci. Instrum.* 70:2438–44.

Cascão-Pereira, L. G., Theodoly, O., Blanch, H. W. and Radke, C. J. 2003a. Dilatational rheology of BSA conformers at the air/water interface. *Langmuir* 19:2349–56.

Cascão-Pereira, L. G., Johansson, C., Radke, C. J. and Blanch, H. W. 2003b. Surface forces and drainage kinetics of protein-stabilized aqueous films. *Langmuir* 19:7503–13.

Damodaran, S. 2004. Adsorbed layers formed from mixtures of proteins. *Curr. Opin. Colloid Interface Sci.* 9:328–39.

Danov, K. D., Radulova, G. M., Kralchevsky, P. A., Golemanov, K. and Stoyanov, S. D. 2012. Surface shear rheology of hydrophobin adsorption layers: laws of viscoelastic behaviour with applications to long-term foam stability. *Faraday Discuss.* 158:195–221.

Díaz, X., Abuin, E. and Lissi, E. 2003. Quenching of BSA intrinsic fluorescence by alkylpyridinium cations Its relationship to surfactant-protein association. *J. Photochem. Photobiol. A*, 155:157–62.

Dickinson, E. 2001. *An Introduction to Food Colloids*. Oxford University Press, Oxford.

Dickinson, E., Horne, D. S., Phipps, J. S. and Richardson, R. M. 1993. A neutron reflectivity study of the adsorption of beta-casein at fluid interfaces. *Langmuir* 9:242–8.

Dimitrova, T. D., Leal-Calderon, F., Gurkov, T. D. and Campbell, B. 2004. Surface forces in model oil-in-water emulsions stabilized by proteins. *Adv. Colloid Interface Sci.* 108:73–86.

Engelhardt, K., Rumpel, A., Walter, J., Dombrowski, J., Kulozik, U., Braunschweig, B. et al. 2012. Protein adsorption at the electrified air–water interface: implications on foam stability. *Langmuir* 28:7780–7.

Fainerman, V. B., Miller, R., Ferri, J. K., Watzke, H., Leser, M. E. and Michel, M. 2006. Reversibility and irreversibility of adsorption of surfactants and proteins at liquid interfaces. *Adv. Colloid Interface Sci.* 123:163–71.

Farrell Jr., H. M., Wickham, E. D., Unruh, J. J., Qi, P. X. and Hoagland, P. D. 2001. Secondary structural studies of bovine caseins: temperature dependence of β-casein structure as analyzed by circular dichroism and FTIR spectroscopy and correlation with micellization. *Food Hydrocolloids* 15:341–54.

Ferri, J. K., Gorevski, N., Kotsmar, Cs., Leser, M. E. and Miller, R. 2008. Desorption kinetics of surfactants at fluid interfaces by novel coaxial capillary pendant drop experiments. *Colloids Surf. A*, 319:13–20.

Ferri, J. K., Kotsmar, Cs. and Miller, R. 2010. From surfactant adsorption kinetics to asymmetric nanomembrane mechanics: Pendant drop experiments with subphase exchange. *Adv. Colloid Interface Sci.* 161:29–47.

Foegeding, E. A., Luck, P. J. and Davis, J. P. 2006. Factors determining the physical properties of protein foams. *Food Hydrocolloids* 20:284–92.

Graham, D. E. and Phillips, M. C. 1979. Proteins at liquid interfaces III. Molecular structures of adsorbed films. *J. Colloid Interface Sci.* 70:427–39.

Grigoriev, D. O., Fainerman, V. B., Makievski, A. V., Krägel, J., Wüstneck, R. and Miller, R. 2002. β-Casein bilayer adsorption at the solution/air interface: experimental evidences and theoretical description. *J. Colloid Interface Sci.* 253:257–64.

Hoorfar, M. and Neumann, A.W. 2006. Recent progress in axisymmetric drop shape analysis (ADSA). *Adv. Colloid Interface Sci.* 121:25–49.

Horne, D. S. 2006. Casein micelle structure: models and muddles. *Curr. Opin. Colloid Interface Sci.* 11:148–53.

Israelachvili, J. N. 2011. *Intermolecular and Surface Forces*, 3rd ed. Academic Press, New York.

Ivanov, I. B. Danov, K. D., Ananthapadmanabhan, K. P. and Lips, A. 2005. Interfacial rheology of adsorbed layers with surface reaction: on the origin of the dilatational surface viscosity, *Adv. Colloid Interface Sci.* 114:61–92.

Kotsmar, Cs., Pradines, V., Alahverdjieva, V.S., Aksenenko, E.V., Fainerman, V.B., Kovalchuk, V.I. et al. 2009. Thermodynamics, adsorption kinetics and rheology of mixed protein–surfactant interfacial layers. *Adv. Colloid Interface Sci.* 150:41–54.

Kralchevsky, P. A. and Nagayama, K. 2001. Particles at fluid interfaces and membranes. Chapter 14. Elsevier, Amsterdam.

Lucassen, J. and van den Tempel, M. 1972. Dynamic measurements of dilatational properties of a liquid interface, *Chem. Eng. Sci.* 27:1283–91.

Mackie, A. R., Gunning, A. P., Ridout, M. J., Wilde, P. J. and Morris, V.J. 2001. Orogenic displacement in mixed β-lactoglobulin/β-casein films at the air/water interfac. *Langmuir* 17:6593–8.

MacRitchie, F. 1998. Reversibility of protein adsorption. In: *Proteins at Liquid Interfaces*, eds. D. Möbius and R. Miller, 149–177. Elsevier, Amsterdam.

Maldonado-Valderrama, J., Martín-Molina, A., Martín-Rodriguez, A., Cabrerizo-Vílchez, M. A., Gálvez-Ruiz, M. J. and Langevin, D. 2007. Surface properties and foam stability of protein/surfactant mixtures: theory and experiment. *J. Phys. Chem. C*, 111: 2715–23.

Marinova, K. G., Gurkov, T. D., Velev, O. D., Ivanov, I. B., Campbell B. and Borwankar, R. P. 1997. The role of additives for the behaviour of thin emulsion films stabilised by proteins. *Colloids Surf. A*, 123, 155–67.

Marinova, K. G., Basheva, E. S., Nenova, B., Temelska, M., Mirarefi, A. Y., Campbell, B. et al. 2009. Physico-chemical factors controlling the foamability and foam stability of milk proteins: Sodium caseinate and whey protein concentrates. *Food Hydrocolloids* 23:1864–76.

McClements, D. J. 2005. *Food Emulsions: Principles, Practices, and Techniques*; 2nd edition. CRC Press, Boca Raton.

Merz, J., Burghoff, B., Zorn, H. and Schembecker, G. 2001. Continuous foam fractionation: Performance as a function of operating variables. *Sep. Purif. Technol* 82:10–18.

Nikolov, A. D. and Wasan, D. T. 1989. Ordered micelle structuring in thin films formed from anionic surfactant solutions. I. Experimental. *J. Colloid Interface Sci.* 133:1–12.

Peters, T., Jr. 1985. Serum albumin. *Adv. Protein Chem.* 37:161–245.

Portnaya, I., Cogan, U., Livney, Y. D., Ramon, O., Shimoni, K., Rosenberg, M. and Danino, D. 2006. Micellization of bovine β-casein studied by isothermal titration microcalorimetry and cryogenic transmission electron microscopy. *J. Agric. Food Chem.* 54:5555–61.

Rotenberg, Y., Boruvka, L. and Neumann, A. W. 1983. Determination of surface tension and contact angle from the shapes of axisymmetric fluid interfaces. *J. Colloid Interface Sci.* 93:169–83.

Russev, S. C., Arguirov, T. V. and Gurkov, T. D. 2000. Beta-casein adsorption kinetics on air-water and oil-water interfaces studied by ellipsometry. *Colloids Surf. B* 19:89–100.

Russev, S. C., Alexandrov, N., Marinova, K. G., Danov, K. D., Denkov, N. D. Lyutov, L. et al. 2008. Instrument and methods for surface dilatational rheology measurements. *Rev. Scientific Instrum.* 79:104102.

Saint-Jalmes, A., Peugeot, M.-L., Ferraz, H. and Langevin, D. 2005. Differences between protein and surfactant foams: microscopic properties, stability and coarsening. *Colloids Surf. A*, 263:219–25.

Scheludko, A. and Exerowa, D. 1959. Device for interferometric measurement of the thickness of microscopic foam films. *Comm. Dept. Chem. Bulg. Acad. Sci.* 7:123–32.

Sengupta, T. and Damodaran, S. 2000. Incompatibility and phase separation in bovine serum albumin/b-casein/water ternary film at the air–water interface. *J. Colloid Interface Sci.* 229:21–28.

Sheludko, A. 1967. Thin liquid films. *Adv. Colloid Interface Sci.* 1:391–464.

Sutherland, K. L. 1952. The kinetics of adsorption at liquid surfaces. *Aust. J. Sci. Res.* 5:683–96; http://dx.doi.org/10.1071/CH9520683.

Svitova, T. F., Wetherbee, M. J. and Radke, C. J. 2003. Dynamics of surfactant sorption at the air/water interface: continuous-flow tensiometry. *J. Colloid Interface Sci.* 261:170–9.

Svitova, T. F. and Radke, C. J. 2005. AOT and Pluronic F68 coadsorption at fluid/fluid interfaces: A continuous-flow tensiometry study. *Ind. Eng. Chem. Res.* 44:1129–38.

Tcholakova, S., Mitrinova, Z., Golemanov, K., Denkov, N. D., Vethamuthu, M. and Ananthapadmanabhan, K. P. 2011. Control of Ostwald ripening by using surfactants with high surface modulus. *Langmuir* 27:14807–19.

Wierenga, P. A., Basheva, E. S. and Denkov, N. D. 2009a. Modified capillary cell for foam film studies allowing exchange of the film-forming liquid. *Langmuir* 25:6035–9.

Wierenga, P. A., van Norél, L. and Basheva, E. S. 2009b. Reconsidering the importance of interfacial properties in foam stability. *Colloids Surf. A* 344:72–8.

Wright, A. K., Thompson, M. R. 1975. Hydrodynamic structure of bovine serum albumin determined by transient electric birefringence. *Biophys. J.* 15:137–41.

Yampolskaya, G. and Platikanov, D. 2006. Proteins at fluid interfaces: adsorption layers and thin liquid films. *Adv. Colloid Interface Sci.* 128:159–183.

Zhang, J., Gao, Y., Su, F. Gong, Zh. and Zhang, Y. 2011. Interaction characteristics with bovine serum albumin and retarded nitric oxide release of $ZCVI_4$-2, a new nitric oxide-releasing derivative of oleanolic acid. *Chem. Pharm. Bull.* 59(6) 734–41.

Zhang, Z., Dalgleish, D. G. and Goff, H. D. 2004. Effect of pH and ionic strength on competitive protein adsorption to air/water interfaces in aqueous foams made with mixed milk proteins. *Colloids Surf. B* 34:113–21.

19 New Trends in Phospholipid Research

Pierre-Léonard Zaffalon and Andreas Zumbuehl

CONTENTS

19.1 MECHANOSENSITIVE VESICLES

Phospholipids contain both a hydrophilic headgroup and a hydrophobic tail. In water, these molecules typically form closed spheres (liposomes or vesicles) of bilayer membranes in order to expose all polar headgroups to the water phase and exclude all hydrophobic carbohydrate chains from contacting water.

The enormous potential of these water-filled vesicles for medical and cosmetic applications led to an explosion of the field of "liposomology" with publication counts still rising. Today, more than 15 vesicle-based drug formulations are approved by the Food and Drug Administration such as Doxil for the treatment of various cancers and AmBisome against fungal infections (Torchilin 2005).

Once injected into a patient's bloodstream, the drug molecules are protected from the hostile environment and can, in theory, be delivered safely to the targeted tissue. This targeting of vesicles has been a major branch of nanotechnology research in recent years (Torchilin 2005). There are, however, many parts of the human body that cannot be reached by the classical biological targeting because, for example, no biomarkers are found to be upregulated. This is the case in stenosed segments of the human cardiovascular system. In a collaboration with the physicist Prof. Dr. Bert Müller and the physician Dr. Till Saxer, we asked ourselves, if not shear forces, what might be used as a purely physical means of targeting drugs to stenosed artery segments?

The shear forces found in a healthy human artery typically do not exceed 1.5 Pa (Cheng et al. 2007; Johnston et al. 2004, 2006). This value increases at least by an order of magnitude when going into a critically stenosed artery. Therefore, theoretically, it should be possible to exploit this difference and target a shear-sensitive nanocontainer to the site of a stenosis.

Twenty years ago, Giorgio et al. made a large unilamellar vesicle with a diameter of 100 nm by the extrusion technique (LUVET100). The egg phosphatidylcholine (eggPC) vesicles entrapped Ca^{2+} ions, and no leaking of these ions was found over time. When shear rates from 27 to 2700 s^{-1} were applied, the passive transbilayer Ca^{2+} transport increased linearly with increasing shear rates (Chakravarthy and Giorgio 1992). The authors predicted an increase in transbilayer permeability of two orders of magnitude if this system is applied in the human cardiovascular system. It was hypothesized that the shearing would increase the vesicle surface but that this force would not be big enough to lead to the disruption of the vesicle. Indeed, a rough calculation for LUVET100 at 37°C puts the critical force way above 40,000 s^{-1} (Bernard et al. 2005). In the case of Giorgio et al., the only liposome parameter that was found to be influenced was a reduction of the number of vesicle aggregates. Therefore, a likely explanation for the measured increase of passive transbilayer Ca^{2+} ion transport would be the attenuation of defects in the bilayer membrane through shear forces (Chakravarthy and Giorgio 1992).

The Ca^{2+} release is strongly dependent on the membrane composition: at room temperature, LUVET100 formulated from pure eggPC showed the biggest shear-induced permeability. If cholesterol was admixed, the intrinsic Ca^{2+} permeability dropped by 55% (Giorgio and Yek 1995). This suggests that the membrane organization has a big influence on the mechanosensitivity of vesicles.

Bernard et al. proposed an elegant solution to take advantage of the geometrical change of vesicles from being spherical under normal conditions to becoming lentil shaped when sheared. The lentil shape consists of essentially two membrane regions with low curvature (at the top and bottom of the vesicle) and high curvature (at the equator). A surfactant with a large headgroup and a small alkyl tail diameter, such as Brij S10, would preferentially partition into regions of high curvature because of the cone shape of the individual molecule. By doing so, regions of higher permeability would be created at the equator and increased passive transbilayer permeability of small molecules should be detectable. The team intended to use this feature in order to develop a postformulation loading protocol for LUVETs.

EggPC LUVET100, LUVET200, and LUVET400 containing 0.1 or 1 mol% of Brij S10 were exposed to shear rates of 0, 2000, 5000, and 10,000 s^{-1}. Shearing at room temperature for 20 min. has an influence on the size distribution of the vesicles. The average diameter of LUVET100 and LUVET200 increased, whereas the average diameter of LUVET400 decreased. This was found to be caused by vesicle fusion: in the presence of 1 mol% Brij S10, up to 10-fold increase of the LUVET100 diameter was measured. No such shear-induced fusion happened in the absence of Brij S10, pointing out the importance of the surfactant for introducing membrane perturbances (Bernard et al. 2005).

We were very pleased to see that our artificial phospholipid Pad-PC-Pad (**1**, see Figure 19.1) would formulate into lentil-shaped vesicles in the resting state (Holme et al. 2012). This morphology represents the geometry of eggPC/Brij S10 vesicles at high shear rates, and therefore, Pad-PC-Pad vesicles should be highly mechanosensitive. We hypothesized that the artificial Pad-PC-Pad would have a near-zero spontaneous curvature, a cylindrical Israelachvili packing factor, and a continuous series of intermolecular hydrogen bonds that would force the molecule

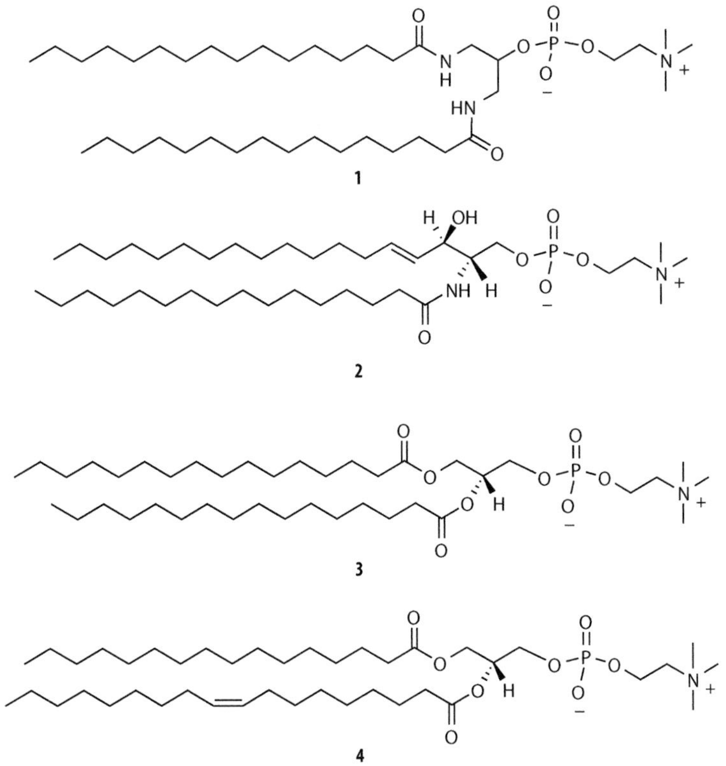

FIGURE 19.1 Chemical structures of the artificial phospholipid Pad-PC-Pad (**1**) bearing two amide bonds at positions 1 and 3 of the original glycerol backbone. The other structures represent the natural lipids N-palmitoyl-D-erythro-sphingosylphosphorylcholine (16:0 SM, **2**), containing one amide bond, DPPC (**3**), and 1-palmitoyl-2-oleoyl-sn-glycero-3-phosphocholine (POPC, **4**), the dominant component of eggPC. **3** and **4** contain no amide bonds, but contain ester bonds instead.

to have a preference for self-assembly into flat surfaces. A vesicle formulated above the phase transition temperature of 35°C would force the lipids into a spherical vesicle geometry that relaxes into a nonspherical shape upon rapid cooling to room temperature.

The self-quenching fluorescent dye 5(6)-carboxyfluorescein was loaded into Pad-PC-Pad LUVET100 and the vesicles were found to not leak their content for at least 1 week (Figure 19.2). However, vigorous shaking on a vortex shaker for 1 min was enough to induce a loss of over 40% of the entrapped fluorescent dye (Holme et al. 2012).

An equally high dye release upon shaking was found when LUVET100 were investigated, which had been formulated from pure eggPC (**4**). But even when the eggPC vesicles were resting on the bench, they spontaneously released over 80% of their content in 1 week. Other natural lipids such as 16:0 SM (**2**) or DPPC (**3**) formed vesicles that did not release any entrapped dye either spontaneously or when shaken vigorously.

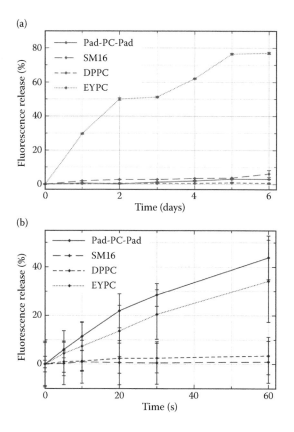

FIGURE 19.2 (a) Spontaneous leaking of 5(6)-carboxyfluorescein from LUVET100. (b) Leaking of the entrapped dye from LUVET100 induced by mechanical shaking.

We therefore suggest that there exist three different types of liposomes: (i) vesicles that are tight and do not release their content either spontaneously or when mechanically stressed (e.g., 16:0 SM or DPPC), (ii) vesicles that do not leak their content spontaneously but are leaky when mechanically stressed (e.g., Pad-PC-Pad), and (iii) vesicles that leak both spontaneously and when mechanically stressed (e.g., eggPC).

Clearly, the newly discovered vesicle type ii bears potential for new treatments for cardiovascular diseases on the basis of shear-induced, targeted drug delivery. In a first proof of concept, we have shown that indeed Pad-PC-Pad (**1**) vesicles released more of the entrapped 5(6)-carboxyfluorescein when passing through a plastic model of a heavily stenosed artery than when the vesicles passed through the model of a healthy artery.

19.2 PHOSPHOLIPASE A$_2$–TRIGGERED DRUG DELIVERY

Each year, cancer affects more than 10 million new patients worldwide. Several critical mutations on normal cells may lead to uncontrolled and rapid proliferation of malignant tumors. Because of poor selectivity and high toxicity of the large number of cytotoxic drugs available, the treatment of the disease is still challenging (Kaasgaard and Andresen 2010).

As early as the 1970s, the first liposomal drug carriers were developed by Gregoriadis et al. (Pedersen et al. 2009). The idea was to minimize the side effects of chemotherapy and to improve the therapeutic effect by delivering the drug specifically to the tumor. This could be achieved if the particle was stable in the bloodstream and could accumulate in the cancerous tissue by exploring the enhanced permeability and retention effect. Doxil, a sterically protected, stealth liposome containing doxorubicin, is a prominent example. The therapy could be enhanced even more if the drug would be released upon a specific trigger. Therefore, recently, Andresen et al. explored the idea of using phospholipase A_2, which is present in high concentration in human tumors, as a drug delivery trigger (Andresen et al. 2010).

The secretory enzyme phospholipase A_2 ($sPLA_2$) consist of a family of enzymes that catalyze the hydrolysis of the *sn*-2 ester bond of the glycerol backbone of phospholipids to provide the free fatty acids as lysophospholipids (Rosseto and Hajdu 2010). $sPLA_2$ are 14 kDa lipases with several subtypes, of which $sPLA_2$ IIA is the most studied. The mechanism of the enzyme requires Ca^{2+} as a co-factor and highly depends on the composition, morphology, and physicochemical properties of the phospholipid membrane (Arouri and Mouritsen 2012). Elevated levels of secretory phospholipase $sPLA_2$ IIA were found in different pathological tissues such as cancers, atherosclerosis, rheumatoid arthritis, and coronary heart disease.

The subtype $sPLA_2$ IIA is overexpressed in cancer cells; is detected in breast, stomach, colorectal, pancreatic prostate, and liver tumors (Pedersen et al. 2010a); and is found in concentrations "one or two orders of magnitude larger than serum levels and much higher than in the healthy tissue lining the tumor" (Mouritsen 2011).

The first ideas were to use $sPLA_2$ in order to destroy the drug-loaded liposomes by hydrolysis of the membrane phospholipid. The shape of the lipids would be changed, leading to a release of the drug. That would change the shape of the liposome and release the drug. Furthermore, lysoetherphospholipids (known for their effect on blood hemolysis) would be produced and could also act on the tumor cells (Mouritsen 2011).

The first application of drug release using secretory phospholipase PLA_2 was published by Andresen et al. in 2009 when a prodrug of the anticancer agent capsaicin (**5**, Figure 19.3) was covalently attached at the *sn*-2 position of a phospholipid (Linderoth et al. 2009). Upon the addition of snake venom $sPLA_2$ (*Agkistrodon piscivorus piscivorus*), a complete release of capsaicin was found after 5000 s (via matrix-assisted laser desorption/ionization time-of-flight mass spectrometry [MALDI-TOF MS]). A similar experiment with human sPLA2 IIA went to completion after 24 h. The mechanism of action for releasing the drug is slightly different from the ones described later because here the palmitic acid was hydrolyzed first followed by an attack of the freed *sn*-2 alcohol upon the *sn*-1 ester.

In 2010, Hajdu followed (Rosseto and Hajdu 2010) with the synthesis of *sn*-2 cleavable prodrugs of several nonsteroidal anti-inflammatory drugs (NSAID) like indomethacin (**6**), sulindac (**7**), and diclofenac (**8**, Figure 19.3). The corresponding drugs are well-established inhibitors of cyclooxygenases and are active against neurodegenerative diseases such as Alzheimer's disease. But similar to doxorubicin,

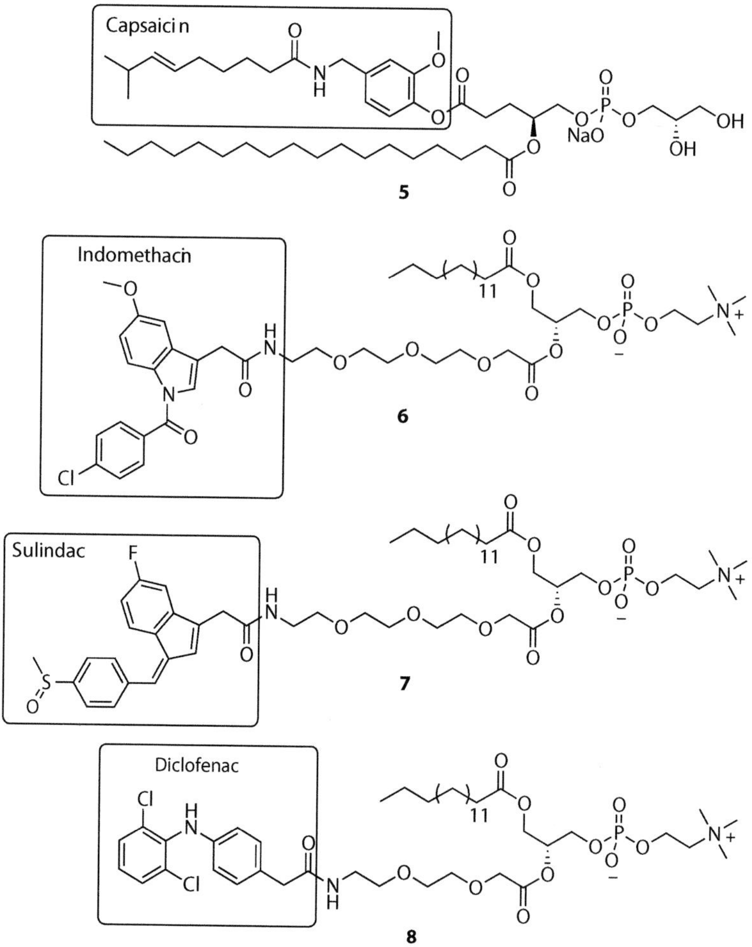

FIGURE 19.3 Phospholipase A$_2$ cleavable prodrugs of anticancer agents 5 and NSAIDs (**6–8**).

prolonged exposure to NSAIDs is characterized by gastrointestinal and renal toxicity. An improved delivery system would therefore significantly reduce the side effects of the drugs.

More complex systems were developed requiring multistep chemical syntheses (Pedersen et al. 2009): Pedersen et al. designed a prodrug of chlorambucil (**9**), a chemotherapeutic agent in use for the treatment of lymphocytic leukemia (Figure 19.4). The enzymatic hydrolysis (24 h at 37°C) of LUVET100 with snake venom sPLA$_2$ (*Naja mossambica mossambica*) led to the completed disintegration of the vesicles. However, no free chlorambucil was observed because of its low stability (half-life of 15 min in a HEPES buffer at 37°C). The liposomal formulation, on the other hand, proved to be very stable (6 weeks at 20°C), which should open new possibilities for storing sensitive drugs.

The antiproliferative prostaglandins were the third class of anticancer drugs evaluated (Pedersen et al. 2010b). 15-Deoxy-$\Delta^{12,14}$-PGJ$_2$ (used in molecule **10a**, Figure 19.4) was chosen because of its activity against leukemia cells and its high lipophilicity. After synthesis and hydration in HEPES, the molecules were formulated into LUVET100. Purified snake venom sPLA$_2$ from *N. mossambica mossambica* or *A. piscivorus piscivorus* was used and MALDI-TOF MS analysis showed complete degradation of **10b** after 24 h whereas **10a** was incompletely degraded

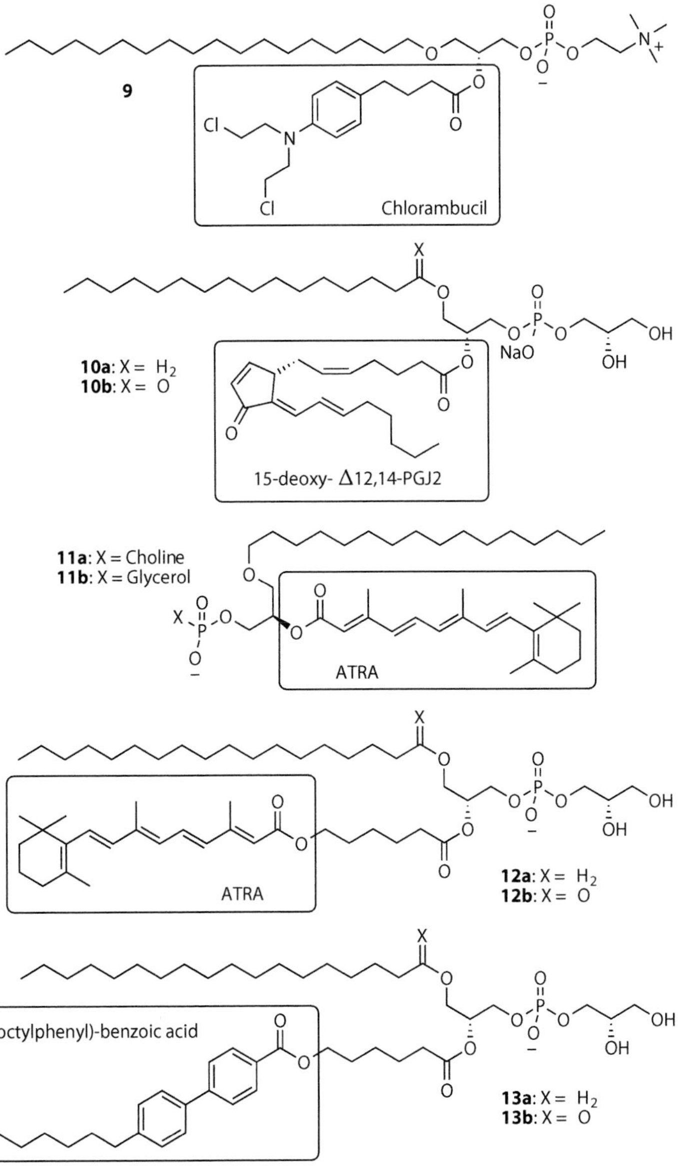

FIGURE 19.4   Additional phospholipase A$_2$ cleavable cancer prodrugs.

(70% of conversion) after 48 h. Furthermore, it was shown that the conjugates **7a** and **7b** were somewhat active even in the absence of PLA$_2$. A possible explanation was the cellular uptake of the prodrug lipid and its metabolic hydrolysis in the cytosol.

For a possible treatment of leukemia, the release of *all-(E)*-retinoic acid (ATRA) was investigated (Pedersen et al. 2010b). Christensen et al. introduced ATRA at the *sn*-2 position of the lipid before attaching a phosphocholine (molecule **11a**, Figure 19.4) or phosphoglycerol (molecule **11b**) headgroup (Christensen et al. 2010). Contrary to the findings with the previous substrates, both the PLA$_2$ activity of *N. mossambica mossambica* or *A. piscivorus piscivorus* was inhibited by the presence of ATRA in **11b**.

ATRA is a molecule rigidified by five double bonds with a methyl next to the carboxylic acid group. In comparison, natural fatty acids such as palmitic acid are saturated and highly flexible. In order to regain PLA$_2$ activity, an aliphatic C$_6$-linker (**12a–b**) was incorporated between the backbone of the phospholipid and the ATRA (Pedersen et al. 2010). The glycerol headgroup was preferred because sPLA$_2$ IIA has a strong affinity for negatively charged substrates. Furthermore, in order to avoid side effects due to the unselectivity of ATRA toward the retinoic receptors (RAR), 4-(4-octylphenyl)-benzoic acid (**13a–b**) was investigated, which acts as agonist of RARβ2 and has a suppressive effect for human cancer.

In conclusion, Andresen et al. introduced a fresh type of liposomal drug delivery on the basis of the activity of PLA$_2$ on phospholipids. A prodrug at the *sn*-2 position of an artificial phospholipid can be cleaved upon the hydrolytic action of the enzyme. As PLA$_2$ is upregulated in cancer tissue, the approach is a promising new tool in nanomedicine.

19.3 POLYMERIZABLE PHOSPHOLIPIDS

Natural plasma membranes are supported by the cellular cytoskeleton. This stabilization is missing in drug delivery liposomes, and therefore, these systems suffer from poor mechanical stability. If the individual phospholipid molecules could be polymerized, the rigidity of natural bilayers might be recreated. Therefore, phospholipid researchers have been interested in such systems for over 30 years.

The field started in the early 1980s with phospholipids containing polydiacetylene-modified chains. In the following years, the molecules were diversified by attaching various polymerizable groups (diacetylene, methacryloyl, butadiene, dienoyl, sorboyl) on the headgroup, near the backbone, and in the chains of phospholipids (Mueller and O'Brien 2002; Okada et al. 1998; Ringsdorf et al. 1988).

Today, the synthesis of artificial polymerizable phospholipids is mainly focused on hydrogel-forming molecules. The researchers are using already existing, commercially available molecules elaborated during the last decades, and only very few new syntheses were carried out as will be highlighted in the coming paragraphs (Monge et al. 2011; Puri and Blumenthal 2011).

In order to have a straightforward access route to polymerizable phospholipids, Singh et al. decided to start with a natural phosphatidylethanolamine and attach a styryl headgroup (Lawson et al. 2003a,b, 2005). Several lipids were prepared containing unsaturated (**14**, Figure 19.5) and saturated (**15**) chains. They were polymerized in the presence of an initiator at 65°C. The mechanical stability of the membrane and its resistance against detergent-induced lysis were enhanced but the membrane permeability (controlled by a calcein release experiment in vesicles) remained unchanged.

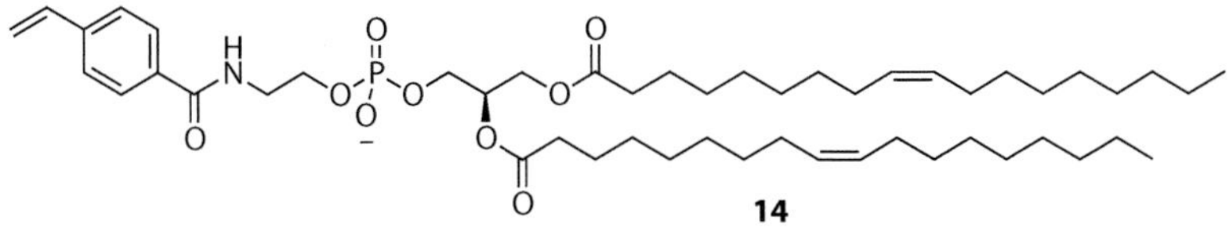

14

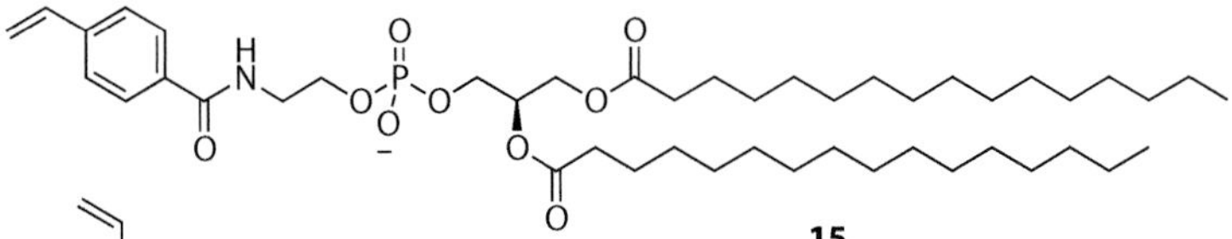

15

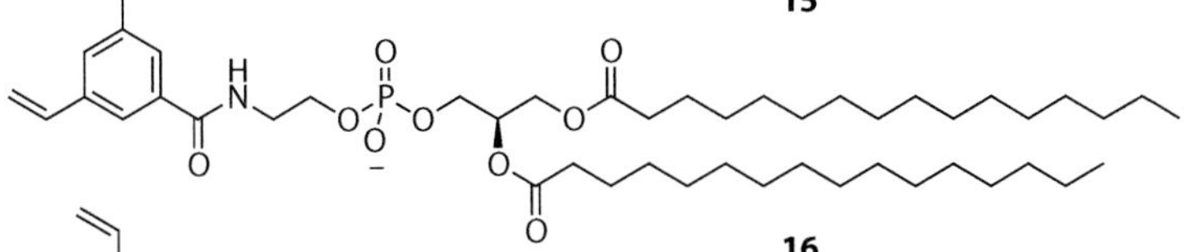

16

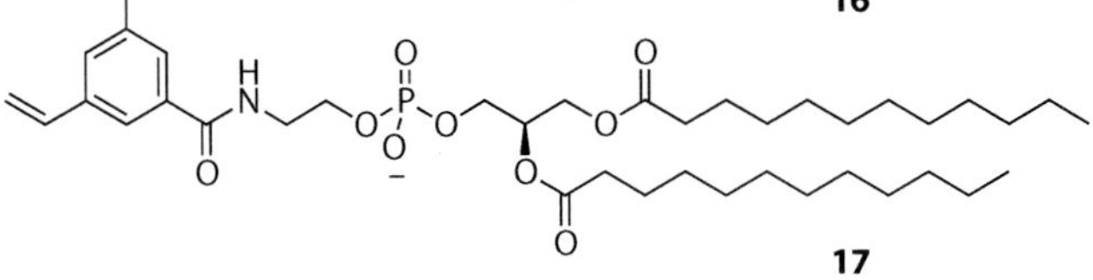

17

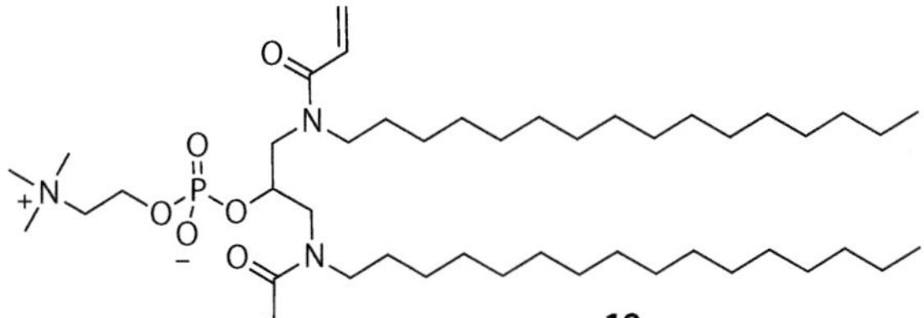

18

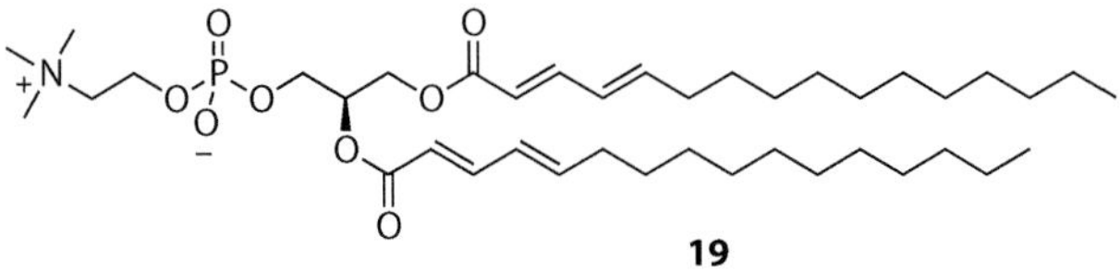

19

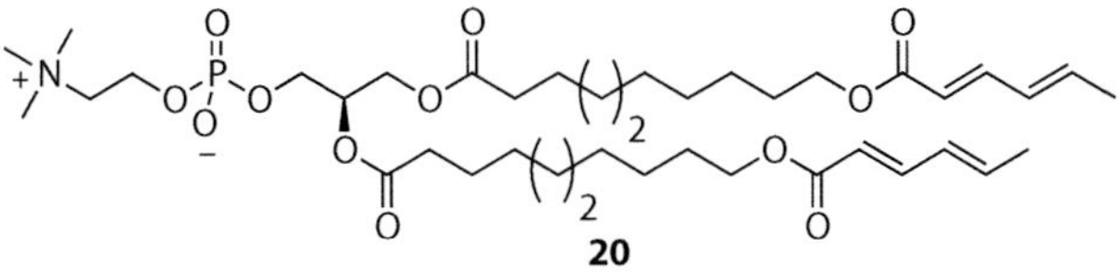

20

FIGURE 19.5 Recently synthesized polymerizable phospholipids.

In order to improve the stability, Singh et al. synthesized the lipids **16** and **17** by coupling dipalmitoyl and dilauroyl phosphoethanolamine with 3,5-divinylbenzoic acid. Polymerization was achieved with the addition of a radical initiator, 2,2′-azobis (2-aminodipropane)hydrochloride, followed by UV irradiation at 254 nm. The best results were obtained for **16** with 100% polymerization after 5 min of irradiation time at 25°C without any change in vesicle diameter during the polymerization process. Several types of enzymes were then successfully encapsulated and protected in such vesicles, opening new possibilities for enzyme stability and transportation.

Placing the polymerizable groups directly in the interface region between the hydrophobic tails and the hydrophilic headgroups should preserve both the headgroup chemistry and the tail fluidity. Therefore, we have synthesized a 1,3-diaminophospholipid containing two acrylamide moieties (**18**, Zaffalon et al. 2011). LUVET100 were polymerized with Irgacure 2959 under UV radiation. We did not obtain single polymerized vesicles; instead, we obtained large aggregates. Scanning electron microscopy revealed that the polymer consisted of intact vesicles linked together by polymerized molecules coming from destroyed vesicles. If drug-loaded vesicles would be used in the polymerization process, such a material might have potential in drug-eluting polymeric materials.

Recently, for reasons of absence of commercial suppliers, Jones et al. revived the field of phospholipids containing polymerizable tails and presented the synthesis of bisDenPC **19** and bisSorbPC **20** (Jones and Hall 2011). These types of molecules were first synthesized by O'Brien starting from glycerophosphocholine that was coupled with a long tail acid (Mueller and O'Brien 2002). Jones and Hall reverted the synthesis and attacked the hydrophobic core first. Encountering difficulties with the attachment of the unsaturated dienoyl tails for **19**, they used the Yamaguchi protocol with trichlorobenzoyl chloride to form an ester anhydride with (E,E)-2,4-octadecadienoic acid reacting with DMAP and (R)-(+)-3-benzyloxy-1,2-propanediol. The phosphocholine headgroup was then introduced using 2-chloro-2-oxo-1,3,2-dioxaphospholane and the Chabrier reaction with trimethylamine. The same procedure was applied for bisSorbPC **20** except that a dicyclocarbodiimide coupling was sufficient for the attachment of the tails. This small paragraph should stand as a reminder that the synthesis of phospholipids is not straightforward and ample possibilities exist for improving previous synthetic routes.

In conclusion, the polymerization of phospholipids has been extensively studied from the standpoint of both chemical synthesis and engineering. For the past 10 years, only few new syntheses have been reported. Challenges, however, remain in order to get access to cheap, synthetically straightforward and abundant polymerizable phospholipids.

19.4 "CLICKING" PHOSPHOLIPIDS

Among the way over 100 reaction types developed by chemists, only a handful are used by biochemists as of today. It is exceedingly difficult to engineer a reaction to be run with nontoxic, cheap chemicals, in environmentally benign solvents (e.g., water), with high selectivity and high yields. Nonetheless, modern chemical biology depends on reactions that are bioorthogonal: reactions that allow the "labeling of a target biomolecule[s] under ambient conditions within the complex environments of live cells and extracts" (Best et al. 2011).

One very popular reaction is the 1,3-dipolar cycloaddition between an alkyne (**21**, Figure 19.6) and an azide (**22**), better known as the Huisgen–Sharpless–Meldal (HSM) click reaction (Huisgen 1963; Lewis et al. 2002; Meldal and Tornøe 2008). The term *click* depicts the fact that two chemical reaction partners are essentially inert toward the environment but react readily with one another. The "click" concept allows to easily build complex molecules by combining small molecular building blocks.

The HSM click reaction was rapidly adopted in the field of soft matter research. And from the very beginning, the reaction was used to label phospholipids and as biomarkers in cell-based assays. An additional important advantage of the HSM click reaction was that it can be performed in aqueous buffer and it therefore was now

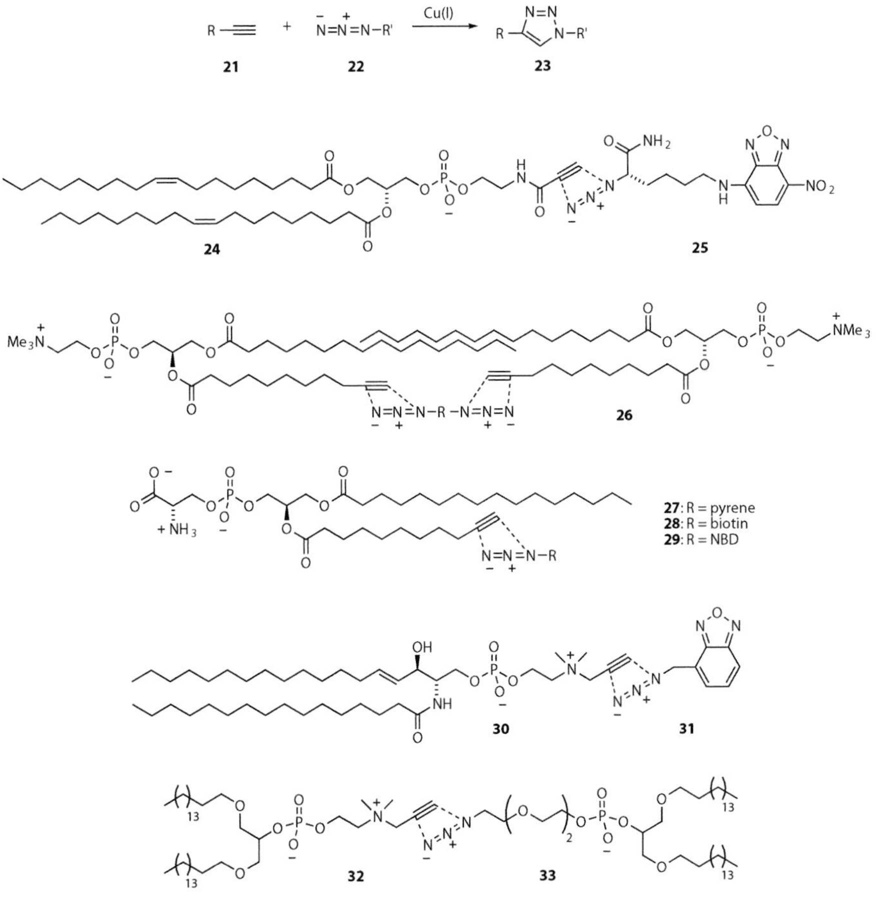

FIGURE 19.6 HSM click reaction between various alkynes and azides.

possible to modify phospholipids that were already formulated in a LUVET100. The first to use this concept were Kros et al. who introduced an alkyne-labeled DOPE (**24**) that they were able to formulate into DOPC liposomes. Then, they were able to click an externally added NBD-azide (**25**) onto the outer membrane layer of the liposomes (Cavalli et al. 2006). Since the inner layer of the vesicle bilayer membrane is protected from the environment, only phospholipids on the outer membrane are available for a reaction with a dye added externally. The reaction was cleverly monitored by Förster resonance energy transfer between the clicked NBD-phospholipid and a Lissamine Red–labeled phospholipid that had already been present in the liposomal membrane, and it was found that the reaction takes approximately 4 h in order to run to completion (Cavalli et al. 2006).

Bolaamphiphiles are amphophilic lipids containing two hydrophilic headgroups connected by ether linkages to a hydrophobic spacer. Making bolaamphiphiles is interesting because the corresponding liposomes should have high phase transition temperatures and should be less permeable to ions and small polar molecules compared to liposomes formulated from standard phospholipids (O'Neil et al. 2007). Indeed, in nature, bolaamphiphiles are found, for example, in *archaea*, living in hostile environments such as hot springs. In these environments, it is important to have a very tight membrane and therefore the membranes of *archaea* consist of lipids that span across both sheets of a membrane bilayer.

The chemical synthesis of bolaamphiphiles is complicated and therefore Smith et al. suggested linking together the hydrophobic parts of two normal phospholipids with an HSM click reaction (O'Neil et al. 2007). However, the resulting bolaamphiphiles (**26**) were found to disturb the integrity of the bilayer membrane when they were added to POPC/cholesterol LUVET100. This led to an increased leakage of 5(6)-carboxyfluorescein from the liposomes most probably due to the fact that the two polar 1,4-triazole units prefer a hydrophilic environment and therefore the bolaamphiphiles were not spanning both membrane layers but would fold upon themselves.

The beauty of the HSM click reaction lies in the many combinations that are possible. In an early example, Smith et al. synthesized a phosphatidylserine containing one alkyl terminated acyl chain (Lampkins et al. 2008). This single starting material could easily be labeled either with a pyrene group (**27**), a biotin tag (**28**), or an NBD label (**29**), providing a flexible library of labeled phospholipids.

Focusing on headgroup modifications, Cairo et al. were able to synthesize a sphingomyelin containing an alkyne (**30**) and preserving the trialkyl ammonium moiety (Sandbhor et al. 2009). The alkyne was clicked with a benzoxadiazole azide (**31**) to provide a fluorescently labeled sphingomyelin. In the presence of sphingomyelinase, this labeled headgroup was cleaved by hydrolysis of the phosphodiester and the released alcohol could be easily detected. Such a straightforward test for sphingomyelinase could be very beneficial for the early detection of sphingomyelinase D activity after spider bites, which can lead to massive tissue damage.

The technology was applied to living cells by Salic et al. (Jao et al. 2009): NIH 3T3 cells were incubated with propargyl choline, and this alcohol was successfully incorporated into natural phospholipids as seen by electrospray ionization tandem mass spectrometry. Adding an Alexa568-azide, the alkyne-terminated sphingomyelins underwent a HSM click reaction and the fluorescently labeled phospholipids could be tracked *in vivo*.

We were interested to see if it was possible to use the HSM reaction in order to click together vesicles and make three-dimensional vesicle networks (Loosli et al. 2012). We thought it would be nice to link together vesicles via a Huisgen–Sharpless click reaction. Therefore, we have synthesized an alkyno phospholipid (**32**) and an azido phospholipid (**33**) using the reactive P(III) reagent (benzyloxy)dichlorophosphine (Loosli et al. 2012; Zaffalon and Zumbuehl 2011). Unfortunately, neither of the two phospholipids formed vesicles by themselves, which is a standard problem in vesicle research.

Then, two populations of liposomes were formulated: Mix A contained 80 mol% eggPC, 10 mol% of the alkyno phospholipid, and 10 mol% of the positively charged (di-*n*-dodecyl)dimethylammonium bromide. Mix B was overall negatively charged and contained 90 mol% eggPC and 10 mol% of the azido-terminated phospholipid. The opposite charges on the vesicles were not enough of a driving force to induce vesicle aggregation. But the transient intervesicle contact could be trapped by the addition of CuBr, which would promote an HSM click reaction between the alkyno and the azido phospholipids. Finally, the resulting vesicle aggregates would fuse and form giant unilamellar vesicles. Overall, about 67 vesicles fused in this way. If the vesicles were formulated lacking one of the HSM click partners, only about 7 vesicles were found to aggregate and fuse upon the addition of the Cu(I) salt. This shows that without a proper intervesicle spacer, it is not possible to aggregate vesicles and always a fusion of vesicles is to be expected.

In conclusion, the HSM click reaction proves to be a valuable tool in phospholipid research. The reaction is compatible with the rather labile chemical groups present in phospholipids (esters and phosphodiesters), and because of their bioorthogonality, click reactions can be performed in living organisms.

Overall, the field of phospholipid research remains attractive for many scientific disciplines from biology, chemistry, and physics and many new discoveries can be expected to come from this truly interdisciplinary research.

REFERENCES

Andresen, Thomas L, David H Thompson, and Thomas Kaasgaard. 2010. Enzyme-Triggered Nanomedicine: Drug Release Strategies in Cancer Therapy. *Mol. Membr. Biol.* 27 (7): 353–363.

Arouri, Ahmad, and Ole G Mouritsen. 2012. Phospholipase A_2-Susceptible Liposomes of Anticancer Double Lipid-Prodrugs. *Eur. J. Pharm. Sci.* 45 (4): 408–420.

Bernard, Anne-Laure, Marie-Alice Guedeau-Boudeville, Valérie Marchi-Artzner, Thadeus Gulik-Krzywicki, Jean-Marc di Meglio, and Ludovic Jullien. 2005. Shear-Induced Permeation and Fusion of Lipid Vesicles. *J. Colloid Interface Sci.* 287 (1) (July): 298–306.

Best, Michael D, Meng M Rowland, and Heidi E Bostic. 2011. Exploiting Bioorthogonal Chemistry to Elucidate Protein–Lipid Binding Interactions and Other Biological Roles of Phospholipids. *Acc. Chem. Res.* 44 (9): 686–698.

Cavalli, Silvia, Alicia R Tipton, Mark Overhand, and Alexander Kros. 2006. The Chemical Modification of Liposome Surfaces via a Copper-Mediated [3 + 2] Azide–Alkyne Cycloaddition Monitored by a Colorimetric Assay. *Chem. Commun.* 3193–3195.

Chakravarthy, Srinivasa R, and Todd D Giorgio. 1992. Shear Stress-Facilitated Calcium Ion Transport Across Lipid Bilayers. *Biochim. Biophys. Acta* 1112 (2): 197–204.

Cheng, Caroline, Frank Helderman, Dennie Tempel, Dolf Segers, Beerend Hierck, Rob Poelmann, Arie van Tol, Dirk Duncker, Danielle Robbers-Visser, and Nicolette Ursem. 2007. Large Variations in Absolute Wall Shear Stress Levels Within One Species and Between Species. *Atherosclerosis* 195 (2): 225–235.

Christensen, Mikkel S, Palle J Pedersen, Thomas L Andresen, Robert Madsen, and Mads H Clausen. 2010. Isomerization of All-(*E*)-Retinoic Acid Mediated by Carbodiimide Activation—Synthesis of ATRA Ether Lipid Conjugates. *Eur. J. Org. Chem.* 2010 (4): 719–724.

Giorgio, Todd D, and S H Yek. 1995. The Effect of Bilayer Composition on Calcium Ion Transport Facilitated by Fluid Shear Stress. *Biochim. Biophys. Acta* 1239 (1): 39–44.

Holme, Margaret, Illya A Fedotenko, Daniel Abegg, Jasmin Althaus, Lucille Babel, France Favarger, Renate Reiter, Radu Tanasescu, Pierre-Léonard Zaffalon, André Ziegler, Bert Müller, Till Saxer, and Andreas Zumbuehl. 2012. Shear-Stress Sensitive Lenticular Vesicles for Targeted Drug Delivery. *Nature Nanotechnology* 7 (8): 536–543.

Huisgen, Rolf. 1963. 1,3-Dipolar Cycloadditions. Past and Future. *Angew. Chem., Int. Ed. Engl.* 2 (10): 565–598.

Jao, Cindy Y, Mary Roth, Ruth Welti, and Adrian Salic. 2009. Metabolic Labeling and Direct Imaging of Choline Phospholipids in Vivo. *Proc. Natl. Acad. Sci. U.S.A.* 106 (36): 15332–15337.

Johnston, Barbara M, Peter R Johnston, Stuart Corney, and David Kilpatrick. 2004. Non-Newtonian Blood Flow in Human Right Coronary Arteries: Steady State Simulations. *J. Biomech.* 37 (5): 709–720.

Johnston, Barbara M, Peter R Johnston, Stuart Corney, and David Kilpatrick. 2006. Non-Newtonian Blood Flow in Human Right Coronary Arteries: Transient Simulations. *J. Biomech.* 39 (6): 1116–1128.

Jones, Ian W, and H K Hall Jr. 2011. Demonstration of a Convergent Approach to UV-Polymerizable Lipids bisDenPC and bisSorbPC. *Tetrahedron Lett.* 52 (29): 3699–3701.

Kaasgaard, Thomas, and Thomas L Andresen. 2010. Liposomal Cancer Therapy: Exploiting Tumor Characteristics. *Expert Opin. Drug Deliv.* 7 (2): 225–243.

Lampkins, Andrew J, Edward J O'Neil, and Bradley D Smith. 2008. Bio-Orthogonal Phosphatidylserine Conjugates for Delivery and Imaging Applications. *J. Org. Chem.* 73 (16): 6053–6058.

Lawson, Glenn E, J J Breen, Manuel Marquez, Alok Singh, and Bradley D Smith. 2003a. Polymerization of Vesicles Composed of *N*-(4-Vinylbenzoyl)Phosphatidylethanolamine. *Langmuir* 19 (8): 3557–3560.

Lawson, Glenn E, Yongwoo Lee, and Alok Singh. 2003b. Formation of Stable Nanocapsules from Polymerizable Phospholipids. *Langmuir* 19 (16): 6401–6407.

Lawson, Glenn E, Yongwoo Lee, Frank M Raushel, and Alok Singh. 2005. Phospholipid-Based Catalytic Nanocapsules. *Adv. Funct. Mater.* 15 (2): 267–272.

Lewis, Warren G, Luke G Green, Flavio Grynszpan, Zoran Radić, Paul R Carlier, Palmer Taylor, M G Finn, and K Barry Sharpless. 2002. Click Chemistry in Situ: Acetylcholinesterase as a Reaction Vessel for the Selective Assembly of a Femtomolar Inhibitor From an Array of Building Blocks. *Angew. Chem., Int. Ed. Engl.* 41 (6): 1053–1057.

Linderoth, Lars, Günther H Peters, Robert Madsen, and Thomas L Andresen. 2009. Drug Delivery by an Enzyme-Mediated Cyclization of a Lipid Prodrug with Unique Bilayer-Formation Properties. *Angew. Chem., Int. Ed. Engl.* 48 (10): 1823–1826.

Loosli, Frédéric, David Alonso Doval, David Grassi, Pierre-Léonard Zaffalon, France Favarger, and Andreas Zumbuehl. 2012. Clickosomes—Using Triazole-Linked Phospholipid Connectors to Fuse Vesicles. *Chem. Commun.* 48 (10): 1604–1606.

Meldal, Morten, and Christian Wenzel Tornøe. 2008. Cu-Catalyzed Azide-Alkyne Cycloaddition. *Chem. Rev.* 108 (8): 2952–3015.

Monge, Sophie, Benjamin Canniccioni, Alain Graillot, and Jean-Jacques Robin. 2011. Phosphorus-Containing Polymers: A Great Opportunity for the Biomedical Field. *Biomacromolecules* 12 (6): 1973–1982.

Mouritsen, Ole G. 2011. Lipids, Curvature, and Nano-Medicine. *Eur. J. Lipid Sci. Technol.* 113 (10): 1174–1187.

Mueller, Anja, and David F O'Brien. 2002. Supramolecular Materials via Polymerization of Mesophases of Hydrated Amphiphiles. *Chem. Rev.* 102 (3): 727–758.

Okada, Sheldon, Susan Peng, Wayne Spevak, and Deborah Charych. 1998. Color and Chromism of Polydiacetylene Vesicles. *Acc. Chem. Res.* 31 (5): 229–239.

O'Neil, Edward J, Kristy M DiVittorio, and Bradley D Smith. 2007. Phosphatidylcholine-Derived Bolaamphiphiles via Click Chemistry. *Org. Lett.* 9 (2): 199–202.

Palle J Pedersen, Mikkel S Christensen, Tristan Ruysschaert, Lars Linderoth, Thomas L Andresen, Fredrik Melander, Ole G Mouritsen, Robert Madsen, and Mads H Clausen. 2009. Synthesis and Biophysical Characterization of Chlorambucil Anticancer Ether Lipid Prodrugs. *J. Med. Chem.* 52 (10): 3408–3415.

Palle J Pedersen, Sidsel K Adolph, Arun K Subramanian, Ahmad Arouri, Thomas L Andresen, Ole G Mouritsen, Robert Madsen, Mogens W Madsen, Günther H Peters, and Mads H Clausen. 2010a. Liposomal Formulation of Retinoids Designed for Enzyme Triggered Release. *J. Med. Chem.* 53 (9): 3782–3792.

Pedersen, Palle J, Sidsel K Adolph, Thomas L Andresen, Mogens W Madsen, Robert Madsen, and Mads H Clausen. 2010b. Prostaglandin Phospholipid Conjugates with Unusual Biophysical and Cytotoxic Properties. *Bioorg. Med. Chem. Lett.* 20 (15): 4456–4458.

Puri, Anu, and Robert Blumenthal. 2011. Polymeric Lipid Assemblies as Novel Theranostic Tools. *Acc. Chem. Res.* 44 (10): 1071–1079.

Rosseto, Renato, and Joseph Hajdu. 2010. Synthesis of Oligo(Ethylene Glycol) Substituted Phosphatidylcholines: Secretory PLA$_2$-Targeted Precursors of NSAID Prodrugs. *Chem. Phys. Lipids* 163 (1): 110–116.

Ringsdorf, Helmut, Bernhard Schlarb, and Joachim Venzmer. 1988. Molecular Architecture and Function of Polymeric Oriented Systems: Models for the Study of Organization, Surface Recognition, and Dynamics of Biomembranes. *Angew. Chem., Int. Ed. Engl.* 27 (1): 113–158.

Sandbhor, Mahendra S, Jessie A Key, Ileana S Strelkov, and Christopher W Cairo. 2009. A Modular Synthesis of Alkynyl-Phosphocholine Headgroups for Labeling Sphingomyelin and Phosphatidylcholine. *J. Org. Chem.* 74 (22): 8669–8674.

Torchilin, Vladimir P. 2005. Recent Advances with Liposomes as Pharmaceutical Carriers. *Nat. Rev. Drug Discov.* 4 (2): 145–160.

Zaffalon, Pierre-Léonard, and Andreas Zumbuehl. 2011. BODP—A Versatile Reagent for Phospholipid Synthesis. *Synthesis* 2011 (05): 778–782.

Zaffalon, Pierre-Léonard, Etienne Stalder, Illya A Fedotenko, France Favarger, and Andreas Zumbuehl. 2011. The Synthesis of an Amine-Bearing Polymerizable Phospholipid. *Tetrahedron Lett.* 52 (32): 4215–4217.

20 Thermodynamics and Specific Ion Effects in Connection with Micellization of Ionic Surfactants

Ana Kroflič, Bojan Šarac, and Marija Bešter-Rogač

CONTENTS

20.1 INTRODUCTION

Surfactants are widely used in many fields, ranging from the cosmetic, food, and pharmaceutical industries to the petroleum industry and oil recovery. In everyday life, their important role can be attributed especially to applications as cleaning, wetting, dispersing, emulsifying, foaming, antistatic, and bactericidal agents. Since all of these properties can be assigned to the surfactant characteristic amphophilic character, their behavior in aqueous solutions became a topic of interest for many researchers.

A surfactant molecule typically consists of a hydrophilic and a hydrophobic part, usually denoted as a polar head and non-polar tail. Therefore, surfactant monomers in water always tend to migrate to interfaces and orientate in such a manner that the polar group lies in the water or in aqueous solution and the non-polar part of the molecule lies in the oil phase or in the air. Above characteristic concentration, denoted as the critical micelle concentration (cmc), amphophilic monomers start to self-assemble in solution. The simplest self-assembled structures formed above cmc are spherical micelles, but at higher surfactant concentrations or at salt addition, more entangled and higher organized structures can also be found in solution such as prolonged and threadworm micelles, vesicles, liposomes, lamellar structures, and bilayers. Because of their simple structure and resemblance to the biological self-assemblied structures, surfactants are often taken as model systems for cell membranes and are also commonly used in biochemistry for membrane protein isolation and purification. Furthermore, liposome formation and microemulsion phenomenon became two general approaches for assessing drug delivery in pharmaceuticals and are nowadays also frequently used in cosmetic products.

The most accepted and scientifically sound surfactant classification is based on the nature of the polar headgroup and its dissociation in water. Anionic surfactants, as the most commonly used surface-active substances, dissociate in water on an amphophilic anion and a more or less simple cation (usually alkaline metal or quaternary ammonium). In contrast, cationic surfactants are dissociated into an amphophilic cation and mostly a halogen or organic-type anion in aqueous solution. If a single surfactant molecule exhibits both anionic and cationic group, it is zwitterionic. These surfactants are usually positively charged at low pH values and negatively charged at high pH values, although some of them do not dissociate in water in a wide pH range. Because of their high cost, as well as their high biological compatibility and low toxicity, the use of zwitterionic surfactants is limited to special applications. Further, as a consequence of possessing a non-dissociable hydrophilic group, non-ionic surfactants do not ionize in aqueous solutions at all, and polymeric surfactants exhibit a polyionic structure and represent the newest class of surface-active compounds. Typical members of anionic, cationic, zwitterionic, and non-ionic surfactants are shown in Figure 20.1.

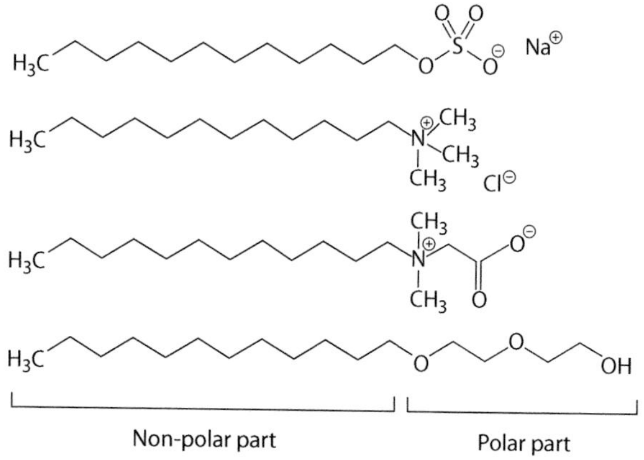

FIGURE 20.1   Schematic representation of anionic (sodium dodecyl sulfate), cationic (dodecyltrimethylammonium chloride), zwitterionic (dodecyldimethyl betaine), and non-ionic (diethylene glycol monododecyl ether) surfactants.

The above-mentioned structural discrepancies among surfactants also lead to differences in controlling forces of the micellization process in aqueous solution. The micellization of classical surfactants with a relatively small polar head and a long aliphatic tail is primarily driven by the hydrophobic effect, which is denoted as a positive entropy change accompanying the release of ordered water molecules from hydrophobic surfaces in the bulk. On the other hand, micellization is hindered by conformational restrictions upon aggregate formation. Besides weak dispersion forces that are of lesser importance for amphiphile self-assembly, electrostatic interactions also play an important role in ionic surfactant micellization. An intricate balance consists between repulsive electrostatic forces among equally charged headgroups of surfactant monomers upon micelle formation and favorable attractive interactions upon binding of ions from the bulk solution to the oppositely charged micelle surface. Furthermore, added salt in solution screens the repulsive electrostatic forces between surfactant monomers and together with bound counterions also lowers the effective net charge of the micelle.

In order to evaluate the importance of electrostatic interactions and specific ion effect at the ionic surfactant self-assembly, several attempts have been made to correlate the micellization parameters of distinct amphophilic ion and different types of counterions with a Hofmeister series of ions, which is partially presented in Figure 20.2. On the basis of protein (polymer) solubility, the Hofmeister series divides ions into two categories: (i) kosmotropes or water structure makers and (ii) chaotropes or water structure breakers. In general, relatively small polarizability, high surface charge density, strong electrostatic interactions, and large hydrated radii are typical for kosmotropes. Since their interaction with the charged micelle is often relatively weak, it is believed that they retain the hydrating water upon adsorbing to the micelle surface, which prevents their stronger binding. In contrast, chaotropes are characterized by significant polarizability, low surface charge density, and poor electrostatic interactions, but are loosely hydrated, so they lose their water of hydration easily and are usually strongly involved in ion binding to the micelle surface.

Furthermore, based on the Hofmeister series, (Collins 2004; Collins et al. 2007) proposed a simple law of matching water affinities and concluded that short-range forces are mostly important for interacting ions in water. Highly hydrated small ions (kosmotropes) obviously bind nearby water molecules so tightly that they can merely share their water shell with another kosmotrope and form a contact ion pair in solution.

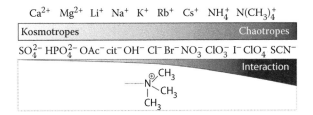

FIGURE 20.2 Typical ordering of ions in the Hofmeister series; kosmotropic and chaotropic ions and strength of interaction between anions and quaternary ammonium group. (Redesigned from Kunz, W., *Curr. Opin. Colloid Interface Sci.* 15, 34–39, 2010.)

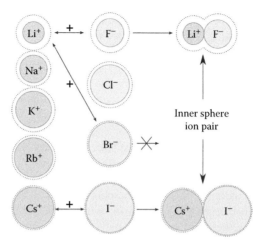

FIGURE 20.3 Ion size controls the tendency of oppositely charged ions to form contact ion pairs in aqueous solution. Dotted circles represent the hydrating water around ions. (Redesigned from Kunz, W., *Curr. Opin. Colloid Interface Sci.* 15, 34–39, 2010.)

Vice versa, chaotropes can only interact favorably with other loosely hydrated chaotrope ions in solution. Collins' law is schematically represented in Figure 20.3. Many biological processes involving ions in the surrounding medium should be thus dominated by their hydration–dehydration, which is a non-electrostatic component, and not by long-range electrostatic forces that were found to be of lesser importance. The law of matching water affinities will be verified in this chapter later on.

20.2 MICELLIZATION OF IONIC SURFACTANTS

Some molecular–thermodynamic theories for modeling the micellization of ionic surfactants in aqueous solution with added electrolytes have been proposed by different authors (Evans et al. 1984; Heindl and Kohler 1996; Nagarajan and Ruckenstein 1991; Srinivasan and Blankschtein 2003a,b). According to all these models, the fraction of counterions in solution binds onto the micelle surface and effectively reduces the micelle surface charge while the rest of the counterions are distributed in the diffuse region around the micelle. This molecular model of micellization takes into account various contributions to the free energy associated with surfactant self-assembly processes in salt aqueous solution, which can be divided into two groups, mainly: (i) the micelle core formation and (ii) the micelle–water interfacial shell formation. The free energy associated with the formation of the micelle core is related first to the transition of the lipophilic moieties from bulk water to a bulk hydrocarbon phase (favorable), then to a hydrocarbon–water interface formation (unfavorable), and finally to a loss of conformational degrees of freedom of non-polar species in solution (unfavorable). Furthermore, in terms of the micelle–water interfacial shell formation, the micellization is hindered by steric interactions between surfactant heads and adsorbed counterions, but favored by an entropic gain associated with mixing of the surfactant headgroups and bound counterions and by the most important reduction of electrostatic repulsion upon surfactant heads as a consequence of counterion binding to the micelle surface ("screening effect").

Counterions in solution either released by the surfactant ionic head or added by any electrolyte obviously play an important role in controlling the micellization process. Salt addition normally not only drastically decreases the cmc of ionic amphiphiles but also affects the size and shape of the self-assembled structures in solution. It should be the size and polarizability of counterions that determine their influence on the micelle formation besides the possible additional effect of a hydrophobic moiety if organic counterions are present in the surrounding medium. In the case of electrolyte addition composed of counterions that differ from the surfactant counterions, competitive condensation must also be considered in solution; therefore, we focus our attention mostly on the single-counterion systems. Otherwise, ion-specific effects are even more complex to study. Since the Hofmeister effect is known to be stronger at anions than at cations (regarding ion–water interactions, anions interact stronger with water molecules than cations as reported by Kunz 2010), we decided to pay greater attention to the micellization of cationic surfactants.

20.2.1 ALKYLTRIMETHYLAMMONIUM SURFACTANT MICELLIZATION: CMC, DEGREE OF MICELLE IONIZATION, AND AGGREGATION NUMBER

Alkyltrimethylammonium surfactants are composed of a long aliphatic tail and a positively charged quaternary ammonium headgroup (see Figure 20.1). They are usually used in the form of chloride or bromide salts, even though other alkyltrimethylammonium salts can be also synthesized. Most commonly studied cationic surfactants are composed of a linear 8- to 16-carbon-atom-long hydrophobic chain, whereas the ability of surfactant to self-assemble in aqueous solution improves substantially with an increase in the aliphatic tail as will be discussed later on. As regards Collins' law of matching water affinities (Figure 20.3), alkyltrimethylammonium cation is considered as a chaotrope and is thus expected to form contact ion pairs with other chaotropic counterions in solution (Vlachy et al. 2009). On the other hand, small kosmotropic anions are not foreseen to strongly interact with the surfactant quaternary ammonium headgroup as it is also demonstrated in Figure 20.2.

In the past few years, we have made an intense investigation on decyl-, dodecyl-, and tetradecyltrimethylammonium chloride surfactants (DeTAC, DTAC, and TTAC, respectively) in water and salt aqueous solutions. The temperature dependence of cmc and the degree of micelle ionization, β, is presented in Figure 20.4. The same micellization parameters (cmc and β) obtained by several researchers on a similar system at room temperature are gathered in Table 20.1 together with the reported aggregation numbers, n_{agg}. As already mentioned, here, only the data with a single counterion present in the surfactant surroundings are taken into consideration owing to the additional competition effect in the case of diverse ions in the solution. At this point, we should also stress out that all of the data obtained by different experimental techniques are in good agreement.

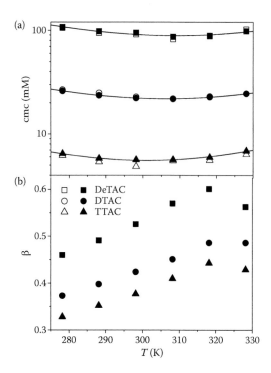

FIGURE 20.4 Temperature dependence of (a) cmc and (b) degree of micelle ionization, β, for DeTAC, DTAC, and TTAC in water as estimated by isothermal titration calorimetry (open symbols; Šarac and Bešter-Rogač 2009 and Kroflič et al. 2011) and conductivity measurements (solid symbols; Kroflič et al. 2012). Solid lines represent the second-order polynomial fits through the experimental data points from conductivity measurements.

A plot of temperature-dependent cmc typically exhibits a U-shaped form (Figure 20.4a), and several attempts have already been made to properly describe the observed dependence with a mathematical polynomial function (González-Pérez et al. 2005; Kim and Lim 2004). A temperature of the minimum, T_{min}, is usually close to room temperature and lowers slightly with an increase in the hydrophobicity of the surfactant as reported by Chen et al. (1998). The same trend was observed for alkyl-trimethylammonium chlorides in water by Perger and Bešter-Rogač (2007) and can be also seen in Figure 20.4a for the presented systems. In fact, the T_{min} decrease with the lengthening of the alkyl chain of the amphiphile may be connected to decreased solubility of the surfactant in water. Moreover, the characteristic T_{min} is related to the enthalpy–entropy compensation effect, which is typical for the micellization process and also agrees well with a temperature where the Gibbs free energy of micellization is of purely entropic contribution (the micellization enthalpy, $\Delta H_M = 0$), T_0, which will be discussed in more detail in the next paragraph.

The water of hydration of classical surfactant monomers is believed to play the most important role in their micellization process. At low temperature, water molecules are oriented at non-polar surfaces of surfactant monomers. The destruction of the water structure surrounding the hydrophobic part of the surfactant when it escapes from water and aggregates at micellization is an endothermic process (unfavorable). On the other hand, micellization is driven by an entropic gain upon release of ordered water molecules in the bulk when aggregates are formed. Furthermore, at high temperature or upon salt addition, the water structure around non-polar surfaces of surfactant monomers is disturbed; thus, the unfavorable enthalpic contribution upon micellization diminishes together with a less expressed increase in entropy. Afterward, the micellization becomes exothermic and is thus accompanied by the favorable enthalpic contribution at high temperature. As a matter of fact, the effect of polar group hydration–dehydration and electrostatic interactions are discussed later on.

In connection to the hydrophobic hydration effect, Zieliński (2001) also found the characteristic T_{min} to decrease with salt addition into octyltrimethylammonium bromide (OTAB) aqueous solutions. Evidently, this can be at least partially explained by the effect of electrolyte addition on the hydration of non-polar surfaces, since ions "disturb" the water shell around surfactant alkyl chains and thus favor the micellization at lower temperature.

Furthermore, it can be deduced from the collected data in Table 20.1 that cmc decreases (see also Figure 20.4a) and n_{agg} increases substantially with the lengthening of the surfactant hydrophobic tail. In fact, the short alkyl chain surfactants are more rigid in comparison to the long alkyl chain ones; hence, they are believed to form smaller and loosely packed aggregates in aqueous solution. Besides, water molecules are presumably present in such loose micelle interiors even at surfactant concentrations much above the cmc (Baar et al. 2001; Buchner et al. 2005). On the contrary, longer alkyl chain surfactants can form denser micelles of a higher aggregation number with water molecules completely expelled from their interior. The hydrophobic effect as a driving force of the micellization is thus more pronounced at the long hydrophobic tail surfactants; thus, the micelles can be formed at a lower surfactant concentration in solution compared to the shorter alkyl chain ones.

Additionally, the aggregation number of ionic surfactants is usually found to decrease with an increase in temperature. This trend has been reported by several authors for the ionic alkyltrimethylammonium surfactants (Bales and Zana 2002; Malliaris et al. 1985; Roelants and De Schryver 1987; Zieliński 2001), and it is believed that the dehydration of the polar groups with increasing temperature plays a crucial role. Consecutively, the repulsive electrostatic forces between the equally charged surfactant headgroups are strengthening and hinder denser micelle formation at higher temperature. At this point, it should be stressed out that n_{agg} usually increases with an increase in surfactant concentration in solution as well (Hansson et al. 2000), which indicates that the represented data for n_{agg} should be treated with care. Namely, it is often found very problematic to determine the micelle aggregation number around cmc and, even more, higher aggregated structures as prolonged micelles frequently appear in solution at increased surfactant concentration.

Moreover, the cmc decrease and n_{agg} increase can be observed in Table 20.1 at Cl^- or Br^- addition, which was also confirmed by the atomistic simulations of Jusufi

TABLE 20.1

Micellization Parameters (cmc, Degree of Micelle Ionization [β] and Aggregation Number [n_{agg}]) for Alkyltrimethylammonium Surfactants in Water and Salt Aqueous Solutions Obtained by Different Analytical Procedures and Measured or Modeled around Room Temperature

Literature	c_{salt} (mM)	T (K)	cmc (mM)	β	n_{agg}
OTAB					
Baar et al. 2001; Buchner et al. 2005; Gaillon et al. 1997	Water	298.15	284	0.35	20
Zieliński 2001	Water	298.15	293		
	0.25 M NaBr	298.15	228		
	0.50 M NaBr	298.15	195		
	0.75 M NaBr	298.15	159		
	1 M NaBr	298.15	135		
DeTAC					
Perger and Bešter-Rogač 2007; Roger et al. 2008	Water	298.15	94.7	0.53, 0.54	26
Kroflič et al. 2011 (Kroflič et al. 2012)	Water	298.15	86.4 (94.3)	0.50	
	0.01 M NaCl	298.15	87.4 (94.0)	0.52	
	0.1 M NaCl	298.15	67.0 (71.2)	0.57	
	1 M NaCl	298.15	18.0 (–)		
DeTAB					
Chakraborty and Moulik 2007; Ribeiro et al. 2004	Water	298.15	57.5, 65.6, 71	0.28, 0.30	40
Ray et al. 2005	Water	303.15	66.3, 66.9	0.31	
DTAOH					
Gaillon et al. 1999; Llanos and Zana 1983	Water	298.15	29.5, 30.5	0.63, 0.68, 0.74	20
DTAF					
Gaillon et al. 1999	Water	298.15		0.60	
DTAAc					
Gaillon et al. 1999	Water	298.15		0.58	
DTAC					
Baar et al. 2001; Buchner and Baar 2005; Gaillon et al. 1997, 1999; Perger and Bešter-Rogač 2007; Roger et al. 2008; Bales and Zana 2002	Water	298.15	20.3, 22.2, 22.7	0.37, 0.42, 0.46	32, 35, 45
Roelants and De Schryver 1987	Water	296.15	17		47
	0.02 M NaCl	296.15	12.8		48

(continued)

TABLE 20.1 (Continued)
Micellization Parameters (cmc, Degree of Micelle Ionization [β] and Aggregation Number [n_{agg}]) for Alkyltrimethylammonium Surfactants in Water and Salt Aqueous Solutions Obtained by Different Analytical Procedures and Measured or Modeled around Room Temperature

Literature	c_{salt} (mM)	T (K)	cmc (mM)	β	n_{agg}
	0.155 M NaCl	296.15	4.8		56
	0.5 M NaCl	296.15	2.4		65
Šarac and Bešter-Rogač 2009 (Kroflič et al. 2012)	Water	298.15	21.4 (22.6)	0.41	
	0.01 M NaCl	298.15	18.9 (19.35)	0.48	
	0.1 M NaCl	298.15	8.37 (8.55)	0.52	
	1 M NaCl	298.15	1.42 (–)		
DTAB					
Baar et al. 2001; Gaillon et al. 1997, 1999; Hansson et al. 2000; Ribeiro et al. 2004	Water	298.15	13, 15.0, 15.34, 15.5	0.23, 0.27	47, 55, 56
Ray et al. 2005	Water	303.15	14.8, 15.7	0.28	63
Bales and Zana 2002	Water	298.15	14.9	0.26	55
	0.008 M NaBr	298.15			63
	0.074 M NaBr	298.15			72
DTANO₃					
Gaillon et al. 1999	Water	298.15		0.22	
DTABen					
Gaillon et al. 1999	Water	298.15		0.20	
DTASal					
Srinivasan and Blankschtein 2003b	Water	298.15	1.68	0.23	
TTAOH					
Gaillon et al. 1999; Llanos and Zana 1983	Water	298.15	5, 7.2	0.66, 0.69	42
TTAF					
Gaillon et al. 1999	Water	298.15		0.58	
TTAAc					
Gaillon et al. 1999	Water	298.15		0.54	
TTAC					
Gaillon et al. 1999; Perger and Bešter-Rogač 2007; Roger et al. 2008	Water	298.15	5.63	0.37, 0.38, 0.40	52
Roelants and De Schryver 1987	Water	297.15			66

(*continued*)

TABLE 20.1 (Continued)
Micellization Parameters (cmc, Degree of Micelle Ionization [β] and Aggregation Number [n_{agg}]) for Alkyltrimethylammonium Surfactants in Water and Salt Aqueous Solutions Obtained by Different Analytical Procedures and Measured or Modeled around Room Temperature

Literature	c_{salt} (mM)	T (K)	cmc (mM)	β	n_{agg}
Kroflič et al. 2011 (Kroflič et al. 2012)	Water	298.15	5.06 (5.54)	0.36	
	0.01 M NaCl	298.15	2.90 (3.14)	0.40	
	0.1 M NaCl	298.15	0.80 (0.98)	0.33	
	1 M NaCl	298.15	0.14 (–)		
TTAB					
Gaillon et al. 1999; Hansson et al. 2000; Lah et al. 2000; Ribeiro et al. 2004; Ruiz et al. 2007	Water	298.15	3.5, 3.5, 3.6, 3.75, 3.94	0.20, 0.23, 0.27	60, 70, 74
Ray et al. 2005	Water	303.15	3.94, 4.08	0.27	
Dahirel et al. 2010	0.01 M NaBr	298.15	1.0		30
	0.05 M NaBr	298.15	0.5		52
	0.2 M NaBr	298.15	0.1		70
TTANO$_3$					
Gaillon et al. 1999; González-Pérez et al. 2004	Water	298.15	3.38	0.19, 0.24	
TTABen					
Gaillon et al. 1999	Water	298.15		0.18	
TTASal					
Srinivasan and Blankschtein 2003b	Water	298.15	0.35	0.22	
CTAOH					
Gaillon et al. 1999; Llanos and Zana 1983	Water	298.15	1.8	0.64, 0.7	46
CTAF					
Gaillon et al. 1999; Jiang et al. 2005	Water	298.15	1.54, 1.62	0.48, 0.56	
CTAAc					
Gaillon et al. 1999	Water	298.15		0.51	
CTAC					
Gaillon et al. 1999; Jiang et al. 2005	Water	298.15	1.15, 1.16	0.35, 0.38	

(continued)

TABLE 20.1 (Continued)
Micellization Parameters (cmc, Degree of Micelle Ionization [β] and Aggregation Number [n_{agg}]) for Alkyltrimethylammonium Surfactants in Water and Salt Aqueous Solutions Obtained by Different Analytical Procedures and Measured or Modeled around Room Temperature

Literature	c_{salt} (mM)	T (K)	cmc (mM)	β	n_{agg}
Roelants and De Schryver 1987	Water	296.15			89
	0.050 M NaCl	296.15			107
	0.104 M NaCl	296.15			136
	0.155 M NaCl	296.15			176
Wang and Larson 2009	Water	300		0.37	100
	0.13 M NaCl	300		0.32	94
	0.54 M NaCl	300		0.25	98
	1.08 M NaCl	300		0.21	105
CTAB					
Baar et al. 2001; Gaillon et al. 1997, 1999; Lah et al. 2000; Hansson et al. 2000; Jiang et al. 2005; Ribeiro et al. 2004	Water	298.15	0.78, 0.9, 0.94, 0.95, 0.96, 0.99	0.17, 0.24, 0.33	104, 90, 95
Ray et al. 2005	Water	303.15	0.93, 0.92	0.27	61
Jakubowska 2010	Water	298.15	0.92	0.35	
	0.002 M KBr	298.15	0.41	0.47	
CTANO$_3$					
Gaillon et al. 1999; Jiang et al. 2005	Water	298.15	0.89	0.15, 0.31	
CTASO$_4$					
Gaillon et al. 1999; Jiang et al. 2005	Water	298.15	0.54, 0.61	0.29	
CTABen					
Gaillon et al. 1999	Water	298.15		0.14	
DTASal					
Srinivasan and Blankschtein 2003b	Water	298.15	0.07	0.29	

Note: The data are presented for octyl- (OTAX), decyl- (DeTAX), dodecyl- (DTAX), tetradecyl- (TTAX), and hexadecyl- (CTAX) trimethylammonium hydroxide (X = OH), fluoride (X = F), acetate (X = Ac), chloride (X = Cl), bromide (X = Br), nitrate (X = NO$_3$), sulfate (X = SO$_4$), benzoate (X = Ben), and salicylate (X = Sal).

et al. (2009) recently. As a matter of fact, this effect is much more pronounced at ionic as at non-ionic surfactants, where lower solubility of the hydrophobic moiety in salt solution results in a slight cmc decrease. Evidently, electrostatic interactions are of great importance at the micellization of ionic surfactants. However, the intricate balance between opposing effects of salt addition on the micellization is assumed since added electrolyte screens all of the electrostatic interactions present in the solution (both repulsive and attractive ones). Added salt thus favors the micellization because of screening of the repulsive forces between the equally charged headgroups of surfactant monomers upon micelle formation. It lowers the cmc consequently and also enables the formation of denser aggregates since it simultaneously lowers the effective surface charge of the micelle. In addition, counterion condensation on the charged micelle surface is accompanied by an exothermic contribution that favors the micellization, although the counterion binding $(1-\beta)$ does not always increase at higher salt concentration, which can arise from the lowered micelle surface charge at higher ionic strength as will be discussed later. On the other hand, electrolyte in solution disturbs the ordered water structure at non-polar surfaces and thus weakens the hydrophobic effect upon micelle formation, which hinders the micellization process of ionic surfactants as it has been mentioned already.

A simple empirical law that describes the effect of added salt on the cmc was formerly developed for non-ionic (Ray and Nemethy 1971) and also applied on ionic (Kroflič et al. 2011) surfactants:

$$\log(\text{cmc/mM}) = \text{constant} - k_s \cdot (c_{salt}/\text{M}), \tag{20.1}$$

where k_s represents a salt–surfactant–specific parameter and c_{salt} is the salt concentration. Recently, the relation has been found applicable for DeTAC ($k_s \approx 70$ in the temperature range between 278.15 and 328.15 K) but not for DTAC and TTAC (Kroflič et al. 2011).

Whereas the cmc and n_{agg} behavior with increased alkyl chain length, temperature, and salt addition can be somehow predicted for ionic surfactants, this cannot be claimed for β dependence on electrolyte addition. As a matter of fact, it was found that β either decreases (Wang and Larson 2009; Zieliński 2001) or increases (Jakubowska 2010; Jusufi et al. 2009; Kroflič et al. 2012) with an increase in salt concentration or even stay at the same level (Kroflič et al. 2012). Moreover, the observed trends obviously depend on the type and concentration range of the electrolyte added (Dutkiewicz and Jakubowska 2002). At first glance, the decrease in β upon salt addition can be easily explained, since the greater number of counterions is believed to bind to the micelle surface at higher ionic strength and thus the repulsion between headgroups is screened. On the contrary, the increased β could also be assigned to the increased charge screening by the diffuse layer around the micelle at higher electrolyte concentration. Consequently, the micelle's effective charge lowers and fewer counterions bind to the aggregate surface. Further, β usually decreases with the lengthening of the surfactant's non-polar tail (Baar et al. 2001; Ray et al. 2005; Ribeiro et al. 2004), which can be ascribed to higher surface charge density of more compact micelles with higher aggregation number. Inversely, β increases with an increase in temperature (Bales and Zana 2002; Perger and Bešter-Rogač 2007;

Roger et al. 2008; Zieliński 2001) since the formation of looser micelles with a lower aggregation number and less net surface charge is assumed at higher temperature. The β dependence on alkyl chain length, temperature, and salt addition is also shown in Figure 20.4b and Table 20.1.

20.2.2 ALKYLTRIMETHYLAMMONIUM SURFACTANT MICELLIZATION: THERMODYNAMIC ANALYSIS

Micellization is a favorable process usually accompanied by a large negative Gibbs free-energy change. It was also found that the standard Gibbs free energy of micellization, ΔG_M^0, does not vary drastically with temperature or salt addition, which is believed to be a direct consequence of the enthalpy–entropy compensation effect typical for the micellization process. As it has already been briefly discussed before, the standard micellization enthalpy, ΔH_M^0, decreases with an increase in temperature (thus, the standard heat capacity change upon micellization, $\Delta c_{p,M}^0$, is always negative, which is usual for the hydrophobic effect as a major driving force of the micellization process, which will be discussed later), and the standard entropy contribution to the micellization, ΔS_M^0, diminishes simultaneously, which results in a slight temperature dependence of ΔG_M^0 values. At T_0 (or T_{min}), the contribution of the enthalpy to the Gibbs energy is neglected compared to the entropy term.

The standard thermodynamic parameters of micellization process (ΔG_M^0, ΔH_M^0, and ΔS_M^0) for several alkyltrimethylammonium amphiphiles at different temperatures are gathered in Table 20.2. Evidently, these values confirmed the already mentioned assertion that micellization is more favorable for the long alkyl chain surfactants in comparison to the short ones (the same trend was also observed by Beyer et al. 2006). This is indeed in agreement with the assumption of denser aggregate formation of the former where more water molecules are obviously expelled from the micelle interior (positive entropy change) and more counterions are bound to the micelle surface owing to its higher surface charge density (negative enthalpy change). Moreover, a consistent change in all of the determined thermodynamic parameters of micellization was actually observed for every CH_2 group addition to the surfactant tail at any temperature. Stodghill et al. (2004) reported a 6.3 kJ mol^{-1} decrease in ΔG_M^0 and a 3.8 kJ mol^{-1} decrease in ΔH_M^0 upon every CH_2 addition to the surfactant alkyl chain for dodecyltrimethylammonium bromide (DTAB), tetradecyltrimethylammonium bromide (TTAB), and cetyltrimethylammonium bromide (CTAB) at 301.15, 303.15, and 308.15 K. In parallel, they also observed a 7.9 J mol^{-1} K^{-1} increase in ΔS_M^0, which is in accordance with the enthalpy–entropy compensation effect. Similarly, Ray et al. (2005) studied the micellization of decyltrimethylammonium bromide (DeTAB), DTAB, TTAB, and CTAB at 303 K and reported the succeeding contributions of the CH_2 addition to the thermodynamic parameters: $\Delta G_M^0(CH_2) = -3.26$ kJ mol^{-1}, $\Delta H_M^0(CH_2) = -1.04$ kJ mol^{-1}, and $\Delta S_M^0(CH_2) = 7.23$ kJ mol^{-1} K^{-1}. Beyer et al. (2006) also published comparable increments for DTAB, TTAB, and CTAB in 0.1 M NaCl obtained by isothermal titration calorimetry (ITC) in the temperature range from 293.15 to 333.15 K [$\Delta G_M^0(CH_2) = -2$ kJ mol^{-1}, $\Delta H_M^0(CH_2) = -1.2$ kJ mol^{-1}, and $\Delta S_M^0(CH_2) = 10$ kJ mol^{-1} K^{-1}].

TABLE 20.2

Standard Thermodynamic Parameters (Standard Gibbs Free Energy [ΔG_M^0], Enthalpy [ΔH_M^0], and Entropy [ΔS_M^0]) for Micellization of Octyl-, Decyl-, Dodecyl-, Tetradecyl-, and Cetyltrimethylammonium Chloride, Bromide, Fluoride, Nitrate, and Sulfate in Water at Different Temperatures

T (K)	ΔG_M^0 (kJ mol^{-1})	ΔH_M^0 (kJ mol^{-1})	ΔS_M^0 (J mol^{-1} K^{-1})	ΔG_M^0 (kJ mol^{-1})	ΔH_M^0 (kJ mol^{-1})	ΔS_M^0 (J mol^{-1} K^{-1})
	OTAC			**OTAB** Zieliński 2001		
293.15	No data found in the literature			−13.3	5.9	66
298.15				−13.6	4.9	62
303.15				−13.9	3.8	59
308.15				−14.2	2.7	55
313.15				−14.5	1.5	51
318.15				−14.7	0.1	47
323.15				−14.9	−1.3	42
328.15				−15.1	−2.8	38
	DeTAC Perger and Bešter-Rogač 2007 (Kroflič et al. 2011)			**DeTAB** Ray et al. 2005		
278.15	−22.26 (−22.31)	10.34 (12.6)	117.2 (125.5)			
283.15	−22.73	9.42	113.5			
288.15	−22.9 (−23.05)	8.26 (9.1)	108.2 (111.5)			
293.15	−23.18	7.13	103.4			
298.15	−23.28 (−23.41)	5.90 (3.8)	97.9 (91.3)			
303.15	−23.54	4.63	92.9	−28.0	−1.62	87.0
308.15	−23.68 (−23.87)	3.29 (2.0)	87.5 (84.0)			
313.15	−23.79	1.91	82.1			
318.15	−23.82 (−23.86)	0.48 (−1.5)	76.4 (70.3)			
328.15	−24.70	−2.7	66.9			
	DTAC Perger and Bešter-Rogač 2007 (Šarac and Bešter-Rogač 2009)			**DTAB** Ray et al. 2005 (Stodghill et al. 2004)		
278.15	−28.84 (−28.75)	13.99 (12.44)	153.9 (148.1)			
283.15	−29.05 (−28.81)	11.68 (9.22)	143.8 (134.3)			

(continued)

TABLE 20.2 (Continued)
Standard Thermodynamic Parameters (Standard Gibbs Free Energy [ΔG_M^0], Enthalpy [ΔH_M^0], and Entropy [ΔS_M^0]) for Micellization of Octyl-, Decyl-, Dodecyl-, Tetradecyl-, and Cetyltrimethylammonium Chloride, Bromide, Fluoride, Nitrate, and Sulfate in Water at Different Temperatures

T (K)	ΔG_M^0 (kJ mol⁻¹)	ΔH_M^0 (kJ mol⁻¹)	ΔS_M^0 (J mol⁻¹ K⁻¹)	ΔG_M^0 (kJ mol⁻¹)	ΔH_M^0 (kJ mol⁻¹)	ΔS_M^0 (J mol⁻¹ K⁻¹)
288.15	−29.79	9.42	136.1			
	(−29.60)	(6.24)	(124.4)			
293.15	−30.12	6.89	126.2			
	(−30.06)	(5.77)	(122.2)			
298.15	−30.56	4.27	116.8			
	(−30.53)	(4.34)	(117.0)			
303.15	−30.71	1.47	106.2	−35.1 (−42.00)	−2.18 (−3.46)	109 (109.0)
	(−30.75)	(1.79)	(107.3)			
308.15	−31.10	−1.45	96.2	− (−37.01)	− (−5.33)	− (102.8)
	(−31.12)	(−1.28)	(96.8)			
313.15	−31.29	−4.52	85.5			
	(−31.09)	(−3.08)	(89.5)			
318.15	−31.25	−7.69	74.2			
	(−31.19)	(−4.92)	(82.6)			

| | **TTAC** | | | **TTAB** | | |
| | **Perger and Bešter-Rogač 2007** | | | **Ruiz et al. 2007** | | |
	(Kroflič et al. 2011)			**(Ray et al. 2005; Stodghill et al. 2004)**		
278.15	−35.04	12.50	170.9			
	(−35.19)	(13.93)	(176.6)			
283.15	−35.70	10.11	161.8			
288.15	−36.19	7.54	151.7			
	(−36.49)	(7.61)	(153.0)			
293.15	−36.66	4.83	141.5	−41.7	−0.83	140
298.15	−37.01	1.95	130.7	−42.1	−4.08	128
	(−37.60)	(1.96)	(132.7)			
303.15	−37.27	−1.08	119.4	−42.5	−7.51	116
				(−41.3, −42.63)	(−6.14, −6.98)	(116, 117.6)
308.15	−37.43	−4.24	107.7	−42.6	−11.05	103
	(−37.56)	(−3.97)	(109.0)	(−, −43.28)	(−, −9.24)	(−, 110.5)
313.15	−37.57	−7.55	95.9	−42.8	−14.83	89.5
318.15	−37.64	−10.97	83.8			
	(−37.91)	(−8.99)	(90.9)			
328.15	−38.89	−12.72	79.7			

| | **CTAF** | | | | | |
	Jiang et al. 2005					
278.15	−24.4	−0.8	79.1			

(continued)

TABLE 20.2 (Continued)

Standard Thermodynamic Parameters (Standard Gibbs Free Energy [ΔG_M^0], Enthalpy [ΔH_M^0], and Entropy [ΔS_M^0]) for Micellization of Octyl-, Decyl-, Dodecyl-, Tetradecyl-, and Cetyltrimethylammonium Chloride, Bromide, Fluoride, Nitrate, and Sulfate in Water at Different Temperatures

T (K)	ΔG_M^0 (kJ mol⁻¹)	ΔH_M^0 (kJ mol⁻¹)	ΔS_M^0 (J mol⁻¹ K⁻¹)	ΔG_M^0 (kJ mol⁻¹)	ΔH_M^0 (kJ mol⁻¹)	ΔS_M^0 (J mol⁻¹ K⁻¹)
					CTAB	
		CTAC			Jiang et al. 2005	
		Jiang et al. 2005		Ray et al. 2005	(Stodghill et al. 2004)	
298.15	−27.6	−4.4	77.8	−28.9	−6.9	73.8
303.15				−47.3	−7.26	132
				(−48.96)	(−11.11)	(124.9)
308.15				−(49.65)	−(14.41)	−(114.4)
		CTANO₃			**CTASO₄**	
		Jiang et al. 2005			Jiang et al. 2005	
278.15	−29.4	−7.8	72.4	−24.8	3.1	93.6

As previously mentioned, ΔG_M^0 was found to be more or less constant upon increased temperature. Just a slight decrease was actually observed by Zieliński 2001 at OTAB. Beyer et al. (2006), Perger and Bešter-Rogač (2007), and Stodghill et al. (2004) also reported a similar course for DeTAX, DTAX, TTAX, and CTAX as it is partially presented in Table 20.2. Moreover, these thermodynamic parameters also show only slight dependence on salt addition [Zieliński (2001) for OTAB and Šarac and Bešter-Rogač (2009) for DTAC]. The effect of ionic strength on the micellization parameters is again more pronounced at ΔH_M^0 and ΔS_M^0 compared to ΔG_M^0 and is illustrated for DeTAC, DTAC, and TTAC in water and NaCl solutions in terms of the enthalpy–entropy compensation effect in Figure 20.5.

The enthalpy–entropy compensation has already been observed for the transfer of a variety of small solutes into water (Lumry and Rajender 1970) as well as for the micellization of diverse surfactants in aqueous solutions (Chen et al. 1998; Perger and Bešter-Rogač 2007). In general, the compensation effect can be described by the relation

$$\Delta H_M^0 = \Delta H_M^* + T_c \Delta S_M^0, \tag{20.2}$$

where T_c, the so-called compensation temperature, is the slope of the compensation plot and ΔH_M^* is the corresponding intercept. ΔH_M^* provides information on the solute–solute interactions and is considered as an index of a "chemical part" of the micellization process (aggregation of the surfactant monomers to form the micellar unit). On the other hand, T_c has been proposed as a measure of a "solvation part" of micelle formation (desolvation of the hydrophobic chains). For a variety of processes

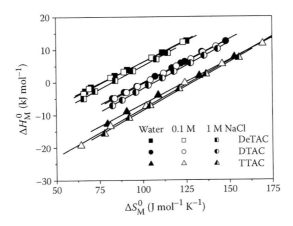

FIGURE 20.5 Enthalpy–entropy compensation effect for DeTAC, DTAC, and TTAC in water (solid symbols), 0.1 M NaCl (open symbols), and 1 M NaCl (black and white symbols). Solid lines represent the linear fits through the experimental data points and the slopes correspond to the compensation temperature, T_c. (Data from Kroflič, A., B. et al., *Chem. Thermodyn.* 43:1557–1563, 2011; Šarac, B., M. Bešter-Rogač, *J. Colloid Interface Sci.* 338:216–221, 2009.)

involving aqueous solutions of small molecules and biological systems, Lumry and Rajender (1970) showed that around room temperature (298.15 K), the T_c values lay in a narrow range between 250 and 320 K. Moreover, for the processes dominated by dehydration, a value of $T_c \approx 280$ K should be assumed as it was summarized from Lumry's law.

Evidently, the enthalpy–entropy compensation plots for DeTAC, DTAC, and TTAC in water and NaCl solutions, presented in Figure 20.5, are linear and almost parallel. The T_c values obtained (259, 269, and 271 K in water; 268, 272, and 294 K in 0.1 M NaCl; 282, 284, and 297 in 1 M NaCl, respectively) are in reasonable agreement with Lumry's law. With an increase in the alkyl chain length, the ΔH_M^* determined in water decreases from −19.9 kJ mol^{-1} at DeTAC to −27.1 kJ mol^{-1} at DTAC and −33.9 kJ mol^{-1} at TTAC. The ΔH_M^* values are found slightly lower in 0.1 M NaCl (−20.7 kJ mol^{-1}, −27.8 kJ mol^{-1}, and −37.6 kJ mol^{-1} at DeTAC, DTAC, and TTAC, respectively) and even lower in 1 M NaCl (−23.3 kJ mol^{-1}, −30.5 kJ mol^{-1}, and −38.4 kJ mol^{-1}). In general, T_c and ΔH_M^* for all of the three surfactants in water and salt solutions are in good agreement. Therefore, the micellization processes in all investigated solvents can be assumed to be similar and mainly governed by the hydrophobic dehydration.

20.2.3 ALKYLTRIMETHYLAMMONIUM SURFACTANT MICELLIZATION: COUNTERION CONDENSATION PHENOMENON

From the literature survey, it is evident that the degree of micelle ionization is strongly dependent on the type of the added electrolyte. Therefore, next to the electrostatic interactions, ion-specific effects are believed to play an important role in counterion binding to the micelle surface. The micelle can be actually treated as a highly charged particle in solution and a counterion condensation phenomenon describes accumulation of ions in its vicinity. Although several attempts have been made to test the validity of the Hofmeister series in ionic surfactant micellization, the specific effect of ions in charged amphiphile aqueous solution is not yet fully understood.

For an easier understanding of the data, we first divided the studied counterions into three groups, similar as Leontidis (2002) did. Class I includes most of the typical Hofmeister ions (OH$^-$, F$^-$, Cl$^-$, Br$^-$, I$^-$, CO$_3^-$, NO$_3^-$, N$_3^-$, ClO$_2^-$, SCN$^-$, SO$_4^{2-}$, and PO$_4^{3-}$), Class II ions are large and complex (acetate, oxalate, tartrate, citrate, S$_2$O$_3^{2-}$, S$_2$O$_7^{2-}$, and B$_4$O$_7^{2-}$), whereas Class III consists of organic ions with large hydrophobic moiety (benzoate and salicylate). Further, we focus our attention on the micellization parameters of different alkyltrimethylammonium salts and even extend the study on data where the excess of diverse electrolytes was added into an alkyltrimethylammonium chloride or bromide aqueous solution so that the exchange of counterions can also be predicted.

Gaillon et al. (1999) prepared seven different salts of dodecyl-, tetradecyl-, and cetyltrimethylammonium amphiphiles (DTAX, TTAX, and CTAX; X = OH, F, Ac, Cl, Br, NO$_3$, and Ben) and studied the influence of different counterions on the cmc and β by electromotive force measurements at 298 K. The research was indeed concluded with an observation that the nature of counterions tremendously affects the properties of the surfactant solution. However, both of the micellization parameters investigated (cmc and β) were found to follow the Hofmeister series and decrease in the order OH$^-$ > F$^-$ > acetate > Cl$^-$ > Br$^-$ > NO$_3^-$ > benzoate. A similar trend was also observed by Jiang et al. (2005) for CTAX (X = F, Cl, Br, NO$_3$, and SO$_4$) at 298.15 K using ITC and conductivity measurements to determine the cmc, β, and ΔH_M^0. Again, the cmc and β decrease within the Hofmeister series for monovalent ions F$^-$ > Cl$^-$ > Br$^-$ > NO$_3^-$ > SO$_4^{2-}$, but an unpredictable change in the position of the divalent sulfate ion can be observed. Therefore, it is believed that divalency plays an important role in the counterion condensation phenomenon owing to stronger electrostatic interactions of such an ion with the surfactant headgroups. Although the large hydrated radius hinders the ability of SO$_4^{2-}$ to bind to the micelle surface, the divalency and greater distance of hydrating water from the center of the ion obviously enable the sulfate to neutralize the micelle charge effectively. On the other hand, micellization was found less exothermic in the order NO$_3^-$ > Br$^-$ > Cl$^-$ > F$^-$ > SO$_4^{2-}$ and even slightly endothermic at CTASO$_4$. However, the observed trend can be explained by the unfavorable enthalpy contribution of counterion dehydration upon binding to the micelle surface that hinders its interaction with the oppositely charged aggregate. The dehydration of a highly hydrated kosmotropic ion (see Figure 20.2) is obviously a more endothermic process but is additionally accompanied by the greater positive entropy change that favors micellization simultaneously. Moreover, binding of the counterion to the micelle surface also lowers the entropy of the system, which is more pronounced at adsorption of two single charged ions in comparison with one divalent ion. The sulfate ion binding upon micellization thus comprises the higher entropy gain; hence, it favors micellization at a lower surfactant concentration despite the endothermic contribution of its dehydration to the process investigated.

Furthermore, ion binding to the micelle surface can also be estimated by the differences in the micelle microenvironment in the vicinity of distinct counterions. Electron spin resonance measurements were used by Jiang et al. (2005) to characterize the microviscosity and micropolarity of the CTAX micelle surroundings, which should be related to the surfactant headgroup packing and, thus, to the counterion condensation as well. As a matter of fact, F^- and SO_4^{2-} counterions in solution result in the highest micropolarity and the lowest microviscosity, indicating the loosely packed aggregate formation, which also assumes the poorest counterion binding to the micelle surface. Sulfate ions are thus believed not to be as efficient as monovalent ions in promoting micelle formation, since loose aggregates are formed despite the lowest cmc and β determined. From the above data, it can be concluded that the micellization parameters at monovalent counterions in solution are in good correlation with the Hofmeister series. However, divalent counterion behavior in the surfactant solution is more complex and its influence on the micellization process should also be related to its valency besides changes in hydration.

As already mentioned, the excess of counterions obtained by salt addition to the surfactant solution is commonly used to study the ion-specific effect on the micellization process, but the competition for counterion condensation to the micelle surface is often neglected. However, Chakraborty and Moulik (2007) studied the self-aggregation of DeTAB in different sodium salts (0.3 M aqueous solutions) at 303.15 K and also observed the cmc dependence within the Hofmeister series for monovalent counterions $F^- > SO_4^{2-} > Cl^- > Br^- > S_2O_7^{2-} > I^- >$ benzoate > salicylate. Further, ΔH_M^0 was found to become less negative in the reverse order: $I^- >$ salicylate $> Br^- >$ benzoate $> F^- \approx S_2O_7^{2-} \approx Cl^- > SO_4^{2-}$. In contrast to the behaviors of CTAX discussed in previous paragraphs, the addition of SO_4^{2-} does not shift cmc drastically toward lower cmc and an endothermic micellization was even observed at F^- and Cl^- counterion addition, which could be the consequence of surfactant rigidity and its loose aggregate formation. Moreover, aromatic counterions (benzoate and salicylate) lower the cmc strongly because of the pronounced hydrophobic effect upon micellization owing to the incorporation of their hydrophobic moiety into the micelle core. Additionally, Hsiao et al. (2005) also reported the decreasing β within the Hofmeister series $F^- > Cl^- > Br^-$ upon sodium salt addition into the TTAB solution determined by direct measurement of the counterion concentration with an ion-selective electrode at 298 K.

More discrepancies from the Hofmeister series were actually reported by Maiti et al. (2009) upon investigation of the micellization process of TTAB at 303 K in the presence of different sodium salts (ionic strength of 0.01 M). Here, first, the reported cmc, β, and ΔH_M^0 dependence on the Class I ion addition will be presented. cmc lowers in the order $F^- > PO_4^{3-} > CO_3^- > Cl^- > Br^- > SO_4^{2-} > N_3^- > NO_3^- > ClO_2^- > SCN^-$, β decreases in the order $ClO_2^- > PO_4^{3-} \approx NO_3^- \approx F^- > N_3^- > SO_4^{2-} \approx SCN^- \approx Br^- > Cl^- > CO_3^-$, and ΔH_M^0 becomes less negative in the order $SCN^- > Br^- \approx NO_3^- > N_3^- > PO_4^{3-} \approx SO_4^{2-} > F^- \approx Cl^- > CO_3^- > ClO_2^-$. Furthermore, for the Class II ions, the cmc lowers in the order tartrate > oxalate $> S_2O_3^{2-} >$ citrate > acetate $> B_4O_7^{2-}$, β decreases in the order oxalate > tartrate $> S_2O_3^{2-} >$ acetate $> B_4O_7^{2-} >$ citrate, and the micellization process is less exothermic in the order citrate > oxalate $> S_2O_3^{2-} >$ tartrate > acetate $> B_4O_7^{2-}$. Again, salicylate and benzoate addition was found to lower the cmc very efficiently and ΔH_M^0 is also strongly negative for both of the aromatic ions, which is in agreement with the trend observed for DeTAB and has already been discussed above. In contrast to previously reported data, this study shows that the counterion effect on the TTAB micellization loosely follows the Hofmeister series, which could be the consequence of the competitive ion condensation to the micelle surface.

Naskar et al. (2012) recently studied the influence of sodium salts (ionic strength of 0.01 M) on the CTAB micellization at 303.15 K. The cmc was found to decrease in the order $F^- > Cl^- >$ acetate $\approx Br^- >$ oxalate $\approx SO_4^{2-} >$ citrate $\approx PO_4^{3-} \approx$ salicylate and the determined ΔH_M^0 was less negative in the order $Br^- > Cl^- >$ acetate $> F^- >$ oxalate $> SO_4^{2-} >$ citrate. Both series are indeed similar to that reported by Gaillon et al. (1999) and Jiang et al. (2005) for CTAX and agree well with the observed importance of counterion valency and hydrophobicity at the micellization process. Further, the zeta potential of the micelle at a concentration approximately 10-fold higher than that from cmc decreases in the order $F^- > Br^- \approx Cl^- > PO_4^{3-} > SO_4^{2-} >$ salicylate $>$ acetate $>$ citrate, which even supports the role of electrostatic interactions at multivalent counterion condensation upon micellization. Indeed, the efficacy of the micelle surface charge screening by counterions in the solution decreases from monovalent to bivalent and trivalent anions, although salicylate and acetate show additional ion specificity.

Moreover, they also observed a salt-induced transition phenomenon (minimum of surface tension observed just below cmc), which was attributed to additional interactions of added counterions (except acetate, sulfate, and phosphate) with cetyltrimethylammonium cation (CTA^+). Although spherical micelles were confirmed at the cmc, sphere-to-prolate ellipsoid transition was found at higher surfactant concentration and in the presence of salt, which was also confirmed by Wang and Larson (2009) with molecular dynamics simulations for cetyltrimethylammonium chloride in NaCl and sodium salicylate solutions. Šarac et al. (2011) reported a similar transition for DTAC upon sodium salicylate addition, and other authors also observed elongated micelles of alkyltrimethylammonium bromide, iodide, nitrate, salicylate, and so on.

In contrast to all these data, it was reported by Jakubowska (2008) that the ability of anion to bind the CTA^+ monomer and the stability of the ion pair formed in solution decrease in the order following the Hofmeister series $F^- > Cl^- > NO_3^- > Br^- > I^-$ (except for the inversion of NO_3^- and Br^- positions). Further, she reported the Hofmeister series $F^- > Cl^- > Br^- > NO_3^-$ for the cmc dependence on sodium salt addition into the CTAB solution at 298.15 K and also observed the above-mentioned inversion at the β dependence on the type of counterion added $F^- < Cl^- < NO_3^- < Br^-$ (Jakubowska 2010). It follows logically that the greater is the affinity of ion to the micelle surface, the lower is the micelle ionization degree. Nevertheless, it is quite unusual that the most hydrated kosmotropic anion interacts most strongly with the chaotropic alkyltrimethylammonium cation, and on the other hand, it results in the highest cmc.

In fact, the represented data are mainly in accordance with the law of matching water affinities, which means that the counterions in solution spontaneously form inner-sphere ion pairs (see Figure 20.3) with the oppositely charged alkyltrimethylammonium micelles only when they have equal water affinities (if they are both chaotropes). In addition, Moreira and Firoozabadi (2010) proposed a thermodynamic model to predict the inorganic salt addition effect on the micellization of ionic surfactants, which was also found in line with Collins' concept. Still, the ion-specific behavior was found dependent on the interacting surface, which can be noticed from the divalent counterion shift along the series reported at the rigid decyltrimethylammonium amphiphile compared to the longer alkyl chain surfactants. The mentioned correlation was also observed by Maiti et al. (2009) at several surfactants. However, despite all the reported divergences from the Hofmeister series [which could also be attributed to the counterion competition for electrostatic interactions with the oppositely charged micelle surface, which was reported by Kunz (2010) to be important especially at low salt concentrations], the counterion effect on the micelle formation can have an apparent generalization.

20.2.4 ALKYLTRIMETHYLAMMONIUM SURFACTANT MICELLIZATION: HEAT CAPACITY CHANGE

The heat capacity change upon micellization, $\Delta c_{p,M}^0$, is always highly negative and can be ascribed to the removal of large non-polar surface from contact with water upon micelle formation. The polar parts of surfactant monomers are actually believed to freely interact with water molecules even after the self-association occurs since they are always oriented toward bulk in aqueous solution. According to Kresheck (2009), $\Delta c_{p,M}^0$ can thus be directly related to the changes in a water-accessible surface area upon burial of the non-polar groups in the micelle interior, ASA_{np}. In fact, he experimentally established a relationship

$$\Delta c_{p,M}^0 = -1.46(J\ mol-1\ K-1\ \text{\AA}-2) \cdot ASA_{np}, \qquad (20.3)$$

on whose basis the number of buried methylene groups ("theoretical" accesible surface area, $ASA_{th} = 29\ \text{\AA}^2$) along with the methyl group ($ASA_{th} = 89\ \text{\AA}^2$) upon micellization has already been determined for a series of non-ionic (Kresheck 2009) and ionic (Kroflič et al. 2011) surfactants. For the semi-long and long alkyl chain ionic amphiphiles, a good correlation was found: an average burial of 8- and 11-carbon-atom-long alkyl chain for DTAC and TTAC, respectively, has been reported recently (Kroflič et al. 2011). Nevertheless, some discrepancies were observed at the short alkyl chain surfactant DeTAC. Although $\Delta c_{p,M}^0$ is usually found constant in the temperature range investigated (it can thus be represented by a slope of the temperature-dependent enthalpy of micellization as it is shown for DTAC and TTAC in Figure 20.6a), it was observed to diminish substantially with temperature at DeTAC (Figure 20.6a). Since only 1- to 2-carbon-atom-long alkyl chain burial was determined for DeTAC at 318.15 K (Table 20.3), it can be concluded that the above correlation between $\Delta c_{p,M}^0$ and ASA_{np} could not be used for this surfactant. The proposed relationship can thus be applied on the classical surfactants where the hydrophobic effect is the main driving force of the micellization process, but the electrostatic interactions are of greater importance at the short alkyl chain ionic surfactants especially at high temperature.

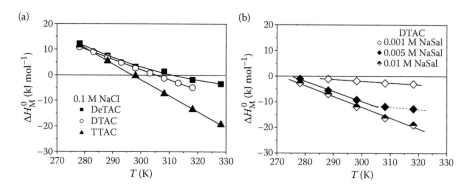

FIGURE 20.6 Temperature dependence of enthalpy of micellization, ΔH_M^0, for (a) DeTAC, DTAC, and TTAC in 0.1 M NaCl (Kroflič et al. 2011) and (b) DTAC in 0.001, 0.005, and 0.01 M NaSal (Šarac et al. 2011). The slopes of solid lines represent $\Delta c_{p,M}^0$, which either is constant or changes with the increase in temperature.

TABLE 20.3
Heat Capacity Changes upon Micellization ($\Delta c_{p,M}^0$), Calculated ASA$_{np}$ Values, and Estimated Burial of Methylene Groups ($n_{burried}$) for Alkyltrimethylammonium Surfactants in Water and Salt Solutions

T (K)	$\Delta c_{p,M}^0$ (J mol^{-1} K^{-1})	ASA$_{np}$ (Å^2)	$n_{burried}$	$\Delta c_{p,M}^0$ (J mol^{-1} K^{-1})	ASA$_{np}$ (Å^2)	$n_{burried}$
		OTAC		**OTAB** Zieliński 2001		
Water	No data found in the literature			−248	170	3
0.25 M NaBr				−335	229	5
0.5 M NaBr				−425	291	7
0.75 M NaBr				−408	279	7
1 M NaBr				−348	238	5
	DeTAC Kroflič et al. 2011 (Perger and Bešter-Rogač 2007)			**DeTAB** Wooley and Burchfield 1985 (Dearden and Wooley 1987)		
Water	−480[a]	328	8	−335.8	230	5
	−150[b]	103	2	(−302)	(207)	(4)
	(−247.8)	(170)	(3)			
0.01 M NaCl	−629[a]	430	12			
	−57[b]	39	1			
0.1 M NaCl	−488[a]	334	8			
	−133[b]	91	2			
1 M NaCl	−291[a]	199	4			
	−291[b]	199	4			

(continued)

TABLE 20.3 (Continued)
Heat Capacity Changes upon Micellization ($\Delta c_{p,M}^0$), Calculated ASA$_{np}$ Values, and Estimated Burial of Methylene Groups ($n_{burried}$) for Alkyltrimethylammonium Surfactants in Water and Salt Solutions

T (K)	$\Delta c_{p,M}^0$ (J mol⁻¹ K⁻¹)	ASA$_{np}$ (Å²)	$n_{burried}$	$\Delta c_{p,M}^0$ (J mol⁻¹ K⁻¹)	ASA$_{np}$ (Å²)	$n_{burried}$
	DTAC			**DTAB**		
	Kroflič et al. 2011 (Perger and Bešter-Rogač 2007; Šarac and Bešter-Rogač 2009)			**Dearden and Wooley 1987**		
Water	−418[a]	286	7	−406	278	7
	−418[b]	286	7			
	(−541.4, −417.9)	(371, 286)	(10, 7)			
0.01 M NaCl	−447[a]	306	8			
	−384[b]	262	6			
	(−, −411.6)	(−, 282)	(−, 7)			
0.1 M NaCl	−422[a]	288	7			
	−381[b]	260	6			
	(−, −401.4)	(−, 275)	(−, 6)			
1 M NaCl	−467[a]	319	8			
	−409[b]	280	7			
	(−, −438.0)	(−, 300)	(−, 7)			
	Šarac et al. 2011					
0.001 M NaSal	−77	53	0			
0.005 M NaSal	−407	279	7			
0.01 M NaSal	−419	287	7			
	TTAC			**TTAB**		
	Kroflič et al. 2011 (Perger and Bešter-Rogač 2007)			**Dearden and Wooley 1987 (Ruiz et al. 2007)**		
Water	−678[a]	464	13	−499	342	9
	−458[b]	314	8	(−699)	(479)	(13)
	(−587.8)	(403)	(11)			
0.01 M NaCl	−574[a]	393	10			
	−422[b]	289	7			
0.1 M NaCl	−627[a]	429	12			
	−623[b]	427	12			
1 M NaCl	−495[a]	339	9			
	−435[b]	298	7			
	TTANO$_3$			**CTAB**		
	González-Pérez et al. 2004			**Dearden and Wooley 1987**		
Water	−692	474	13	−573	393	10

Note: Values at several temperatures are reported where temperature-dependent $\Delta c_{p,M}^0$ was observed.

[a]　Value at 278.15 K.
[b]　Value at 318.15 K.

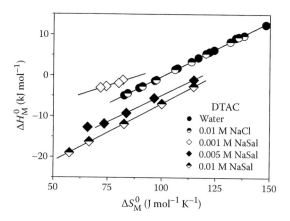

FIGURE 20.7 Enthalpy–entropy compensation effect for DTAC in water and 0.01 M NaCl (Šarac and Bešter-Rogač 2009) and 0.001, 0.005, and 0.01 M NaSal (Šarac et al. 2011). Solid lines represent the linear fits through the experimental data points and the slopes correspond to the compensation temperature, T_c.

Interestingly, similar temperature dependence of $\Delta c_{p,M}^0$ as at DeTAC was also observed at DTAC in 0.005 M NaSal solution by Šarac et al. (2011) and is presented in Figure 20.6b. In addition, the permanently lower slope observed at DTAC in 0.001 M NaSal (Figure 20.6b) indicates loose aggregate formation within the concentration range investigated, which also agrees with the lowest T_c obtained from the enthalpy–entropy compensation plot for the system studied as will be discussed later on. All the $\Delta c_{p,M}^0$ values at different temperatures are gathered in Table 20.3.

The enthalpy–entropy compensation plots for DTAC in water, 0.01 M NaCl, and sodium salicylate solutions are shown in Figure 20.7 for comparison. All of them are once again linear and, with the exception of DTAC in 0.001 M NaSal, mostly parallel. The T_c values obtained for DTAC in water, 0.01 M NaCl, 0.005 M NaSal, and 0.01 M NaSal are 269, 268, 261, and 282 K, respectively, whereas a lower slope at DTAC in 0.001 M NaSal (T_c = 177 K) can be observed. In contrast to all other values presented, the latter T_c is also not in agreement with Lumry's law. ΔH_M^* values decrease from –15.8 kJ mol^{-1} in 0.001 M NaSal to –27.1 kJ mol^{-1} in water and 0.01 M NaCl and –30.9 kJ mol^{-1} in 0.005 M NaSal. The lowest ΔH_M^* is found in 0.01 M NaSal (–35.1 kJ mol^{-1}), where the strongest chemical interactions are predicted upon micellization. On the other hand, the least negative ΔH_M^* and the lowest T_c in 0.001 M NaSal reveal that the process observed cannot be treated as classical micellization and support the prediction of very loose aggregate formation in the investigated system.

20.3 CONCLUSIONS

Micellization is often used as a model system to study the hydrophobic effect, which is believed to play an important role in many biological self-assembly processes, for instance protein or nucleic acid folding, ligands binding to macromolecules, drugs entering the biomembranes, and others. Nevertheless, as far as ionic surfactant micellization is concerned, it is obviously found to be a complex process and the conclusions about the leading forces cannot be drawn straightforward. Besides hydrophobic hydration, electrostatic interactions are also important in ionic surfactant micellization and the intricate balance between opposing effects should be considered to explain the observed dependencies.

The hydrophobic dehydration effect (and consequently entropy rise) is obviously the main leading force of micelle formation especially at low temperature and favors micellization with the lengthening of the surfactant alkyl chain. At higher temperature, the water lattice at non-polar surfaces is disturbed and enthalpy contribution becomes more important. Micelles with a lower aggregation number but with a higher degree of ionization are also predicted at higher temperature.

Moreover, added salt in an ionic surfactant solution affects its micellization considerably but the effect of electrolyte addition cannot be easily predicted. Although counterions in the solution are expected to interact with the oppositely charged micelle surface via electrostatic interactions in the first place, the diffuse layer around the micelle also screens the repulsive forces between equally charged surfactant headgroups, which indeed hinders counterion condensation on the micelle surface.

Besides, the ion-specific effect that is attributed to ion–water interactions must also be considered at the micellization process. In general, Collins' concept of matching water affinities on the basis of the Hofmeister series was found applicable on alkyltrimethylammonium surfactants. Nevertheless, the effect of micelle surface and electrostatic interactions cannot be neglected at the process investigated. Therefore, the low-surface-charge tetraalkylammonium cation is believed to fundamentally form contact ion pairs with other chaotropic counterions in solution and not with kosmotropic ones. As a matter of fact, this concept is mostly applicable on monovalent counterions, while electrostatics is more important at multivalent ions. Additionally, hydrophobic dehydration should also be taken into account upon hydrophobic counterion addition; thus, it favors micellization in comparison to the simple inorganic counterions. It should also be stressed that more deviations from the Hofmeister series were observed with the excess salt addition into the ionic surfactant solution. Such an experiment thus seems to be inappropriate for testing the law of matching water affinities, since counterion competition should also be considered.

The thermodynamics of micelle formation is somehow easier to generalize. The negative heat capacity change is typical for the micellization process and can be related to the changes in the non-polar surface area, which is in contact with water molecules upon micellization at least in classical long hydrophobic tail surfactants in water and inorganic electrolyte solutions. It is indeed the measure of the hydrophobic effect at the process investigated, but it can only be determined if the experiments are carried out at a broad temperature range. The enthalpy of micellization thus decreases with an increase in temperature. Even more, micellization is usually an endothermic process at lower temperatures and exothermic at higher temperatures. However, the enthalpy–entropy compensation effect is also typical for micellization (micellization enthalpy and entropy decrease simultaneously with an increase in temperature); hence, $\Delta G_{\mathrm{M}}^{0}$ is nearly constant in the temperature range investigated.

REFERENCES

Baar, C., R. Buchner, W. Kunz 2001. Dielectric relaxation of cationic surfactants in aqueous solution. *J. Phys. Chem. B* 105:2906–2922.

Bales, B.L., R. Zana 2002. Characterization of micelles of quaternary ammonium surfactants as reaction media I: Dodecyltrimethylammonium bromide and chloride. *J. Phys. Chem. B* 106:1926–1939.

Beyer, K., D. Leine, A. Blume 2006. The demicellization of alkyltrimethylammonium bromides in 0.1 M sodium chloride solution studied by isothermal titration calorimetry. *Colloids Surf. B* 49:31–39.

Buchner, R., C. Baar, P. Fernandez, S. Schrödle, W. Kunz 2005. Dielectric spectroscopy of micelle hydration and dynamics in aqueous ionic surfactant solutions. *J. Molec. Liquids* 118:179–187.

Chakraborty, I., S.P. Moulik 2007. Self-aggregation of ionic C_{10} surfactants having different headgroups with special reference to the behavior of decyltrimethylammonium bromide in different salt environments: A calorimetric study with energetic analysis. *J. Phys. Chem. B* 111:3658–3664.

Chen, L.-J., S.-Y. Lin, C.-C. Huang 1998. Effect of hydrophobic chain length of surfactants on enthalpy–entropy compensation of micellization. *J. Phys. Chem. B* 102:4350–4356.

Collins, K.D. 2004. Ions from the Hofmeister series and osmolytes: Effects on proteins in solution and in the crystallization process. *Methods* 34:300–311.

Collins, K.D., G.W. Neilson, J.E. Enderby 2007. Ions in water: Characterizing the forces that control chemical processes and biological structure. *Biophys. Chem.* 128:95–104.

Dahirel, V., B. Ancian, M. Jardat, G. Mériguet, P. Turq, O. Lequin 2010. What can be learnt from the comparison of multiscale Brownian dynamics simulations, nuclear magnetic resonance and light scattering experiments on charged micelles? *Soft Matter* 6:517–525.

Dearden, L.V., E.M. Wooley 1987. Heat capacities of aqueous decyl-, dodecyl-, tetradecyl-, and hexadecyltrimethylammonium bromides at 10, 25, 40, and 55°C. *J. Phys. Chem.* 91:4123–4127.

Dutkiewicz, E., A. Jakubowska 2002. Effect of electrolytes on the physicochemical behaviour of sodium dodecyl sulphate micelles. *Colloid Polym. Sci.* 280:1009–1014.

Evans, D.F., D.J. Mitchell, B.W. Ninham 1984. Ion binding and dressed micelles. *J. Phys. Chem.* 88:6344–6348.

Gaillon, L., J. Lelièvre, R. Gaboriaud 1999. Counterion effects in aqueous solutions of cationic surfactants: Electromotive force measurements and thermodynamic model. *J. Colloid Interface Sci.* 213:287–297.

Gaillon, L., M. Hamidi, J.- Lelievre, R. Gaboriaud 1997. Influence of the counterion in aqueous solutions of cationic surfactants. Second part: Conductivity and thermodynamic model. *J. Chim. Phys.* 94:728–749.

González-Pérez, A., J. Czapkiewicz, J.L. Del Castillo, J.R. Rodríguez 2004. Micellar properties of tetradecyltrimethylammonium nitrate in aqueous solutions at various temperatures and in water-benzyl alcohol mixtures at 25°C. *Colloid Polym. Sci.* 282:1359–1364.

González-Pérez, A., J.M. Ruso, M.J. Romero, E. Blanco, G. Prieto, F. Sarmiento 2005. Application of thermodynamic models to study micellar properties of sodium perfluoroalkyl carboxylates in aqueous solutions. *Chem. Phys.* 313:245–259.

Hansson, P., B. Jönsson, C. Ström, O. Söderman 2000. Determination of micellar aggregation numbers in dilute surfactant systems with the fluorescence quenching method. *J. Phys. Chem. B* 104:3496–3506.

Heindl, A., H.-H. Kohler 1996. Rod formation of ionic surfactants: A thermodynamic model. *Langmuir* 12:2464–2477.

Hsiao, C.C., T.-Y. Wang, H.-K. Tsao 2005. Counterion condensation and release in micellar solutions. *J. Chem. Phys.* 122:144702–144712.

Jakubowska, A. 2008. Stability of anion binding with monomers of a cationic surfactant. *Chem. Phys. Chem.* 9:829–831.

Jakubowska, A. 2010. Interactions of different counterions with cationic and anionic surfactants. *J. Colloid Interface Sci.* 346:398–404.

Jiang, N., P. Li, Y. Wang, J. Wang, H. Yan, R.K. Thomas 2005. Aggregation behavior of hexadecyltrimethylammonium surfactants with various counterions in aqueous solution. *J. Colloid Interface Sci.* 286:755–760.

Jusufi, A., A.-P. Hynninen, M. Haataja, A.Z. Panagiotopoulos 2009. Electrostatic screening and charge correlation effects in micellization of ionic surfactants. *J. Phys. Chem. B* 113:6314–6320.

Kim, H.-U., K.-H. Lim 2004. A model on the temperature dependence of critical micelle concentration. *Colloids Surf. A* 235:121–128.

Kresheck, G.C. 2009. Isothermal titration calorimetry studies of neutral salt effects on the thermodynamics of micelle formation. *J. Phys. Chem. B* 113:6732–6735.

Kroflič, A., B. Šarac, M. Bešter-Rogač 2011. Influence of the alkyl chain length, temperature, and added salt on the thermodynamics of micellization: Alkyltrimethylammonium chlorides in NaCl aqueous solutions. *J. Chem. Thermodyn.* 43:1557–1563.

Kroflič, A., B. Šarac, M. Bešter-Rogač 2012. What affects the degree of micelle ionization: Conductivity study of alkyltrimethylammonium chlorides. *Acta. Chim. Slov.* 59:564–570.

Kunz, W. 2010. Specific ion effects in colloidal and biological systems. *Curr. Opin. Colloid Interface Sci.* 15:34–39.

Lah, J., C. Pohar, G. Vesnaver 2000. Calorimetric study of the micellization of alkylpyridinium and alkyltrimethylammonium bromides in water. *J. Phys. Chem. B* 104:2522–2526.

Leontidis, E. 2002. Hofmeister anion effects on surfactant self-assembly and the formation of mesoporous solids. *Curr. Opin. Colloid Interface Sci.* 7:81–91.

Llanos, P., R. Zana 1983. Micellar properties of alkyltrimethylammonium hydroxides in aqueous solution. *J. Phys. Chem.* 87:1289–1291.

Lumry, R., S. Rajender 1970. Enthalpy–entropy compensation phenomena in water solutions of proteins and small molecules: A ubiquitous property of water. *Biopolymers* 9:1125–1227.

Maiti, K., D. Mitra, S. Guha, S.P. Moulik 2009. Salt effect on self-aggregation of sodium dodecylsulfate (SDS) and tetradecyltrimethylammonium bromide (TTAB): Physicochemical correlation and assessment in the light of Hofmeister (lyotropic) effect. *J. Mol. Liquids* 146:44–51.

Malliaris, A., J. Le Moigne, J. Sturm, R. Zana 1985. Temperature dependence of the micelle aggregation number abd rate of intramicellar excimer formation in aqueous surfactant solutions. *J. Phys. Chem.* 89:2709–2713.

Moreira, L., A. Firoozabadi 2010. Molecular thermodynamic modeling of specific ion effects on micellization of ionic surfactants. *Langmuir* 26:15177–15191.

Nagarajan, R., E. Ruckenstein 1991. Theory of surfactant self-assembly: A predictive molecular thermodynamic approach. *Langmuir* 7:2934–2969.

Naskar, B., A. Dan, S. Ghosh, V.K. Aswal, S.P. Moulik 2012. Revisiting the self-aggregation behavior of cetyltrimethylammonium bromide in aqueous sodium salt solution with varied anions. *J. Mol. Liquids* 170:1–10.

Perger, T.-M., M. Bešter-Rogač 2007. Thermodynamics of micelle formation of alkyltrimethylammonium chlorides from high performance electric conductivity measurements. *J. Colloid Interface Sci.* 313:288–295.

Ray, A., G. Nemethy 1971. Effects of ionic protein denaturants on micelle formation by nonionic detergents. *J. Am. Chem. Soc.* 93:6787–6793.

Ray, G.B., I. Chakraborty, S. Ghosh, S.P. Moulik, R. Palepu 2005. Self-aggregation of alkyltrimethylammonium bromides (C_{10}-, C_{12}-, and C_{16}TAB) and their binary mixtures in aqueous medium: A critical and comprehensive assessment of interfacial behavior and bulk properties with reference to two types of micelle formation. *Langmuir* 21:10958–10967.

Ribeiro, A.C.F., V.M.M. Lobo, A.J.M. Valente, E.F.G. Azevedo, M. da G. Miguel, H.D. Burrows 2004. Transport properties of alkyltrimethylammonium bromide surfactants in aqueous solutions. *Colloid Polym. Sci.* 283:277–283.

Roelants, E., F.C. De Schryver 1987. Parameters affecting aqueous micelles of CTAC, TTAC, and DTAC probed by fluorescence quenching. *Langmuir* 3:209–214.

Roger, G.M., S. Durand-Vidal, O. Bernard, P. Turq, T.-M. Perger, M. Bešter-Rogač 2008. Interpretation of conductivity results from 5 to 45°C on three micellar systems below and above the cmc. *J. Phys. Chem. B* 112:16529–16538.

Ruiz, C.C., L. Díaz-López, J. Aguiar 2007. Self-assembly of tetradecyltrimethylammonium bromide in glycerol aqueous mixtures: A thermodynamic and structural study. *J. Colloid Interface Sci.* 2007:293–300.

Šarac, B., J. Cerkovnik, B. Ancian, G. Mériguet, G. M. Roger, S. Durand-Vidal, M. Bešter-Rogač 2011. Thermodynamic and NMR study of aggregation of dodecyltrimethylammonium chloride in aqueous sodium salicylate solution. *Colloid Polym. Sci.* 289:1597–1607.

Šarac, B., M. Bešter-Rogač 2009. Temperature and salt-induced micellization of dodecyl-trimethylammonium chloride in aqueous solution: A thermodynamic study. *J. Colloid Interface Sci.* 338:216–221.

Srinivasan, V., and D. Blankschtein 2003a. Effect of counterion binding on micellar solution behavior: 1. Molecular-thermodynamic theory of micellization of ionic surfactants. *Langmuir* 19:9932–9945.

Srinivasan, V., and D. Blankschtein 2003b. Effect of counterion binding on micellar solution behavior: 2. Prediction of micellar solution properties of ionic surfactant—Electrolyte systems. *Langmuir* 19:9946–9961.

Stodghill, S.P., A.E. Smith, J.H. O'Haver 2004. Thermodynamics of micellization and adsorption of three alkyltrimethylammonium bromides using isothermal titration calorimetry. *Langmuir* 20:11387–11392.

Vlachy, N., B. Jagoda-Cwiklik, R. Vácha, D. Touraud, P. Jungwirth, W. Kunz 2009. Hofmeister series and specific interactions of charged headgroups with aqueous ions. *Adv. Colloid Interface Sci.* 146:42–47.

Wang, Z., R.G. Larson 2009. Molecular dynamics simulations of threadlike cetyltrimethyl-ammonium chloride micelles: Effects of sodium chloride and sodium salicylate salts. *J. Phys. Chem. B* 113:13697–13710.

Wooley, E.M., T.E. Burchfield 1985. Model for thermodynamics of ionic surfactant solutions. 3. Enthalpies, heat capacities, and volumes of other surfactants. *J. Phys. Chem.* 89:714–722.

Zieliński, R. 2001. Effect of temperature on micelle formation in aqueous NaBr solutions of octyltrimethylammonium bromide. *J. Colloid Interface Sci.* 235:201–209.

21 Stiff and Flexible Water-Poor Microemulsions

Disconnected and Bicontinuous Microstructures, Their Phase Diagrams, and Scattering Properties

Magali Duvail, Jean-François Dufrêche, Lise Arleth, and Thomas Zemb

CONTENTS

21.1 INTRODUCTION

Microemulsions are stable dispersions of two immiscible fluids separated by a layer of surfactant, called the surfactant film (Eicke 1979; Chevalier and Zemb 1990; Wennerström and Lindman 1979). The ternary solutions called, for historical reasons, reverse micelles or swollen micelles correspond to water/oil (w/o) and oil/water (o/w) microemulsions. Rationale in terms of thermodynamics has been started with the early recognition by Winsor (1948) that four types of thermodynamical equilibria can be found. These four types were also found in a simple model for flexible interfaces by Andelman et al. (1987). Nearly a dozen structural models have been developed (Zemb et al. 1990).

Polar and apolar parts of the amphophilic film are separated in a water-soluble head-group part and a solvent-soluble tail part. This allows deriving from the three known water, oil, and film volume fractions, the important variables as the apolar volume fraction ϕ_a and the polar volume fraction $(1 - \phi_a)$. The surface separating polar and apolar volumes is the neutral surface and the area per unit volume can be measured without assumption by scattering experiment or derived from the known surfactant content (Zemb 2002). The second important quantity is the area per unit volume of this neutral interface. The last important quantity is the interfacial film thickness. In incompressible fluids, the thickness of the surfactant film l_s is linked to the area per unit volume and the surfactant volume fraction x_s

$$l_s = x_s \frac{S}{V}. \tag{21.1}$$

The interfacial film, considered as a thin solid shell, has some spontaneous average curvature H_0 and some spontaneous Gaussian curvature K_0 considered here to be negligible (Hyde et al. 1997). The expression for solids has been used by Helfrich for the surfactant film and the cost in free energy per unit area of interfacial film associated to bending away from spontaneous curvature is

$$F_{\text{bending}} = 2k(H - H_0)^2 + \bar{\kappa}(K - K_0), \tag{21.2}$$

where κ and $\bar{\kappa}$ are the bending and Gaussian elastic constants, respectively, and H and K are the mean average and Gaussian curvatures. The curvatures H and K are given by $\frac{1}{2}(c_1 + c_2)$ and $(c_1 \times c_2)$, respectively, where c_1 and c_2 are the local principal curvatures. If one now takes into account that the interfacial film cannot be torn and has no edge, differential geometry shows that there is a constraint relating a scalar p linking the integral of curvatures and called the coverage relation

$$p = 1 + H l_s + \frac{1}{3} K l_s^2 .$$ (21.3)

Now, if one uses as differential thickness the film l_s of molecular length, simple geometrical considerations link the film thickness to molecular volume and area per molecule a_0 at equilibrium (Israelachvili et al. 1976)

$$p = \frac{V_m}{a l_s} ,$$ (21.4)

where p is the molecular volume of the surfactant in the interfacial film. Molecular considerations show that all surfactants (except bolaforms) in a film have a well-defined configuration of minimal free energy

$$p_0 = \frac{V_m}{a_0 l_s} .$$ (21.5)

In physical chemistry and engineering literature, the spontaneous packing is defined in arbitrary units and called HLB for nonionic surfactants and HDL for ionic surfactants (Kunz et al. 2009). Identification with Equation 21.3 gives

$$p_0 = 1 + H_0 l_s + \frac{1}{3} K_0 l_s^2 .$$ (21.6)

Using this formalism taking into account the incompressibility of the surfactant molecules and the condition of continuity of the surfactant film gives the general expression of the free energy of a microemulsion

$$F_{bending} = \frac{1}{2} \kappa * (p - p_0)^2 .$$ (21.7)

The Helfrich–Gauss expression of free energy and the relation valid in case of incompressible monolayers above are equivalent at first order if

$$\kappa* = 2\kappa + \bar{\kappa}.$$ (21.8)

In some cases, using the Helfrich expression instead of the general one gives divergence of the free energy or inconsistencies in solutions (Fogden et al. 1991). It is more convenient and fast in evaluation to use the two-term expression in Equation 21.7 instead of the general expression (Equation 21.2). However, the free energy of bending has been calculated according to the general equation. It should be noticed that neglecting $\bar{\kappa}$ is related to the next neighbors in the film plane, fixed to be six in preferred local packing, which implies a 2- to 3-fold or a 6-fold axis symmetry in the molecule as represented by a flexible body perpendicular to the interface (Charvolin and Sadoc 1990).

Thermodynamics and structural studies have focused on a peculiar point: least possible concentration of surfactant and equal amount of water and oil. For applications, minimizing the amount of surfactant is crucial. However, the point with equal volumes of water and oil is much easier to model since several terms vanish.

A general model, able to capture together phase behavior and microstructure on the basis of wavelets and two level-cuts, has been proposed in 2001 (Arleth et al. 2001). The present study focuses on all possible behaviors concerning the domain of stability and the microstructures with lowest free energy. These are characterized in real and reciprocal space and compared to x-ray or neutron scattering experimental results. Special attention is given far from the most studied systems where the flexibility of the surfactant is vanished, that is, low bending constant and equal volume fraction of oil and water (Pieruschka and Marčelja 1994). In most real applications, either w/o or o/w microemulsions are considered as microemulsions giving the best efficiency. Therefore, we consider here the whole domain of polar volume fraction, that is, from 5% to 95%.

For the first time, we present in this work practical examples corresponding to "molten" lyotropic liquid crystals. To our best knowledge, there is no such study available in literature, that is, structures far from the symmetric point corresponding to equal volume fractions of water and oil.

21.2 METHODS

21.2.1 HELFRICH FORMALISM AND GAUSSIAN RANDOM FIELDS

As already mentioned, microemulsions are commonly modeled as two-domain systems composed of water and oil separated by a self-assembled monolayer of surfactant. The typical size of such interfaces is typically nanometric and a continuous approach is assumed to be valid. Thus, the free energy of the system depends on the topology of the interface. Within the Helfrich formalism, it is calculated through an expansion with respect to the curvatures of the interfaces from a flat reference state. The resulting Hamiltonian reads (Helfrich 1973)

$$\mathcal{H} = \int_s dS[2\kappa(H - H_0)^2 + \bar{\kappa}K], \tag{21.9}$$

where H is the mean curvature, K is the Gaussian curvature, and H_0 is the spontaneous curvature, which represents the tendency of the film to bend preferably toward the water or the oil domain. The Gaussian random field model is based on the idea of Cahn (1965) and Berk (1991), which suggested that the structure of bicontinuous microemulsions could be described by means of a Gaussian distribution of wavelengths having random amplitude, direction, and phase. In the 1990s, Teubner pointed out that analytical expressions of the mean H, Gaussian K, and mean squared H_2 curvatures could be derived from the procedure suggested by Cahn and Berk, since the structures generated using such an approach were realizations of Gaussian random fields (Teubner 1991). Shortly later, Piersuschka and Safran showed that the

Helfrich free energy could be expressed in terms of level-cut Gaussian random fields (Pieruschka and Safran 1993, 1995). They also pointed out that the structure factor $v(k)$ that generates the Gaussian random field has the same functional form as the one defined in the Teubner and Strey model (Teubner and Strey 1987). However, this model took into account only the curvatures and the elastic constants. Therefore, Marčelja introduced an additional term in the free energy representing the entropy (Pieruschka et al. 1994). Because we consider that the physics of a system composed of soft interfaces (low bending energy, negligible long-range electrostatic and steric interactions), such as the systems we modeled, is governed by the Hamiltonian of the surfactant interface, the use of such approach allows to provide results on the basis of the physics contained in the Helfrich Hamiltonian. Originally, this model was developed for only one level-cut, the level-cut allowing actually to define the polar and apolar domains in the microemulsion. In such a model, the concentrations of the polar and apolar phases are controlled by their volume fractions, ϕ_a and ϕ_p where subscripts "a" and "p" denote apolar and polar, respectively. The polar phase is composed of the entire water phase and the polar head of the surfactant, whereas the apolar one contains both the entire oil phase and the hydrophobic chain of the surfactant, implying thus $\phi_a + \phi_p = 1$. Later, Arleth and Marčelja improved this model by adding a second level-cut, allowing therefore to simulate a continuous transition from positive to negative mean curvature of the surfactant film (Arleth et al. 2001), which was not possible when only one level-cut is used. In both cases (one or two level-cuts), a variational minimization of the Helfrich free energy is used

$$f_{\text{free}} = \frac{S}{V}\left[2\kappa\langle(H-H_0)^2\rangle + \bar{\kappa}\langle K\rangle\right] - \frac{1}{2\pi^2}\int_k dk\, k^2 \ln v(k), \qquad (21.10)$$

where S/V is the surface-to-volume ratio of the surfactant, and $S/V = \phi_s/l_s$, where ϕ_s is the volume fraction of surfactant and l_s is the surfactant layer thickness ($l_s = 1$ nm). The last term of Equation 21.10 is an ansatz for the entropy term. Furthermore, it has been shown that the microemulsion structure factor $v(k)$ (which minimizes the free energy) generated using this approach has the same functional form as the one defined by Teubner and Strey (1987),

$$v(k) = \frac{a}{k^4 - bk^2 + c}, \qquad (21.11)$$

where the constants a, b, and c are calculated using the same procedure as done by Arleth et al. (2001). The structure factor $v(k)$ allows then to obtain a representation of the microemulsion structure in the direct (three-dimensional [3D] visualizations) and inverse (scattering functions) spaces.

21.2.2 PHASE DIAGRAM CONSTRUCTION

A complete ternary phase diagram of water–oil–surfactant can be obtained theoretically by calculating the binodal lines that distinguish the stable states from the unstable one. The latter are calculated by writing the various chemical equilibria between the possible phases, that is, by searching corresponding states for which the first derivatives of the free energy (i.e., chemical potentials) are equals. However, the calculation of the binodal lines is far from obvious since it is the solution of coupled nonlinear equations. Another way of investigating the phase diagram of complex systems consists in calculating the spinodal lines, which correspond to the limit of local stability in the phase diagram. Spinodal lines are not equivalent to binodal lines, but they allow the evaluations of unstable domains of the phase diagram—no distinction was made between two or three phases. They provide important informa-tion needed to understand the complex phase diagrams of ternary systems. Thus, we have chosen to determine the geometry of the spinodal lines. To this end, we calculated the determinant of the Hessian matrix, defined as the matrix of the second derivatives of the Gibbs potential. Its sign provides information on the stability (>0) or instability (<0) of the calculated point at the phase diagram. More precisely, let us define the free energy g as a function of the volume fractions of surfactant ϕ_s and apolar phase ϕ_a, that is, $g(\phi_s, \phi_a)$. The points of the spinodal line are calculated each time the determinant of the Hessian matrix is equal to zero:

$$\begin{vmatrix} \dfrac{\partial^2 g(\phi_s, \phi_a)}{\partial \phi_s^2} & \dfrac{\partial^2 g(\phi_s, \phi_a)}{\partial \phi_s \partial \phi_a} \\[2ex] \dfrac{\partial^2 g(\phi_s, \phi_a)}{\partial \phi_a \partial \phi_s} & \dfrac{\partial^2 g(\phi_s, \phi_a)}{\partial \phi_a^2} \end{vmatrix} = 0. \tag{21.12}$$

The lamellar instability, that is, the case where the trivial solution with all inter-facial curvatures is zero with lower free energy, has also been calculated in order to determine an eventual overlap between these two kinds of instabilities. The lamellar instability was solely calculated from the Helfrich free energy without considering the entropic term. Indeed, in the case of lamellar structure, the spontaneous curva-ture H_0 and the average Gaussian curvature $\langle K \rangle$ are everywhere equal to zero:

$$f_{\text{free}}^{\text{lam}} = 2\frac{S}{V}\langle H^2 \rangle. \tag{21.13}$$

In this study, we have chosen to model the behavior of the water/n-octane/$C_{10}E_5$, a widely studied "flexible" microemulsion (Chevalier and Zemb 1990). In order to plot the ternary phase diagram, we have expressed the volume fraction of the polar and apolar partitions of total space and the volume-to-surface ratio of the surfactant in terms of volume fraction of oil, water, and surfactant taking into account the density of all the components. Therefore, to determine the volume fraction of surfactant, we took into account the percentage of the surfactant chain in the apolar phase (75%) and the head area of the polar group of a typical oil-soluble surfactant, that is, 50 Å^2 (Banc et al. 2010).

21.2.3 SYSTEMS STUDIED

In the study, we generated different types of microemulsions far away from the "symmetrical" composition 50% polar, 50% apolar case, where most interesting terms disappear. For clarity, we report in Table 21.1 all the studied microemulsions as a function of the bending elastic constant κ^*, the Gaussian elastic constant κ, the effective rigidity constant $\bar{\kappa}$ (see Section 21.5.1), the topological parameter τ (see Section 21.5.2), and the volume fractions of surfactant ϕ_s and oil ϕ_a.

TABLE 21.1
Parameters Used for the Different Studied Microemulsions

Figure	κ^a	$\bar{\kappa}^b$	κ^{*c}	τ^d	S/V^e	ϕ_a^f	x_w^g	x_o^h	x_s^i	Microstructure Type
21.2a	1.00	−1.00	1.00	1.00	0.26	0.57	0.34	0.32	0.34	Flexible bicontinuous
21.2b	2.00	−1.00	3.00	0.50	0.26	0.57	0.34	0.32	0.34	Rigid bicontinuous
21.2c	3.00	−1.00	5.00	0.33	0.26	0.57	0.34	0.32	0.34	Ultra-rigid bicontinuous
21.2d	1.00	−1.00	1.00	1.00	0.19	0.85	0.09	0.67	0.24	Frustrated connected droplets
21.2e	2.00	−1.00	3.00	0.50	0.19	0.85	0.09	0.67	0.24	Frustrated locally lamellar
21.2f	3.00	−1.00	5.00	0.33	0.19	0.85	0.09	0.67	0.24	Frustrated locally lamellar
21.6 (1)	2.00	0.00	4.00	0.00	0.14	0.91	0.05	0.77	0.18	Frustrated locally lamellar
21.6 (2)	2.00	0.00	4.00	0.00	0.16	0.71	0.24	0.55	0.21	Oil rich near the plait point
21.6 (3)	2.00	0.00	4.00	0.00	0.31	0.85	0.05	0.55	0.40	Unfrustrated locally lamellar

[a] Bending elastic constant (in $k_B T$).
[b] Gaussian elastic constant (in $k_B T$).
[c] Effective rigidity constant (in $k_B T$).
[d] Topological parameter.
[e] Surface to volume ratio of the surfactant (in nm^{-1}).
[f] Apolar volume fraction.
[g] Water volume fraction.
[h] Oil volume fraction.
[i] Surfactant volume fraction.

21.3 INFLUENCE OF THE BENDING ELASTIC CONSTANT κ

The bending elastic constant is related to the incompressibility of the surfactant in terms of area per molecule or, in other words, to the rigidity of the surfactant film. Increasing (decreasing) the rigidity of the surfactant film can be interpreted in terms of raising (diminishing) the number of carbon atom in the surfactant chain. Indeed, the study of Sottmann et al. on 19 nonionic D_2O/n-alkane/C_iE_j ternary microemulsions pointed out that the larger the number of carbon atoms in the surfactant chain was, the larger was the rigidity of the surfactant (Sottmann et al. 1997). It is expected from macroscopic body physics that rigidity goes like the cube of thickness. This has been shown to be surprisingly valid (Roux et al. 1992). Most studies with single-chain nonionic linear ethoxylates deal with 10 or 12 carbons in the chain, and the value of κ is close to 1 $k_B T/a_0$. For double-chain, ionic, or large head-groups, the range is 1–5 $k_B T/a_0$. For branched surfactants such as the widely studied AOT or HDEHP, the bending constants are less than 1 $k_B T/a_0$. Here, we investigate the influence of the bending elastic constant κ on the calculated ternary phase diagram and the scattering functions predicted by the Gaussian random field model. To this end, we kept a fixed value for the Gaussian elastic constant $\bar{\kappa}$ to -1 $k_B T$.

21.3.1 TERNARY PHASE DIAGRAMS

Increasing the κ value from 1 to 3 $k_B T$ changes slightly the domains where the spinodal instabilities are located in the ternary phase diagrams (Figure 21.1).

Although the "height" of the instability is the same whatever the value of κ, we observe a broader instability domain when decreasing the bending elastic constant, especially in the water and oil corners of the ternary phase diagram. Indeed, for a κ value equal to 1 $k_B T$, an instability is observed in the ($x_o < 0.10$, $x_s \sim 0.15$) region, whereas for a 3 $k_B T$ bending elastic constant, the instability has almost disappeared. The height of the shaded domain at the bottom of the phase triangle is referred to in classical engineering as the "efficiency" of the surfactant. Note that for κ = 3 $k_B T$, another instability domain is observed from $x_o \sim 0.50$ and $x_s = 0.20$. Therefore, the influence of the bending elastic constant κ—keeping the $\bar{\kappa}$ parameter constant—is very weak on the ternary phase diagrams, which means that the surfactant efficiency is not affected.

21.3.2 MICROSTRUCTURES AND SCATTERING PROPERTIES

21.3.2.1 Representation in the Direct Space

Although weak changes are observed on the ternary phase diagrams as a function of the bending elastic constant, this latter parameter has a significant influence on the microemulsion structures. Indeed, for both bicontinuous and frustrated microemulsions, the persistence length of the system, as defined by de Gennes and Taupin (1982), depends on the bending elastic constant κ (Figure 21.2).

Increasing the κ parameter and, thus, the rigidity of the surfactant film promotes the formation of larger droplets. This is easily understood, since the larger the κ parameter is, the harder it is to bend the surfactant film to form small droplets.

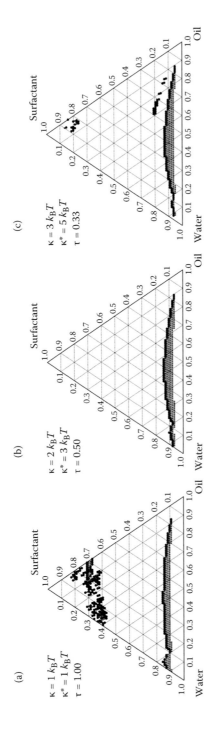

FIGURE 21.1 Ternary phase diagrams calculated for $\bar{\kappa} = -1\,k_{\mathrm{B}}T$ and $\kappa = 1\,k_{\mathrm{B}}T$ (a), $2\,k_{\mathrm{B}}T$ (b), and $3\,k_{\mathrm{B}}T$ (c) and a zero spontaneous curvature H_0. Points represent the spinodal instabilities. The shaded areas at the bottom of the triangles are two- or three-phase domains. The unshaded area is the area where a defined minimum of free energy for a bicontinuous structure without long-range order exists.

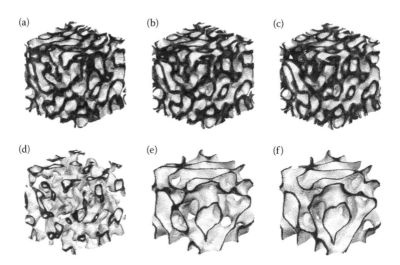

FIGURE 21.2 3D field projections of the direct space representation of nonsymmetrical microemulsions obtained for a zero spontaneous curvature as predicted by the Gaussian random field. For the illustrations, water is black, oil is light gray, and the surfactant is gray. The side lengths of the figures are 20π nm ~ 63 nm. (Top) Bicontinuous microemulsions $\phi_a = 0.57$, $Sl_s/V = 0.26$, and $\bar{\kappa} = -1\,k_BT$: (a) flexible $\kappa = 1\,k_BT$, (b) rigid $\kappa = 2\,k_BT$, (c) ultra-rigid $\kappa = 3\,k_BT$. (Bottom) Frustrated microemulsions $\phi_a = 0.85$, $Sl_s/V = 0.19$, and $\bar{\kappa} = -1\,k_BT$: (d) connected droplets $\kappa = 1\,k_BT$, (e) rigid locally lamellar $\kappa = 2\,k_BT$, (f) ultra-rigid locally lamellar $\kappa = 3\,k_BT$. All the characteristics of these microemulsions are reported in Table 21.1.

However, it seems that the influence of the bending elastic constant is less pronounced in the case of bicontinuous microemulsions compared to the frustrated ones. Indeed, looking at the 3D field projections, no difference is observed between the rigid and the ultra-rigid bicontinuous microemulsions (Figure 21.2b and c) since microemulsions begin to become locally lamellar. In the case of the flexible bicontinuous microemulsion (Figure 21.2a), the locally lamellar microstructure is not yet observed, since the structure is intermediate between a connected droplet-like and a locally lamellar-like structure. On the other hand, this effect is more accentuated for frustrated microemulsions since we observed the transition from a connected droplet structure to a locally lamellar structure when the bending elastic constant κ is increased (Figure 21.2d–f). This can be explained by the volume ratio between the water and the oil. Indeed, in the case of the well-studied symmetric bicontinuous microemulsions, the amount of water and oil is almost the same, whereas in the case of frustrated microemulsions, the amount of oil is much larger than the one of water, while there is no spontaneous curvature. Therefore, the structure of the bicontinuous microemulsions is more "frozen" than the frustrated one, explaining the difficulty to change the microstructures (even when the κ value is increased) without increasing the free energy linked to the microstructure. Since the spontaneous curvature of the system is zero in these cases, the only energy that the system can have comes from the bending and Gaussian elastic constants. Changing the spontaneous curvature of the surfactant film will certainly induce different microstructures when increasing κ

even in the case of bicontinuous microemulsions. Note that, although we have represented here only oil-rich frustrated microemulsion, the same effect is observed for water-rich frustrated microemulsions.

21.3.2.2 Scattering Properties

The microstructures shown in Figure 21.2 are indeed different by local geometry, as well as local connectivity. But it is difficult, using only the naked eye, to determine if they correspond to coalescence droplets or connected cylinders, or if they are locally lamellar (foam- or sponge-like structures)—the three possible topologies of dispersions based on Voronoï tessellation (Barnes et al. 1988; Zemb 1997).

In order to approach this, we generated all the scattering functions in the reciprocal q-space, as already done in two dimensions (Welberry and Zemb 1988). These are shown in Figure 21.3.

First of all, all the three classes of scattering patterns observed up to now for microemulsions are produced by our model:

- Scattering as a broad peak, which can be roughly fitted at first by three phenomenological parameters, such as the Teubner–Strey formula (Teubner and Strey 1987) or at higher order to four parameter fits (Choi et al. 1997).

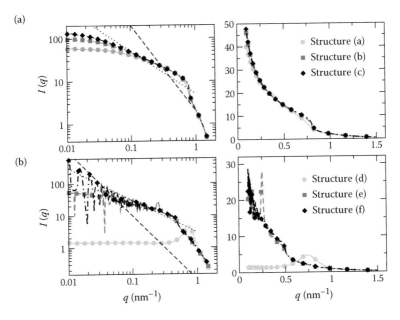

FIGURE 21.3 Predictions of the scattering functions calculated for the microemulsion structures presented in Figure 21.2. (a) Bicontinuous microemulsions: flexible (light gray circles), rigid (gray squares), and ultra-rigid (black diamonds). (b) Frustrated microemulsions: connected droplets (light gray circles), rigid locally lamellar (gray squares), and ultra-rigid locally lamellar (black diamonds). Left and right panels correspond to a representation of the scattering function predictions in logarithm and linear scales, respectively. Note that the q^{-2} (black dashed line) and q^{-1} (black dotted line) are also shown for the logarithm scale.

Microstructures associated can go from disconnected droplets to locally lamellar microstructures (Barnes et al. 1990).

- Scattering with a low-q behavior similar to a living polymer, usually designed as cylindrical or wormlike micelle swollen by the internal phase. This microstructure is topologically close to droplets coalescing with nearest neighbors and designed as connected cylinders (Ninham et al. 1987).
- The intermediate case, when a more or less defined peak or shoulder separates a high-q domain with q^{-2} to q^{-4} behavior from weakly varying intensities at low q (Auvray et al. 1986). This intermediate case was used to propose a general relation linking the position of this broad peak position to the volume fraction ϕ_a and area per unit volume S/V (de Gennes and Taupin 1982) $D^* \times S/V = 6\phi_a(1 - \phi_a)$.

More precise evaluation in the case of flexible films gives the numerical constant 5.87 instead of 6. This prediction has been confronted to several experimental results (Testard and Zemb 2002). These scattering functions allow us to better identify the structures of our systems. Indeed, for the bicontinuous microemulsions, we observe that the general appearance of the spectra is almost the same (Figure 21.3, top). For large q values we observe the "Porod" limit with a q^{-4} dependence (Porod 1951)

$$\lim_{q \to \infty} I(q) = \frac{2\pi(\Delta\rho)^2}{q^4} \frac{S}{V}, \tag{21.14}$$

where $\Delta\rho$ is the difference between the densities of both phases (polar and apolar). This means that the interface between these two phases is clearly defined and smooth (Spalla 2002). This was indeed observed on the 3D field projections in the direct space. Look now at the behavior of $I(q)$ at small q values. It is difficult to determine any general power-law q dependence of the spectra since several slopes can be calculated for different small q ranges. However, we can compare the q dependence as a function of the bending elastic constants used (from 1 to 3 $k_B T$). Indeed, it is well known that the q dependence at small q values allows determining, the local structure of the microemulsion, i.e. q^{-2} corresponds to a locally lamellar structure, $q^{-1.6}$ to $q^{-1.0}$, and to a locally cylindrical one.

When increasing the κ value, we observe an increase of the q dependence of the spectra. Therefore, one may conclude that the more rigid the surfactant is, the more locally lamellar-like the structure is. This was actually observed in the direct space representations. Note that when looking at the 3D field representations, we were able to distinguish the structures between the flexible and the rigid and ultra-rigid bicontinuous microemulsions. Nevertheless, it was hard to see the difference between the rigid and the ultra-rigid ones. Examining the scattering functions, we are now able to discriminate between the structures corresponding to these two microemulsions and conclude that the ultra-rigid bicontinuous microemulsion is more locally lamellar than the rigid one.

As already mentioned for the 3D field projections, the transition from the coalesced droplet to the locally lamellar structure is easier observed for frustrated oil-rich (or water-rich) microemulsions. Also, this phenomenon is clearly observed on the scattering functions (Figure 21.3, bottom). Again the difference between flexible and rigid (or ultra-rigid) frustrated microemulsions is clearly observed since the scattering functions have a totally different global aspect. Indeed, at small q values, the behavior of the flexible microemulsion is close to the one observed for droplets, that is, q^0, whereas for the rigid one, a q^{-n} dependence can be calculated with $n > 0$. Now, look at the difference between rigid and ultra-rigid frustrated microemulsions. As already observed for the bicontinuous microemulsions, the scattering properties of rigid and ultra-rigid microemulsions are similar compared to the flexible ones. However, some changes are observed. Indeed, the q^{-n} dependence at small q values is different, and it seems that the n value of this dependence is higher in the case of the ultra-rigid microemulsion, as already mentioned. In any case, the q^{-1} to q^{-2} dependence is found in a large number of experiments and relates to the closest approximation of local microstructure as locally lamellar or locally cylindrical connected structures. It must be noted that when ϕ_a is close to 0.5, the low-q part is not informative because the scattering plateau observed at low q mainly reflects the osmotic compressibility. Moreover, in the case of the rigid frustrated microemulsion, a broad correlation peak appears at about 0.25 nm^{-1}, whereas for the ultra-rigid one, no peak is observed. The vanishing of the peak enlightens the trend to form more locally lamellar structures when going from rigid to ultra-rigid systems.

Note that, in most cases—bicontinuous or frustrated—the scattering functions of these microemulsions cannot be fitted by the Teubner and Strey formula (Teubner and Strey 1987)

$$I(q) = \frac{1}{a_2 + c_1 q^2 + c_2 q^4}. \tag{21.15}$$

Figure 21.4 shows a scattering pattern produced by our model, as fitted to a Teubner–Strey three-parameter expression. The fit is fair around the bump or in the

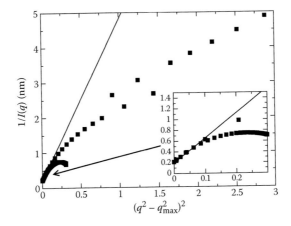

FIGURE 21.4   Variation of $I_{max}/I(q)$ as a function of $\left(q^2 - q_{max}^2\right)^2$ calculated using the Teubner and Strey parameters a_2, c_1, and c_2 (line) and the calculated scattering function (symbol).

Porod regime: most used in literature is fitting in linear scale and avoiding the Porod domain. However, a great advantage of the Teubner–Strey fitting is that it allows determining the main spatial frequency D^* and the persistence length ξ

$$\xi = \left[\frac{1}{2} \sqrt{\frac{a_2}{c_2}} + \frac{1}{4} \frac{c_1}{c_2} \right]^{-1/2} \tag{21.16a}$$

$$D^* = 2\pi \left[\frac{1}{2} \sqrt{\frac{a_2}{c_2}} - \frac{1}{4} \frac{c_1}{c_2} \right]^{-1/2}. \tag{21.16b}$$

In most cases studied so far, that is, for fluid microemulsions and close to symmetry between polar and apolar volume ($\phi_a = 0.5$), a natural ratio D^*/ξ close to 1 has been reported experimentally. This experimental finding is also reproduced by our model ($\xi = 7.1$ nm and $D^* = 8.3$ nm). The deep reason is that Voronoï tessellation, with random seed points, is an efficient basis for description of fluid dispersion (Hyde et al. 1989).

For a good visualization of the validity of the inverse quadratic expression by Teubner and Strey, one needs to plot the inverse of scattering versus quadratic expression in q: a perfect fit would mean a straight line only since

$$\frac{1}{I(q)} = c_2 (q^2 - q_{max}^2)^2 + a_2 + c_2 q_{max}^4. \tag{21.17}$$

Deviation versus straight line illustrates the approximations in the fit (Figure 21.4).

A good agreement between our scattering function and the Teubner and Strey formula is observed near the maximum of the peak $(q^2 - q_{max}^2)^2$ values. However, a quite important difference between both is observed relatively quickly. This means that, when a correlation peak is observed on the scattering functions, the Teubner and Strey formula is valid only for a short range of q value (centered at the maximum of the peak). Hence, it seems that the Teubner and Strey formula cannot be used in an attempt to model the scattering function for the whole q value range.

As a conclusion, we can say that, although the bending elastic constant κ has no (or very few) influence on the oil–water–surfactant ternary phase diagrams, it modifies the structure of the system and the scattering spectra. Indeed, it seems that increasing the bending elastic constant promotes the formation of locally lamellar structures for both bicontinuous ($x_o \sim x_w \sim x_s$) and frustrated ($x_o \gg x_w$ or $x_o \ll x_w$) microemulsions. This phenomenon is observed on both 3D field projections in the direct space and scattering functions (in the inverse space). Moreover, it appears that the Teubner and Strey formula—defined to reproduce the scattering properties—can be used in very few cases, since the presence of a "visible" correlation peak is not obvious, contrary to a commonly accepted assumption (Teubner and Strey 1987).

21.4 INFLUENCE OF THE GAUSSIAN ELASTIC CONSTANT $\bar{\kappa}$

Although the role of the bending elastic constant κ is relatively well understood for microemulsions, much less is known for the Gaussian elastic constant (Safran 1991), which is linked to the number of nearest neighbors with the surfactant film (Charvolin and Sadoc 1990). Indeed, this parameter is linked to the topology of the microemulsions and more precisely to the polydispersity (shape and size fluctuations of the droplets), as a consequence of the Gauss–Bonnet theorem. Here, we look at the influence of the Gaussian elastic constant $\bar{\kappa}$ on the ternary phase diagrams and the microemulsion structures (3D visualizations and scattering functions). To investigate this point, we have increased the $\bar{\kappa}$ parameter (from –2 to 0 $k_B T$), keeping fixed the bending elastic constant κ (2 $k_B T$).

21.4.1 TERNARY PHASE DIAGRAMS

Contrary to what was observed for the influence of κ, the ternary phase diagrams depend strongly on the $\bar{\kappa}$ parameter (Figure 21.5).

Indeed, the height of the spinodal instability in the ternary phase triangle increases with the $\bar{\kappa}$ parameter, involving therefore a decrease of the surfactant efficiency. However, in the case of nonzero $\bar{\kappa}$, the two phase diagrams calculated have a similar appearance with only one instability domain for low surfactant volume fraction. For the zero $\bar{\kappa}$ value, in addition to the "central" instability domains already observed for the nonzero $\bar{\kappa}$ values, two other instability domains are observed on each side of the ternary phase diagram for surfactant volume fraction between 0.2 and 0.4 (Figure 21.5, right). These two instability domains, located on each side of the triangle, bring up two ultra-stable corridors. These two corridors allow the transition from the frustrated zone (below the instability) to the unfrustrated zone (above the instability). Looking at the ternary phase diagram calculated for a κ value between –1 and 0 $k_B T$ (not shown), no additional instability domains have been observed. However, an instability line almost parallel to the surfactant axis can be seen. Is this line the precursor of additional instability domain? This is not known at the present.

21.4.2 MICROSTRUCTURES AND SCATTERING PROPERTIES

Contrary to the influence of the bending elastic constant κ, the Gaussian elastic constant $\bar{\kappa}$ has no clear influence on the microemulsion structures. Indeed, only the scattering function calculated for $\bar{\kappa} = 0\ k_B T$ differs from the others—the others being exactly the same. This is certainly due to the undefined system of underlying relations, when Gaussian elastic constant is zero. Indeed, when one of the elastic constants is zero, the Helfrich decomposition with only first- and second-order terms in curvature is inconsistent (Fogden et al. 1991). Therefore, we focus here on the influence of the oil and water volume fractions on the 3D field projections and scattering properties. Since, for $\bar{\kappa} = 0\ k_B T$, the ternary phase diagram calculated is totally different from all other phase diagrams calculated, we focus now on understanding the different microstructures for this special case, labeled (1), (2), and (3) on Figure 21.5 (right). These three microemulsions correspond to a frustrated one (1),

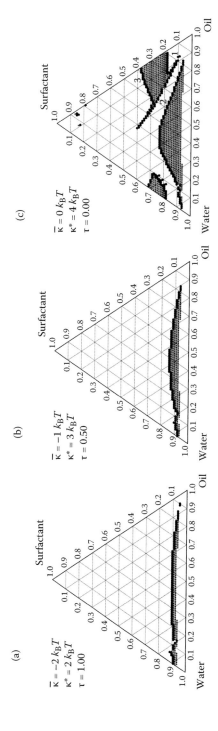

FIGURE 21.5 Ternary phase diagrams calculated for $\kappa = 2\, k_BT$ and $\bar{\kappa} = -2\, k_BT$ (a), $-1\, k_BT$ (b), and $0\, k_BT$ (c) and a zero spontaneous curvature H_0. Points represent the spinodal instabilities.

an oil-rich near the plait point (which corresponds to the point where the binodal and the spinodal should meet) located in the ultra-stable corridor (2), and an unfrustrated one (3). As already mentioned, the microemulsion structure depends on the concentration of oil, water, and surfactant (Figure 21.6).

Indeed, looking at the 3D field projections in the direct space, we can notice that, although the frustrated [Figure 21.6 (1)] and the unfrustrated [Figure 21.6 (3)] structures are locally lamellar, the transition from one state to another (via the ultra-stable corridor) goes through a bicontinuous-like structure [Figure 21.6 (2)]. This is observed on the scattering functions. Indeed, for the bicontinuous structure, we calculated a scattering function close to the one calculated for the droplet structure (Figure 21.3, bottom), that is, a q^{-4} and a q^0 dependence at large and small q values, respectively. Although these two systems (connected droplets and bicontinuous) have similar scattering functions, their 3D field projections look totally different. The link between "typical" scattering function and the various kinds of microemulsion is not bijective. This means that scattering alone cannot reflect properties such as connectivity and tortuosity. Supplementary information can be obtained via perturbation. Perturbations can be either a dilution at a constant ratio of two of the components or an addition of a solute of known volume and interfacial area. This determination from dilution curves has been done experimentally in several cases (Testard and

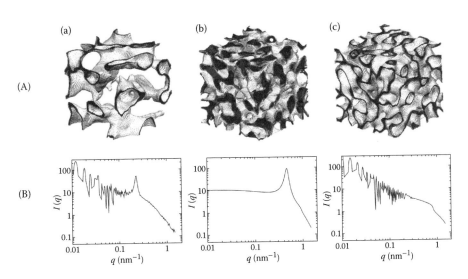

FIGURE 21.6 (A) 3D field projections of the direct space representation of microemulsions obtained for a zero spontaneous curvature and the set of parameters ($\kappa = 2\,k_B T$, $\bar{\kappa} = 0\,k_B T$) as predicted by the Gaussian random field. For the illustrations, water is black, oil is light gray, and the surfactant is gray. The side lengths of the figures are 20π nm ~ 63 nm. (a) Frustrated microemulsion $\phi_a = 0.91$ and $Sl_s/V = 0.14$, (b) oil-rich near the plait point $\phi_a = 0.71$ and $Sl_s/V = 0.16$, and (c) unfrustrated microemulsion $\phi_a = 0.85$ and $Sl_s/V = 0.31$. (B) Corresponding predictions of the scattering functions calculated for the previous microemulsion structures (A). All the characteristics of these microemulsions are reported in Table 21.1.

Zemb 2002), and initially using peptides and proteins (Huruguen et al. 1992; Pileni et al. 1985; Zemb et al. 1990) and lindane (Testard et al. 1997; Testard and Zemb 1998) as well as model aroma (Tchakalova et al. 2008a,b). The crucial information about the type of microstructures with very similar microstructures comes from the variation of slopes and peak position, as well as the variation of the S/V ratio induced by the addition of one component. Concerning the two locally lamellar structures (frustrated and unfrustrated), we found again a q^{-n} dependence at small q values, with n between 1 and 2. Note that in the case of the frustrated structure, in addition to the decrease of the scattering intensity, a correlation peak appears (at about 0.25 nm^{-1}). This peak was not observed for the other frustrated locally lamellar structures. We can notice here that, when looking at the transition from the unfrustrated state to the frustrated one, we observe a displacement of the correlation peak toward the small q values. This effect has already been observed (Testard et al. 1997). This means that the persistence length of the system, through this transition, tends to increase. This is in agreement with the observations made on the microemulsion swelling.

As a conclusion, we can say that the influence of the Gaussian elastic constant $\bar{\kappa}$ is totally different from that of the bending elastic constant κ. Indeed, although only the microstructures and the scattering functions (for a given composition of oil, water, and surfactant) depend on κ, the shape of the ternary phase diagram is strongly related to the $\bar{\kappa}$ value. For a fixed value of the bending elastic constant, increasing the $\bar{\kappa}$ diminishes the surfactant efficiency since larger unstable domains are calculated. In the "special" case of a zero Gaussian elastic constant, two ultra-stable domains are observed. These domains, through which the transition between the frustrated and unfrustrated states is made, are located on each side of the ternary phase diagram. It seems that the universal microstructure in these ultra-stable domains is bicontinuous, whereas the frustrated and unfrustrated structures are closer to locally lamellar.

21.5 SHOULD WE CONSIDER κ AND $\bar{\kappa}$ SEPARATELY?

As seen previously, the bending κ and Gaussian $\bar{\kappa}$ elastic constants have different influences on the ternary phase diagram, the microemulsion structures, and the scattering functions. The question we address here is the following: Is it wise to decorrelate the influence of κ without taking account of $\bar{\kappa}$, and vice versa? To answer this question, we will consider here the effect of two parameters depending on both κ and $\bar{\kappa}$, that is, the effective rigidity constant κ^* and a topological parameter named τ.

21.5.1 EFFECTIVE RIGIDITY CONSTANT

In the Helfrich formalism, the free energy is minimized as a function of the spontaneous H_0, the mean H, and the Gaussian κ curvatures for given bending κ and Gaussian $\bar{\kappa}$ elastic constants. In the 1970s, Israelachvili et al. suggested that the surfactant shape was driven by the so-called surfactant packing parameter p (Israelachvili et al. 1976). Therefore, the relevant parameters that we should consider are the spontaneous p_0 and mean p packing parameters, instead of H_0, H, and K (Equations 21.3 and 21.6). Considering now the effective bending constant κ^* (Equation 21.8), and not κ

and $\bar{\kappa}$ separately anymore, one may wonder if this unique scale parameter controls the phase diagrams and the microstructures.

Looking first at the ternary phase diagrams, we observed no uniform variation of the phase diagram as a function of κ^*. Indeed, compiling the results presented in Figures 21.1 and 21.5, we notice that the ternary phase diagram calculated for $\kappa^* = 4\ k_BT$ (Figure 21.5, right) does not correspond to an intermediate phase diagram between $\kappa^* = 3\ k_BT$ (Figure 21.1, middle) and $\kappa^* = 5\ k_BT$ (Figure 21.1, right). This is also the case for $\kappa^* = 2\ k_BT$ (Figure 21.5, left), since, for this value, the calculated spinodal instability is flatter than the ones calculated for $\kappa^* = 1\ k_BT$ (Figure 21.1, left) and $\kappa^* = 3\ k_BT$ (Figure 21.1, middle).

Concerning the variation of the scattering properties as a function of κ^*, no clear influence of $\bar{\kappa}$ is observed, since, as mentioned above, it is quite difficult to determine the influence of $\bar{\kappa}$ on the scattering functions. Indeed, when looking only at the influence of the bending elastic constant κ variation, one may conclude that the effective bending constant κ^* has the same influence as κ. However, when looking at the influence of the Gaussian elastic constant $\bar{\kappa}$, exactly the same scattering functions have been generated for two different $\bar{\kappa}$ values.

Therefore, it seems that neither the shape of the ternary phase diagram nor the scattering properties are driven by the effective bending constant κ^*. This is quite surprising since κ^* represents the rigidity of the surfactant film, as seen from a molecular point of view: incompressibilities, no tearing. This points out that other parameters—depending on κ and $\bar{\kappa}$—may control the ternary phase diagram and the scattering properties.

21.5.2 TOPOLOGICAL PARAMETER

Another parameter, depending on both κ and $\bar{\kappa}$, can be defined, that is, a topological parameter named τ. This dimensionless parameter was introduced in the 1990s by Pieruschka et al. and corresponds to the ratio between the Gaussian and the bending elastic constants (Pieruschka and Marčelja 1994)

$$\tau = -\bar{\kappa}/\kappa. \tag{21.18}$$

In their study, they looked at the influence of the τ parameter on the microemulsion structures. They enlightened four different τ ranges, each domain corresponding to a specific structure. Note that the well-known stability criterion for bending energy is $-2\kappa < \bar{\kappa} < 0$; therefore, no stable structure can be calculated for $\tau < 0$ and $\tau > 2$.

From our results, we observed a well-defined variation of the ternary phase diagram shape as a function of τ. We notice that when the τ parameter decreases, the instability domain becomes larger. Moreover, it seems that decreasing τ until the threshold value zero promotes the formation of other instabilities (compared to the central one) on each side of the triangle. Indeed, for $\tau = 0.33$ (Figure 21.1, right), we observed the beginning of the instability line formation, already described for $\tau = 0$ (Figure 21.5, right). Other calculations performed for $\tau = 0.36$ (not shown) also show this instability line in the oil domain. However, it seems that the ternary phase diagram calculated for this τ value has an intermediate shape between those calculated for $\tau = 0.33$ and 0, which is not coherent when looking at the effect of the parameter on the shape of the phase diagrams. As a conclusion, it seems that the only parameter that drives the shape of the ternary phase diagram might be the Gaussian elastic constant $\bar{\kappa}$. This hypothesis is supported by the fact that a ternary phase diagram calculated for $\kappa = 1\ k_BT$ and $\bar{\kappa} = 0\ k_BT$ ($\kappa^* = 2\ k_BT$ and $\tau = 0$) has a similar shape as the one calculated for $\kappa = 2\ k_BT$ and $\bar{\kappa} = 0\ k_BT$ ($\kappa^* = 4\ k_BT$ and $\tau = 0$).

For the scattering properties, again, since the influence of $\bar{\kappa}$ is not clearly defined, no interpretation of the influence of τ can be made. Indeed, since the same scattering functions are calculated for the ($\kappa = 2\ k_BT$, $\bar{\kappa} = -2\ k_BT$, $\kappa^* = 2\ k_BT$, and $\tau = 1$) and ($\kappa = 2\ k_BT$, $\bar{\kappa} = -1\ k_BT$, $\kappa^* = 3\ k_BT$, and $\tau = 0.5$) systems, no conclusion can be given.

21.6 INFLUENCE OF SPONTANEOUS CURVATURE H_0 SET BY CHEMICAL COMPOSITION

Few experimental results are available in the literature concerning the influence of the spontaneous curvature of the surfactant H_0 on the ternary phase diagrams and the microemulsion structures. However, some studies report the influence of temperature on the phase diagrams (Anderson et al. 1989; Deen and Pedersen 2008) and scattering functions (Chen et al. 1990; Lee et al. 1990), the spontaneous curvature being related to the temperature

$$H_0(T) \propto (\bar{T} - T), \tag{21.19}$$

where T is the temperature of the system and $\bar{T}$ is the phase inversion temperature (Strey 1994). Since, in our model, the free energy of the microemulsion is minimized as a function of the spontaneous curvature, and other parameters, we have the possibility to observe directly the influence of H_0 on microemulsion properties or, more precisely, the influence of the dimensionless parameter $H_0 l_s$ (l_s being the surfactant thickness). Note that since the spontaneous packing parameter p_0 can be defined as a function of $H_0 l_s$ (Equation 21.3), looking at the influence of $H_0 l_s$ means looking at the influence of p_0. Here, we investigate the influence of the spontaneous curvature ($H_0 l_s$ between -0.50 and 0.30) on the ternary phase diagrams and the microemulsion structures—for the ($\kappa = 2\ k_BT$, $\bar{\kappa} = 0\ k_BT$) system—that have a peculiar phase diagram. Note that we use the convention that curvature toward the water is positive.

21.6.1 TERNARY PHASE DIAGRAMS

As mentioned above, this system provides a peculiar ternary phase diagram at zero spontaneous curvature, with two ultra-stable domains on each side of the central instability (Figure 21.5, right). What about the instability domains and the location of these ultra-stable domains when changing the spontaneous curvature? Since the sign of H_0 drives the curvature either toward water or oil, changes are observed on the ternary phase diagrams as a function of $H_0 l_s$ (Figure 21.7).

Indeed, when increasing $H_0 l_s$, we observe a displacement of the instabilities from the water domain to the oil domain. This can be explained as follows. For high negative

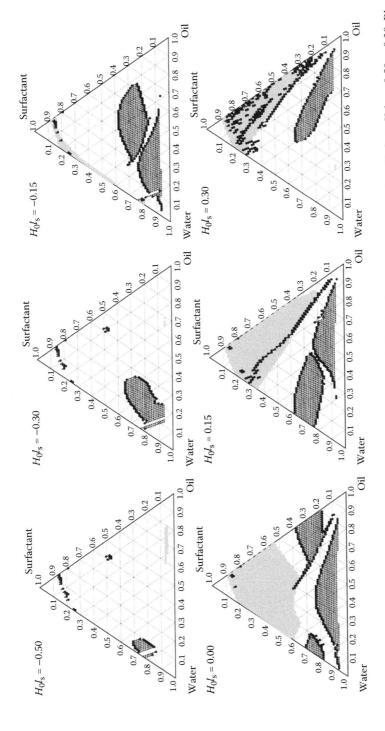

FIGURE 21.7 Ternary phase diagrams calculated for $\kappa = 2\,k_B T$ and $\kappa = 0\,k_B T$ and various spontaneous curvature H_0 from $H_0 l_s = -0.50$ to 0.30. Black points represent the spinodal instabilities, and gray points represent the lamellar instabilities.

spontaneous curvature ($H_0 l_s = -0.50$), the microemulsion is composed of small drop-lets of oil in water (since $H_0 = 1/R$, where R is the droplet radius). The decrease of $|H_0 l_s|$ ($H_0 l_s = -0.30$) tends to enlarge the droplet size, allowing thus a larger incorpora-tion of oil in the droplets. Therefore, we observe the displacement of this instability to the oil domain. Continuing to diminish $|H_0 l_s|$ ($H_0 l_s = -0.15$), a break (into two parts) of the instability is observed, promoting an ultra-stable region between both parts of the instability. Again, since $H_0 l_s$ increases, the size of the droplets grows and, for the same reason as the one argued above, the instability moves to the oil domain. Now, for a zero spontaneous curvature, as already described previously, two ultra-stable domains are formed on each part of the central instability. This means that both oil droplets in water and water droplets in oil are stable. Note that, in this case, the main instability is centered at almost $x_o \simeq x_w \simeq 0.40$, corresponding to an apolar volume fraction of 0.55. Increasing then $|H_0 l_s|$ ($H_0 l_s = 0.15$ and 0.30), the same effects are observed compared to negative $H_0 l_s$ values. Indeed, the ternary phase diagrams cal-culated for $H_0 l_s = -0.15$ and $H_0 l_s = 0.15$ are similar and symmetric compared to $H_0 l_s = 0$. Although a break of the instability was observed for the negative value of $H_0 l_s$, for the positive value, a "fusion" between two instability domains is observed. Note that this "fusion" occurs on the water side, since the curvature is toward water for positive H_0, whereas it was toward oil for negative H_0. Finally, for high positive $H_0 l_s$, it seems that the structure composed of small droplets of water in oil is the stable structure of microemulsion, which is the contrary to what we notice for highly negative sponta-neous curvature. We can notice that for nonzero positive spontaneous curvature, the instability line calculated for $H_0 l_s = 0$—from ($x_w = 0.08$, $x_o = 0.83$) to ($x_w = 0.32$, $x_o = 0.29$)—remains. However, for $H_0 l_s = 0.15$ and 0.30, we cannot say if this line should actually be found since it is coincident with the calculated lamellar instability.

In conclusion, the spontaneous curvature has a strong influence on the general appear-ance of the ternary phase diagrams generated. Moreover, all "typical" phase diagrams have been experimentally observed (Chevalier and Zemb 1990). This implies that, from an experimental point of view, the temperature (driving the surfactant spontaneous cur-vature) is a crucial parameter to control the desired surfactant efficiency. However, we should keep in mind that we presented here a "special" ternary phase diagram (with a zero Gaussian elastic curvature). For nonzero Gaussian elastic constant, the break or fusion of instability domains is not necessarily observed, and only the passage of the instability from the water domain to the oil domain is observed (not shown). Indeed, this passage is the only phenomenon common to all (κ, $\bar{\kappa}$) systems observed on all the ternary phase diagrams when changing the curvature from water to oil, and vice versa.

21.6.2 Scattering Properties

Finally, we are interested in understanding the microemulsion structure variation as a function of the spontaneous curvature H_0 for a given composition of oil, water, and surfactant. As we noticed previously, changes in the ternary phase diagrams do not necessarily lead to changes in microstructures and, therefore, in scattering functions (Section 21.4). For this study, we have chosen to study the influence of $H_0 l_s$ on the scattering properties of the point labeled (2) in Figure 21.5 (right); that is, $\phi_a = 0.71$ and $S l_s/V = 0.16$ (Figure 21.8).

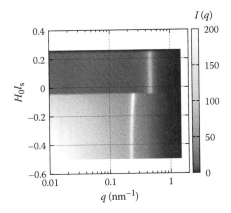

FIGURE 21.8   2D map of the scattering functions as a function of $H_0 l_s$ for the ($\kappa = 2\ k_B T$, $\bar{\kappa} = 0\ k_B T$) system, and $\phi_a = 0.71$ and $Sl_s/V = 0.16$ (corresponding to the point labeled (2) in Figure 21.5, right). Clear domains correspond to high peak intensities and the dark ones correspond to low peak intensities.

The variation of the scattering functions as a function of $H_0 l_s$ clearly shows an abrupt transition in the microemulsion structure for $H_0 l_s \simeq -0.05$. Two different persistence lengths can be therefore calculated. Indeed, above this threshold value of –0.05, the scattering functions calculated have the same behavior as the ones that we could calculate for the droplet microstructure (see Figure 21.6 (2)), since until the correlation peak ($q = 0.47$ nm^{-1}), the peak intensity is weak (dark zone) and constant. On the other hand, below this threshold value, it seems that the microemulsion might be a locally lamellar structure, since for small q the intensity is high and decreases until $q = 0.22$ nm^{-1} (from the white zone to the dark gray zone). Moreover, a slight displacement of the correlation peak is observed below $H_0 l_s = -0.05$, but there is no continuity between these two domains.

The transition from the "frustrated" state to the "unfrustrated" state can be explained as follows: as mentioned above, we adopt the convention that the curvature is toward water when H_0 is positive. Therefore, below the calculated threshold value (which is not exactly zero), only frustrated microemulsions are calculated since the curvature is now toward oil (H_0 negative). Above the threshold value, the microemulsions are not anymore frustrated since the curvature is in the "correct" direction. Note that the transition occurs exactly at the moment when this composition point we have chosen meets the instability. Therefore, because of the presence of the instability (and thus the phase separation), the microemulsion is more stable in a droplet structure than in a locally lamellar one.

21.7   CONCLUSION

Taking into account elastic energy, our model has been able to generate example microstructures of flexible, rigid, frustrated, and unfrustrated microemulsions. All shapes of scattering induced by the three possible types of local microstructures have been obtained theoretically from the model. The effect of bending constants, either

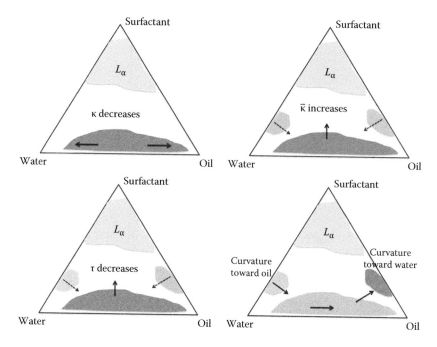

FIGURE 21.9 Schematic representations of the influence of the κ, $\bar{\kappa}$, τ, and H_0 parameters on a ternary phase diagram.

in the Helfrich–Gauss or in the generalized single-parameter form, has been given. In Gauss formulation or in the generalized form taking into account packing, the transition from frustrated to unfrustrated microstructure is near instabilities in phase diagram and is accompanied by a "jump" in scaled peak position.

Finally, it is very important to note that all these conclusions could be drawn (Figure 21.9) since we have examined the model far from a degenerate point, that is, when polar and apolar volume fractions are close to 50%, which has been up to now the object of a vast majority of studies, but all physical properties, including scattering via the Babinet principle, degenerate for symmetry reasons.

REFERENCES

Andelman, D., Cates, M. E., Roux, D. and Safran, S. A. 1987. Structure and phase equilibria of microemulsions. *J. Chem. Phys.*, 87(12):7229–7241.

Anderson, D., Wennerström, H. and Olsson, U. 1989. Isotropic bicontinuous solutions in surfactant—Solvent systems: the L3 phase. *J. Phys. Chem.*, 93(10):4243–4253.

Arleth, L., Marčelja, S. and Zemb, Th. 2001. Gaussian random fields with two level-cuts—Model for asymmetric microemulsions with nonzero spontaneous curvature? *J. Chem. Phys.*, 115(8):3923–3936.

Auvray, L., Cotton, J. P., Ober, R. and Taupin C. 1986. Structure of concentrated Winsor microemulsions by SANS. *Phys. B+C*, 136(13):281–283.

Banc, A., Bauduin, P. and Diat, O. 2010. Tracking an ion complexing agent within bilayers. *Chem. Phys. Lett.*, 494:301–305.

Barnes, I. S., Hyde, S. T., Ninham, B. W., Derian, P.-J., Drifford, M. and Zemb, Th. 1988. Small-angle x-ray scattering from ternary microemulsions determines microstructure. *J. Phys. Chem.*, 92(8):2286–2293.

Barnes, I. S., Derian, P.-J., Hyde, S. T., Ninham, B. W. and Zemb, Th. 1990. A disordered lamellar structure in the isotropic phase of a ternary double-chain surfactant system. *J. Phys. France*, 51:2605–2628.

Berk, N. F. 1991. Scattering properties of the leveled-wave model of random morphologies. *Phys. Rev. A*, 44(8):5069–5079.

Cahn, J. W. 1965. Phase separation by spinodal decomposition in isotropic systems. *J. Chem. Phys.*, 42(1):93–99.

Charvolin, J. and Sadoc, J.-F. 1990. Cubic phases as structures of disclinations. *Colloid Polym. Sci.*, 268(2):190–195.

Chen, S.-H., Chang, S.-L. and Strey, R. 1990. Structural evolution within the one-phase region of a three-component microemulsion system: Water-*n*-decane-sodium-bis-ethylhexylsulfosuccinate (AOT). *J. Chem. Phys.*, 93(3):1907–1918.

Chevalier, Y. and Zemb, Th. 1990. The structure of micelles and microemulsions. *Rep. Prog. Phys.*, 53(3):279–371.

Choi, S. M., Chen, S. H., Sottmann, T. and Strey, R. 1997. Measurement of interfacial curvatures in microemulsions using small-angle neutron scattering. *Phys. B*, 241–243:976–978.

de Gennes, P. G. and Taupin, C. 1982. Microemulsions and the flexibility of oil/water interfaces. *J. Phys. Chem.*, 86(13):2294–2304.

Deen, R. G. and Pedersen, J. S. 2008. Phase behavior and microstructure of $C_{12}E_5$ nonionic microemulsions with chlorinated oils. *Langmuir*, 24(7):3111–3117.

Eicke, H.-F. 1979. On the cosurfactant concept. *J. Colloid Interface Sci.*, 68(3):440–450.

Fogden, A., Hyde, S. T. and Lundberg, G. 1991. Bending energy of surfactant films. *J. Chem. Soc., Faraday Trans.*, 87(7):949–955.

Helfrich, W. 1973. Elastic properties of lipid bilayers—Theory and possible experiments. *Z. Naturforsch. C*, 28:693–703.

Huruguen, J., Zemb, Th. and Pileni, M.-P. 1992. Influence of proteins on the percolation phenomenon in AOT reverse micelles: Structural studies by SAXS. *Progr. Colloid Polym. Sci.*, 89:39–43.

Hyde, S. T., Ninham, B. W. and Zemb, Th. 1989. Phase boundaries for ternary microemulsions: Predictions of a geometric model. *J. Phys. Chem.*, 93(4):1464–1471.

Hyde, S. T., Andersson, S., Larsson, K., Blum, Z., Landh, T., Lidin, S. and Ninham, B. W. 1997. *The Language of Shape: The Role of Curvature in Condensed Matter: Physics, Chemistry and Biology*. Elsevier Science, Amsterdam.

Israelachvili, J. N., Mitchell, D. J. and Ninham, B. W. 1976. Theory of self-assembly of hydrocarbon amphiphiles into micelles and bilayers. *J. Chem. Soc., Faraday Trans. 2*, 72:1525–1568.

Kunz, W., Testard, F. and Zemb, Th. 2009. Correspondence between curvature, packing parameter, and hydrophilic lipophilic deviation scales around the phase-inversion temperature. *Langmuir*, 25(1):112–115.

Lee, L., Langevin, D., Meunier, J., Wong, K. and Cabane, B. 1990. Film bending elasticity in microemulsions made with nonionic surfactants. *Progr. Colloid Polym. Sci.*, 81:209–214.

Ninham, B. W., Barnes, I. S., Hyde, S. T., Derian, P.-J. and Zemb, Th. 1987. Random connected cylinders: a new structure in three-component microemulsions. *Europhys. Lett.*, 4(5):561–568.

Pieruschka, P. and Marčelja, S. 1994. Monte Carlo simulation of curvature-elastic interfaces. *Langmuir*, 10(2):345–350.

Pieruschka, P. and Safran, S. A. 1993. Random interfaces and the physics of microemulsions. *Europhys. Lett.*, 22(8):625–630.

Pieruschka, P. and Safran, S. A. 1995. Random interface model of sponge phases. *Europhys. Lett.*, 31:207.

Pieruschka, P., Marčelja, S. and Teubner, M. 1994. Variational theory of undulating multilayer systems. *J. Phys. II France*, 4(5):763–772.

Pileni, M.-P., Zemb, Th. and Petit, C. 1985. Solubilization by reverse micelles: Solute localization and structure perturbation. *Chem. Phys. Lett.*, 118(4):414–420.

Porod, G. 1951. Die Röntgenkleinwinkelstreuung von dichtgepackten kolloiden Systemen. I. Teil. *Colloid Polym. Sci.*, 124(2):83–114.

Roux, D., Nallet, F., Freyssingeas, E., Porte, G., Bassereau, P., Skouri, M. and Marignan, J. 1992. Excess area in fluctuating-membrane systems. *Europhys. Lett.*, 17(7):575–581.

Safran, S. A. 1991. Saddle-splay modulus and the stability of spherical microemulsions. *Phys. Rev. A*, 43(6):2903–2904.

Sottmann, T., Strey, R. and Chen, S.-H. 1997. A small-angle neutron scattering study of nonionic surfactant molecules at the water–oil interface: Area per molecule, microemulsion domain size, and rigidity. *J. Chem. Phys.*, 106(15):6483–6491.

Spalla, O. 2002. General theorems in small-angle scattering. In *Neutron, X-rays and Light: Scattering Methods Applied to Soft Condensed Matter*, eds. P. Lindner and Th. Zemb, 49–71. Elsevier, North Holland.

Strey, R. 1994. Microemulsion microstructure and interfacial curvature. *Colloid Polym. Sci.*, 272(8):1005–1019.

Tchakalova, V., Testard, F., Wong, K., Parker, A., Benczédi, D. and Zemb, Th. 2008a. Solubilization and interfacial curvature in microemulsions: I. Interfacial expansion and co-extraction of oil. *Colloids Surf. A Physicochem. Eng. Aspects*, 331(12):31–39.

Tchakalova, V., Testard, F., Wong, K., Parker, A., Benczédi, D. and Zemb, Th. 2008b. Solubilization and interfacial curvature in microemulsions: II. Surfactant efficiency and PIT. *Colloids Surf. A Physicochem. Eng. Aspects*, 331(12):40–47.

Testard, F. and Zemb, Th. 1998. Excess of solubilization of lindane in nonionic surfactant micelles and microemulsions. *Langmuir*, 14(12):3175–3181.

Testard, F. and Zemb, Th. 2002. Understanding solubilisation using principles of surfactant self-assembly as geometrical constraints. *C.R. Geosci.*, 334:649–663.

Testard, F., Zemb, Th. and Strey, R. 1997. Excess solubilization of lindane in bicontinuous microemulsions. *Progr. Colloid Polym. Sci.*, 105:332–339.

Teubner, M. 1991. Level surfaces of Gaussian random fields and microemulsions. *Europhys. Lett.*, 14(5):403–408.

Teubner, M. and Strey, R. 1987. Origin of the scattering peak in microemulsions. *J. Chem. Phys.*, 87:3195–3200.

Welberry, T. R. and Zemb, Th. 1988. Scattering of two-dimensional models of microemulsions. *J. Colloid Interface Sci.*, 123(2):413–426.

Wennerström, H. and Lindman, B. 1979. Micelles. Physical chemistry of surfactant association. *Phys. Reports*, 52(1):1–86.

Winsor, P. A. 1948. Hydrotropy, solubilisation and related emulsification processes. *Trans. Faraday Soc.*, 44:376–398

Zemb, Th. 1997. The DOC model of microemulsions: Microstructure, scattering, conductivity and phase limits imposed by sterical constraints. *Colloids Surf. A Physicochem. Eng. Aspects*, 129–130:435–454.

Zemb, Th. 2002. Scattering by microemulsions. In *Neutron, X-rays and Light: Scattering Methods Applied to Soft Condensed Matter*, eds. P. Lindner and Th. Zemb, 317–350. Elsevier, North Holland.

Zemb, Th., Barnes, I. S., Derian, P.-J. and Ninham, B. W. 1990. Scattering as a critical test of microemulsion structural models. *Progr. Colloid Polym. Sci.*, 81:20–29.

Index

Page numbers followed by *f* and *t* indicate figures and tables, respectively.

529

T - #0184 - 251019 - C564 - 234/156/25 - PB - 9780367379759